Energetic Materials Research, Applications, and New Technologies

Rene Francisco Boschi Goncalves
Federal University of Para, Brazil

José Atilio Fritz Fidel Rocco
Aeronautics Institute of Technology, Brazil

Koshun Iha
Aeronautics Institute of Technology, Brazil

A volume in the Advances in Chemical and Materials Engineering (ACME) Book Series

IGI Global
DISSEMINATOR OF KNOWLEDGE

Published in the United States of America by
IGI Global
Engineering Science Reference (an imprint of IGI Global)
701 E. Chocolate Avenue
Hershey PA, USA 17033
Tel: 717-533-8845
Fax: 717-533-8661
E-mail: cust@igi-global.com
Web site: http://www.igi-global.com

Library of Congress Cataloging-in-Publication Data

Names: Goncalves, Rene Francisco Boschi, 1985- editor. | Rocco, Jose Atilio Fritz Fidel, 1956- editor. | Iha, Koshun, editor.
Title: Energetic materials research, applications, and new technologies / Rene Francisco Boschi Goncalves, Jose Atilio Fritz Fidel Rocco, and Koshun Iha, editors.
Description: Hershey, PA : Engineering Science Reference, [2018] | Includes bibliographical references.
Identifiers: LCCN 2017012020| ISBN 9781522529033 (hardcover) | ISBN 9781522529040 (ebook)
Subjects: LCSH: Explosives--Research. | Propellants--Research.
Classification: LCC TP271 .E97 2018 | DDC 662/.2--dc23 LC record available at https://lccn.loc.gov/2017012020

This book is published in the IGI Global book series Advances in Chemical and Materials Engineering (ACME) (ISSN: 2327-5448; eISSN: 2327-5456)

British Cataloguing in Publication Data
A Cataloguing in Publication record for this book is available from the British Library.

All work contributed to this book is new, previously-unpublished material. The views expressed in this book are those of the authors, but not necessarily of the publisher.

For electronic access to this publication, please contact: eresources@igi-global.com.

Advances in Chemical and Materials Engineering (ACME) Book Series

J. Paulo Davim
University of Aveiro, Portugal

ISSN:2327-5448
EISSN:2327-5456

Mission

The cross disciplinary approach of chemical and materials engineering is rapidly growing as it applies to the study of educational, scientific and industrial research activities by solving complex chemical problems using computational techniques and statistical methods.

The **Advances in Chemical and Materials Engineering (ACME) Book Series** provides research on the recent advances throughout computational and statistical methods of analysis and modeling. This series brings together collaboration between chemists, engineers, statisticians, and computer scientists and offers a wealth of knowledge and useful tools to academics, practitioners, and professionals through high quality publications.

Coverage

- Materials to Renewable Energies
- Ceramics and glasses
- Biomaterials
- Industrial chemistry
- Coatings and surface treatments
- Artificial Intelligence Methods
- Multifuncional and Smart Materials
- Fatigue and Creep
- Metallic Alloys
- Wear of Materials

IGI Global is currently accepting manuscripts for publication within this series. To submit a proposal for a volume in this series, please contact our Acquisition Editors at Acquisitions@igi-global.com or visit: http://www.igi-global.com/publish/.

The Advances in Chemical and Materials Engineering (ACME) Book Series (ISSN 2327-5448) is published by IGI Global, 701 E. Chocolate Avenue, Hershey, PA 17033-1240, USA, www.igi-global.com. This series is composed of titles available for purchase individually; each title is edited to be contextually exclusive from any other title within the series. For pricing and ordering information please visit http://www.igi-global.com/book-series/advances-chemical-materials-engineering/73687. Postmaster: Send all address changes to above address.

Titles in this Series

For a list of additional titles in this series, please visit: www.igi-global.com/book-series

3D Printing and Its Impact on the Production of Fully Functional Components Emerging Research and Opportunities
Petar Kocovic (Union – Nikola Tesla University, Serbia)
Engineering Science Reference • copyright 2017 • 115pp • H/C (ISBN: 9781522522898) • US $150.00 (our price)

Advanced Applications of Supercritical Fluids in Energy Systems
Lin Chen (Tohoku University, Japan & Japan Society for the Promotion of Science (JSPS), Japan) and Yuhiro Iwamoto (Nagoya Institute of Technology, Japan)
Engineering Science Reference • copyright 2017 • 682pp • H/C (ISBN: 9781522520474) • US $235.00 (our price)

Innovative Applications of Mo(W)-Based Catalysts in the Petroleum and Chemical Industry Emerging Research and Opportunities
Hui Ge (Chinese Academy of Sciences, China) Xingchen Liu (Chinese Academy of Sciences, China) Shanmin Wang (Oak Ridge National Laboratory, USA) Tao Yang (China University of Petroleum, China) and Xiaodong Wen (Synfuels China, China)
Engineering Science Reference • copyright 2017 • 162pp • H/C (ISBN: 9781522522744) • US $135.00 (our price)

Sustainable Nanosystems Development, Properties, and Applications
Mihai V. Putz (West University of Timişoara, Romania & Research and Development National Institute for Electrochemistry and Condensed Matter (INCEMC) Timişoara, Romania) and Marius Constantin Mirica (Research and Development National Institute for Electrochemistry and Condensed Matter (INCEMC) Timişoara, Romania)
Engineering Science Reference • copyright 2017 • 794pp • H/C (ISBN: 9781522504924) • US $245.00 (our price)

Computational Approaches to Materials Design Theoretical and Practical Aspects
Shubhabrata Datta (Calcutta Institute of Engineering and Management, India) and J. Paulo Davim (University of Aveiro, Portugal)
Engineering Science Reference • copyright 2016 • 475pp • H/C (ISBN: 9781522502906) • US $215.00 (our price)

Green Approaches to Biocomposite Materials Science and Engineering
Deepak Verma (Graphic Era Hill University, Dehradun, India) Siddharth Jain (College of Engineering Roorkee, India and University of Alberta, Canada) Xiaolei Zhang (Queens University, Belfast, UK) and Prakash Chandra Gope (College of Technology, G.B.Pant University of Agriculure and Technology, Pantnagar, India)
Engineering Science Reference • copyright 2016 • 322pp • H/C (ISBN: 9781522504245) • US $165.00 (our price)

IGI Global
DISSEMINATOR OF KNOWLEDGE

701 East Chocolate Avenue, Hershey, PA 17033, USA
Tel: 717-533-8845 x100 • Fax: 717-533-8661
E-Mail: cust@igi-global.com • www.igi-global.com

Table of Contents

Detailed Table of Contents

Luigi T. DeLuca, Politecnico di Milano, Italy
Manfred A. Bohn, Fraunhofer-Institut für Chemische Technologie (ICT), Germany
Volker Gettwert, Fraunhofer-Institut für Chemische Technologie (ICT), Germany
Volker Weiser, Fraunhofer-Institut für Chemische Technologie (ICT), Germany
Claudio Tagliabue, Fraunhofer-Institut für Chemische Technologie (ICT), Germany

Solid rocket propulsion enjoys unique properties favoring its use in space exploration and military missions still for decades to come. Yet, it also suffers a limited performance especially in terms of gravimetric specific impulse. Although new high-energy materials have been identified, most of them are far from being practically usable in the short range. Presently, no integrated vehicle designs make use of these new ingredients. A broad overview is discussed in this paper and attention is paid to Ammonium Dinitramide, ADN to overcome the current limitations of Ammonium Perchlorate, AP. The latter imply not only a limited gravimetric specific impulse but also a negative impact on the environment and personal health. ADN-based dual-oxidizer formulations, with Al-based dual-metal fuels and inert or energetic binders, are promising solutions for a variety of solid rocket propulsion missions aiming respectively at minimizing environmental impact (ADN + AN) or maximizing performance (ADN + AP).

Rene Francisco Boschi Goncalves, Federal University of Pará, Brazil
Koshun Iha, Aeronautics Institute of Technology, Brazil
José A. F. F. Rocco, Aeronautics Institute of Technology, Brazil

Fuels and energetic materials enhancements have been focus of several studies in the past few years, as the sole development of the engines (design, structure of the combustion chambers and stronger materials) leads to a limited efficiency. One of the most effective ways of enhancing the energetic capabilities of a fuel is via enrichment with a highly reactive species, such as hydrogen and ozone, as these materials significantly increase the reaction mechanism kinetics, enabling the occurrence of faster reactions and increasing the temperature of the chamber. The present chapter aims to summarize some of the applications of fuel enrichment and the so far achieved results, focusing in internal combustion engines and systems for automotive, energy-generation or aerospace applications.

This chapter adopts the Chapman-Jouguet approach to derive jump conditions across combustion waves propagating in Van der Waals and Noble-Abel gases. The steady one-dimensional balance equations of mass, momentum and energy, assuming different properties of reactants and products, are applied to obtain the main properties of combustion waves, including velocities, Mach numbers, pressures and temperatures, in terms of the covolumes and intermolecular force parameters. In general, the effects of covolumes are more significant than the effects of the intermolecular attraction forces on Hugoniot curves and on properties of combustion waves. However, theoretical results using the Van der Waals equation of state matched more closely the experimental results for detonations of mixtures of propane and diluted air at high initial pressures.

This Chapter presents a theoretical analysis of radiation interaction with a semi-transparent metalized energetic material. Main regularities of the laser pulse interaction with metalized compositions are considered within the framework of non-resonant interaction of radiation with matter. The large variety of metalized composite propellants with different properties of the components, their ratio and dispersion can be divided into two classes, depending on the ratio of the laser irradiation's characteristic time (tr) and the thermal relaxation time of the propellant characteristic cell containing one metal particle (tm). Analysis of the role of metallic particles shape shows that in the case of spherical metal particles, duration of the laser pulse corresponds to the optimal size of particles, heated to a maximum temperature. In the case of flat metallic particles and constant pulse duration, the critical radiation flux and the critical density of ignition energy significantly decrease with decreasing thickness of the particle.

This work presents the simulation of the internal flow in a swirl atomizer. The geometry of the atomizer is calculated by analytical equations used in engineering. The numerical simulation of the two-phase flow is performed by using two equations k-ε turbulence model. The fluids are presented as two-fluid homogeneous model. The interface between two phases is calculated by free surface model. The distribution fields of the axial and tangential velocities, pressures and air core are obtained. The aim of this work is to compare the results obtained by numerical simulation with ones obtained analytically. Also, to study the internal fluids flow inside the atomizer.

The development of liquid sheets that emerges from nozzles is influenced mainly by their initial speed and by the physical properties of the liquid and the ambient gas. A minimum speed of the sheet is necessary for its enlargement against the superficial tension that tends to contract the surface. As this speed increases, the sheet expands until a main extremity is formed, where balance exists among the superficial tension and the inertial forces. The form and regularity of the sheet's disintegration process has influence in the size distribution of the produced drop and in the Sauter mean diameter (SMD). The initial thickness of the produced liquid sheet is important to determine the medium size of obtained drops. It was observed that thicker films produce thicker ligaments and larger drops. The medium drop diameter produced in conical sheets of pressurized swirl atomizers is calculated according to the thickness of the sheets and in the wavelength for the maximum growth tax.

Solid rocket motors (SRM) are extensively employed in satellite launchers, missiles and gas generators. Design considers propulsive parameters with dimensional, manufacture, thermal and structural constraints. Solid propellant geometry and computation of its burning rate are essential for the calculation of pressure and thrust vs time curves. The propellant grain geometry changes during SRM burning are also important for structural integrity and analysis. A computational tool for tracking the propagation of tridimensional interfaces and shapes is then necessary. In this sense, the objective of this work is to present the developed computational tool (named RSIM) to simulate the burning surface regression during the combustion process of a solid propellant. The SRM internal ballistics simulation is based on 3D propagation, using the level set method approach. Geometrical and thermodynamic data are used as input for the computation, while simulation results of geometry and chamber pressure versus time are presented in test cases.

Friendly or green compositions are being sought in many kinds of applications, such as fertilization, building materials, energy generation, and so on. Thus, this classification (green) can be established after the subjection of the compound to a thorough toxicity study. The research for low toxicity and no damage to the environment has stimulated the development of specific investigation lines in many areas. Inevitably, the criteria for safe handling, sensitivity and, mainly, specific impulse (efficiencies) of propellant compositions is still superior compared to ecological appeals. Nowadays, however, the solid or liquid propulsion, as aerospace as military, has already compounds to efficiency and eco-friendly characteristics. Thus, the chapter proposed will show a review in the energetic materials area, aiming at the most promising materials to be used as oxidizers and combustibles in a green propulsion system.

This research aims to provide a methodology for the project of labscale hybrid motors. This development began with the thermal analysis of the fuel grain using the Flynn, Wall and Ozawa method, generating simulation entry data to maximize the motor performance. The simulation was performed with the Chemical Equilibrium Specific Impulse Code. Based on the optimum oxidizer to fuel ratio, the literature was used to supply the mathematical background to calculate the motor geometrical parameters whose operating conditions were determined throughout the simulation. Finally, firing tests were conducted to verify the reliability of the project methodology. The firing tests were performed with three injectors: two swirling and one axial. The tests showed that the higher the operating pressure the more suitable is the project, meaning the methodology developed works best in hybrid rocket motors with high operating pressures. Additionally, it was shown that the swirling flow injectors produce higher efficiency.

The thermal decomposition kinetics of ammonium perchlorate (AP)/hydroxyl-terminated-polybutadiene (HTPB) samples, with Iron Oxide catalyst at nano and micro scale were studied by thermal analysis techniques at different heating rates in dynamic nitrogen atmosphere. The exothermic reaction kinetics was studied by differential scanning calorimetry (DSC) in isothermal conditions. The Arrhenius kinetic parameters were obtained by Flynn-Wall and Ozawa Kissinger and Starink methods. The propellant samples thermal decomposition was studied simultaneously by TG-DTA. For this purpose, solid propellant grains containing nano and micro scale iron oxide were formulated. The effect of catalysts on the propellant burning rate and the propellant initiation sensitivity were also evaluated by friction and impact. The effect of the catalyst in the propellant binder reaction was evaluated by viscosity and mechanical properties. SEM/EDS technique was used to evaluate the iron oxide morphology. Three bench firing tests were performed with rockets motor in order to know the ballistics parameters.

Storage of rocket motors loaded with composite solid propellant for long periods may change the propellant properties, thus causing failure and affecting the safety during launch. In this study, an accelerated aging assay was carried out, in order to predict the useful lifetime and to evaluate variations on the propellant properties with time by means of thermal analysis (TG/DSC). The aging temperatures used were 65°C, and samples were withdrawn after 3 months. Aging was also carried out at room temperature. There was significant variation in the activation energy of the solid propellant samples thermal decomposition in the two kinetic methods used – Ozawa or model-free isoconversional method and Kissinger method – during the aging period. There was significant decrease of enthalpy of aged propellant enthalpy causing changes in ballistics parameters of the solid propellant grain affecting the rocket's performance.

In the late twentieth century, liquid and solid propulsion technologies have been integrated into hybrid engines currently apllied in propulsion launch vehicles and missiles. The reaction of polyol (HTPB) and diisocyanate (IPDI) provides the most versatile of the binders in the production of solid propellants due to its ability to withstand high loads combined with low cost and ease of processing. A propellant based on HTPB obtained in this study was submitted to natural and accelerated aging tests, seeking to evaluate the modifications of mechanical properties as tensile strength, elongation and hardness up to 360 days. The mechanism considered in the aging process is the increase of crosslink density by breaking the double bond contained in the HTPB molecule, which causes the instability of the propellant, increasing its handling risk. Samples of these propellants subjected to aging presented variations in their properties that match the values available in the literature.

A large scale experimental program consisting of testing 10 RC slabs with different variations of concrete compressive strength, reinforcement ratio and retrofit was conducted in Brazil. As part of that test program, a small-scale blast pre-test setup and associated dynamic analysis were conducted in order to confirm the proper functioning of the blast test sensors (pressure gages, displacement meter and accelerometers). The results of the pre-test were compared with theoretical blast wave parameter predictions using established equations and maximum displacement predictions using simplified dynamic analysis. The pre-test experiment provided useful insights and was shown to be critical for the success of the subsequent large scale blast tests.

High explosives are tools used in different areas of research and are critical components in conventional and nuclear weapons. PBXs (plastic explosives) were developed to reduce the sensitivity of explosives, and are widely used in both applications, civil and military, where high performance is required. As the weapons may be used in aggressive thermal and mechanical environments, it is important to characterize the PBXs, in order to know its behavior and properties. Therefore, contributing for research of PBXs, is discussed in this chapter a methodology to characterize and to quantify the polymer content in PBX by Fourier Transform Infrared Spectroscopy (FT-IR), using the High-Performance Liquid Chromatography (HPLC) and Thermogravimetry (TG) as reference techniques to quantitative method. The proposed methodology for quantification of polymers in PBX uses the technique FT-IR of reflection ATR and UATR, being faster than the usual methodologies and eliminating the disadvantage of chemical residues generation.

Electroexplosive devices, EEDs or squibs (an electric resistance encapsulated by a primary explosive), fundamentally convert electrical energy into heat, solely to start off an explosive chemical reaction. Obviously, the EED activation shall not happen by accident or, even worse, by intentional exogenous influence. From an ordinary differential equation (ODE), which describes this device thermal behaviour for both continuous and pulsed electrical excitation, a remarkable, but certainly not intuitive, dependence of the temperature response on the heat transfer process time-constant is verified: the EED temperature profile dramatically changes as the time-constant spans a wide range of values, from much lesser than the pulse width to much greater than the pulse period. On the basis of this dependence, important recommendations, concerning the EED safety (and efficient) operation, are presented.

Preface

Energetic materials are known since antiquity and are characterized by their ability to release large amounts of energy in a short time and consequent production of heat, light, smoke and noise; a shock-wave can produce up to 500,000 atm of pressure and 5,500 K, while traveling at 10 km/s. The first explosive synthesized by man was black gunpowder, initially used in the making of fireworks by the Chinese (ninth century). However, its formula was first quoted in the writings of English monks in the thirteenth century. From the determination of the components of black powder and the development of new formulations of other types of gunpowder it was possible to improve the products of the industries of defense materials and fireworks.

With the evolution of science and technology, new types of explosives have been developed, increasing their applicability in different areas, such as: mining, opening of canals, construction of highways and railroads, deepening of seaports, seismic prospecting, petrochemical industry, propulsion by Rocket engines, among others. At the same time, several laboratories around the world have spared no effort in research into the synthesis of energy materials with improved properties and improved handling, transport and storage safety.

Explosives can be defined as substances or materials that undergo highly exothermic decomposition reactions, with great formation and consequent gas displacement (shock wave). They can be classified into two main types: explosive (or low-power) explosives and high-powered explosives.

In the case of high-explosives, they have a higher rate of decomposition, resulting in higher pressures (when confined) and, consequently, resulting in greater destructive power if their final application is, for example, -war. Also, they are used in perforations of road and underground tunnels, military explosives and other applications. This type of explosive suffers the effect called detonation, that is, the thermal degradation reaction is fed back by the shock wave coming from the reaction itself, exerting great pressure on the material.

Low explosives are usually used for ballistic works, such as propellants for bullets, mortars or rocket engines. They are also used in controlled explosions of mines, among others. The constituent substances of this type of material react with each other, releasing large amounts of energy or forming new compounds that undergo highly exothermic decomposition reactions. After ignition, by the phenomenon of deflagration (propagation of combustion through thermal conductivity), chain reactions occur that spread throughout the material in a period of time of the order of fraction of a second.

In the case of rocket motors powered by solid propellant, the performance in flight depends directly on the characteristics of the propellant used. These characteristics determine the distance to be traveled in the trajectory described by the mill, the maximum speed reached, the acceleration, the stability of the system, among others. Full knowledge of the formulation and characteristics of the propellants prior to their use in these engines is necessary.

The processing of these energy materials varies according to the type of compound to be produced. Solid propellants produced with oxidants in the form of crystals and agglomerates of resins, for example, can be obtained by mechanical extrusion process (with or without solvent), pressing and molding or direct polymerization.

These synthetic processes require standards and procurement procedures that vary according to the material to be obtained. For some of these materials, strict temperature and pressure control, as well as friction and impact care are required, due to the high instability of the compound (nitroglycerin, for example), while other materials are highly stable and not reactive, only entering into process by means of a detonator (trinitrotoluene, for example).

Another class of energetic materials is the pyrotechnics, which may be considered a mixture of materials suitable for combustion, producing some special effects. Usually, the composition is similar to propellants, i.e. contain fuel, oxidizer and a binder for structural integrity. The 'special effects' produced by pyrotechnics can be the emission of heat, light, smoke or sound; main applications of these effects include igniters, illumination devices, signaling, decoy and distraction devices.

In the process of characterization of energy materials, several techniques of analysis can be employed. Among these, the following stand out: differential scanning calorimetry (DSC), Fourier transform infrared spectroscopy (FTIR), high performance liquid chromatography (HPLC), and sensitivity to friction, impact, sparks and heat analysis.

Differential Scanning Calorimetry (DSC) is a widely used thermal analysis technique in these cases because it allows the use of a reduced amount of sample, milligram, speed of execution of the test and the possibility of determining the kinetic parameters of a Thermal decomposition reaction.

Fourier transform infrared spectroscopy (FTIR) provides structural information regarding the substances analyzed, depending on the specific vibration frequencies of chemical bonds (vibrational levels), making it possible to determine the composition and location of substituents and, in some Determination of the total structure of the molecule.

High performance liquid chromatography is one of the most commonly used techniques for substance identification and purity checking. The HPLC technique employs small columns, filled with specially prepared materials and a mobile phase, eluted under high pressures, making it possible to carry out quantitative separations and analyzes of a large number of compounds present in several types of samples, in a time scale of few minutes, With high resolution, efficiency and sensitivity.

The sensitivity of an explosive is the ability to react to a particular external excitation, and each type of excitation requires a different technique. For example, friction sensitivity analysis is determined by friction of a small sample of energetic material on a porous porcelain plate, while the impact sensitivity analysis is measured by a "falling hammer".

Using the combination of results from the different materials characterization techniques, it is possible to determine the structure of the synthesized compound, the degree of purity, the identification of possible intermediates and interferents and the determination of speed and energy parameters of Activation of reactions (including excitation sensitivity).

ROCKET PROPULSION

One of the major applications for energetic materials, propulsion by rocket engines followed two distinct paths throughout the twentieth century. the first was the one aimed at allowing access to space in the direction of space travel according to the view of Tsiokolsky in Russia, Goddard in the United States of America, Esnault-Pelterie in France and Oberth in Germany who readily understood that motor propulsion - rocket was the key to the conquest of outer space. liquid propellant rocket propulsion was chosen to achieve the space since its performance is superior when compared to the propulsion by solid propellant rocket motor. the other path, in the direction of the development of armament systems, was rocket propulsion with solid propellant because they had to combine good performance with other characteristics such as immediate use, compact dimensions and long service life. in the late twentieth century, both liquid and solid propulsion technologies were integrated and can be found both in propulsion of launch vehicles and in missiles, among others.

when compared to other types of engines, the advantages of solid rocket propulsion of aerospace systems, especially in the case of tactical missiles and even in satellite launch vehicles, outweigh its disadvantages and for this reason remain employees on current days. the greatest of these advantages is that the solid propellant grain has a high density when compared, for example, to the liquid propellant used in liquid propellant rocket engines, among others. this can be a limiting factor in the choice of the propellant to be used in the engine according to the characteristics of the propulsive design. in addition, the solid propellant grain can be considered, depending on the formulation used to obtain it, "chemically" stable, keeping its physicochemical characteristics, little altered for a long period of time (useful life). at the current stage of development in which these solid propellant rocket motor-based propulsion systems are integrated into tactical missiles this useful life is estimated at 15 years. another important feature of solid propellant rocket motor can be described as its "relative" simplicity when compared to propellant systems by rocket propellant or even gas turbines (turbojet). again, considering warlike artifacts such as tactical missiles, this simplicity can be a determining factor when high scales of production and issues such as reproducibility of ballistic characteristics become imperative. also, because of this simplicity, solid propellant rocket engines have their development and production costs significantly lower compared to others as mentioned above. exception made for the propulsion in the subsonic regime of medium-range missiles where the turbojet is the option of lower cost. this same line of reasoning can be applied to the maintenance costs of the mills when the solid propellant rocket motor is adopted as a propulsion system in relation to others. as the main disadvantages, we can mention the low value of the specific impulse (isp), important ballistic property of a rocket motor that will be characterized later, and the difficulty in modulating the thrust in large solid propellant rocket motor. however, great progress has been achieved in the last decades and tend to minimize these two limitations of solid propellant rocket motor. thus, rocket engines based on solid propellant grain firing are the simplest way to generate thrust from chemical reactions (chemical propulsion) and meet, among others, requirements such as: sustainer phase of rockets and missiles; "booster" (acceleration) phase in missile propulsion in the "ramjet" and "scramjet" cycles; phase "and" sustainer "in satellite launch vehicles, such as the Brazilian Space Program.

ORGANIZATION OF THE BOOK

The book is organized into 15 chapters. A brief description of each of the chapters follows:

Chapter 1 presents the novel formulations for energetic materials, focusing on the solid rocket propellants, which has been extensively studied in the last decades. The chapter presents new ideas and innovative formulations applicable in rocket motors for space propulsion.

Chapter 2 establishes the current studies in enhancement of energetic materials using additives and/or different fuels for increasing some combustion characteristics. This chapter reviews several studies developed over the past few years and shows the capability of each additive for improvement of different characteristic of engines and combustion processes.

Chapter 3 adopts the Chapman-Jouguet approach to derive jump conditions across combustion waves propagating in Van der Waals and Noble-Abel gases. Based on analytical analyses, and using some physical characteristics and balance equations, the main properties of combustion waves are characterized.

Chapter 4 presents a theoretical analysis of radiation interaction with a semi-transparent metalized energetic material. Based on different propellant structures and formulations, type, content and granulometry of metallized propellants, the ignition process can be extremely distinct, which may cause some significant disadvantages when using the materials in large scale.

Chapter 5 analyzes swirl atomizers by computational fluid dynamics simulations; In this work, some swirl atomizers are designed and have the fuel/oxidizer flow analyzed. The simulation results are compared to analytical calculations, also present in the chapter. Standard turbulence models were applied to this simulation, in order to evaluate mostly the interactions to the surface and geometry of the system.

Chapter 6 presents experimental results of the disintegration of liquid paraffin sheets produced in swirl injectors. Liquid paraffin is an interesting liquid to evaluate, as it melts in low temperatures and provides low-cost trials for evaluating the injector properties (mechanical, structural). These results are possible to be compared to simulated ones in similar injectors.

Chapter 7 discusses the solid propellant ballistic properties when inside the rocket engine. The tridimensional granulometry of the formulation by itself causes the development of different ballistic properties by the system, i.e. specific impulse, thrust etc. The chapter shows simulation results for the geometry and chamber pressure according to the time after ignition of the system.

Chapter 8 presents the current trend of green propellants. As rockets are charged with huge quantities of propellant materials, during the past years the emissions have become a major concern to the environment. Novel materials, different combustion systems, and improved formulations are some of the 'green' ways to reduce hazard to the environment. The author reviews several materials which are being employed as substitution to halogenated or toxic materials.

Chapter 9 shows a complete analytical and experimental test of a lab-scale hybrid rocket engine, using a swirl injector for improving the mixing of the reactants. The authors present a very detailed description and several pictures of the system, as well as the combustion analysis. Prior to the experiments, several simulations were developed to find the optimum parameters for a better performance of the system.

Chapter 10 discusses the use of nanotechnology in solid propellant formulations. All the formulations were evaluated by thermal analysis methods and the effect of using different iron oxide granulometries was studied (micro and nano scale). Kinetic results obtained by the methods are important parameters, especially due to the possibility of calculating the velocity for the global combustion process.

Chapter 11 discusses the shelf life of rocket motors, filled with solid propellants. For several reasons, many places store several rocket motors for long periods, which can cause an aging of the propellant formulations. The authors show the effects on the physical-chemical properties of the propellant formulations after different storage times, by using thermal analysis techniques.

Chapter 12 presents some experimental tests for a concise evaluation of energetic materials aging process. In this chapter, the author submits a solid propellant formulation to accelerated aging procedures and analyses the effects by thermal analysis methods.

Chapter 13 shows the pre-test and analysis of the structural and integrity of reinforced concrete slab after suffering blasts from non-confined explosives. These pre-tests can be as important as precise simulations, as they provide information on the performance of the system with a reduced cost. After optimization, the pretest and analysis procedure can be applied to large scale trials.

Chapter 14 discusses the application of infrared techniques for characterize and quantify the polymer (binder) content in explosives, which can be also applied to other energetic materials. The polymer content in plastic explosives can determine the performance or the expected detonation, as the internal structure is a major aspect in the ballistic properties, so the authors present a methodology for fast evaluation.

Chapter 15 shows an analytical analysis of the initiation process that occur in electroexplosive devices, which are commonly used to ignite other energetic materials or to start some chemical reactions. A precise performance of these materials is required to achieve the full potential of the working system.

Acknowledgment

The editors would like to acknowledge the help of all the people involved in this project and, more specifically, to the authors and reviewers that took part in the review process. Without their support, this book would not have become a reality.

First, the editors would like to thank each one of the authors for their contributions. Our sincere gratitude goes to the chapter's authors who contributed their time and expertise to this book.

Second, the editors wish to acknowledge the valuable contributions of the reviewers regarding the improvement of quality, coherence, and content presentation of chapters. Most of the authors also served as referees; we highly appreciate their double task.

Rene F. B. Gonçalves
Aeronautics Institute of Technology, Brazil

José A. F. F. Rocco
Aeronautics Institute of Technology, Brazil

Koshun Iha
Aeronautics Institute of Technology, Brazil

Chapter 1
Innovative Solid Rocket Propellant Formulations for Space Propulsion

Luigi T. DeLuca
Politecnico di Milano, Italy

Manfred A. Bohn
Fraunhofer-Institut für Chemische Technologie (ICT), Germany

Volker Gettwert
Fraunhofer-Institut für Chemische Technologie (ICT), Germany

Volker Weiser
Fraunhofer-Institut für Chemische Technologie (ICT), Germany

Claudio Tagliabue
Fraunhofer-Institut für Chemische Technologie (ICT), Germany

ABSTRACT

Solid rocket propulsion enjoys unique properties favoring its use in space exploration and military missions still for decades to come. Yet, it also suffers a limited performance especially in terms of gravimetric specific impulse. Although new high-energy materials have been identified, most of them are far from being practically usable in the short range. Presently, no integrated vehicle designs make use of these new ingredients. A broad overview is discussed in this paper and attention is paid to Ammonium Dinitramide, ADN to overcome the current limitations of Ammonium Perchlorate, AP. The latter imply not only a limited gravimetric specific impulse but also a negative impact on the environment and personal health. ADN-based dual-oxidizer formulations, with Al-based dual-metal fuels and inert or energetic binders, are promising solutions for a variety of solid rocket propulsion missions aiming respectively at minimizing environmental impact (ADN + AN) or maximizing performance (ADN + AP).

DOI: 10.4018/978-1-5225-2903-3.ch001

NOMENCLATURE

Roman Symbols

ADN: Ammonium DiNitramide
AN: Ammonium Nitrate
AP: Ammonium Perchlorate
ARC: Atlantic Research Corporation
BAMO: 3,30-Bis(AzidoMethyl) Oxetane
BC: Before Christ
BFF: 3,4-bis(3-fluorodinitro-methyl furazan-4-oxy)furazan
BHEGA: N, N-bis(2-hydroxyethyl)glycol amide, bonding agent also known as HX-880; BHEGA may act also as a cross-linker
BPS: Bis-Propargyl Succinate
Bu-NENA: n-Butyl-NitratoEthylNitrAmine
ca: circa
CalTech: CALifornia institute of TEChnology
CCP: Condensed Combustion Products
CL-20: 2,4,6,8,10,12-hexanitro-2,4,6,8,10,12-hexaazaisowurtzitane, also known as HNIW
CMDB: Composite Modified Double-Base
CTPB: Carboxyl-Terminated PolyButadiene
DB: Double-Base
EMCDB: Elastomer Modified Cast Double-Base
ESA: European Space Agency
ESD: ElectroStatic Discharge
FOI: Swedish Defence Research Agency
GALCIT: Guggenheim Aeronautical Laboratory at California Institute of Technology
GAP: Glycidyl Azide Polymer
GUDN: GuanylUrea-DiaNitramide, also called FOX-12
HADNMNT: HydroxylAmmonium 2-DiNitroMethyl-5-NitroTetrazolate
HATO: diHydroxylAmmonium 5,5'-bisTetrazole-1,1'-diOlate (also known as TKX-50)
HBIW: HexaBenzylhexaazaIsoWurtzitane
HEM: High-Energy Materials
HMX: cyclotetramethylenetetranitramine, High Melting Explosive (also known as Octogen, Her Majesty's Explosive)
HNF: Hydrazinium nitroformate
HNIW: 2,4,6,8,10,12-HexaNitro-2,4, 6,8,10,12-HexaazaIsoWurtzitane, commonly known as CL-20
HTPB: Hydroxyl-Terminated PolyButadiene
HTPE: Hydroxyl-Terminated PolyEther
ICBM: InterContinental Ballistic Missile
ICP: Institute of Chemical Physics
ICT: Fraunhofer Institute for Chemical Technology
IM: Insensitive Munitions
Is: Gravimetric Specific Impulse

JATO: Jet Assisted Take Off
JPL: Jet Propulsion Laboratory
KP: Potassium Perchlorate
NASA: National Aeronautics and Space Administration
nAl: nano-sized Al particles
NC: NitroCellulose
NEPE: Nitrate Ester Plasticized polyEther (or Nitrate Ester PolyEther)
NG: NitroGlycerine
OB: Oxygen Balance
PBAA: PolyButadiene Acrylic Acid
PBAN: PolyButadiene AcryloNitrile
PDL: Pressure Deflagration Limit
PETN: PentaErythritol Tetra Nitrate
PGN: Poly-Glycidyl Nitrate
PU: Polyurethane
RDX: cyclotrimethylenetrinitramine, Research Department explosive (also known as Hexogen, Cyclonite, T4, Royal Demolition Explosive)
SL: Skeleton Layer
SNPE: Société Nationale des Poudres et des Explosifs (1971–2013)
T_f: Adiabatic flame temperature
THF: TetraHydroFuran
USA: United States of America
XLDB: Cross-Linked Double-Base

Other Symbols

ΔH_f: Heat of Formation
μAl: Micron-sized Al particles
ρ: Density
$\mathcal{M}$: Molar mass

INTRODUCTION

Solid rocket propulsion provides a very large thrust with relatively small volume and contained costs. Solid propellants additionally offer the advantages of large density, well extended storage time, and immediate readiness. In addition to military applications, these unique features make solid propulsion particularly suitable for first stage or add-on boosters of heavy space launchers as well as for boost stages of small space launchers. However, solid propellants suffer a limited performance in terms of gravimetric specific impulse, Is. Moreover, the composite formulations based on the matrix AP-HTPB, currently the most used one especially for space exploration missions, also suffer a negative impact on the environment (because of copious emission of hydrochloric acid HCl) and personal health (because of perchlorate competition with iodide in entering the thyroid gland). Further difficulties stem from the disposal of AP-based HEM, for all sorts of military applications, after service.

To overcome the above difficulties, a variety of new ingredients is under examination worldwide at many advanced laboratories. Several families of high-energy materials propellants were identified, such as formulations based on high-nitrogen compounds, octanitrocubane $C_8(NO_2)_8$, metallic hydrogen, atomic radicals, metastable helium, and so on. Likewise, metal oxides like NiO and CuO are being studied as new catalysts and allow the accurate adjustment of the burning rate of propellant formulations. However, most of these compounds are far from being practically usable in the short-range due to a variety of severe difficulties in large scale manufacture, ground processing, personnel safety, handling and transportation safety, shock sensitivity, prolonged storage, and cost considerations. Notwithstanding continuous work, presently no integrated vehicle designs make use of these radically new potential propellant ingredients or formulations.

A broad overview of the status is presented in this paper based on a joint international editorial effort just completed (DeLuca et al., 2016). After a quick historical survey, several promising modern ingredients in terms of solid oxidizers, metal fuels, binder systems, and additives are scrutinized. Among the oxidizers, attention is paid to ADN-based systems. However, the intrinsic properties of ADN (physical, chemical, mechanical, ballistic, and so on) are far from the values ideally desired for applications in solid rocket propulsion, especially for space exploration. Dual-oxidizer systems, such as (ADN+AN) or (ADN+AP), are probably preferable solutions for a variety of solid rocket propulsion missions.

The objective of this paper is to show that by properly combining these three inorganic oxidizers, ADN and the two well-known salts AN and AP, a wide range of interesting applications beyond the current limitations appear feasible. Examples of formulations are discussed in terms of thermochemistry and ballistics. However, upgrading from laboratory to motor applications needs full-scale ballistic testing together with adequate analyses of manufacture, mechanical, aging, and hazard properties as well as cost.

A TIMELINE OF SOLID ROCKET PROPELLANTS DEVELOPMENT

For centuries, rudimental forms of solid rocket propulsion were based on the use of black powder, following the fortuitous alchemist activities in China dating as early as ca 220 BC. Significant advances were made in Europe, introducing NC-based smokeless propellants, only during the years 1863-1888. Further decisive advances were made in the USA, introducing castable composite propellants in the framework of the GALCIT program dedicated to JATO rockets. This project, 1934-1943, was directed by Theodore Von Karman and included other prominent figures such as Martin Summerfield, Frank Malina, and Jack (John) Parsons. In June 1942 Parsons, a self-educated and very eccentric but also imaginative chemist, combining an organic matrix (asphalt) with a crystalline inorganic oxidizer (KP) succeeded in making the first castable composite solid propellant. Just at the end of the GALCIT project, Parsons also occasionally employed AP as crystalline inorganic oxidizer, instead of KP, to reduce smoke emission. Systematic testing of AP was conducted by Aerojet Engineering Corporation (now Aerojet Rocketdyne) starting in 1948 (Umholtz, 1999). The GALCIT team created Aerojet Engineering Corporation in 1942 (sold to General Tire and Rubber Company in 1944) and finally transformed into what is today JPL in 1943. Composite propellants eventually replaced DB propellants, based on NC and NG mixtures, in most rocket applications. The GALCIT project was the start of modern solid propellant formulations.

Following the breakthrough by Parsons in 1942, composite solid propellants were actively developed in the USA along different directions. The need for more volume-efficient propellant for the launch of ICBM's from submarines (US Navy) or silos (US Air Force) was the strongest driver for advancing solid propellant formulations. But this push toward more and more performance seems now at its end, as shown by the flattening curve of delivered Is during the last decades; see Figure 1. For high-energy solid propellants, two main families can be distinguished essentially based on the kind of the organic binder system and this implies an important consequence in terms of hazard classification, in turn a discriminating factor for applications.

XLDB propellants, using energetic binders or at least plasticizers, generally belong to class 1.1 hazards and conventional composite propellants using inert binders belonging to class 1.3 hazards. High-energy XLDB formulations in a way represent a combination (Davenas, 2003) of the historical class of homogeneous formulations used as gun propellants and the castable heterogeneous formulations first devised by Parsons. Yet, definitions are somewhat loosely applied and the nomenclature is confusing. Despite intrinsic differences, the various expressions CMDB, XLDB, EMCDB, and NEPE can be considered broadly equivalent.

Highlights in the worldwide history of HEM developments are summarized in Table 1.

At least in the Western world, the state-of-the-art in solid propulsion for space exploration is represented by AP/HTPB/Al formulations. Advanced ingredients under study are discussed in (DeLuca et al., 2016). Sinditskii (DeLuca et al., 2016, Ch. 3) focused his attention on high-nitrogen compounds. In fact, HEM that burn at the expense of energy derived from positive heat of formation rather than oxidation of the carbon backbone have recently attracted attention not only as explosives but also as components of gas-generating and propellant compositions, because they can produce relatively high specific impulse despite low combustion temperature. 1,2,4,5-Tetrazine (also known as s-tetrazine) derivatives are among these compounds. 1,2,4,5-Tetrazines are unique heterocyclic compounds containing the maximum number of nitrogen atoms in the cycle among the stable azines. Among polynitrogen HEM, 1,2,4,5-tetrazine

Figure 1.

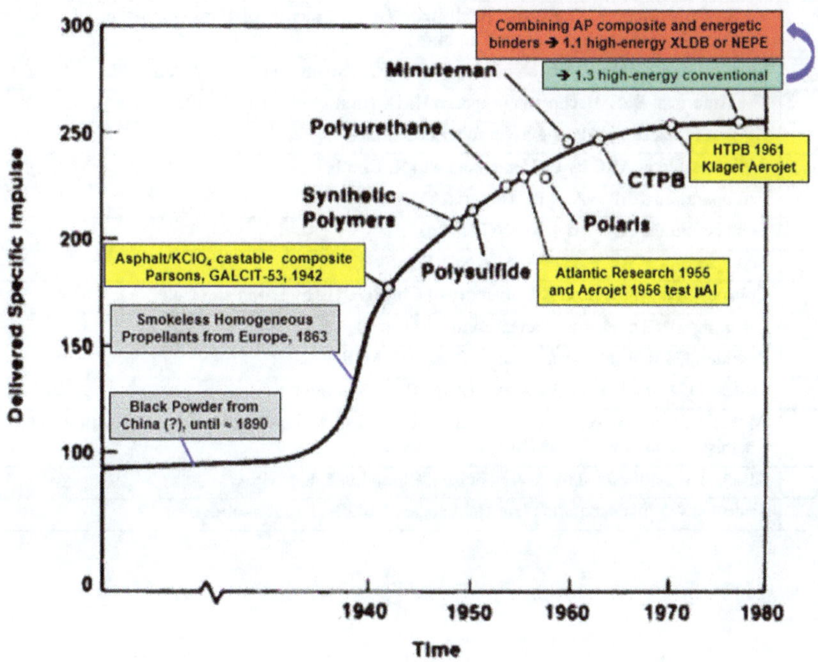

Table 1. Highlights in the history of high-energy materials (DeLuca, 2016, Ch 42)

Date	Event
220 BC	Chinese alchemists accidentally made a solid mixture based on nitrates close to black powder
85 ca	Roman emperor Domitianus alerted to the use of incendiary arrows.
690	Arabs used black powder at the siege of Mecca.
940	The Chinese invented the 'Fire Ball' using an explosive composition similar to black powder.
1232	Battle of Kai-Fung-Fu: rockets or incendiary arrows repelled Mongols.
1379	Battle for Isle of Chiozza: devastating powder rockets used as "rocchetta"
1500 ca	The Chinese Wàn Hoo, during the Ming dynasty, first recorded astronaut.
1659	The German J.R. Glauber synthesized AN
1792 1799	Combustion chambers with metallic walls allowed the Indian troops of Tipu Sultan to twice defeat the British army at Seringapatam.
1805	After defeats in India, the British Colonel William Congreve extended the range of powder rockets from 600 to 2000 yard (ca 550 to 1829 m).
1807	Most of Copenhagen burned by 14,000 powder rockets by Royal Navy.
1812	Congreve's rockets used by Royal Navy at siege of Fort McHenry, Baltimore (rockets' red glare in 'The Star-Spangled Banner').
1816	The Austrian Count Frederick von Stadion synthesized KP in Germany.
1831	The French G.S. Serullas discovered AP.
1846	The Italian A. Sobrero discovered liquid nitroglycerine.
1863	The Swede A. Nobel manufactured nitroglycerine.
1863	The Prussian F.E. Schultze prepared smokeless NC propellant.
1864	The French P. Vieille prepared smokeless NC propellant.
1875	Nobel mixed nitroglycerine with nitrocellulose to form a gel.
1877	The French L. Jousselin prepared nitroguanidine NQ
1888	Nobel invented the smokeless double-base Ballistite.
1889	The British F. Abel and J. Dewar patented smokeless double base Cordite
1899	The German G.F. Henning prepared RDX for medicinal use.
1938	A.F. Belyaev and Ya.B. Zel'dovich, ICP, Russia, proposed "Theory of combustion of volatile explosives" (leading reaction in the gas phase).
1942	Ya.B. Zel'dovich, ICP, Russia, published "On the theory of the combustion of powders and explosives" (transient solid propellant burn).
1942	First castable KP composite solid propellant by J. Parsons at GALCIT (later JPL), CA, USA.
1943	The American W.E. Bachmann prepared HMX (first synthesized in 1930).
1948	KP first replaced by AP at Aerojet to reduce smoke.
1955	Al first tested at ARC by C. Henderson and K. Rumbel.
1960	first operational AP + Al + PU (Polaris A1).
1962	first operational AP + Al + NC/NG (Polaris 2).
1964	first operational AP + HMX + Al + NC (Polaris 3).
1964	Cubanes first synthesized at University of Chicago, IL, by Eaton and Cole.
1968	Patent for HNF preparation released to J.R. Lovett, USA
1970	Poseidon C3 first submarine launch, 2 stages CMDB.
1977	Trident I C4 first launch, 3 stages XLDB-70 (70% solids).
1971	ADN first synthesized by V.A. Tartakovsky and O.A. Luk'yanov, Zelinsky Institute of Organic Chemistry, Moscow, Russia (SRI, USA, 1988).
1987	CL-20 first synthesized by Arnie Nielsen, China Lake, CA, USA.
1990	Trident II D5 first submarine launch, 3 stages XLDB-75 (75% solids).

derivatives are of interest for their high density, thermostability, and remarkable insensitivity to ESD, friction, and impact. High enthalpy of formation and good thermal stability of tetrazine cycle allow producing tetrazine-based HEM, which can be used as insensitive, thermostable, environmentally friendly ingredient in many applications. Studies of combustion behavior show that most tetrazines are low-volatile substances with high surface temperatures, which determines a dominant role of the condensed phase in combustion of many tetrazine derivatives. The tetrazine cycle shows low reactivity in redox reactions, leading to rather moderate burning rates of its salts with oxidizing acids and coordination compounds despite high energetic characteristics.

According to Singh (DeLuca et al., 2016, Ch. 4), use of new powerful and green oxidizers like ADN and ammonium ozonide along with energetic binders like GAP, BAMO and BAMO-THF copolymer can boost Is above 320 s. Inclusion of energetic plasticizers like tetra azido malonate and tetra azido glutarate can further enhance Is. Likewise, inclusion of nitrogen-rich compounds like nitro guanidinium azides, N_8, N_{10}, etc., can boost energy content further on the higher side. Addition of nano-sized metal powders and nano-sized metal hydrides is likely to enhance burning rates significantly. Yet, most of these new eco-friendly ingredients and formulations still must go through the long and uncertain process of upgrading from laboratory to industrial level.

ADVANCED OXIDIZERS FOR SOLID ROCKET PROPELLANTS

The main target for advanced formulations is to identify a green and highly performing replacement for AP. The demands for this ideal solid oxidizer are chlorine-free, as low toxic as possible, thermally stable, compatible with the commonly used propellant ingredients, and a good aging behavior during in-service time (DeLuca et al., 2016, Ch. 1, 30, 31 and 32). In this respect, new oxidizers were studied in (DeLuca et al., 2016, Ch. 2 and 6). Kettner and Klapötke (DeLuca et al., 2016, Ch. 2) underline that the quest for novel oxygen sources for chemical propulsion is the focus for research activities all around the world. The strategy these authors follow is the synthesis of neutral CHNO-based materials for full or partial replacement of AP. Several new compounds with positive oxygen balance, that were synthesized in their laboratory over the last few years, are described. Interest is on the syntheses of molecules containing the trinitromethyl functionality due to the high oxygen content of this group. In their systematic and long-range investigation, the authors present four different molecule classes as backbones for diverse polynitro moieties. The resulting neutral compounds are discussed in detail concerning their properties as oxygen carriers in solid rocket propellants. They all show up advantages and disadvantages for the suitability in such applications. The fluorine-containing materials were proven to be inferior compared to their analogue trinitromethyl derivatives in terms of Is, but they meet the requirements of high thermal stabilities and low sensitivities better than the trinitromethyl compounds.

Fan et al. (DeLuca et al., 2016, Ch. 6) characterized two new tetrazole salts, HADNMNT and HATO. Theoretically, HADNMNT is a potential oxidizer to replace AP, RDX, HMX, and CL-20 in composite propellants. Safety tests showed that HATO exhibits excellent thermal stability and low mechanical sensitivities. The compatibilities of HATO with HTPB, AP, RDX, and Al powder in vacuum stability tests were good. Results from comparative studies of HATO and RDX as ingredients for composite propellants showed that HATO formulations offer the advantages of high burning rate and low mechanical sensitivities. The higher combustion rate of propellants based on HATO is in a good agreement with the greater burning rate of neat HATO compared with RDX (Sinditskii et al., 2015).

However, these and other promising suggestions are still far from flight applications. Today, possible candidates as primary oxidizers for solid rocket propulsion are mainly AN, AP, HNF, ADN, and HNIW (commonly known as CL-20). Moreover, RDX and HMX are often used as secondary oxidizers in high performance military propulsion missions. While AN and AP are well-known solid inorganic oxidizers and have been used at all levels for several decades, RDX and HMX are well-known solid organic explosives. On the opposite, HNF ADN, and CL-20 are relatively new solid HEM characterized by high-energy content, chlorine-free formulation, and smokeless exhausts. All of them are in principle potential candidates to replace AP. At any rate, castable propellants demand reasonable viscosity and high solid fraction loading. In this respect, particles with minimum spatial extension are sought for, favoring a low aspect ratio and preferably a spherical shape.

Main features of the above solid oxidizers as monopropellants are listed in Table 2 (DeLuca et al., 2013); T_f is the self-deflagration adiabatic temperature at 6.8 MPa. To assure data consistency, when possible, one common reference was selected for the same property (for example, O_2 balance and T_f). A comparative experimental analysis of ballistic properties of all mentioned oxidizers (except AN) - as monopropellants - was performed, under strand burner conditions, by (Atwood et al., 1999). For ADN the self-deflagration rate is the highest, while its initial temperature sensitivity is higher than that of AP for most of the commonly used pressure range (say, 0.1 – 7.0 MPa for space launchers). A value of self-deflagration is around 30 mm/s at 6 MPa with pressure sensitivity $n \approx 0.6$, while over the pressure interval 0.2 - 4 MPa sensitivity is slightly reduced to n=0.46 - 0.6; the yellow area is where data scattering is concentrated. The observed high self-deflagration rate, well evidenced in the comparative plot of Figure 2, is a drawback for space exploration applications. Note also the much lower deflagration rates and much larger PDL of AN as a monopropellant, while intermediate values of deflagration rates and PDL are observed for AP as a monopropellant. In addition, "burning rates of ADN-based propellants are characterized by an inverse relationship with oxidizer dispersity, as compared to the AP-based ones" (Pak, 1993).

HNF is ruled out as AP replacement because of compatibility problems with ingredients and a decomposition behavior difficult to stabilize (Bohn, 2015; Tummers et al., 2012; Lowers et al. 1999; Sinditskii et al., 2002; Sinditskii et al., 2002). Ballistic tests of HNF-based solid propellants have often revealed an exceedingly large pressure dependence (with the pressure exponents near one), or severe

Table 2. Properties of current solid propellant oxidizers (DeLuca et al., 2013).

Denomination	Chemical Formula	Oxygen % content/balance	Molar mass $\mathcal{M}$, g/mol	Density ρ, g/cm³	ΔH_f kJ/mol	T_f K	Cost	Environmental Impact
Ammonium Dinitramide (ADN)	$NH_4N(NO_2)_2$	51.6 / +25.8	124.1	1.81	−134.6	2051	–	+ (HNO)
Ammonium Nitrate (AN)	NH_4NO_3	60.0 / +20.0	80.0	1.73	−365.7	1247	+	+ (HNO)
Ammonium Perchlorate (AP)	NH_4ClO_4	54.5 / +34.0	117.5	1.95	−295.8	1406	=	– (ClHNO)
CL-20 or HNIW	$(NNO_2)_6(CH)_6$	43.8 / -10.9	438.2	2.04	+454.0	3591	– –	= (CHNO)
Tetranitramine (HMX)	$C_4H_8N_4(NO_2)_4$	43.2 / -21.6	296.2	1.90	+75.0	3278	≈	= (CHNO)
Hydrazinium Nitroformate (HNF)	$N_2H_4HC(NO_2)_3$	52.4 / +25.0	183.0	1.86	−72.0	3082	NAp	= (CHNO)
Trinitramine (RDX)	$C_3H_6N_3(NO_2)_3$	43.2 / -21.6	222.1	1.82	+61.5	3286	≈	= (CHNO)

= is the reference value; ≈ about equal to reference; – less good than reference; + better than reference (qualitative scale).

Figure 2.

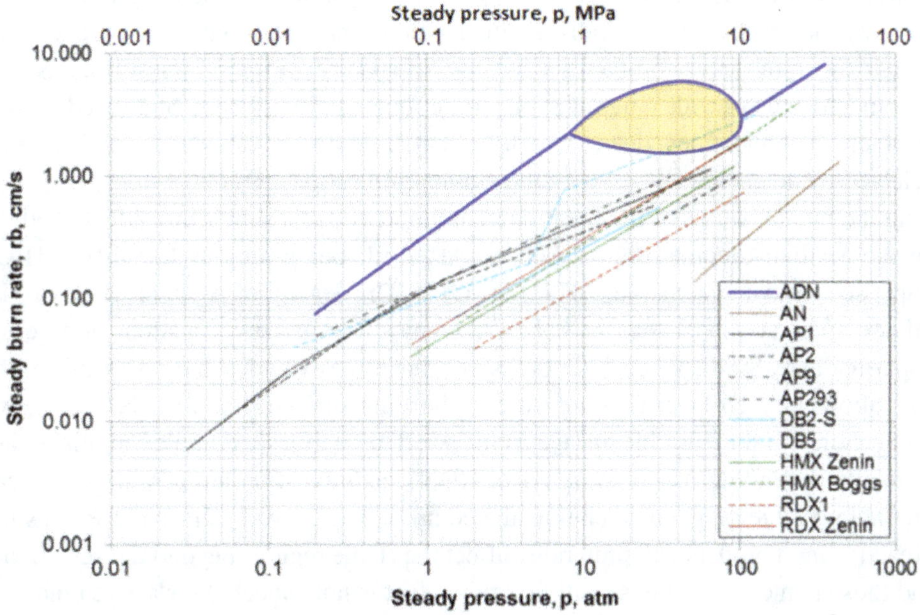

incompatibilities when the pressure exponent could be reduced to more desirable values, say below 0.6. In Europe, HNF was actively produced in the Netherlands; today, mainly India and China continue efforts.

CL-20 is the highest energy as well as the highest density compound known among organic chemicals, albeit some concerns exist about its susceptibility to impact and friction (DeLuca et al., 2016, Ch. 29 by Simakova and Parmon). All known methods to synthesize CL-20 are based on HBIW conversion, but this is a major challenge. A low conversion factor and high costs of the platinum catalysts commonly used require significant improvements to reduce CL-20 production costs. CL-20 is manufactured in the ε crystal phase (highest density) by Thiokol in USA and SNPE, now Eurenco, in France; manufacturing is also available in Russia, India, and China.

Although RDX and HMX are not employed as monopropellants because of their explosive nature, their use as co-oxidizers in high-performance propulsion systems is widespread. Several examples of synergistic effect in solid rocket propulsion are discussed in (DeLuca et al., 2016, Ch. 13, 15, 16 and 17); for details see next sections. Pang et al. (DeLuca et al., 2016, Ch. 17) studied the synergistic effect of AP on HMX in formulations including in mass 65% oxidizer (AP+HMX), 18% μAl, 12% HTPB binder, 2% catocene, and 3% additives. Under the explored experimental conditions, up to 15 MPa, AP+HMX dual-oxidizer formulations result in heat of explosion and density decrease as well as burning rate decrease of 12.3–14.5% compared with that of the reference AP-based mono-oxidizer propellant. The mechanical properties of these dual-oxidizer compositions are lower than those of the AP-based mono-oxidizer propellant.

In principle, the main shortcomings of currently used AP-based propellant formulations can be overcome by replacing AP with more energetic and chlorine free solid oxidizers. But using RDX, HMX, HNF, and CL-20 as main oxidizers for space exploration missions is not recommended for safety and performance reasons. In detail, HNF experimentally revealed too hazardous, while HNIW or CL-20 still appears way too expensive and maybe hazardous as well for large-scale industrial applications. The remaining candidate oxidizers are the ever green AN, the ubiquitous AP, and the promising ADN. AP is in many other ways an excellent solid oxidizer (considered a "miracle of nature" by propellant chemists) and still today is by far the most commonly used one for space exploration missions. AN is a low-cost green oxidizer much used in a wide range of applications, but mainly its Phase-Stabilized version (PSAN) is viable for practical applications, due to a series of crystalline transitions near ambient temperatures. The three selected oxidizers are all based on the ammonium cation $(NH4)^+$.

Comparing the main features of the mentioned monopropellants shows that ADN ideal gravimetric specific impulse neatly overcomes that of AN and AP, while being roughly like that of HNF and CL-20, for example see (Kuo and Acharya, 2012, Fig. 1.45). The density of ADN (1.81 g/cm^3) is less than that of all other common oxidizers except AN (1.73 g/cm^3). Detailed investigations of pure ADN flame structure, in particular by Sinditskii et al. (2006), show that three extended and essentially gaseous zones (like for DB) follow the melt layer floating over the burning surface (like for AN). Thus, ADN flame is not tightly anchored to the burning surface as for AP. This implies a complex multistep heat release mechanism mainly depending on pressure. Also, ADN decomposition temperatures are scattered but overall lie in the low value range, much below the commonly observed values of AP. This fact implies that slowly reactive inert binders will preferably affect the flame regions beyond the aerosol decomposition zone and thus augment pressure sensitivity; this effect is not expected for active binders.

Systematic studies about the aging of ADN containing formulations were conducted by Bohn and Cerri (DeLuca et al., 2016, Ch. 32). Several ADN-based propellants were tested including different prepolymers (GAP, Desmophen® D2200), curing agents (BPS, Desmodur® N3400), and filler types (Al, HMX). AP-based formulations were also manufactured for comparisons. The glass-rubber transition temperature was found systematically lower with the AP formulations compared to ADN. The Desmophen®-based propellant formulations were subjected to an extensive aging program. The aging temperature ranged between 60 and 85 °C, with aging times adjusted to a thermal equivalent load of 15 years at 25 °C. Whereas the AP-based materials did not change by aging, the ADN-based ones showed significant degradation effects.

ADN, as a monopropellant, features high burning rates and fair pressure sensitivities, low surface temperatures, low signature, and full environmental respect. ADN also possesses several unique physical and chemical properties as compared to other solid oxidizers, such as high condensed phase heat capacity of 0.49 cal/g/K, surface temperatures like AN but less than AP, and PDL around 0.2 MPa. But the observed high burning rate may be a drawback for space exploration applications. Moreover, the needle shaped ADN crystals hinder manufacture of propellant samples combining suitable ballistic performance and good mechanical properties. In Europe, manufacture of ADN was licensed by FOI to Eurenco Bofors in Sweden in 1996. Production of ADN prills from needle shaped crystals was developed at FOI (by spraying molten ADN through a nozzle) and ICT in Germany (by an emulsion method) (DeLuca et al., 2013).

METALLIC FUELS FOR SOLID ROCKET PROPELLANTS

Metals are the favorite ingredient to impart energy to solid rocket propellants. Important properties of relevant metals (Al, B, Mg, Zr) and of their most common oxides are summarized in (DeLuca et al., 2016, Ch. 8). The volumetric heat release is maximum for B, followed by Al and Zr; while the gravimetric heat release is maximum for B, followed by Al and Mg. The metal melting temperature is much less than that of the corresponding oxide for Al and Mg, while the opposite is true for B. The peculiar thermodynamic properties of B hamper its efficient combustion and nozzle expansion. The interest for Zr in propulsion applications is explained by its large value of density (6.52 g/cm³), making possible remarkable ΔV increments of the vehicle velocity. The density of all other commonly employed metalized fuels is much less, including AlH_3 (1.477 g/cm³). The α crystal phase of AlH_3 can be very stable in time, compatible with most ingredients of common use, conducive to superior ballistic properties, and prone to excellent performance of solid rocket motor upper stages. However, stabilization of α-AlH_3 is an essential pre-requisite for practical applications. As of today, the proper stabilization technique is known only to Russia and not shared. Al and even more AlH_3, like most metallic additives, lead to an Is increase mainly because Al is strongly oxidized by H_2O as combustion product, releasing large energy and the light molecular mass H_2.

Updated discussions about metallic ingredients, including nano-sized metals and agglomeration effects, are presented in several papers of (DeLuca et al., 2016). DeLuca et al. (2016, Ch. 8) tested a range of modified Al powders, from uncoated to coated nAl particles and from chemically to mechanically activated μAl. These variants were comparatively tested under laboratory burning conditions. Mainly the class of aluminized composite propellants (AP/inert binder) and operating conditions typical of space applications were investigated. Each of the tested Al variants has its own properties, and implementation in full-scale propulsive systems needs to be carefully evaluated for an overall assessment. For example, a promising technique proposed to make B a useful ingredient, despite its unfavorable thermodynamic properties, resorts to dual Al and B metal fuels (Chan et al., 2004).

In this respect, several Mg_xB_y composite dual-metal powders were tested under laboratory conditions differing by the Mg coating amount (ranging from 10 to 60% in mass) and B purity (90 or 95%). The theoretical ideal Is, as well as the adiabatic flame temperature, decrease whenever Al is replaced, totally or partially, by B, Mg, or Mg_xB_y composite metals. Experimental results show that the addition of Mg_xB_y composite metal powders leads to a measurable increase of steady burning rate, with respect to the corresponding μAl baseline, with both 90% and even more 95% purity. In either case, the precise content of Mg coating has a minor effect on the propellant burning rate if it is in the range 10–25%. For increasing Mg content up to 60%, the burning rate gain fades away. Experimental results (Dossi et al., 2012) also point out the beneficial effect of Mg in lowering the Mg_xB_y ignition temperature to the level of nAl. Further experimental results (DeLuca et al., 2012) indicate a general decrease of the incipient average agglomerate size, especially for 25% Mg coating. The combustion process of the B compounds is essentially characterized by the formation of thin flakes at or below the burning surface.

Vorozhtsov et al. (DeLuca et al., 2016, Ch. 9) describe the production of 5 μm metal borides by self-propagating high temperature synthesis (SHS). The sharp size distribution curve of particles and purity are enough for use as fuel in high-energy materials.

Weiser et al. (DeLuca et al., 2016, Ch. 10) investigated the combustion behavior of Al particles in an ADN/GAP matrix at various pressures up to 15 MPa. The temperature measurements close to the propellant surface indicate higher values near the Al boiling point that accelerate the melting of Al particles and influence the agglomeration process. At higher pressures, the temperatures are in the magnitude of Al_2O_3 evaporation and decomposition close to 3000 K. The agglomeration process of aluminum particles is like that of the AP/HTPB/Al system. The incipient average agglomerate size is mainly influenced by the oxidizer particle size and decreases with increasing pressure.

Zhao et al. (DeLuca et al., 2016, Ch. 11) tested ignition of nAl coated with Oleic Acid (nAl@OA), Perfluorotetradecanoic Acid (nAl@PA) and Nickel Acetylacetonate (nAl@NA) by laser ignition. A critical value of laser heat flux was found in the ignition process of the coated nAl powders. The ignition delay time of nAl@NA is shorter than those of nAl@PA and nAl@OA, owing to catalysis effects of nickel acetylacetonate. The burning rate of the propellant sample containing nAl@NA is the highest among all the tested formulations and it reaches 26.13 mm/s at 15 MPa.

Babuk (DeLuca et al., 2016, Ch. 13) reviewed the state-of-the-art about the influence of various formulation factors on the CCP, i.e., agglomerates and smoke oxide particles, formed at the burning propellant surface. The influence of binder, oxidizer, and metal fuel was investigated considering active and inactive binders, AP, AN (pure and phase stabilized), ADN, HMX, μAl and nAl powder, Al powder with a polymeric and refractory covering. Within the framework of a general physical picture, the CCP properties (in terms of size, chemical composition, and internal structure) were seen to depend on the burning propellant surface layer properties, which in turn depend on the propellant ingredients. The importance of the SL nature was underlined.

The results allow making reasonable choices when creating new solid propellants. Overall, based on a wealth of experimental investigations, the recommended strategy for best results is a dual-mode Al mixture, (μAl+nAl) or (μAl+Mg_xB_y) or (μAl+AlH^3), synergistically exploiting each component. The well-known technology of nano-metal powders, successful in many other fields, is not yet suitable for rocket propulsion for a series of drawbacks including clustering of particles (DeLuca et al., 2016, Ch. 27), decreased active metal content, increased end of mix viscosity as well as propellant hazard sensitivities, and dependence on passivation technique. Both for (μAl+nAl) and (μAl+MgxBy) dual-mode Al mixture, a good approach appears that of a partial replacement of μAl (say, around 1/5 to 1/6 of the total μAl commonly loaded in composite propellants for space propulsion).

BINDERS FOR COMPOSITE SOLID ROCKET PROPELLANTS

Binders are an essential ingredient for castable composite propellants. They allow binding the propellant solid particles (oxidizers and metal, if present) and attaining the wanted mechanical properties. Starting in the late 1970's, HTPB became the material of choice for a variety of commercial applications; a "propulsion grade" HTPB is being tested in the USA as of this writing. HTPE is another hydroxylated binder primarily used today for rocket propellants that contain energetic plasticizer, usually Bu-NENA, providing formulations with relatively high elongations. Propellants based on HTPE binders essentially feature mechanical properties like those that use HTPB, but show a less severe response to slow cook-off, ESD, and bullet impact tests for IM compliance. The HTPE density is larger than that of HTPB (1.04 vs. 0.92 g/cm^3). In addition, OB for HTPE is larger than for HTPB (-220.5% vs. -323.8%), thus allowing more Al to be loaded and further increasing density. Full details about the polyester diol Desmophen® D2200, recently tested as inert prepolymer for elastomeric binder at ICT, are reported in (Gettwert, Tagliabue, Weiser et al., 2015; Gettwert, Fischer & Menke, 2013). A summary of relevant binder properties is shown in Table 3. The nitrate ester PGN featuring large values of OB (-60.5%) and density (1.39-1.45 g/cm^3), stands as one of the most energetic nitrate ester pre-polymers for potential use in binder systems. However, when cured using aliphatic isocyanate agents, PGN ages poorly and this hinders its applicability. Also, the glass-rubber transition temperature, being within about -35 °C to -30 °C, is somewhat high.

Updated discussions about binders, including both inert and energetic binders, are dealt with in (DeLuca et al., 2016). Babuk (DeLuca et al., 2016) clarifies how, for propellants containing an active binder, the formation of SL and CCP depends on the combustion pressure. In the low-pressure domain, a SL is formed but this practically disappears when moving to high pressures, which leads to a sharp falling of the agglomerate mass fraction in the CCP composition. The specific laws depend on the combustion features of the active binder. In general, the domain of high pressures is found above 5.0 MPa for all propellants of this type.

Pei et al. (DeLuca et al., 2016, Ch. 14) examined the combustion characteristics of propellants based on (BAMO-GAP) copolymer. Their results show that while for ideal energetic properties there is an optimum ratio between AP and RDX or HMX or CL-20, performance improves linearly if AP is replaced by ADN. When AlH_3 replaces Al in the propellant formulation, energetic properties are much improved. Propellants based on (BAMO-GAP) copolymer have excellent properties, such as low flame temperature, high gravimetric specific impulse, good mechanical properties, and stable combustion. The (BAMO-GAP) copolymer appears a most promising energetic binder for solid propulsion applications.

Pivkina et al. (DeLuca et al., 2016, Ch. 15) studied the synergistic effect of AP on HMX and showed that formulations with active binder and AP-coated HMX provide higher burning rate than those ones with mechanical mixtures of HMX with fine AP. This implies the possibility to use considerably less amount of AP to achieve the same level of burning rate.

Rashkovskiy et al. (DeLuca et al., 2016) proposed a theoretical combustion model, for mixtures of energetic binders with inert and active fillers, which considers the curvature of the binder layer burning surface and ignition delay of the filler. The results of the model were successfully compared with experimental data for mixtures of energetic binders with several oxidizers. An alternative combustion mechanism was proposed in (Sinditskii et al., 2012).

In conclusion, HTPB and HTPE (for IM applications) are today the recommended inert binder systems, while GAP and BAMO-GAP appear the most viable for active binders.

ADN-BASED SINGLE- OXIDIZER SOLID ROCKET PROPELLANTS

Based on the work under progress, the basic propellant ingredients considered in this paper are the following:

- ADN, AP, AN, and PSAN as solid crystalline inorganic oxidizers.

Table 3. Properties of some prepolymers used for current binder systems.

Denomination	Chemical Formula	Oxygen balance, %	Molar mass $\mathcal{M}$, g/mol	Density ρ, g/cm^3	ΔH_f kJ/mol	T_f K
Desmophen® D2200	$C_{10}H_{16.678}O_{5.267}$	-166.9	221.2	1.18	-976.1	-
GAP DIOL	$C_3H_5N_3O$	-121.1	99.1	1.28	+117.2	1570
HTPB-R45T	$C_{10}H_{15.4}O_{0.07}$	-323.8	136.8	0.92	-62.0	-
HTPE	$C_6H_{12}O_2$	-220.5	116.1	1.04	-485.3	-
PGN	$C_3H_5NO_4$	-60.5	119.1	1.39-1.45	-322.8	1465

- Al (of different size and variants by proprietary chemical/physical treatments) and AlH_3 as solid metallic fuels. This list includes on one side propulsion grade micron-sized Al (μAl), activated Al (actAl), amorphous Al (amAl), aluminum hydride (AlH_3) and on the other side nano-sized Al (nAl).
- HTPB, HO-[(CH_2=CH-CH=CH_2)]n-OH, and a polyester diol-based, Desmophen® D2200, were employed as prepolymers for the inert propellant binders. The energetic polyether polyol-based GAP, HO-(CH_2-CH(CH_2N_3)O)n-H, was employed as prepolymer for the active propellant binder (DeLuca et al., 2016, Ch. 32).

Detailed properties of most of the ingredients under consideration (oxidizers, metallic fuels, and binder prepolymers) were presented elsewhere (DeLuca et al., 2013; DeLuca et al., 2014; DeLuca, 2016; Palmucci, 2014). Comparing the ideal gravimetric specific impulse shows that ADN again neatly overcomes that of AN and AP using either inert (HTPB) or energetic (GAP) binder system.

Bonding agents, for example TEPANOL (or HX-878), TEPAN (or HX-879) and BHEGA (or HX-880), were mainly developed for AP/HTPB-based formulations. In some cases, typically CTPB- or PBAN-based formulations, bonding agents also serve as curing agents. In the open literature, much is reported for AP bonding agents but little for ADN.

A wide variety of nanomaterials has been developed over the years (Gromov et al., 2016; Yan et al., 2016), including nano-metal particles, metal oxides, metal salts, metallic composites, organometallic compounds, energetic nano-catalysts, and carbon nano-materials. These additives could increase both the decomposition and the burning rate as well as enhance the combustion efficiency of the corresponding solid propellants. In the open literature, much is reported about AP, little about AN, and virtually nihil for ADN.

Experimental testing at several international sites pointed out the inadequacy of ADN as single oxidizer. For example, the European project HISP (High specific impulse Propellant for In-Space Propulsion, www.hispfp7.eu) focused on ADN as a single-oxidizer and analyzed two metallized formulations, based on either the standard AP/HTPB matrix or ADN with inert or active binder. Under the explored operating conditions, none of the tested ADN-based formulations looked ready to immediately replace AP for space propulsion. This is shown in Figure 3, evidencing the strong effects of the implemented binder (inert HTPB or energetic GAP) on the steady burning rates of ADN-based propellants, with respect to a flight proven industrial AP/HTPB/Al propellant taken as reference. GAP yields a reasonable pressure dependence but with excessively large burning rates (Wingborg, 2010); HTPB yields reasonable burning rates (at relatively low pressures) but with excessively large pressure dependence (de Flon et al., 2011). As usual, the presence of μAl does not imply a significant effect on the ballistic properties of ADN/GAP, in terms of both burning rates and pressure dependence (Wingborg, 2014).

In addition to several Al variants, stabilized alane AlH_3 was also tested, in view of its highly attractive propulsive features (rumored indeed to be used in upper stages of Russians military vehicles) (DeLuca et al., 2009; Wingborg and Calabro, 2016). Under all explored circumstances, AlH_3 appreciably mitigates the burning rate pressure dependence. Alane shows fast dehydrogenation at the burning surface and leaves behind porous Al crystals ready to react, removing heat release from the gas phase to near the burning surface and thus decreasing pressure sensitivity. See DeLuca et al. (2013, Figure 6) for AP/HTPB formulations and Figure 4 in this paper for ADN/GAP formulations (Wingborg, 2014).

Figure 3.

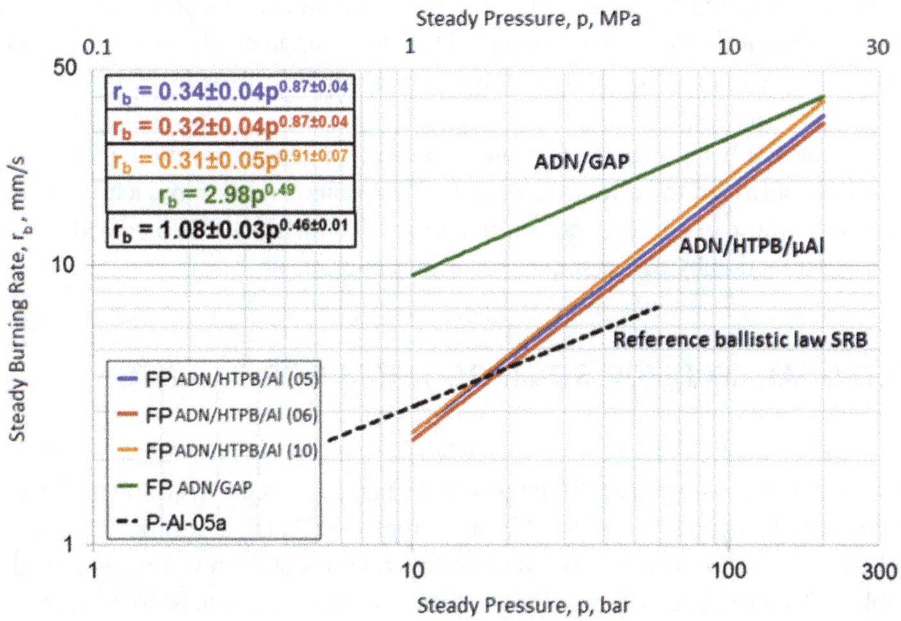

Figure 4.

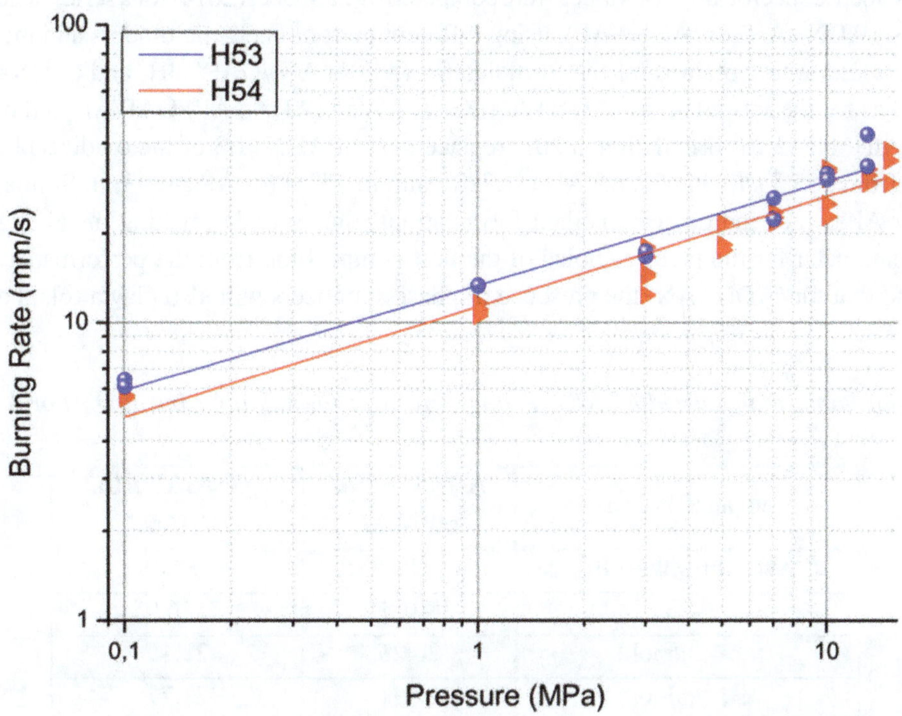

In the case of AP/HTPB, also an impressive increase of burning rate by a factor of 2 is achieved. For ADN/GAP, samples H53 and H54, containing AlH_3 with a bimodal ADN prill (70% 208 μm + 30% 55 μm), reveal only a minor influence on burning rate. Despite the appreciable change in formulation from 16% for H53 to 26% alane for H54, steady burning rates are quite close (24-26 mm/s at 7 MPa), with the 16% sample a bit faster.

In conclusion, despite being high-performance and fully green, as from other programs, ADN as a single-oxidizer is confirmed to be less than ideal. The recommended approach is to use ADN with a proper co-oxidizer whose choice depends on the propulsive mission. This updated approach will be illustrated in the next section by two examples of current interest.

ADN-BASED DUAL-OXIDIZER SOLID ROCKET PROPELLANTS

Compositions based on dual-oxidizers, such as (ADN+AN) or (ADN+AP), allow a partial replacement of ADN and thus mitigate its intrinsically large self-deflagration rates. Slowdown effects on burning rate were experimentally reported, for either AN or AP replacing ADN, in several sources from the open literature (DeLuca et al., 2014). But the overall ballistic properties of the resulting dual-oxidizer systems are also sensibly influenced, depending on the ADN replacement fraction. Moreover, in both cases the ideal specific impulse is penalized, although only slightly as discussed next. At any rate, the possible presence of AP again implies the presence of some chlorine in the exhausts, while AN requires a phase stabilizer for high performance propulsive missions.

Ideal gravimetric specific impulse values were computed by Palmucci (2014) for a series of dual-oxidizer formulations, (ADN+AN) or (ADN+AP), using different inert or energetic binders and metallic fuels.

Detailed results were obtained for the formulations (ADN+AN)/GAP/AlH_3 and (ADN+AP)/GAP/AlH_3 as well as for the formulations (ADN+AN)/GAP/Al and (ADN+AP)/GAP/Al. In the case of the ADN/GAP/AlH_3 60/12/28 formulation, a 20% replacement of ADN (1/3 of the oxidizer) implies a decrease of Is from 355.3 s to 349.2 s for AN (-1.72%) and to 352.0 for AP (-0.93%). Similar trends are observed for Al as fuel and other propellant formulations, thus actually making metal agglomeration the crucial parameter for the precise choice of the best composition from the performance viewpoint. Keep in mind that for (ADN+AN) the presence of phase stabilizers will also play a role. Paying atten-

Table 4. Performance of metalized ADN/GAP single-oxidizer formulations burning Al or AlH_3 fuel

Computed Property	ADN/GAP/Al 60/18/22	ADN/GAP/*AlH*3 62/10/28
I_s (vacuum with ϵ=40), s	331.50	355.88
T_c, K	3801.12	3556.05
$\mathcal{M}_c$, g/mol	26.26	21.15
$\mathcal{M}_e$, g/mol	27.31	21.77
CCP (chamber), molar %	9.40	9.01
CCP (nozzle exit), molar %	11.13	10.16

tion to maximum performance, some results from thermochemical computations are reported in Table 4. Alane as a fuel gives higher performance in terms not only of theoretical specific impulse (almost +8% increase), but also of adiabatic flame temperature, molecular mass of the combustion products, and amount of condensed products. Indeed, the combustion temperature is lower, resulting in less need of thermal protection, as well as the average molecular mass is reduced. Their trends are reported in Figure 5, showing that the molecular mass decreases faster than the temperature, thus implying a higher T_c/M_c. The condensed products amount is also smaller leading to reduced losses by two-phase flow during the nozzle expansion of the mixture. Another advantage of adding AlH_3 to a GAP matrix is the pressure sensitivity reduction of the burning propellant. This is very useful in improving internal ballistics of ADN-based propellants. Although formulations with only 10% by mass of GAP are difficult to manufacture, more suitable formulations can be tested with only minor changes of the performance level.

Relevant properties of both (ADN+AN)- or (ADN+AP)-based dual-oxidizer propellants are discussed next.

(ADN+PSAN)-BASED DUAL-OXIDIZERS

As stressed in Wingborg and Calabro (2016), sustainable development for the space industry in Europe is reflected by ESA´s Clean Space Initiative and the Green Propulsion Harmonization Process. In this respect, the unsatisfactory experimental results of the HISP project, see Figure 3, motivated another joint effort focused on directly affecting the oxidizer behavior. The new project called GRAIL (GReen Advanced hIgh energy propellant for Launchers, www.grail-h2020.eu) is again coordinated by FOI for three-years, ca 2015-2018, and includes a number of European partners. GRAIL resorts to ADN as a dual-oxidizer and precisely seeks for a replacement of AP by using a green mixture of the high-performance ADN and the low-cost AN. ADN has a dual advantage over the workhorse oxidizer AP in terms of clean combustion and superior heat of combustion. But due to lower oxygen content, it is not possible to replace AP by ADN one-to-one (Wingborg and Calabro, 2016; Tagliabue, 2015; Gettwert et al., 2015; Tagliabue et al., 2016).

Figure 5.

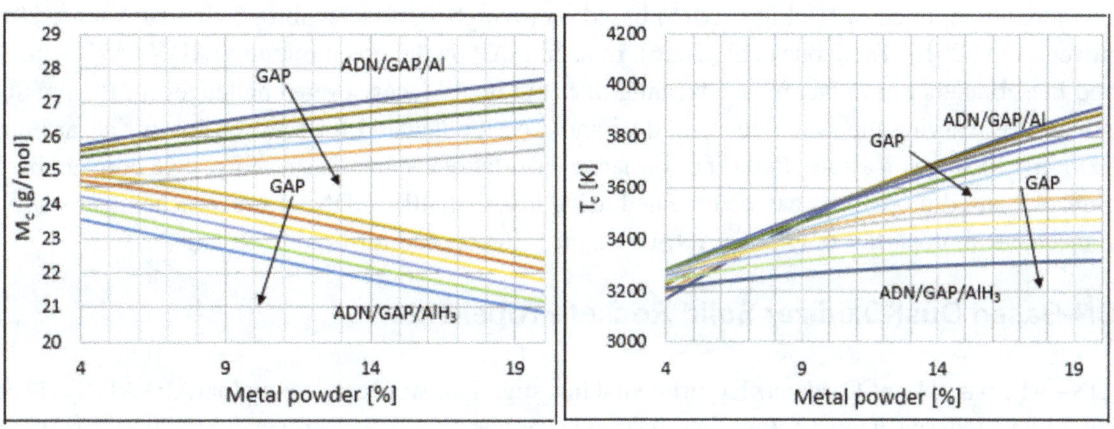

In a preliminary series of tests conducted at ICT, a total of 24 different formulations were investigated, changing the PSAN/ADN ratio GAP, HTPB, Desmophen® D2200, and a mixture GAP/ Desmophen® D2200 (80/:20) were tested as binder systems. This led to different amount of oxidizer for each binder system due to processibility issue, but the ratio of coarse/fine particle distribution was kept constant (70/30). ADN was used in a bimodal distribution of prills with 48 μm and 212-218 μm nominal particle size. Al powder was used in two particle sizes: 4 μm and 20 μm nominal diameter. For all formulations, Al was 18% of the total mass.

AN was used only its Phase-Stabilized version (PSAN). Spherical particles of PSAN were produced at ICT stabilized with either 7% KNO3 (KNO_3-PSAN, 30 μm size) or 3% NiO (NiO-PSAN, 120 μm size). Solid mixtures of ADN and PSAN proved compatible only for KNO_3-PSAN, whereas NiO-PSAN promoted ADN decomposition.

Ignition revealed difficult for both inert binders at the low range of the tested pressures. For HTPB-based samples, PDL would increase with a larger content of ADN: from 1 MPa with AN to 3 MPa with ADN. For Desmophen®-based samples, PDL would increase with a larger content of AN: from 2 MPa with pure ADN to 10 MPa for AN/ADN ratio 70/30. Desmophen®-based samples with pure AN oxidizer could be ignited only occasionally.

Partially replacing ADN with AN led to a monotonically gradual and fair reduction of the ideal Is going from pure ADN to pure AN. Within the explored operating conditions, steady burning rate was also reduced but in a less regular way: the decrease became considerable only above, say 30% ADN replacement. In detail, the partial replacement of GAP by Desmophen® could halve burning rates over most of the AN/ADN range.

Unfortunately, the pressure sensitivity was also dramatically affected. The pressure exponent was already too large for pure ADN (above 1 for inert binders and around 0.7 for active binders) and with ADN replacement it further increased for inert binders, while it remained essentially constant for active binders; see Figure 6 for both trends. Within the explored operating conditions, at least 70% AN of the total oxidizer content is required to achieve a reasonable value of pressure exponent with inert binder.

The comparison between active (GAP) and inert (HTPB) binder systems was extended to friction and impact sensitivities as well. Experimental testing pointed out that altogether insensitive, or reduced sensitivity, propellants can be obtained only with pure AN as oxidizer and the effect of ADN/AN ratio is quite low. Moreover, friction sensitivity is somewhat worsened by GAP binder, while impact sensitivity appears about the same for both active and inert binders.

As already reported in (DeLuca, 2016) based on (Wingborg and Calabro, 2016; Tagliabue, 2015; Gettwert et al., 2015; Tagliabue et al., 2016), replacing AP by the green mixture (ADN+AN) with the tested inert binders can in fact reduce burning rates but it still faces severe challenges for the pressure exponent. By varying the ratio between ADN and AN, the demand for low sensitivity (hazard classification 1.3) seems feasible. Initial performance calculations show that performance comparable to current AP based propellants can be obtained using less than 40% ADN. However, chemical stability and mechanical properties of (ADN+AN)-based propellants need further investigations.

ADN-Based Dual-Oxidizer Solid Rocket Propellants

(ADN+AP)-based Dual-Oxidizers Experimental investigations were carried out for (AP+ADN)/HTPB/Al at Xi'an Modern Chemistry Research Institute (Pang et al., 2013; DeLuca et al., 2016, Ch. 17) by testing formulations including 64% oxidizer (AP+ADN), 18 μAl, 13% HTPB binder, and 5% additives

Figure 6.

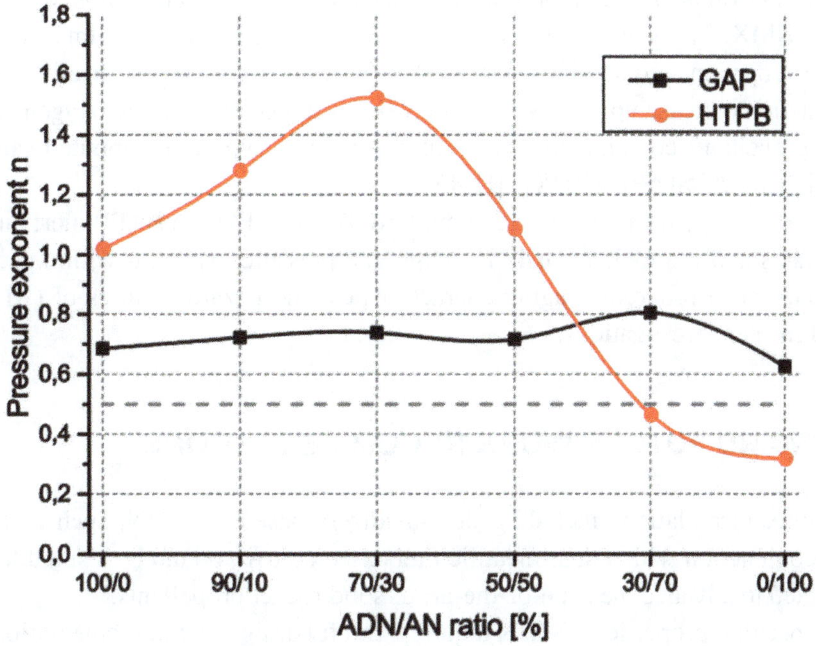

in mass. Four (AP+ADN) mixtures were implemented ranging from 64+0 (pure AP) to 54+10, 49+15, and 44+20 mass fractions. AP was used in a bimodal grain distribution: the average size was 105-147 μm for coarse particles and 1-5 μm for fine particles. ADN prills were obtained by spraying molten ADN through a nozzle and the average diameter was ranged 147 to 205 μm. The solidified ADN drops were then coated by a polymer binder.

Little difference could be observed for the tested four formulations of (AP+ADN)/Al/HTPB in terms of computed thermochemical properties, under the ideal operating conditions of 7.0 MPa combustion pressure and 0.1 MPa exit pressure: specific impulse Is (from 256.8 s to 260.1 s) and characteristic velocity c* (from 1518.5 m/s to 1540.3 m/s) slightly increase with increasing AP replacement by ADN. On the contrary, the adiabatic flame temperature (from 2895 K to 2851 K) and density (from 1.718 to 1.674 g/cm^3) slightly decrease by increasing the AP (ρ=1.95 g/cm^3) replacement by ADN (ρ = 1.81 g/cm^3).

The burning response is similar for most of the tested dual-oxidizer formulations: (AP+HMX), (AP+CL-20), (AP+GUDN). But for (AP+ADN), experimental results point out a sensible increase of burning rate and pressure exponent (from 0.41 to 0.71) for increasing AP replacement by ADN; see DeLuca et al. (2016, Figure 13). Experimental hazard results reveal that using the tested coated ADN particles involves a considerable increase of the propellant friction and impact sensitivities. This requires attention and warrants further work in possible ADN-based propellant applications. Both the friction and impact sensitivities increase for increasing ADN fraction.

Further analyses were indirectly carried out by Pei et al. (DeLuca et al., 2016, Ch. 14) to investigate the effects of Al mass fraction on the performance of propellants made of (BAMO-GAP) copolymer 15%, GAP 5%, and oxidizer 70%. The oxidizers were gradually replaced by Al up to 20%. Ideal performance was calculated by a Chinese thermochemical code at 7 MPa. The results show that the performance corresponding to each oxidizer can be quite different. For the formulations containing RDX, HMX, or

CL-20, the propellant Is first rises and then reduces when the Al mass fraction replacing the oxidizer goes beyond a certain value. Thus, an optimum value exists for the Al amount, say 8% for CL-20 and 10% for RDX or HMX. The value of T_c vs. the Al mass fraction shows the same trend as that of Is. However, for the propellants containing ADN or AP, the value of Is monotonically increase with the oxidizer mass fraction increase up to 20%. This may be attributed to the large oxygen content of ADN and AP. For the propellants containing different oxidizers, the maximum computed values of Is are in the order [ADN]>[CL-20]>[AP]>[RDX]>[HMX].

Overall, replacing AP by the more energetic mixture (ADN+AP) with HTPB inert binder can in fact reduce burning rates and it simultaneously increases the pressure exponent within acceptable limits. However, the mechanical properties, aging characteristics, and hazard features of (ADN+AP)-based propellants need further investigations.

CONCLUSION AND FUTURE WORK RECOMMENDATIONS

For the near future, formulations including dual-oxidizers based on ADN, such as (ADN+AN) or (ADN+AP), in conjunction with a dual-metallic fuel based on μAl, bound by a suitable active or inert binder, are expected to advance the state-of-the-art of solid rocket propellants.

However, the peculiar properties of ADN require a careful design and implementation of the propellant formulation. Most investigations so far accomplished did not fully exploit the unusual dependence of ADN ballistic properties on particle size (Shang and Huang, 2010; Franzin, 2013; Pang et al., 2014). Small ADN prills are available, down to 2 μm at ICT (Roßmann et al., 2014), involving the potential dual advantage of reduced slope and burning rate (above the crossover pressure observed around 2 - 3 MPa for monomodal ADN/GAP 70/30) as well as reduced agglomeration. Appropriate burning rate modifiers or coolants may be needed to fully control the propellant ballistic properties. As of now, in view of the experimental results already obtained, the mixture (ADN+AP) seems easier to master than (ADN+AN).

Overall, when dealing with ADN, it is the binder system still demanding a decisive effort to manufacture wellbehaving formulations; other critical areas are the implementation of a suitable phase-stabilizer for AN (only KNO_3-PSAN was compatible with ADN (Tagliabue et al., 2016)) and bonding agent for the oxidizer (BHEGA is compatible with AP but not with ADN (DeLuca et al., 2016, Ch. 32)). Concerning AN phase stabilizer, it was shown that AN can be phase stabilized over the temperature range -50 °C to + 100 °C by adding small amounts of organic compounds (Golovina et al., 2006). Concerning bonding agents, specific products were already tested for ADN-based formulations with either uncertain (Gettwert et al., 2013) or unconfirmed results (Consaga, 1990; MACH I, 2009).

At any rate, the mechanical, aging, and hazard properties of the resulting formulations must be closely monitored and possibly improved to achieve the target of 1.3 hazard rating.

ACKNOWLEDGMENT

Sincere thanks are due to Prof. Valery P. Sinditskii, Dr. Thomas Heintz, and Mr. Andrea Franzin for useful comments.

REFERENCES

Atwood, A. I., Boggs, T. L., Curran, P. O., Parr, T. P., Hanson-Parr, D. M., Price, C. F., & Wiknich, J. (1999). Burning Rate of Solid Propellant Ingredients, Part 1: Pressure and Initial Temperature Effects. *Journal of Propulsion and Power, 15*(6), 740–747. doi:10.2514/2.5522

Bohn, M. A. (2015). Review of some peculiarities of the stability and decomposition of HNF and ADN. In *Proceedings of the 18th seminar new trends in research of energetic material*, University of Pardubice, Czech Republic, April 15–17.

Calabro, M. (2012). Evaluation of the Interest of New ADN Solid Propellants for the Vega Launch Vehicle. In Proceedings of 12-IWCP, Politecnico di Milan, Milan, Italy, June 9-10.

Cerri, S. (2011). Characterization of the Ageing of Advanced Solid Rocket Propellants and First Step Design of Green Propellants [PhD. Thesis]. Politecnico di Milano, Milan, Italy.

Cerri, S., Bohn, M. A., Menke, K., & Galfetti, L. (2014). Characterization of ADN/GAP-Based and ADN/Desmophen®-Based Propellant Formulations and Comparison with AP Analogues. *Propellants, Explosives, Pyrotechnics, 39*(2), 192–204. doi:10.1002/prep.201300065

Chan, M. L., Parr, T., Hanson-Parr, D., Bui, T., & Mason, M. (2004). Characterization of a boron containing propellant, In L.T. DeLuca, L. Galfetti, R.A. Pesce-Rodriguez (Eds.), Novel energetic materials and applications. Grafiche GSS, Bergamo.

Consaga, J. P. (1990). Bonding Agents for Composite Propellants. U.S. Patent 4,944,815.

Davenas, A. (2003). The Development of Modern Solid Propellants. *Journal of Propulsion and Power, 19*(6), 1108–1128. doi:10.2514/2.6947

de Flon, J., Andreasson, S., Liljedahl, M., Oscarson, C., Wanhatalo, M., & Wingborg, N. (2011). Solid Propellants Based on ADN and HTPB. *AIAA Paper.*

DeLuca L. T. (2016) Innovative Solid Formulations for Rocket Propulsion. *Eurasian Chemico-Technological Journal, 18*(3), 181-196.

DeLuca L. T., Galfetti L, Maggi F, Colombo G, Reina A, Dossi S, Consonni D, Brambilla M (2012). Innovative metallized formulations for solid or hybrid rocket propulsion. *Hanneng Cailiao [Chin. J. Energetic Mater.], 20*(4), 465–474.

DeLuca, L. T., Galfetti, L., Severini, F., Rossettini, L., Meda, L., Marra, G., & Pavlovets, G. J. et al. (2007). Physical and ballistic characterization of AlH_3-based space propellants. *Aerospace Science and Technology, 11*(1), 18–25. doi:10.1016/j.ast.2006.08.010

DeLuca, L. T., Palmucci, I., Dossi, S., Maggi, F., Franzin, A., Trache, D., & Weiser, V. et al. (2014). *Combustion Features of ADN-Based Solid Rocket Propellants for Space Applications, Zel'dovich Memorial* (Vol. 1, pp. 108–116). Moscow, Russia: Torus Press.

DeLuca, L. T., Palmucci, I., Franzin, A., Weiser, V., Gettwert, V., Wingborg, N., & Sjöblom, M. (2014, February). *New Energetic Ingredients for Solid Rocket Propulsion.* In Proceedings of *HEMCE '14* (pp. 13–15).

DeLuca, L.T., Rossettini, L., Kappenstein, C., Weiser, V. (2009). Ballistic Characterization of AlH_3-Based Propellants for Solid and Hybrid Rocket Propulsion (AIAA Paper 2009-4874).

DeLuca, L. T., Shimada, T., Sinditskii, V. P., & Calabro, M. (Eds.). (2016). Chemical rocket propulsion: A comprehensive survey of energetic materials. doi:. Cham, Switzerland: Springer International Publishing. doi:10.1007/978-3-319-27748-6

Dossi, S., Reina, A., Maggi, F., & DeLuca, L. T. (2012). Innovative metal fuels for solid rocket propulsion. *Int. J. Energy Mater. Chem. Propuls.*, *11*(4), 299–322.

Franzin, A. (2013). *Burning and Agglomeration of Aluminized ADN/GAP Solid Propellants* [MSc. Thesis]. Politecnico di Milano, Milan, Italy.

Gettwert, V., Fischer, S., & Menke, K. (2013). Aluminized ADN/GAP Propellants – Formulation and Properties. In *Proceedings of the 44th Int'l Annual Conference of ICT*, Karlsruhe, Germany.

Gettwert, V., Tagliabue, C., Weiser, V., & Imiolek, A. (2015). Green Advanced High Energy Propellants for Launchers (GRAIL) - First results on the Burning Behavior of AN/ADN Propellants. In *Proceedings of the 6th European Conference for Aeronautics and Space Sciences* (EuCASS 2015), Krakow, Poland.

Golovina, N. I., Nechiporenko, G. N., Lempert, D. B., & Lempert, D.B. et al. (2006). Phase stabilization of ammonium nitrate by means of determined classes of nitrate heterocyclic compounds. In Condensed Energetic Systems.

Gromov, A. A., Korotkikh, A. G., Il'in, A., DeLuca, L. T., Arkhipov, V. A., Monogarov, K. A., & Teipel, U. (2016). Nanometals: Synthesis and Application in Energetic Systems. In V.E, Zarko & A.A. Gromov (Eds.), Energetic Nanomaterials: Synthesis, Characterization, and Application (Ch. 3). Elsevier. doi:10.1016/B978-0-12-802710-3.00003-9

DeLuca, L. T., Maggi, F., Dossi, S., Weiser, V., Franzin, A., Gettwert, V., & Heintz, T. (2013). High-energy metal fuels for rocket propulsion: Characterization and performance. *Chin. J. Explos. Propellants*, *36*, 1–14.

Kuo, K. K., & Acharya, R. (2012). Fundamentals of Turbulent and Multiphase Combustion. In Solid Propellants and Their Combustion Characteristics (Ch. 9). John Wiley & Sons, Inc.

Louwers, J., Gadiot, G. M. H. J. L., Landman, A. J., Peeters, T. W. J., van der Meer, Th. H., & Roekaerts, D. (1999). Combustion and Flame Structure of HNF Sandwiches and Propellants (*AIAA Paper* 99–2359).

MACH I, Inc. (2009). A New Class of Bonding Agents for High Energy Propellants (Technical Bulletin).

Pak, Z. P. (1993). Some Ways to Higher Environmental Safety of Solid Rocket Propellant Application. (*AIAA Paper* 93–755).

Palmucci, I. (2014, July 25). ADN-Based Double Oxidizer Solid Propellants Formulations [MSc. Thesis]. Politecnico di Milano, Milan, Italy.

Pang, W. Q. (2012). Effects of ADN on the Properties of Nitrate Ester Plasticized Polyether (NEPE) Solid Rocket Propellants. In Proceedings of 12-IWCP, Politecnico di Milan, Milan, Italy, June 9-10.

Pang, W. Q., DeLuca, L. T., Fan, X. Z., & Liu, F. L. (2014). Effects of ADN on the Properties of NEPE Solid Propellant. In *Proceedings of the International Workshop on New Energetic Materials and Propulsion Techniques for Space Exploration*, Milan, Italy, June 9-10.

Pang, W. Q., Fan, X. Z., Zhang, W., Xu, H. X., Wu, S. X., Liu, F. L., & Yan, N. et al. (2013). Effect of ADN on the Characteristics of HTPB Based Composite Solid Propellant. *Journal of Chemical Science and Technology*, *2*(2), 53–60.

Roßmann, Ch., Heintz, T., & Herrmann, M. (2014) Generation of Very Fine ADN Particles by Bead Milling Technology. In *Proceedings of the 45th International Annual Conference of ICT*, Karlsruhe, Germany.

Shang, Dong-qin, & Huang, Hong-yong (2010). Combustion Properties of PGN/ADN Propellants. Chinese Journal of Energetic Materials, 18(4), 372–376.

Sinditskii, V. P., Egorshev, V. Y., Levshenkov, A. I., & Serushkin, V. V. (2006). Combustion of Ammonium Dinitramide, Part 1: Burning Behavior. *Journal of Propulsion and Power*, *22*(4), 769–776. doi:10.2514/1.17950

Sinditskii, V. P., Egorshev, V. Y., Levshenkov, A. I., & Serushkin, V. V. (2006). Combustion of Ammonium Dinitramide, Part 2: Combustion Mechanism. *Journal of Propulsion and Power*, *22*(4), 777–785. doi:10.2514/1.17955

Sinditskii, V. P., Egorshev, V. Yu., Serushkin, V. V., Filatov, S. A., & Chernyi, A. N. (2012). Combustion Mechanism of Energetic Binders with Nitramines. *Int. J. Energy Mater. Chem. Propuls.*, *11*(5), 427–449.

Sinditskii, V. P., Egorshev, V. Y., Serushkin, V. V., & Levshenkov, A. I. (2002) Chemical peculiarities of combustion of solid propellant oxidizers. In *Proc. of the 8th Int. Workshop on Rocket Propulsion: Present and Future*, Pozzuoli, Italy, June 16–20 (pp. 1-20).

Sinditskii, V. P., Filatov, S. A., Kolesov, V. I., Kapranov, K. O., Asachenko, A. F., Nechaev, M. S., & Shishov, N. I. et al. (2015). Combustion behavior and physico-chemical properties of dihydroxylammonium 5,5-bistetrazole-1,1-diolate (TKX-50). *Thermochimica Acta*, *614*, 85–92. doi:10.1016/j.tca.2015.06.019

Sinditskii, V. P., Serushkin, V. V., Filatov, S. A., & Egorshev, V. Y. (2002). Flame structure of hydrazinium nitroformate. *International Journal of Energetic Materials and Chemical Propulsion*, *5*(1-6), 576–586. doi:10.1615/IntJEnergeticMaterialsChemProp.v5.i1-6.600

Tagliabue, C. (2015). Burning Behavior of ADN/AN based Solid Rocket Propellants [MSc. Thesis]. Politecnico di Milano, Milan, Italy.

Tagliabue, C., Weiser, V., Imiolek, A., Bohn, A. M., Heintz, T., & Gettwert, V. (2016) Burning Behavior of AN/ADN Propellants. In *Proceedings of the 47th International Annual Conference of ICT*, Karlsruhe, Germany.

Tummers, M. J., van der Heijden, A. E. D. M., & van Veen, E. H. (2012). Selection of burning rate modifiers for hydrazinium nitroformate. *Combustion and Flame*, *159*(2), 882–886. doi:10.1016/j.combustflame.2011.08.010

Umholtz, P. D. (1999). The History of Solid Rocket Propulsion and Aerojet. (*AIAA Paper* 99–2927).

Weiser, W., Franzin, A., De Luca, L. T., Fischer, S., Gettwert, V., & Kelzenberg, S., Knapp, S., Raab, A., & Roth, E. (2012). Burning Behavior of ADN Solid Propellants Filled with Aluminum and Alane. In Proceedings of 12-IWCP, Politecnico di Milan, Milan, Italy, June 9-10.

Wingborg, N. (2012). Status of ADN-Based Solid Propellant Development. In Proceedings of 12-IWCP, Politecnico di Milan, Milan, Italy, June 9-10.

Wingborg, N., Andreasson, S., de Flon, J., Johnsson, M., Liljedahl, M., Oscarson, C., & Wanhatalo, M. et al. (2010). Development of ADN-Based Minimum Smoke Propellants. *AIAA Paper*, 2010–6586.

Wingborg, N., & Calabro, M. (2016). Green Solid Propellants for Launchers. In *Proceedings of the Space Propulsion conference*, Rome, Italy, May 2-6.

Yan, Q.-L., Zhao, F.-Q., Kuo, K. K., Zhang, X.-H., Zeman, S., & DeLuca, L. T. (2016). Catalytic Effects of Nano Additives on Decomposition and Combustion of RDX-, HMX-, and AP-Based Energetic Compositions. *Progress in Energy and Combustion Science*, *57*, 75–136. doi:10.1016/j.pecs.2016.08.002

Chapter 2
Enhancement of Energetic Materials Combustion Process

Rene Francisco Boschi Goncalves
Federal University of Pará, Brazil

Koshun Iha
Aeronautics Institute of Technology, Brazil

José A. F. F. Rocco
Aeronautics Institute of Technology, Brazil

ABSTRACT

Fuels and energetic materials enhancements have been focus of several studies in the past few years, as the sole development of the engines (design, structure of the combustion chambers and stronger materials) leads to a limited efficiency. One of the most effective ways of enhancing the energetic capabilities of a fuel is via enrichment with a highly reactive species, such as hydrogen and ozone, as these materials significantly increase the reaction mechanism kinetics, enabling the occurrence of faster reactions and increasing the temperature of the chamber. The present chapter aims to summarize some of the applications of fuel enrichment and the so far achieved results, focusing in internal combustion engines and systems for automotive, energy-generation or aerospace applications.

INTRODUCTION

During the past decades, propulsive systems have gone through significant modifications and enhancements, aiming higher efficiencies, lower pollutant emissions and lower fuel consumption. The problem resides in the technical limits of these modifications, i.e. physical changes in the system (e.g. dimensions, design, composite materials etc.) shall not impact so much without modifying the reactant mixture (fuel/oxidizer) in parallel. On the other hand, application of other fuel species may increase the power yield of the engine, but generate even worse pollutant emissions (such as halogenated species). In the hypothesis of using new fuels or formulations, the emissions generated must be severely controlled.

DOI: 10.4018/978-1-5225-2903-3.ch002

Despite all of the precautions taken according to the Kyoto Protocol United Nations Framework Convention on Climate Change, an increase of 27% was observed in CO_2 emissions from 1990 to 2004, and transport related CO_2 emissions increased by 37% (Sopena et al., 2010). Therefore, enhancements of the current combustion processes require immediate attention for new technologies and improvements.

One of the most promising ideas to fill that purpose is the use of additives for enhancing the completion of the combustion process. These additives may be oxidizer or fuel species. As an example, the sole use of higher O_2 concentrations causes an increase of ~30% (2200 K to 2800 K) in the temperature of the chamber and in water final content, i.e. improves the combustion process (Figure 1) (Beltrame et al., 2001). This result was achieved experimentally and numerically with a diffusion flame of methane with against current, with oxidant composed by a mixture of air and oxygen (68% enrichment). As the peaks approach the nozzle, it is stated that the combustion is faster (lower duration) and more complete, which generate less intermediates (highly pollutant species) like CO and unburned hydrocarbons (the reactants are completely consumed before the stagnation plane). The observed increase in the combustion chamber temperature enables the use of lean regime (lower concentration of fuels species – lower fuel consumption).

During the past few years, low molecular weight (and high reactive) fuel species have been studied in the presence of different fuels and engines, as an attempt of achieving better efficiencies and conditions. Therefore, this chapter will cover briefly the combustion of several formulations and systems, as well as some enhancement studies and forthcoming technologies.

HIDROGEN ENRICHMENT

The most common species used as additive or secondary fuel for combustion improvement is hydrogen. In small to moderate amounts, H_2 is able to reduce pre-ignitions (due to its high-octane rating), lower NOx emissions, reduce energy losses and increase power output of IC engines (Akansu et al., 2007; Ma et al., 2008; Bauer and Forest, 2001; Kahraman et al., 2009). Figure 2 below, adapted from the work

Figure 1. Computed temperature and major species profiles for: (a) methane/air flame, (b) methane/68% O_2 flame (Sopena et al., 2010)

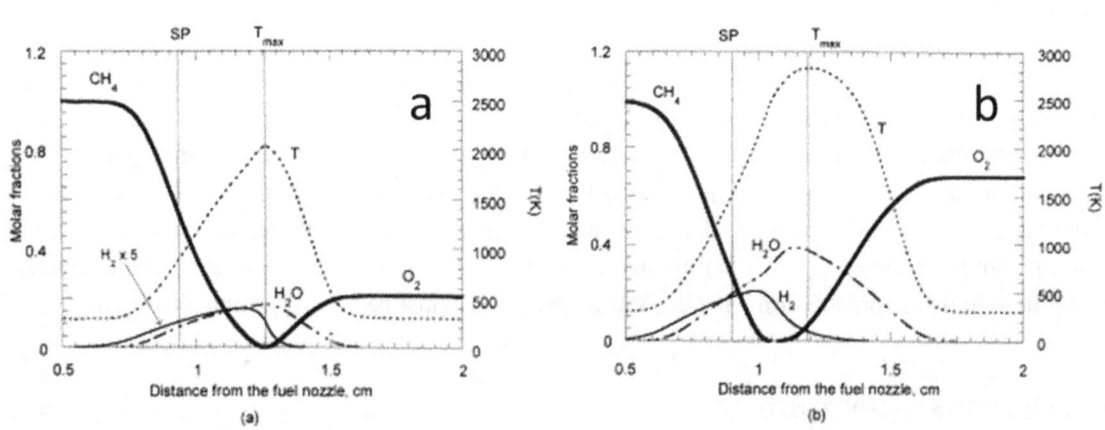

Figure 2. H_2 benefits and challenges

Property	Benefits	Challenges
Wide flammability range	Enables lean or more dilute combustion.	Lean combustion reduces power.
Low ignition energy	Allows lean yet rapid ignition.	Can prematurely ignite on hot spots.
Small quenching distance	Flame can travel closer to walls.	Can lead to flashback.
High auto-ignition temp	Enables higher compression ratio.	Hard to ignite in CI engine.
High diffusivity	Facilitates more uniform mixture (Homogenous).	Can leak to undesirable areas.
High flame speed	More closely approaches ideal Otto engine cycle (isometric heat addition).	N/A
High octane rating	Enables higher compression ratio. Reduces knock & MEP variation.	Hard to ignite in CI engine.
Low density	N/A	Lower energy density per volume.

of Greenwood et al. (2014), presents a comparison between H_2 benefits and challenges, when used as secondary fuel in IC engines.

One of the mentioned challenges of hydrogen regards its low density, which leads to a storage issue, as there are no safe and efficient methods available to all demands. Also, storing H_2 as cryogenic liquid, compressed gas or dissolved in metal hydrides significantly increases the weight of the vehicles, the overall price (Fontana et al., 2002; Whiete et al., 2006) and the risks of explosion due to an easy ignition (Abdel-Aal et al., 2005).

There are on-going studies for hydrogen production from biomass, heat and industrial chemical wastes (Saravanan and Nagarajan, 2008; McLellan et al., 2005), or via sunlight and water (electrolysis) inside the vehicle, so there will be no need for onboard storage (Bari and Esmaeil, 2010). However, there is still much to pursue in this front.

Hydrogen shows better performance in spark ignition engines, compared to compression ignition engines, as it accelerates the flame propagation, enhances the lean operational region, stability and BTE, while reducing emissions (Sandalci and Karagöz, 2014), as the combustion of hydrogen itself does not emit hydrocarbons, sulphur oxide, acids or CO_2 (Al-Baghdadi, 2004; Sastri, 1987). For hydrogen engines, a relative improvement in efficiency of 15-25% over gasoline operation was assumed, whereas fuel cell efficiency was taken to be 45% irrespective of load (Verhelst, 2014). Based on recent studies (Dhole et al., 2016; Dhole et al. 2016), a 10% increase on H_2 percentage increases the flame front speed by 46% and reduces the combustion duration by 24%, decreasing the heat transfer and fuel consumption and enhancing the engine efficiency (Figure 3). Although hydrogen combustion performance and characteristics are well known in theory, it is still difficult to precisely predict the effect of hydrogen addition to a specific combustion process or system in practice. Depending on the injection type, fuel mixture, flow, oxidizer content, internal pressure and other characteristics, the observed behavior changes drastically. In some cases there is great improvement in combustion and reduction in emissions; in others, higher fuel consumption is observed and the generation of toxic gases increases. Therefore, it is extremely important to study the effects generated by the specific characteristics of H_2 when modifying geometries, fuels, injection and cycles.

As an example, Khalil and Gupta (2013) studied the effect of H_2 enrichment in methane flames (Figure 4). A considerable difference is observed only by modifying the injection, especially regarding NO emissions. In non-premixed combustion, there is an approximate increase from 7 to 27% of NO generated with 15% H_2, while the observed increase is just from 5 to 7% for the premixed regime.

One of the most important and desired characteristic of any IC engine is a low fuel consumption. A useful way of reducing the fuel consumption in any system (turbines, CI and SI engines etc.) is the use of lean combustion. Lean air-fuel mixtures reduce significantly the temperature inside the combustion

Figure 3. Properties of fuels (Dhole et al., 2016)

Sr. No.	Properties	Diesel	Producer gas	Hydrogen
1	Lower heating value (kJ/kg)	42,800	6000	1,20,000
2	Minimum ignition energy (mJ)	-	-	0.26
3	Flame speed (cm/s)	2.0-8.0	20-30	265-325
4	Flammability limit (% volume in air)	0.6-7.5	7.0-21.6	4-75
5	Flammability limit (equivalence ratio)	0.6-2.0	-	0.1-7.1
6	Diffusion coefficient (cm^2/s)	-	-	0.61

Figure 4. NO and CO emissions in a gas turbine fueled with hydrogen-enriched methane in (a) premixed and (b) non-premixed combustion (Khalil and Gupta, 2013)

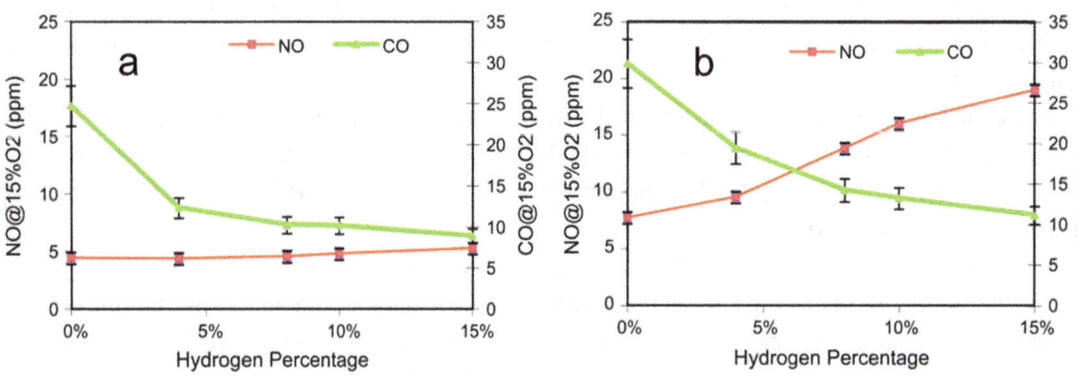

chamber, leading to lower NOx emissions (Correa, 1992). However, lean flames are susceptible to local flame extinction and quenching, leading to undesirable characteristics (flame quenching and acoustic combustion instabilities) and flame flashback.

Extended flammability limits achieved with H_2 enrichment enables the engine to operate in leaner air-fuel conditions, reducing NOx, CO, unburned HC and CO_2 emissions (Moreno et al., 2012), as a direct result of higher OH, H and O radical concentrations in the resulting flame (Schefer et al., 2002), and increases fuel economy. The use of lean hydrocarbon/air mixtures, however, are hard to ignite and have low flame speed, which may lead to increased unburned hydrocarbons, reduced performance and malfunction of combustors (Chenglong et al., 2014). As usual, the presence of hydrogen in the combustion chamber causes an increase in the mean temperature, proportional to H_2 volume fraction (Figure 5).

It is important to mention that, besides all applications of H_2 with liquid and gas fuels, there is the solid fuel combustion used in furnaces, boilers and blast furnaces. Ionel et al. (2014) studied the flame behavior of pulverized coal inside a furnace after having been treated with a hydrogen enriched gas. They observed a significant decrease in the sulfur oxides (35-40%) and a small increase in the nitrogen oxides emissions. As seen in the studies above, the characteristics of hydrogen enrichment are also observed in the solid fuel combustion and require further investigation, both theoretical and experimental.

Figure 5. Temperature according to equivalence ratio and hydrogen fraction. (Riahi et al., 2016)

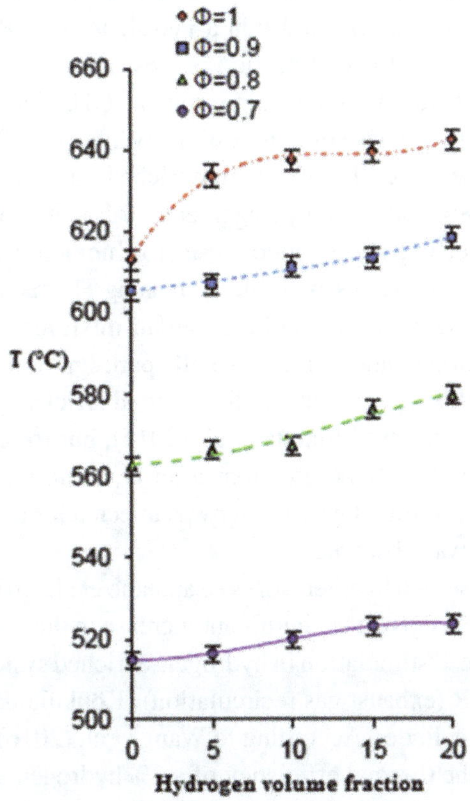

As there is a huge variety of combustion systems and fuels, it is a challenge to mention all applications and possibilities for the use of hydrogen. Therefore, the topics below shall better enlighten the effects of hydrogen enrichment in widely used fuels.

Hydrogen-Enriched Gasoline

Internal combustion engines powered by gasoline (isooctane) has been developed decades ago and are widely used in several applications (Norbye, 1971; Heywood, 1988), such as removable electricity generation, military aircrafts, cars etc. and most of the developments were based on the structure, parts and physical characteristics of the engine. However, there is enough space to add improvements in the combustion by changing or enhancing the air-fuel mixture. One of the most popular substances for enhancing the combustion velocity, temperature and yield is hydrogen. Due to its low ignition energy, high diffusivity, high flame velocity and wide flammability, hydrogen has been widely considered as an alternative fuel for several different engines (Zhen et al., 2013; Changwei et al., 2016; Maher and Haroun, 2003).

Wang et al. (2009 and 2010) have studied hydrogen addition on the basic combustion characteristics of some hydrocarbon fuels, which was very effective for enhancing the laminar burning velocity and shortening Markstain length of natural gas and liquid petroleum gas. Based on their results, H_2 can decrease significantly the duration of the combustion process and also improve the stability of flame structure due to the free radicals present in the flame, such as OH, H and O. This velocity increase is also observed for the flame initiation phase of combustion (Ivanic et al., 2005). Amrouche et al. (2014) analyzed a Wankel rotary engine fueled by hydrogen-enriched gasoline, which resulted in a thermal efficiency increase to about 28.8% with 10% hydrogen energy fraction. When pure gasoline is used in the engine, the highest thermal efficiency is found close stoichiometric conditions near 22%. As seen on Figure 6, the exhaust temperature drops with the increasing H_2 fraction, leading to a shorter burn duration and to a fast and complete combustion of the fuel/air mixtures.

During investigations of hydrogen addition to a small spark-ignition engine, fueled with gasoline, Varde (1981) discovered that hydrogen can improve the thermal efficiency, while reducing cyclic variations. The same effect was published by Chintala et al. (2015), but for compression ignition engines. In their study, there was an increase of NOx emissions after H_2 addition, while decreasing HC and CO emissions; NOx emissions were controlled when water was injected for reducing the temperature during combustion, and 37% reduction was observed.

This increase on NOx was observed in other works (Catapano et al., 2015; Wang et al., 2010; Wang et al., 2013; Ji et al., 2014), and is caused by the significant increase in the in-cylinder temperature, oxygen fraction and turbulence flame speed stimulation in hydrogen-enriched systems (Figure 7). Other possible control for NOx emission is EGR (exhaust gas recirculation) (Kökkülünk et al., 2014), where CO_2 and inert gas are re-injected into the cylinder. According to Wang et al. (2016), the increase on CO_2 volume fraction in the intake increases the thermal efficiency of a 3% hydrogen-enriched gasoline engine, and reduces the NOx emissions by almost 80% (Figure 8). This behavior is similar to the results of Yoshihito and Teruo (2005), where NOx emissions from the engine were almost reduced to zero when fueled with hydrogen-gasoline mixtures at lean conditions.

Figure 6. Brake thermal efficiency (a), exhaust temperature and maximum working chamber temperature (b) for different hydrogen fractions (Amrouche et al., 2014)

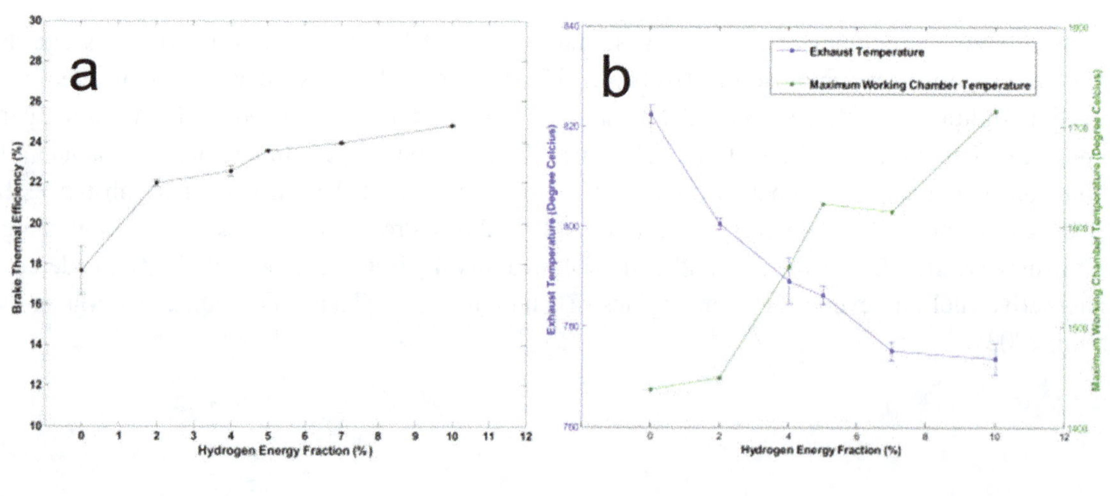

Figure 7. NOx versus hydrogen energy fraction, adapted from Wang et al. (2010)

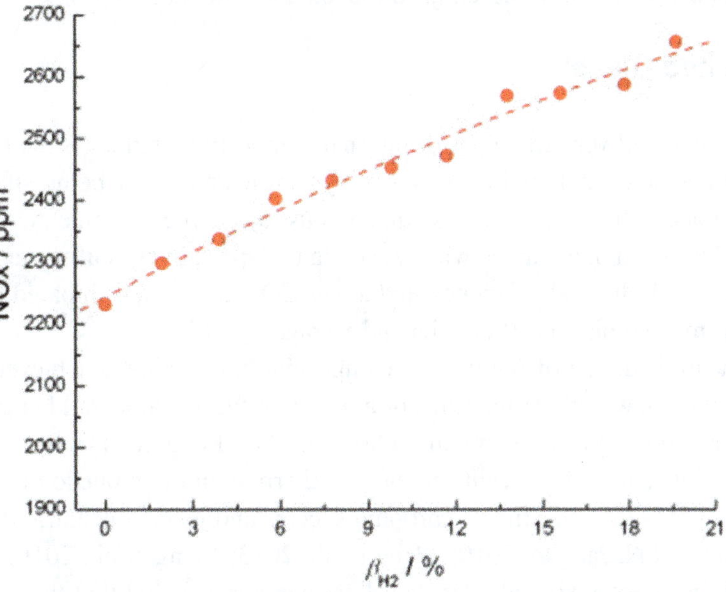

Figure 8. NOx emissions in pure gasoline engine (α_{H2} = 0) and in a 3% hydrogen-enriched gasoline engine (α_{H2} = 3%), according to the volumetric fraction of CO_2 re-injected into the cylinder (α_{CO2}). Adapted from Wang et al. (2016)

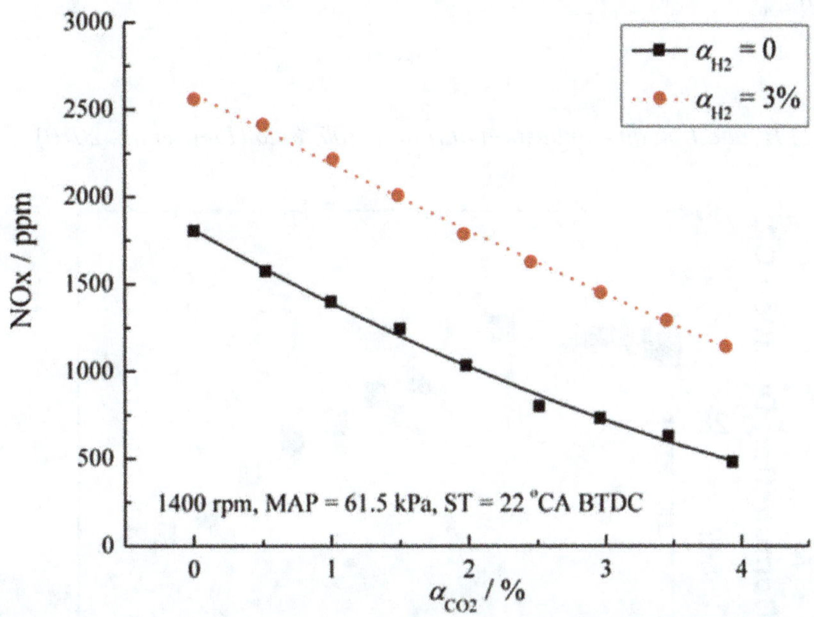

The forthcoming topics shall enlighten the use of H_2 together with the most common fuels widely used in internal combustion engines: diesel, gasoline, ethanol and natural gas.

Hydrogen-Enriched Diesel

One of the most efficient and versatile types of internal combustion engines is the Diesel engine. Used for several different machines and applications, this system is constantly being studied for improving performance and characteristics (injection system, combustion chamber geometry, valves modification, turbocharging etc.). Diesel engines can be widely seen in transport, in agriculture and in electricity production. Besides their high thermal efficiency and lower CO_2 emissions compared to SI engines, there are higher NOx and smoke emissions (Sandalci and Karagöz, 2014).

As previously mentioned, one of the ways for improving the engine is achieved by enhancing the combustion process with new fuels or mixtures of alternative fuels (Sahin and Durgun, 2009; Kilzitan, 1988). Similar to gasoline engines, hydrogen addition in diesel engines has been studied in the past few years. Recent results show significant increase in thermal and combustion efficiency, especially in medium to high-loads, while the smoke, carbonyl species and NOx concentrations decreased in the emissions (Saravanan and Nagarajan, 2010; Wang et al., 2013; Jhang et al., 2016). Liew et al. (2010) used hydrogen on a four-stroke, six-cylinder diesel engine, and verified that the combustion duration was drastically shortened (Figure 9) and the peak rate of heat release value increased.

Saravanan et al. (2007) have verified significant increases in the thermal brake efficiency and decrease in the specific energy consumption (Figure 10) using a single-cylinder, water-cooled, direct-injection compression-ignition engine, and 10 lpm constant hydrogen flow rate. In their study, several loads were applied and several injections time and duration were selected. In one of the tests, the TBE increased from 23.6% to 29.4%.

Figure 9. Effect of H_2 addition on combustion duration, 70% load (Liew et al., 2010)

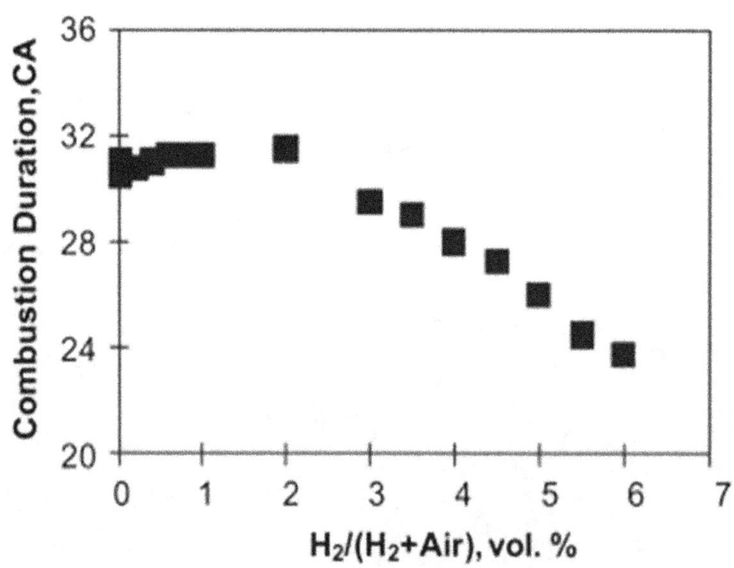

Figure 10. Variation of brake thermal efficiency (a) and specific energy consumption-SEC (b) with load. Adapted from Saravanan et al. (2007)

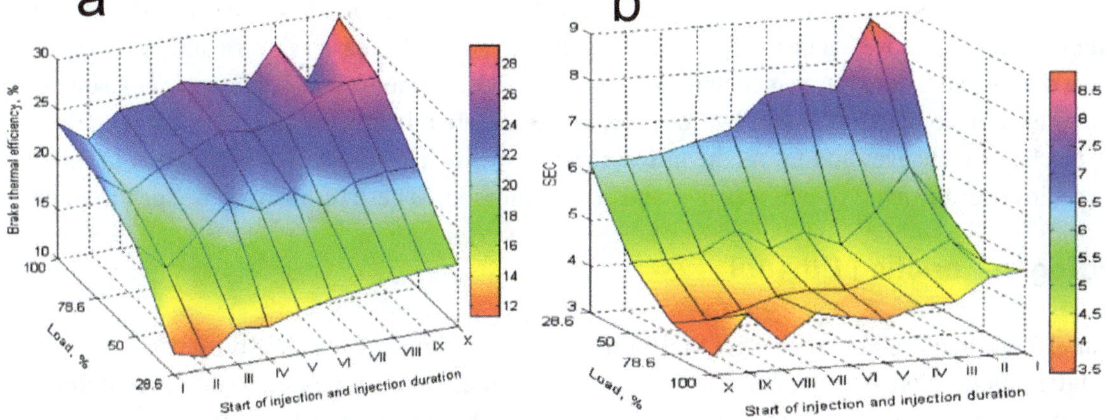

Depending on the engine characteristics, on the speed and on the load, the emissions composition can vary significantly. While in some situations there are improvements in HC, CO, and CO_2 emissions, NOx emissions may increase (Bari and Esmaeil, 2010; Köse and Ciniviz, 2013) (Figure 11), which could be attributed to the increase of HO_2 radical (Lilik et al., 2010). The effect of NOx emission is positive at low engine load and negative at high engine load (Zhou et al., 2014). By increasing the amount of hydrogen in the diesel, NOx can decrease, but with the cost of decreasing the brake thermal efficiency

Figure 11. NOx emissions for different values of hydrogen enrichment (2.5%, 5.0% and 7.5%) at different engine speeds. (Köse and Ciniviz, 2013)

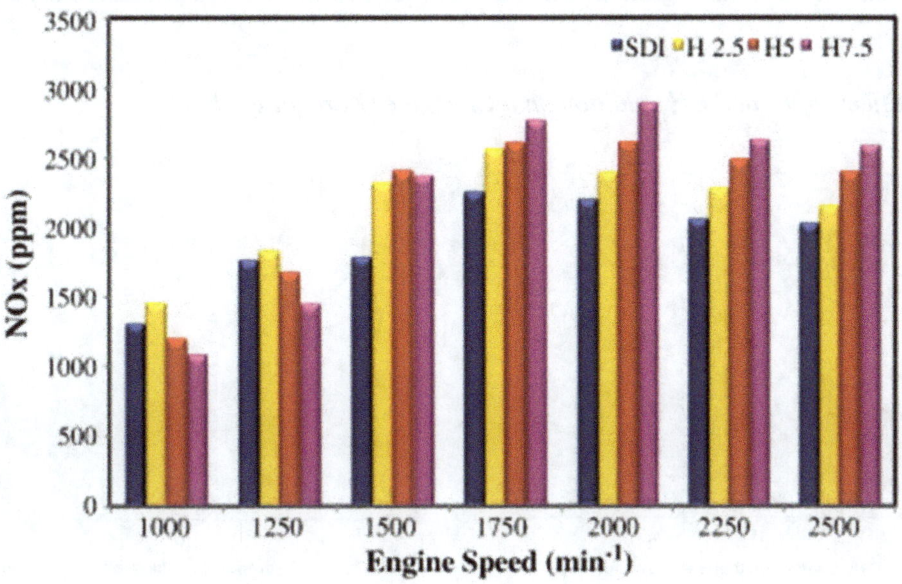

(20% reduction with ~50% H_2) (Karagöz et al., 2016). Hydrogen addition has also the potential to reduce unregulated emissions, such as olefins (C_2H_4 and C_3H_6), BTX (C_6H_6, C_7H_8 and C_8H_{10}) and acetaldehyde (C_2H_5O).

Negative effects were observed for hydrogen-methane mixtures in diesel engines. Different to the studies and results above, when using CH_4/H_2 in diesel, the TBE decreased significantly when compared to neat diesel, and was accompanied with an increase in CO emissions, which demonstrates a decrease in the combustion completion. When hydrogen and methane are mixed with diesel, probably the combustion mechanism changes and favors the fastest reactions, which lead to a high production of intermediate species over the main products (water and carbon dioxide). These results may be seen in Figure 12.

Hydrogen-Enriched Ethanol

Ethanol fuel is a sustainable alternative fuel for spark ignition engines, successfully used in some countries around the globe (Cooney et al., 2009; Schifter et al., 2011). Has also a high-octane rating, enabling higher compression ratios engines without pre-ignition (Radwan, 1985), and with high efficiency. Compared to gasoline, ethanol has a higher heat of vaporization and combustion velocity, which potentially increases the power output, gives a higher charge density through increased charge cooling and may generate less CO, HC and NOx emissions than gasoline (Ayala et al., 2006). Its higher laminar flame speed and lower flammability limit allows it to be run with leaner mixtures. According to Sahin et al. (2005), ethanol has proven itself a good additive for diesel engines, reducing emissions and increasing the mixing velocity of fuel and air, enhancing the combustion process and engine performance.

Among some of the disadvantages in using ethanol, there is the challenge with hydrocarbon emissions during cold start, which can be mitigated using secondary air injection or with fuel additives, like gasoline or hydrogen (H_2 enrichment) (Greenwood et al., 2014).

As hydrogen has the highest mass energy density of any fuel (120.1 MJ/kg vs. 26.9 MJ/kg of ethanol) and the highest stoichiometric air/fuel ratio (34.3 vs. 9 of ethanol), it may reduce the specific fuel consumption of the ethanol engine. Amrouche et al. (2016) found a significant increase in the peak pressure with the advance of the crank angle and an increase of over 14% in TBE of a Wankel rotary engine fueled

Figure 12. Effects of H_2 and CH_4 addition on a CI engine (Karagöz et al., 2016)

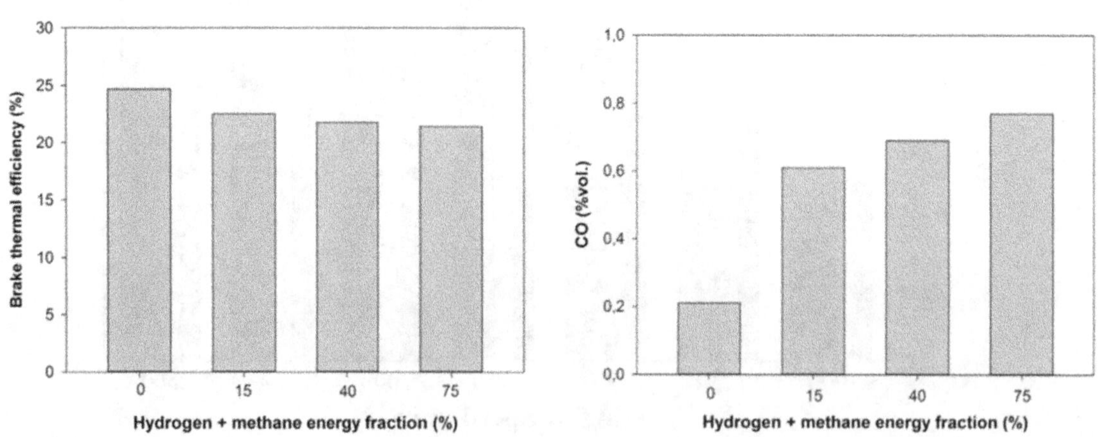

with ethanol enriched with H_2. An increased peak pressure results in an increased temperature inside the chamber. Hydrogen's characteristic high velocity combustion decreases the flame development and the propagation duration, lowering the exhaust losses and fuel consumption.

Hydrogen-Enriched Natural Gas

The use of natural gas in spark ignition engines is not new, and has grown fast in the last years in comparison to gasoline. It is mostly composed of methane, but can present other species in small fractions, such as ethane, propane, nitrogen, among others. When compressed to high pressures (~225 bar), receives the name of compressed natural gas-CNG, and when stored in liquid form (cryogenic cylinders), receives the name of liquefied natural gas-LNG.

The interest in using this type of fuel is due to its cleaner combustion (high H/C ratio) in comparison to other hydrocarbon fuels which, as mentioned before, potentially reduces toxic emissions, such as COx and NOx. The challenge in using natural gas lies in its low flame speed (high combustion duration), high ignition energy and poor stability, leading to lower overall efficiency. Figure 13 summarizes the differences of the main properties for combustion, among CNG, H_2 and gasoline.

The most striking feature of hydrogen for combustion, in comparison to CNG and Gasoline, is the lower heating value (LHV), three times over the others. This high energy allows the use of lower mass for the same output of the engine. However, as hydrogen has a very small density, the volume needed is quite high. Therefore, a way for increasing the LHV of the fuel is mixing the CNG with certain amounts of H_2.

Several studies have been reported about the performance of hydrogen-blended natural gas engines (Gaurav et al., 2016; Park et al., 2014; Hora and Agarwal, 2015; Lee et al., 2014). Park et al. (2013) observed that the lean burn limit was extended and the TBE of the engine increased by 0.74% after the hydrogen addition. Patil et al. (2010) observed an increase in the torque by 3% and in power by 4%, while reducing the fuel consumption by 4% and the emissions by 50% (CO) and 30% (NOx), as the engine can

Figure 13. Properties of CNG, H_2 and Gasoline (Hora and Agarwal, 2016)

Properties	CNG	Hydrogen	Gasoline
Formulae	CH_4	H_2	$C_{7.1}H_{12.56}$
Molar Mass	16	2	98
Octane Number	120	130	70–97
Auto Ignition Temperature (°C)	**540**	**585**	230–500
Flammability Limit (Vol. % air)	**5–15**	**4–75**	1–7.6
Flame Quenching Distance (mm)	**2.03**	**0.64**	2.84
Minimum Ignition Energy (mJ)	**0.29**	**0.02**	0.24
Stoichiometric A/F Ratio	17.16	34.33	14.7
Lower Heating Value (MJ/Kg)	47.3	120	40–45
Diffusion Coefficient (cm^2/sec)	**2.0**	**6.1**	0.5
Laminar Burning Velocity in Air (m/s)	**0.4**	**2.9–3.5**	0.5
Molar Carbon-to-Hydrogen Ratio	0.25	0	0.44
Density (Kg/m^3) at STP	0.717	0.0899	726

operate in a leaner regime (Figure 14). Improvements in thermal efficiency of ~6,5% with hydrogen addition to CNG were also reported by Lim et al. (2014), followed by a 75% reduction in NOx emissions.

NOx emissions tend to increase with hydrogen, due to previously mentioned characteristics, but seems that for CNG the generation of these species do not suffer so much with H_2 addition, probably due to the similar octane number and diffusion coefficient. Some modifications in ignition time or injection time are required for the emission reduction, but overall, they present significant enhancements in the results obtained (Figure 15).

OZONE ENRICHMENT

A substantial increase in the thermal efficiency and decrease in pollutant emissions can be achieved by controlling the point of ignition and delay time (dependent of the chemical species present in the reactants) of a homogeneous charge of fuel and air within the thermodynamic cycle of an engine (Schönborn et al., 2010). By inserting only a tiny quantity of atoms and radicals ($\sim10^{-4\%}$) is possible to shift equilibrium in the system and initiate the combustion process (Xiaoye et al., 2013). These radicals have, therefore, the ability to enhance an engine performance when present in very low concentrations in the fuel mixture. Among all the possible chemicals, ozone is a very reactive specie, which generates radicals almost instantly when in contact with other molecules.

Recently, Wang et al. (2012) proposed a new mechanism compilation for ozone (Figure 16). In this mechanism is possible to observe the effects of O_3 in contact to the most common gas species in a combustion chamber. As observed, there are several reactions with very small (and even negative) activation energies, providing high velocity of the radical chain reactions (production of radicals O, OH, HO_2 etc.), substantially increasing the flame burning velocity, decreasing delay time (Tachibana et al, 1991) and modifying the ignition characteristics of a reactant mixture. Using some simulation methods with this

Figure 14. Comparison of engine performance with CNG and HCNG

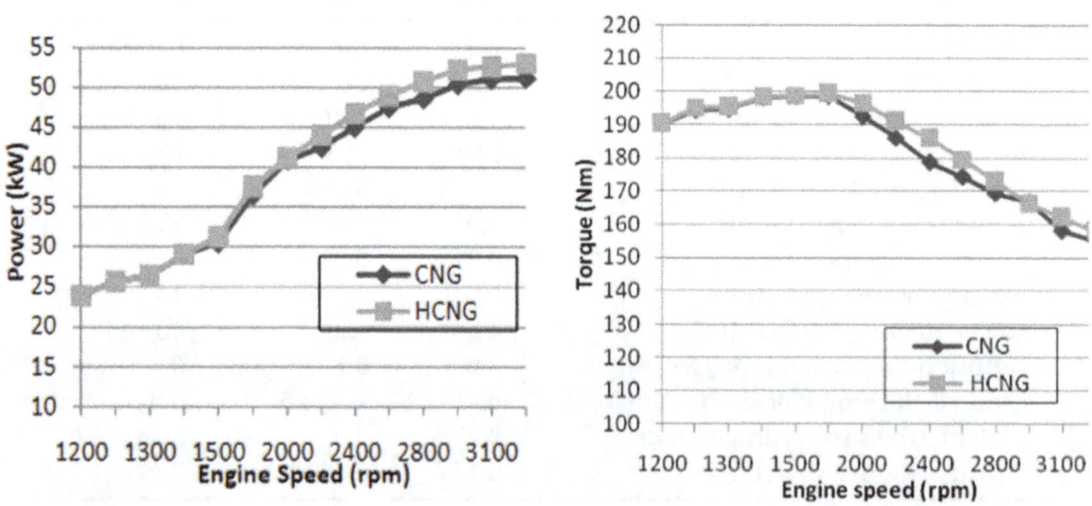

Figure 15. Brake specific NOx emission versus combustion duration for fuel blends of various hydrogen fractions (Ma et al., 2007)

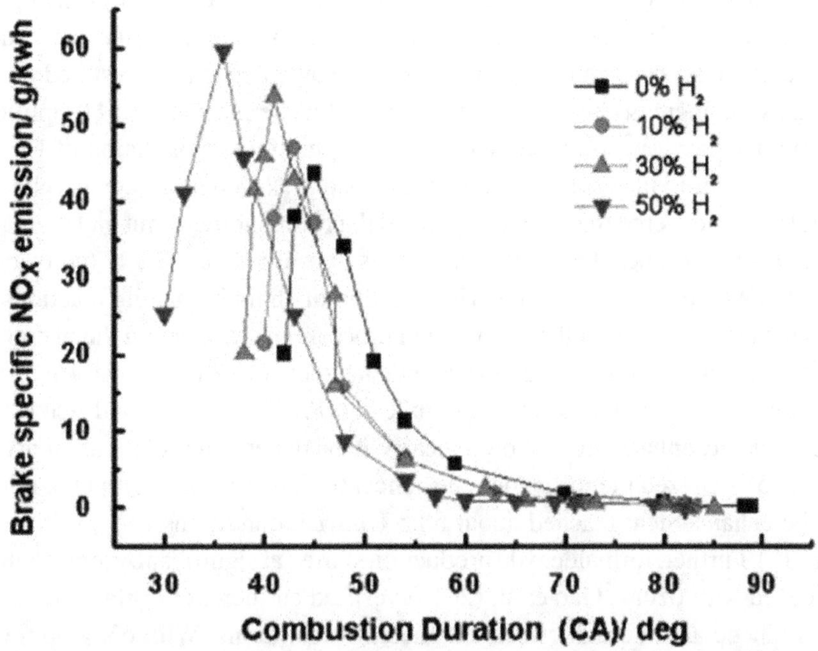

Figure 16. Ozone mechanism with the respective Arrhenius parameters

NO.	Reaction	*A*	*n*	*E*
1	$O_3 + O_2 \rightarrow O_2 + O + O_2$	1.54E+14	0	23,064
2	$O_2 + O + O_2 \rightarrow O_3 + O_2$	3.26E+19	−2.1	0
3	$O_3 + N_2 \rightarrow O_2 + O + N_2$	4.00E+14	0	22,667
4	$O_2 + O + N_2 \rightarrow O_3 + N_2$	1.60E+14	−0.4	−1391
5	$O_3 + O \rightarrow O_2 + O + O$	2.48E+15	0	22,727
6	$O_2 + O + O \Leftrightarrow O_3 + O$	2.28E+15	−0.5	−1391
7	$O_3 + O_3 \rightarrow O_2 + O + O_3$	4.40E+14	0	23,064
8	$O_2 + O + O_3 \rightarrow O_3 + O_3$	1.67E+15	−0.5	−1391
9	$O_3 + H \Leftrightarrow O_2 + OH$	8.43E+13	0	934
10	$O_3 + H \Leftrightarrow O + HO_2$	4.52E+11	0	0
11	$O_3 + OH \Leftrightarrow O_2 + HO_2$	1.85E+11	0	831
12	$O_3 + H_2O \Leftrightarrow O_2 + H_2O_2$	6.62E+01	0	0
13	$O_3 + HO_2 \Leftrightarrow OH + O_2 + O_2$	6.62E+09	0	994
14	$O_3 + O \Leftrightarrow O_2 + O_2$	4.82E+12	0	4094
15	$O_3 + NO \Leftrightarrow O_2 + NO_2$	8.43E+11	0	2603
16	$O_3 + CH_3 \Leftrightarrow CH_3O + O_2$	5.83E+10	0	0

[a] Expressed as $k = AT^n \exp(-E/RT)$ (cal, cm^3, mol, s).

mechanism, burning velocity increases of 8% (rich regime) to 16% (lean regime) were reported (Wang et al, 2012), when using ~7000 ppm of ozone.

Some experimental results (Mohammadi et al., 2006) showed low combustion efficiency of a natural gas CI engine, emitting high HC and CO concentrations when operating at an equivalence ratio of 0.24 using ozone concentrations lower than 40ppm. This lean equivalence ratio resulted in a cold combustion, which explains the increase in the intermediate products emissions. At 80-ppm ozone addition, a decrease in ignition delay was observed, leading to a higher rate of oxidation of HC and CO before exiting the cylinder, resulting in a decrease in those emissions as well as greater combustion efficiency.

Zhang et al. (2016) discovered that ozone expanded the flammability limit and increased the laminar flame speed of $H_2/CO/O_2$ flames. Over 90% of O_3 was consumed at ~700 K (activation of low temperature chemistry) (Won et al., 2015), triggering a group of chain branching reactions and producing active radicals, which interfered with the combustion process and accelerated the overall reaction under near-limit condition. Liang et al. (2013) reported enhancements on burning velocity of $H_2/CO/N_2$/Air premixed flames enriched with O_3 addition on both lean (18.74% increase) and rich (15.78% increase) sides. The burning velocity enhancement showed nearly linear dependency of the amount of ozone seeded.

Weng et al. (2015) analyzed the formaldehyde concentration in flames with O_3 addition. With 4500 ppm of ozone, the enhancement reached about 58.5% at rich conditions and 15.5% at stoichiometric condition (Figure 17). Further, formaldehyde production started at significantly lower temperatures when the flame was seeded with ozone. Gao et al. (2015) verified an increase in flame propagation with O_3 addition, through kinetic and thermal effects at adiabatic conditions. With 6334 ppm O_3 addition, the stoichiometric flame speed (SL) was increased by 8.3% for CH_4, and 7.6% for C_3H_8.

Figure 17. Enhancement rate of a methane-air flame with ozone addition. Adapted from Weng et al. (2015)

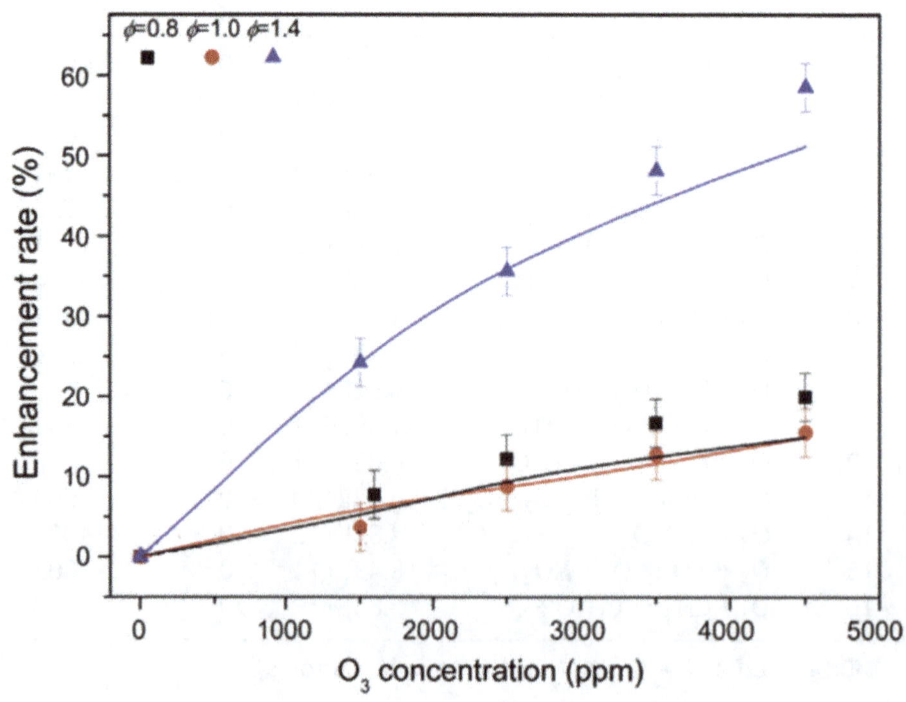

Simple electric discharge devices can produce ozone, which has a half-life of approximately 30-40 minutes in ambient conditions, and may be installed close to the air intake port of an engine. This technique enables a precise controlled addition of O_3 in the oxidizer flow. A similar approach was used by Schönborn et al. (2010) for controlling the point of ignition of the fuel mixture (Figure 18). The point of ignition could be adjusted within the cycle by varying the proportions of fuel oxidized with ozone and not oxidized with ozone. Analysis of the fuel samples suggested that upon contact with ozone, the fuel was partly converted into a peroxide, which is expected to promote earlier autoignition of the charge.

Pinazzi et al. (2016) evaluated the potential of ozone to promote the autoignition reactivity of a 95%-octane gasoline, when direct-injected in a CI engine running at low load, and discovered that at early injection timing, combustion is strongly influenced by residual NO, and at later injection timing, local distribution of the temperature and the equivalence ratio mainly drive the onset of combustion.

Summarizing, ozone has the ability to significantly influence the combustion and emission level of engines (especially CI) due to the increase in the radical species (and extra oxygen compound) in the flame region, greatly increasing the burning velocity of the fuel/oxidizer mixture. This fact enables the combustion to be leaner, contributing to lower fuel consumption. Based on these effects, O_3 addition is responsible of reduced ignition delay, faster combustion process, reduction in CO and HC emissions and increase in NOx emissions. Therefore, due to the reduced ignition time, combustion happens closer to the end of the compression process and the engine experiences higher temperatures, which causes an increase in the efficiency followed by the undesirable increased NOx emissions (Depcik et al., 2014).

ETHANOL ENRICHMENT

In the last decades, the combustion modeling and energetic material ignition evolved enormously (Beckstead, 2005), being applied to several types of explosives, propellants, gas generators or pyrotechnics formulations (Becksted et al., 2007). This procedure considers effects like the distribution of reactional

Figure 18. Scheme of engine running with ozone enriched air

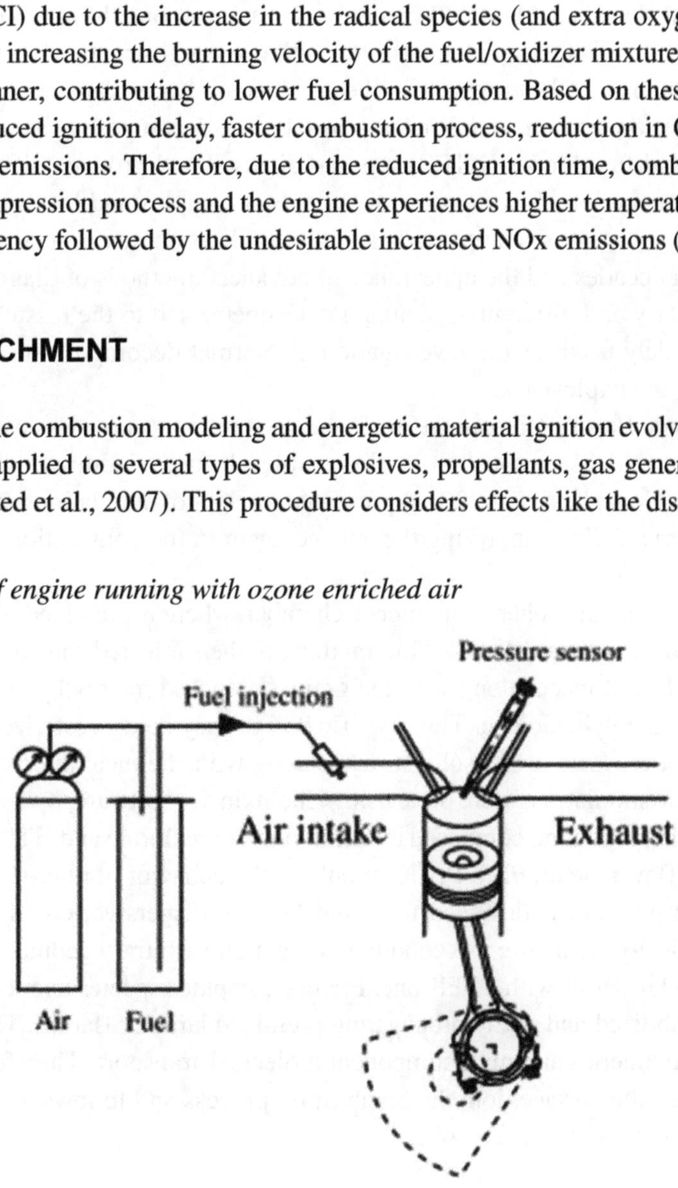

zones in condensed phases, radiation of non-linear specters and non-homogeneity of the propellant (Erikson and Becksted, 1997). Besides all of the previously mentioned engines and combustors, another field where combustion is the most important part of the process is in aerospace propulsion systems. In these devices, highly energetic solid, liquid or hybrid fuel mixtures are employed, in order to achieve extreme velocity, thrust and impulse.

In solid propulsion, the fuel is stored in a chamber (or stage) and, as the material is ignited, the combustion takes place until everything is consumed. Most of the solid propellants used for propulsion (rockets, missiles etc.) were composed by a mixture among several substances, particularly ammonium perchlorate (oxidizer salt) and hydroxyl terminated polybutadiene (binder). Many different formulations were developed and tried, in order to optimize the combustion and increase its efficiency (Kreitz et al., 2012; Ramakrishna et al., 2012; Brill et al., 2000; Krishnan and Jeenu, 2012).

Despite the widespread use and long investigative history of ammonium perchlorate-fuel mixtures, it still can be said that AP alone and AP/HTPB composites remain among the most confounding materials in the research setting (Brill and Budenz, 2000). Since the physical structure of composite propellants like the AP/HTPB composite is heterogeneous, the combustion wave structure appears to be also heterogeneous. During the combustion, at the burning surface, the decomposed gases from the ammonium perchlorate particles and fuel binder (HTPB) are interdiffused and produce diffusion flame streams. Due this, the flame structure of AP composite propellants is complex and locally three-dimensional in shape. The AP/HTPB composite decomposition and the combustion mechanism have been extensively investigated in the last decades and the appearance of advanced methods of diagnostics, like flash pyrolysis, thermogravimetry and differential scanning calorimetry, led to the ressurgence of the interest. These methods are widely used for the investigation of thermal decomposition of organic materials, polymers, composites and explosives.

When traveling inside Earth's atmosphere, the solid mixture may carry less oxidizer and use the atmospheric oxygen to enable the combustion, so this enables the use of additives (enrichment) in the air flow. In a recent work, Gonçalves et al. (2014) presented a series of simulations of AP/HTPB combustion with an atmospheric airflow, analyzing the enhancement in the combustion process with ethanol enrichment of the intake air.

In the evaluated system, atmospheric air enters a chamber, where is premixed with a specific amount of ethanol, provided from onboard tanks. This mixture is then injected into the reactor/combustion chamber and is ignited, so it reacts along with the grain, composed mainly by ammonium perchlorate and hydroxyl terminated polybutadiene. This specific design may be characterized as a pseudo hybrid combustor, as there is a gas flow over a solid surface, along with chemical interactions.

The chemical kinetic simulations were done using Chemkin 4.1 software, by Sandia Labs. The burning behavior of the solid propellant, composed by ammonium perchlorate and HTPB, was analyzed over distance inside a plug flow reactor, PFR. PLUG simulates the behavior of plug-flow chemical reactors. More specifically, the program is designed to model the non-dispersive, one-dimensional flow of a chemically reacting, ideal-gas mixture in a conduit of essentially arbitrary geometry. To do so, a premix reactor model was used in series with a PFR one. Premix computes species and temperature profiles in steady-state burner-stabilized and freely propagating premixed laminar flames. The program accounts for finite-rate chemical kinetics and multicomponent molecular transport. Therefore, the air enters the system, flows over the solid surface, join the combustion process and follows to the PFR reactor until the species leave the system through a nozzle.

A mechanism composed by 76 different chemical species and 483 elementary reactions was assembled using literature data. Among these elementary reactions, the decomposition of ammonium perchlorate, HTPB and ethanol are present. It is interesting that there are several reactions common to each species' decomposition. Although the problem becomes harder to solve and converge (increasing the computational costs), the results may approximate even more to the real case scenario.

In order to broaden the analysis and find optimum conditions, variational studies were developed. Pressure and inlet temperature were varied from 1 to 60 atm and 300 to 2500 K, respectively. When ethanol was inserted, its molar concentration was also varied from 0.01 to 0.6, so several surfaces are generated and specific regions can be used or improved.

Figure 19 shows the combustion behavior of the solid propellant using only dry air as inlet. The molar fractions of the chemical species of interest are shown according to the temperature and distance from the burning surface.

Based on the species variations, it would be wise to say in this case, that the optimum performance of the combustion process occurs with an inlet temperature around 1100 K, as there is low CO, NO and NO_2 concentrations and high contents of water and CO_2 in this region. Hydrochloric acid suffers also a decrease, according to the pressure as well.

High inlet temperatures favor parallel reactions for NOx production, so there's a significant increase in the intermediate species concentration, followed by a consequent decrease on the final products mole fraction.

One of the concerns regarding the combustion gases from the process is the hydrochloric acid concentration, a great responsible for acid rains. It is interesting to find a condition which can sustain the performance for thrust, but also reduces the emission of pollutants. For this specific scenario, the use of ethanol altogether with air should also help decreasing HCl concentration, as there is kinetic increase of some reactions where the species is consumed.

Figure 19. Variational study of AP/HTPB combustion with dry air

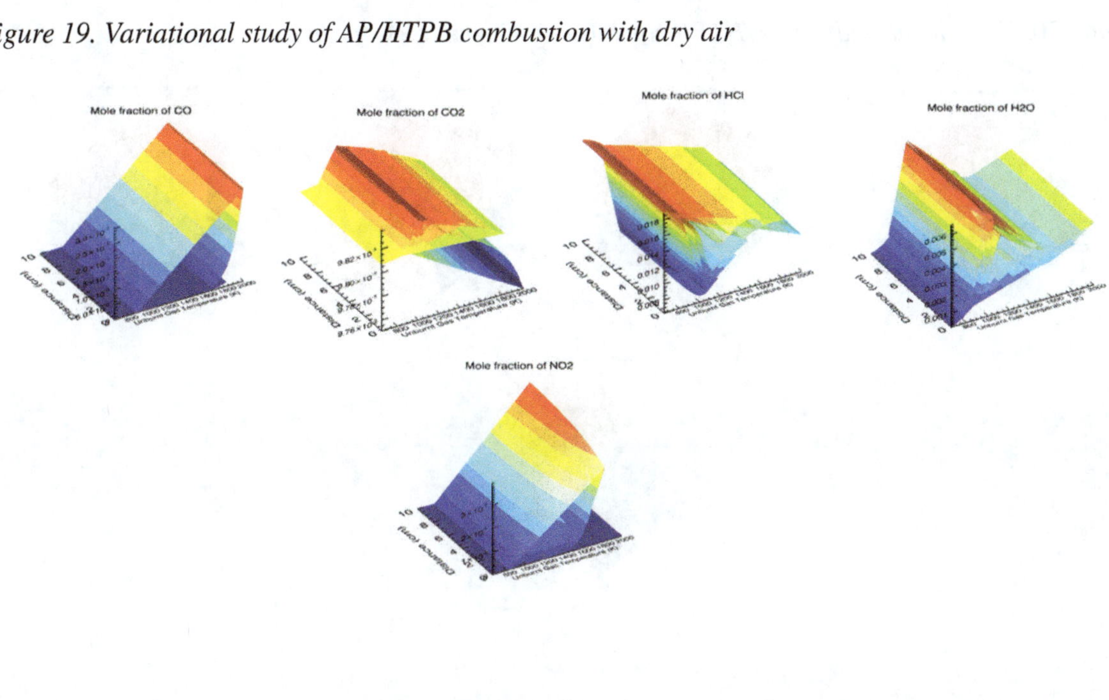

As ethanol is a known liquid fuel, with high heating value and low boiling point, it is assumed that it's presence will increase the burning efficiency and also the whole energy. Figure 20 presents the molar fractions of the species, also according to the temperature and to the distance from the burning surface.

When comparing both results, it is clear that the very presence of ethanol greatly affected the combustion. Regarding the efficiency, the combustion is more complete in the latter, as the water and carbon dioxide concentrations are higher. The molar fraction of carbon dioxide increases very fast in the flame region, due to the consideration of a premixed formulation, and keeps constant to the end of the process. The several layers observed correspond to other parameters variation; as previously mentioned, these are results of a complete variational study, so there are several parameters varying independently at each run.

The behavior at high inlet temperatures was also observed in this case, i.e. there is a significant increase of CO and NO concentration when the inlet temperature approaches 2000 K. These species are characteristic of poor combustion processes.

When inserting ethanol in the system, the hydrochloric concentrations are somewhat more stable according to the inlet temperature. However, the maximum mole fraction observed is about 0.016, while in the first case is about 0.018.

Nitrous oxide showed a very distinct behavior, as its maximum content was reached when the inlet temperature was around 1500 K, and not 2000 K. This indicated that during combustion the temperature inside the chamber increases significantly, being suitable for the NO_2 reactions to occur more frequently. Also, the concentration of both NOx species increase along the reactor, indicating the occurrence of high temperature reactions in the tube.

After observation of a significant evolution on AP/HTPB combustion using air doped with ethanol, there is a need of evaluating the best ethanol content in the inlet air. Therefore, presents the chemical species mole fractions according to the ethanol inlet.

Carbon monoxide is increasingly produced with increasing contents of ethanol. It is likely that the extra molecules of ethanol do not burn completely, so there is a negative effect on the global combus-

Figure 20. Variational study of AP/HTPB combustion with ethanol-doped air

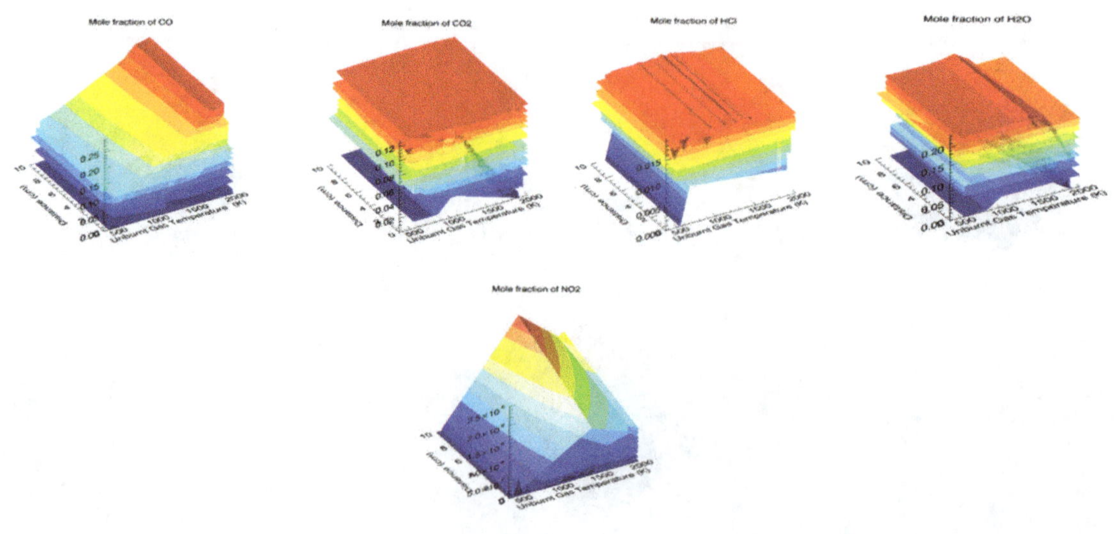

tion process. This effect is proved in the carbon dioxide curve, where the maximum contents are in low ethanol concentrations, around 0.1 and 0.2.

Water production is also increased with ethanol content, reaching its maximum at ethanol mole fraction of 0.4 and beyond. Like the previous species, there are several possibilities of water content, based on the other parameters inside of the combustion chamber.

High values of HCl mole fraction are observed with low ethanol concentrations, around 0.05-0.1. From this region on, the final content decreases significantly, showing that reactions with ethanol and their derivate species are more favorable than HCl production.

NOx species showed an odd behavior. They were only produced when low contents of ethanol were present in the system (up to 0.1). It is possible to say that in that region, ethanol is consumed and causes a great increase on the temperature, also increasing the mean velocity of the chemical reactions. Therefore, with this increase and low ethanol insertion, there is place for the parallel NOx reactions to happen, which are responsible for the continuous growth of the concentrations during the flow in the reactor.

Based on the results observed, the optimum concentration of ethanol in the inlet air would be 0.2, due to the high production of water and carbon dioxide and low emissions of CO, HCl and NOx. This mole fraction is also not so high for avoiding the entrance of oxidizer species to the system.

For an even greater combustion enhancement, instead of atmospheric air, an oxygen tank could be placed. Then, it'd mix with ethanol and enter the system. NOx levels would decay consistently and the global combustion process would be significantly increased.

CONCLUSION

In the present chapter were presented recent studies and technologies for combustion process enhancement. Hydrogen is the most studied species for enrichment of the fuel-air mixture, especially in internal combustion engines (compression ignition and spark ignition). Less common enhancers presented were ozone and ethanol. All of these energetic materials increased the mean temperature of the combustion chamber, decreasing the combustion duration and the fuel consumption, while increasing the combustion efficiency. Several physical adjustments are required for using fuel enrichment, as these advantages come with substantial increases in pollutant emissions, especially NOx. New designs, materials and cycle specifications are needed for guaranteeing the optimum performance of the system, so there will be no risks and further damage to the environment. Therefore, as the observed results are economically interesting, it is a solid possibility the use of this kind of technology in the near future, as the importance of new sustainable fuels and economic vehicles are growing exponentially.

REFERENCES

Abdel-Aal, H. K., Sadik, M., Bassyouni, M., & Shalabi, M. (2005). A new approach to utilize hydrogen as a safe fuel. *International Journal of Hydrogen Energy*, *30*(13-14), 1511–1514. doi:10.1016/j.ijhydene.2005.07.007

Akansu, S. O., Kahraman, N., & Ceper, B. (2007). Experimental study on a spark ignition engine fueled by methane-hydrogen mixtures. *International Journal of Hydrogen Energy*, *32*(17), 4279–4284. doi:10.1016/j.ijhydene.2007.05.034

Al-Baghdadi, M. A. R. S. (2004). Effect of compression ratio, equivalence ratio and engine speed on the performance and emission characteristics of a spark ignition engine using hydrogen as a fuel. *Renewable Energy*, *29*(15), 2245–2260. doi:10.1016/j.renene.2004.04.002

Amrouche, F., Erickson, P., Park, J., & Varnhagen, S. (2014). An experimental investigation of hydrogen-enriched gasoline in a Wankel rotary engine. *International Journal of Hydrogen Energy*, *39*(16), 8525–8534. doi:10.1016/j.ijhydene.2014.03.172

Ayala, F., Gerty, M., & Heywood, J. B. (2006). SI engine performance. Influences of relative air–fuel ratio, load, compression ratio, and combustion phasing (SAE technical paper 2006-01-0229).

Bari, S., & Esmaeil, M. M. (2010). Effect of H2/O2 addition in increasing the thermal efficiency of a diesel engine. *Fuel*, *89*(2), 378–383. doi:10.1016/j.fuel.2009.08.030

Bari, S., & Esmaeil, M. M. (2010). Effect of H2/O2 addition in increasing the thermal efficiency of a diesel engine. *Fuel*, *89*(2), 378–383. doi:10.1016/j.fuel.2009.08.030

Bauer, C. G., & Forest, T. W. (2001). Effect of hydrogen addition on the performance of methane-fueled vehicles. Part I: Effect of hydrogen addition on S.I. engine performance. *International Journal of Hydrogen Energy*, *26*, 55–70. doi:10.1016/S0360-3199(00)00067-7

Beckstead, M. W. (2006, November). Recent Progress in Modeling Solid Propellant Combustion. *Combustion, Explosion, and Shock Waves*, *42*(6), 623–641. doi:10.1007/s10573-006-0096-5

Beckstead, M. W., Puduppakkam, K., Thakre, P., & Yang, V. (2007). Modeling of combustion and ignition of solid-propellant ingredients. *Progress in Energy and Combustion Science*, *33*(6), 497–551. doi:10.1016/j.pecs.2007.02.003

Beltrame, A., Porshnev, P., Merchan, W., Saveliev, A., Fridman, A., & Kennedy, L. A. (2001). Soot and NO formation in methane-oxygen enriched diffusion flames. *Combustion and Flame*, *124*(1-2), 295–310. doi:10.1016/S0010-2180(00)00185-1

Brill, T. B., & Budenz, B. T. (2000). Flash Pyrolysis of Ammonium Perchlorate-Hydroxyl-Terminated-Polybutadiene Mixtures Including Selected Additives. *Progress in Aeronautics and Astronautics*, *185*, 3.

Brill, T. B., Ren, W., Yang, V. (2000) Solid Propellant Chemistry, Combustion, and Motor Interior Ballistics. *Progress in Astronautics and Aeronautics, 185.*

Catapano, F., Iorio, S. D., Magno, A., Sementa, P., & Vaglieco, B. M. (2015). A comprehensive analysis of the effect of ethanol, methane and methane-hydrogen blend on the combustion process in a PFI (portfuel injection) engine. *Energy*, *88*, 101–110. doi:10.1016/j.energy.2015.02.051

Chintala, V., & Subramanian, K. A. (2015). Experimental investigations on effect of different compression ratios on enhancement of maximum hydrogen energy share in a compression ignition engine under dual-fuel mode. *Energy*, *87*, 448–462. doi:10.1016/j.energy.2015.05.014

Cooney, C. P., Worm, Y. J., & Naber, J. D. (2009). Combustion characterization in an internal combustion engine with ethanol gasoline blended fuels varying compression ratios and ignition timing. *Energy & Fuels*, *23*(5), 2319–2324. doi:10.1021/ef800899r

Correa, S. M. (1993, January). A review of NOx formation under gas-turbine combustion conditions. *Combustion Science and Technology*, *87*(1-6), 329–362. doi:10.1080/00102209208947221

Depcik, C., Mangus, M., & Ragone, C. (2014). Ozone-Assisted Combustion—Part I: Literature Review and Kinetic Study Using Detailed n-Heptane Kinetic Mechanism. *Journal of Engineering for Gas Turbines and Power*, *136*(9), 091507. doi:10.1115/1.4027068

Dhole, A. E., Yarasu, R. B., & Lata, D. B. (2016). Investigations on the combustion duration and ignition delay period of a dual fuel diesel engine with hydrogen and producer gas as secondary fuels. *Applied Thermal Engineering*, *107*, 524–532. doi:10.1016/j.applthermaleng.2016.06.151

Dhole, A. E., Yarasu, R. B., & Lata, D. B. (2016). Effect of hydrogen and producer gas as secondary fuels on combustion parameters of a dual fuel diesel engine. *Applied Thermal Engineering*, *108*, 764–773. doi:10.1016/j.applthermaleng.2016.07.157

Erikson, W. W., & Beckstead, M. W. (1997) Modeling unsteady monopropellant combustion with full chemical kinetics. In *Proceedings of the 36th AIAA Aerospace Sciences Meeting and Exhibit, Aerospace Sciences Meetings*, Reno, NV.

Fontana, A., Galloni, E., Jannelli, E., Minutillo, M. (2002). Performance and fuel consumption estimation of a hydrogen enriched gasoline engine at part-load operation (technical paper).

Gao, X., Zhang, Y., Adusumilli, S., Seitzman, J., Sun, W., Ombrello, T., & Carter, C. (2015). The effect of ozone addition on laminar flame speed. *Combustion and Flame*, *162*(10), 3914–3924. doi:10.1016/j.combustflame.2015.07.028

Gonçalves, R. F. B., Iha, K., & Rocco, J. A. F. F. (2015) Enhancement of Ammonium Perchlorate/Hydroxyl-Terminated Polybutadiene Combustion Kinetics Using Ethanol-Doped Air. In *Proceedings of the 51st AIAA/SAE/ASEE Joint Propulsion Conference, Propulsion and Energy Forum*. doi:10.2514/6.2015-3869

Greenwood, J. B., Erickson, P. A., Hwang, J., & Jordan, E. A. (2014). Experimental results of hydrogen enrichment of ethanol in an ultra-lean internal combustion engine. *International Journal of Hydrogen Energy*, *39*(24), 12980–12990. doi:10.1016/j.ijhydene.2014.06.030

Heywood, J. (1988). *Internal Combustion Engine Fundamentals*. McGraw-Hill Education.

Hora, T. S., & Agarwal, A. K. (2015). Experimental study of the composition of hydrogen enriched compressed natural gas on engine performance, combustion and emission characteristics. *Fuel*, *160*, 470–478. doi:10.1016/j.fuel.2015.07.078

Hora, T. S., & Agarwal, A. K. (2016). Effect of varying compression ratio on combustion, performance, and emissions of a hydrogen enriched compressed natural gas fueled engine. *Journal of Natural Gas Science and Engineering*, *31*, 819–828. doi:10.1016/j.jngse.2016.03.041

Ivanic, Z., Ayala, F., & Goldwitz, J. (2005) Heywood JB. Effects of hydrogen enhancement on efficiency and NOx emissions of lean and EGR-diluted mixture in a SI engine (SAE paper 2005-01-0253).

Jhang, S., Chen, K., Lin, S., Lin, Y., & Cheng, W. L. (2016). Reducing pollutant emissions from a heavy-duty diesel engine by using hydrogen additions. *Fuel, 172*, 89–95. doi:10.1016/j.fuel.2016.01.032

Ji, C., Liu, X., Gao, B., Wang, S., & Yang, J. (2014). A laminar burning velocity correlation for combustion simulation of hydrogen-enriched ethanol engines. *Fuel, 133*, 139–142. doi:10.1016/j.fuel.2014.05.013

Ji, C., Su, T., Wang, S., Zhang, B., Yu, M., & Cong, X. (2016). Effect of hydrogen addition on combustion and emissions performance of a gasoline rotary engine at part load and stoichiometric conditions. *Energy Conversion and Management, 121*, 272–280. doi:10.1016/j.enconman.2016.05.040

Kahraman, N., Çeper, B., Akansu, S. O., & Aydin, K. (2009). Investigation of combustion characteristics and emissions in a spark-ignition engine fueled with natural gas-hydrogen blends. *International Journal of Hydrogen Energy, 34*(2), 1026–1034. doi:10.1016/j.ijhydene.2008.10.075

Karagöz, Y., Güler, I., Sandalcı, T., Yüksek, L., & Dalkılıc, A. S. (2016). Effect of hydrogen enrichment on combustion characteristics, emissions and performance of a diesel engine. *International Journal of Hydrogen Energy, 41*(1), 656–665. doi:10.1016/j.ijhydene.2015.09.064

Karagöz, Y., Güler, I., Sandalcı, T., Yüksek, L., Dalkılıc, A. S., & Wongwises, S. (2016). Wongwises, S. (2016) Effects of hydrogen and methane addition on combustion characteristics, emissions, and performance of a CI engine. *International Journal of Hydrogen Energy, 41*(2), 1313–1325. doi:10.1016/j.ijhydene.2015.11.112

Khalil, A. E. E., & Gupta, A. K. (2013). Hydrogen addition effects on high intensity distributed combustion. *Applied Energy, 104*, 71–78. doi:10.1016/j.apenergy.2012.11.004

Kızıltan, E. (1988) *Effect of alcohol addition to motor fuels on engine performance* [MS Thesis]. University of Karadeniz Technical University.

Kökkülünk, G., Parlak, A., Ayhan, V., Cesur, I., Gonca, G., & Boru, B. (2014). Theoretical and experimental investigation of steam injected diesel engine with EGR. *Energy, 74*, 331–339. doi:10.1016/j.energy.2014.06.091

Köse, H., & Ciniviz, M. (2013). An experimental investigation of effect on diesel engine performance and exhaust emissions of addition at dual fuel mode of hydrogen. *Fuel Processing Technology, 114*, 26–34. doi:10.1016/j.fuproc.2013.03.023

Kreitz, K., Petersen, E., Reid, D., & Seal, S. (2012). Scale-Up Effects of Nanoparticle Production on the Burning Rate of Composite Propellant. *Combustion Science and Technology, 184*(6), 750–766. doi:10.1080/00102202.2012.663026

Krishnan, S., & Jeenu, R. (2012). Combustion characteristics of AP/HTPB propellants with burning rate modifiers. *Journal of Propulsion and Power, 8*(4), 748–755. doi:10.2514/3.23545

Lee, S., Kim, C., Choi, Y., Lim, G., & Park, C. (2014). Emissions and fuel consumption characteristics of an HCNG-fueled heavy-duty engine at idle. *International Journal of Hydrogen Energy, 39*(15), 8078–8086. doi:10.1016/j.ijhydene.2014.03.079

Liang, X., Wang, Z., Weng, W., Zhou, Z., Huang, Z., Zhou, J., & Cen, K. (2013). Study of ozone-enhanced combustion in H2/CO/N2/air premixed flames by laminar burning velocity measurements and kinetic modeling. *International Journal of Hydrogen Energy*, *38*(2), 1177–1188. doi:10.1016/j.ijhydene.2012.10.075

Liew, C., Li, H., Nuszkowski, J., Liu, S., Gatts, T., Atkinson, R., & Clark, N. (2010). An experimental investigation of the combustion process of a heavy-duty diesel engine enriched with H2. *International Journal of Hydrogen Energy*, *35*(20), 11357–11365. doi:10.1016/j.ijhydene.2010.06.023

Lilik, G. K., Zhang, H. D., Herreros, J. M., Haworth, D. C., & Boehman, A. L. (2010). Hydrogen assisted diesel combustion. *International Journal of Hydrogen Energy*, *35*(9), 4382–4398. doi:10.1016/j.ijhydene.2010.01.105

Lim, G., Lee, S., Park, C., Choi, Y., & Kim, C. (2014). Effects of compression ratio on performance and emission characteristics of heavy-duty SI engine fueled with HCNG. *International Journal of Hydrogen Energy*, *39*, 8078–8086. doi:10.1016/j.ijhydene.2014.03.079

Ma, F., Liu, H., Wang, Y., Li, Y., Wang, J., & Zhao, S. (2008). Combustion and emission characteristics of a port-injection HCNG engine under various ignition timings. *International Journal of Hydrogen Energy*, *33*(2), 816–822. doi:10.1016/j.ijhydene.2007.09.047

Ma, F., Wang, Y., Liu, H., Li, Y., Wang, J., & Zhao, S. (2007). Experimental study on thermal efficiency and emission characteristics of a lean burn hydrogen enriched natural gas engine. *International Journal of Hydrogen Energy*, *32*(18), 5067–5075. doi:10.1016/j.ijhydene.2007.07.048

Maher, A. R., & Haroun, A. K. (2003). A prediction study of a spark ignition supercharged hydrogen engine. *Energy Conversion and Management*, *44*(20), 3143–3150. doi:10.1016/S0196-8904(03)00127-4

McLellan, B., Shoko, E., Dicks, A. L., & Diniz da Costa, J. C. (2005). Hydrogen production and utilization opportunities for Australia. *International Journal of Hydrogen Energy*, *30*(6), 669–679. doi:10.1016/j.ijhydene.2004.06.008

Mohammadi, A., Kawanabe, H., Ishiyama, T., Shioji, M., & Komada, A. (2006) Study on Combustion Control in Natural-Gas PCCI Engines with Ozone Addition into Intake Gas (SAE Technical Paper No. 2006-01-0419).

Moreno, F., Arroyo, J., Muñoz, M., & Monné, C. (2012). Combustion analysis of a spark ignition engine fueled with gaseous blends containing hydrogen. *International Journal of Hydrogen Energy*, *37*(18), 13564–13573. doi:10.1016/j.ijhydene.2012.06.060

Norbye, J. P. (1971). *The Wankel Engine*. Philadelphia: Chilton Book Company.

Park, C., Choi, Y., & Lee, J. (2013). Operating strategy for exhaust gas reduction and performance improvement in a heavy-duty hydrogen-natural gas blend engine. *Energy*, *50*, 262–269. doi:10.1016/j.energy.2012.10.048

Park, C., Lee, S., Lim, G., Choi, Y., & Kim, C. (2014). Full load performance and emission characteristics of hydrogen-compressed natural gas engines with valve overlap changes. *Fuel*, *123*, 102–106. doi:10.1016/j.fuel.2014.01.041

Patil, K. R., Khanwalkar, P. M., Thipse, S. S., Kavathekar, K. P., Rairikar, S. D. (2010) Development of HCNG blended fuel engine with control of NO x emissions. *Journal of Computer Information Systems and Industrial Management Applications*, *2*, 87–95.

Pinazzi, P. M., & Foucher, F. (in press). Influence of injection parameters, ozone seeding and residual NO on a Gasoline Compression Ignition (GCI) engine at low load. In *Proceedings of the Combustion Institute*.

Pisa, I., Lazaroiu, G., & Prisecaru, T. (2014). Influence of hydrogen enriched gas injection upon polluting emissions from pulverized coal combustion. *International Journal of Hydrogen Energy*, *39*(31), 17702–17709. doi:10.1016/j.ijhydene.2014.08.119

Radwan, M. S. (1985) Performance and knock limits of ethanol-gasoline blends in spark-ignited engines (SAE Paper 850213).

Ramakrishna, P. A., Paull, P. J., & Mukunda, H. S. (2012). Revisiting the Modeling of Ammonium Perchlorate Combustion: Development of an Unsteady Model. *Journal of Propulsion and Power*, *22*(3), 661–668. doi:10.2514/1.15502

Riahi, Z., Bounaouara, H., Hraiech, I., Mergheni, M. A., Sautet, J., & Nasrallah, S. B. (in press). Combustion with mixed enrichment of oxygen and hydrogen in lean regime. *International Journal of Hydrogen Energy*.

Sahin, Z., & Durgun, O. (2009). Numerical investigation of the effects of injection parameters on performance and exhaust emissions in a D I diesel engine. In *Proceedings of the 4th International Exergy, Energy and Environment Symposium*, Sharjah, UAE.

Sahin, Z., Durgun, O., Kurt, M. (2015). Experimental investigation of improving diesel combustion and engine performance by ethanol fumigation-heat release and flammability analysis. *Energy Conversion and Management*, *89*, 175–187.

Sandalcı, T., & Karagöz, Y. (2014). Experimental investigation of the combustion characteristics, emissions and performance of hydrogen port fuel injection in a diesel engine. *International Journal of Hydrogen Energy*, *39*(32), 18480–18489. doi:10.1016/j.ijhydene.2014.09.044

Saravanan, N., & Nagarajan, G. (2008). An experimental investigation of hydrogen-enriched air induction in a diesel engine system. *International Journal of Hydrogen Energy*, *33*(6), 1769–1775. doi:10.1016/j.ijhydene.2007.12.065

Saravanan, N., & Nagarajan, G. (2010). Performance and emission studies on port injection of hydrogen with varied flow rates with diesel as an ignition source. *Applied Energy*, *87*(7), 2218–2229. doi:10.1016/j.apenergy.2010.01.014

Saravanan, N., Nagarajan, G., Dhanasekaran, C., & Kalaiselvan, K. M. (2007). Experimental investigation of hydrogen port fuel injection in DI diesel engine. *International Journal of Hydrogen Energy*, *32*(16), 4071–4080. doi:10.1016/j.ijhydene.2007.03.036

Sastri, M. V. C. (1987). Hydrogen Energy research-and-development in India - an overview. *International Journal of Hydrogen Energy*, *12*(3), 137–145. doi:10.1016/0360-3199(87)90145-5

Schefer, R. W., Wickall, D. M., & Agrawal, A. K. (2002). Combustion of hydrogen-enriched methane in lean premixed swirl stabilized burner. *Proceedings of the Combustion Institute*, *29*(1), 843–851. doi:10.1016/S1540-7489(02)80108-0

Schifter, L., Diaz, R., Rodriguez, J. P., & Gomez Gonzalez, U. (2011). Combustion and emissions behavior for ethanol gasoline blends in a single cylinder engine. *Fuel*, *90*(12), 3586–3592. doi:10.1016/j.fuel.2011.01.034

Schönborn, A., Hellier, P., Aliev, A. E., & Ladommatos, N. (2010). Ignition control of homogeneous-charge compression ignition (HCCI) combustion through adaptation of the fuel molecular structure by reaction with ozone. *Fuel*, *89*(11), 3178–3184. doi:10.1016/j.fuel.2010.06.005

Sopena, C., Diéguez, P. M., Sáinz, D., Urroz, J. C., Guelbenzu, E., & Gandía, L. M. (2010). Conversion of a commercial spark ignition engine to run on hydrogen: Performance comparison using hydrogen and gasoline. *International Journal of Hydrogen Energy*, *35*(3), 1420–1429. doi:10.1016/j.ijhydene.2009.11.090

Tachibana, T., Hirata, K., Nishida, H., & Osada, H. (1991). Effect of ozone on combustion of compression ignition engines. *Combustion and Flame*, *85*(3-4), 515–519. doi:10.1016/0010-2180(91)90154-4

Tang, C., Zhang, Y., & Huang, Z. (2014). Progress in combustion investigations of hydrogen enriched hydrocarbons. *Renewable & Sustainable Energy Reviews*, *30*, 195–216. doi:10.1016/j.rser.2013.10.005

Varde, K. (1981) Combustion characteristics of small spark ignition engines using hydrogen supplemented fuel mixture (SAE paper 810921).

Verhelst, S. (2014). Recent progress in the use of hydrogen as a fuel for internal combustion engines. *International Journal of Hydrogen Energy*, *39*(2), 1071–1085. doi:10.1016/j.ijhydene.2013.10.102

Verma, G., Prasad, R. K., Agarwal, R. A., Jain, S., & Agarwal, A. K. (2016, August). Experimental investigations of combustion, performance and emission characteristics of a hydrogen enriched natural gas fueled prototype spark ignition engine. *Fuel*, *178*, 209–217. doi:10.1016/j.fuel.2016.03.022

Wang, H. K., Chen, K. S., & Lin, Y. C. (2013). Emission reductions of carbonyl compounds in a heavy-duty diesel engine supplemented with H2/O2 fuel. *Aerosol and Air Quality Research-Index*, *13*, 1790–1795.

Wang, J., Huang, Z., Tang, C., Miao, H., & Wang, X. (2009). Numerical study of the effect of hydrogen addition on methane-air mixtures combustion. *International Journal of Hydrogen Energy*, *34*(2), 1084–1096. doi:10.1016/j.ijhydene.2008.11.010

Wang, J., Huang, Z., Tang, C., & Zheng, J. (2010). Effect of hydrogen addition on early flame growth of lean burn natural gas-air mixtures. *International Journal of Hydrogen Energy*, *35*(13), 7246–7252. doi:10.1016/j.ijhydene.2010.01.004

Wang, S., Ji, C., & Zhang, B. (2010). Effects of hydrogen addition and cylinder cutoff on combustion and emissions performance of a spark-ignited gasoline engine under a low operating condition. *Energy*, *35*(12), 4754–4760. doi:10.1016/j.energy.2010.09.015

Wang, S., Ji, C., Zhang, B., Cong, X., & Liu, X. (2016). Effect of CO2 dilution on combustion and emissions characteristics of the hydrogen-enriched gasoline engine. *Energy*, *96*, 118–126. doi:10.1016/j.energy.2015.12.017

Wang, S., Ji, C., Zhang, B., & Liu, X. (2013). Emissions performance of a hybrid hydrogen e gasoline engine-powered passenger car under the new European driving cycle. *Fuel*, *106*, 873–875. doi:10.1016/j.fuel.2013.01.011

Wang, Z. H., Yang, L., Li, B., Li, Z. S., Sun, Z. W., Aldén, M., & Konnov, A. A. et al. (2012). Investigation of combustion enhancement by ozone additive in CH4/air flames using direct laminar burning velocity measurements and kinetic simulations. *Combustion and Flame*, *159*(1), 120–129. doi:10.1016/j.combustflame.2011.06.017

Weng, W., Nilsson, E., Ehn, A., Zhu, J., Zhou, Y., Wang, Z., & Cen, K. et al. (2015). Investigation of formaldehyde enhancement by ozone addition in CH4/air premixed flames. *Combustion and Flame*, *162*(4), 1284–1293. doi:10.1016/j.combustflame.2014.10.021

Whiete, C. M., Steeper, R. R., & Lutz, A. E. (2006). The hydrogen-fueled internal combustion engine: A technical review. *International Journal of Hydrogen Energy*, *31*(10), 1292–1305. doi:10.1016/j.ijhydene.2005.12.001

Won, S. H., Jiang, B., Diévart, P., Sohn, C. H., & Ju, Y. (2015). Self-sustaining n-heptane cool diffusion flames activated by ozone. *Proceedings of the Combustion Institute*, *35*(1), 881–888. doi:10.1016/j.proci.2014.05.021

Yoshihito, S., & Teruo, S. (2005) Effect of hydrogen and gasoline-mixed combustion on spark ignition engine (SAE paper 2005-08-0503).

Zhang, Y., Zhu, M., Zhang, Z., Shang, R., & Zhang, D. (2016). Ozone effect on the flammability limit and near-limit combustion of syngas/air flames with N2, CO2, and H2O dilutions. *Fuel*, *186*, 414–421. doi:10.1016/j.fuel.2016.08.094

Zhen, X., Wang, Y., Xu, S., & Zhu, Y. (2013). Study of knock in a high compression ratio spark ignition methanol engine by multi-dimensional simulation. *Energy*, *50*, 150–159. doi:10.1016/j.energy.2012.09.062

Zhou, J. H., Cheung, C. S., & Leung, C. W. (2014). Combustion, performance, regulated and unregulated emissions of a diesel engine with hydrogen addition. *Applied Energy*, *126*, 1–12. doi:10.1016/j.apenergy.2014.03.089

Chapter 3
Chapman-Jouguet Combustion Waves in Van Der Waals and Noble-Abel Gases

Fernando S. Costa
Istituto Nacional de Pesquisas Espaciais, Brazil

César A. Q. Gonzáles
Universidad Nacional Mayor de San Marcos, Peru

ABSTRACT

This chapter adopts the Chapman-Jouguet approach to derive jump conditions across combustion waves propagating in Van der Waals and Noble-Abel gases. The steady one-dimensional balance equations of mass, momentum and energy, assuming different properties of reactants and products, are applied to obtain the main properties of combustion waves, including velocities, Mach numbers, pressures and temperatures, in terms of the covolumes and intermolecular force parameters. In general, the effects of covolumes are more significant than the effects of the intermolecular attraction forces on Hugoniot curves and on properties of combustion waves. However, theoretical results using the Van der Waals equation of state matched more closely the experimental results for detonations of mixtures of propane and diluted air at high initial pressures.

INTRODUCTION

Combustion waves are flame fronts which can propagate in solid, liquid, gas or multiphase media with speeds that depend on the composition of the mixture and on initial and boundary conditions. Deflagrations are combustion waves with subsonic speeds relative to the reactants and, in general, present velocities of 0.1 to 1 m/s in standard conditions (100 kPa and 298 K) and are, approximately, isobaric. Detonations are combustion waves with supersonic speeds relative to the reactants and, for standard conditions, present burning front velocities of order of 1500-4000 m/s, detonation Mach numbers $4 < M_D < 8$ and detonation pressures varying from 2,5 to 6,0 MPa (Nettleton, 1987).

DOI: 10.4018/978-1-5225-2903-3.ch003

Studies that led to the discovery of detonations in condensed explosives were made by Abel (1869) and in gaseous mixtures by Berthelot and Vielle (1881). However only in the 1950s a comprehensive bibliography of studies on detonations was documented (Bauer et al., 1991; Manson and Dabora, 1993). Current research on detonations involves different research lines defined by the study of a particular phenomenon (Nettleton, 1987; Kuo, 2005; Lee, 2008).

Shortly after the discovery of the detonation phenomenon, Michelson (1893), Chapman (1889) and Jouguet (1905, 1906) proposed a theory to calculate the detonation velocity of an explosive mixture, based on the solutions developed by Rankine (1870) and Hugoniot (1887, 1889) for the shock wave problem. However, Michelson's publications were unknown outside Russia and the detonation theory became known only as the CJ theory (Dremin, 1999).

The classical Chapman-Jouguet (CJ) theory considers a combustion wave as a discontinuity with an infinity reaction rate. A steady one-dimensional flow process is assumed, with chemical reactions proceeding instantaneously just after a shock wave passes through the medium. Based on the CJ theory, it is possible to calculate the properties of the combustion wave and to verify that the combustion products attain a minimum velocity satisfying the conservation equations.

Solution of continuity and momentum equations yields the Rayleigh line that relates pressures and specific volumes across the combustion wave. Solution of the energy, continuity, momentum and equation of state is known as the Rankine-Hugoniot relation. In the case of constant thermodynamic properties, the Rankine-Hugoniot curve takes the form of a hyperbola yielding the equilibrium states of the combustion products, given the initial conditions of the reactant mixture and the heat release from the reaction.

According to the CJ approach, the solution for the combustion wave propagation problem corresponds to the case where the Rayleigh line is tangent to the Rankine-Hugoniot curve. Two points are determined, known as upper CJ point and lower CJ point which are solutions, respectively, for detonation and deflagration waves. The CJ theory does not require the knowledge of the combustion wave structure and reaction kinetics, but only the determination of the thermodynamic equilibrium state of the combustion products.

During the II World War, the CJ model was improved by Zeldovich, Döring and von Neumann which considered the chemical reaction rate. Their model (ZND) describes the detonation wave as a shock wave immediately followed by a reaction zone, whose thickness is determined by the reaction rate. The ZND theory yields the same detonation velocities and pressures that the CJ theory, being the only additional information the determination of the wave thickness (Wingerden et al., 1999; Williams, 1985; Glassman, 1996). Figure 1 shows a scheme of the ZND structure of a detonation wave comprising a shock followed by a deflagration.

In the case of detonations, the real wave structure is proven to be three-dimensional due to hydrodynamic instabilities, however it is apparent that the one-dimensional solution provides good results for average properties, for example, the detonation velocity at the upper CJ point (Fickett and Davis, 2005; Kuo, 2005).

REAL GASES

The equation of state for ideal gases can be used with good precision at low pressures. However, at high pressures the volumes of the molecules and the molecular attraction forces must be considered. Several studies focused on the effects of high pressures on the propagation of detonations and deflagrations.

Figure 1. ZND structure of a detonation wave (adapted from Kuo, 2005)

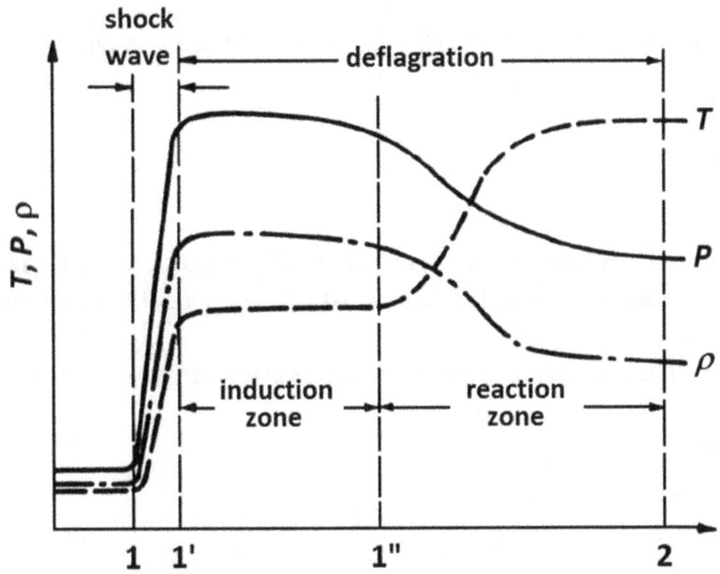

Gealer and Churchill (1960) measured detonation velocities of H_2-O_2 mixtures with initial pressures up to 70 bars, and compared experimental data with results from CJ theory based on the ideal-gas equation of state. Bauer et al. (1979) studied detonation characteristics of gaseous C_2H_4-O_2-H_2 mixtures for detonation pressures of 100-1000 bars. The measured detonation velocities were compared to numerical solutions based on the ideal-gas, BKW and Boltzman equations of state. The theoretical pressures and temperatures of detonations appeared to be less sensitive to the equation of state than the detonation velocities. Astapov et al. (1984) investigated the detonation parameters of H_2-O_2 and H_2-air mixtures at high initial density. The one-dimensional conservation equations and the equilibrium composition based on the Van der Waals equation of state were solved numerically. The authors verified that theoretical results agreed with experiments within the limits of uncertainty. Schmitt and Butler (1995a) investigated theoretically the effects of initial pressures (> 1 bar) and temperatures on CJ detonation properties of several gas systems (H_2, CH_4, C_2H_4, C_2H_6, C_3H_8) using four equations of state: ideal-gas, Redlich-Kwong, Soave and Peng-Robinson. Schmitt and Butler (1995b) extended the ZND theory of gas-phase detonation to use real-gas equations of state, real-gas thermodynamic properties and chemical kinetic reaction rates. Marchionni et al. (2007) adopted the Redlich-Kwong equation of state to simulate freely propagating flames in CH_4-O_2-inert mixtures at pressures up to 150 atm. Modified equilibrium constants were used to obtain reverse reaction rates in a detailed reaction mechanism, considering the high-pressure effects on the enthalpy and specific heat. These works have used the classical solutions for ideal-gas or developed numerical solutions using real gas equations of state.

Therefore, the present work adopts the CJ approach to derive mathematical expressions for the jump conditions in combustion waves propagating in Van de Waals (VDW) and Noble-Abel (NA) gases, assuming different properties of reactants and products. Steady one-dimensional conservation equations of mass, momentum and energy will be applied to obtain the Rayleigh line, Rankine-Hugoniot curve, Mach numbers, CJ propagation velocity, as well as pressure, specific volume and temperature ratios across the combustion wave.

EQUATIONS OF STATE OF VAN DER WAALS AND NOBLE-ABEL

The Van der Waals (VDW) equation of state (EOS) was the first semi-empirical approach used to represent real gas behavior:

$$\left(P + a/v^2\right)\left(v - b\right) = RT \tag{1}$$

where P is pressure, T is temperature, v is specific volume, R is the gas constant, a is the attraction force parameter and b is the covolume, or excluded volume, which corresponds to about four times the volume occupied by the molecules.

Constants a and b can be determined experimentally or estimated from critical parameters:

$$a = \frac{27}{64}\frac{R^2 T_{cr}^2}{P_{cr}}\,;\; b = \frac{RT_{cr}}{8P_{cr}} \tag{2}$$

where T_{cr} and P_{cr} are the critical temperature and pressure, respectively.

Nevertheless, high temperatures can reduce the effects of molecular attraction forces due to the high molecular kinetic energy. For this reason, in some cases, the term concerning the intermolecular forces in VDW-EOS equation of state can be removed without significant loss of precision, thus yielding the Noble-Abel EOS:

$$P\left(v - b\right) = RT \tag{3}$$

CHAPMAN-JOUGUET SOLUTIONS FOR VDW GASES

Considering a coordinate system moving with the combustion wave and assuming a steady one-dimensional adiabatic flow with no body forces, the conservation equations of continuity, momentum and energy are given, respectively, by:

$$\rho_1 u_1 = \rho_2 u_2 = m \tag{4}$$

$$P_1 + \rho_1 u_1^2 = P_2 + \rho_2 u_2^2 \tag{5}$$

$$h_1 + \frac{u_1^2}{2} = h_2 + \frac{u_2^2}{2} \tag{6}$$

where subscripts 1 and 2 denote properties of reactants and products, respectively, m is the mass flow rate per unit area, ρ is density, u is flow velocity and h is specific enthalpy.

The VDW-EOS for reactants and products is:

$$\left(P_i + a_i / v_i^2\right)\left(v_i - b_i\right) = R_i T_i (i = 1, 2) \tag{7}$$

where $v_i = 1/\rho_i$ are the specific volumes.

The enthalpy of VDW gases with constant properties (Gonzáles, 2010) is given by:

$$h_i = h_{f,i} + c_{P,i}^0 T_i + b_i P_i \left(1 + \varepsilon_i\right) - 2 P_i v_i \varepsilon_i \tag{8}$$

where $h_{f,i}$ are the specific formation enthalpies, $c_{P,i}^0$ are the specific heats at low pressure (ideal gas equivalent), and $\varepsilon_i = a_i / P_i v_i^2 << 1$ (i =1, 2).

The specific heats at constant volume and constant pressure of a VDW gas are, respectively, $c_v = c_v^0 = \dfrac{R}{\gamma^0 - 1}$ and $c_P = c_v + \dfrac{(1+\varepsilon)R}{1 - \varepsilon + 2\varepsilon b^*}$.

The ratio of specific heats is $\gamma = \dfrac{\gamma^0\left(1+\varepsilon\right) - 2\varepsilon\left(1 - b^*\right)}{1 - \varepsilon + 2\varepsilon b^*}$ where the superscript 0 designates ideal gas condition.

Rayleigh Line Equation

Substituting the flow velocities $u_1 = m/\rho_1 = m v_1$ and $u_2 = m/\rho_2 = m v_2$ into the momentum equation, Eq. (5), yields the Rayleigh line equation:

$$\frac{P_2 - P_1}{v_2 - v_1} = -m^2 \tag{9}$$

or

$$\frac{p - 1}{v - 1} = -\mu \tag{10}$$

where $p = P_2/P_1$ and $v = v_2/v_1$ are pressure and specific volume ratios, respectively, and $\mu = m^2 v_1 / P_1 = u_1^2 / P_1 v_1$ is a dimensionless mass flow rate which can be expressed in terms of the Mach number of the reactants, $M_1 = u_1 / c_1$, where c_1 is the speed of sound in the reactants.

The speed of sound c_i can be obtained (Oates, 1984) from:

$$c_i = \left[\gamma_i v_i^2 \left(\frac{\partial P_i}{\partial T_i} \right)_{v_i} \left(\frac{\partial T_i}{\partial v_i} \right)_{P_i} \right]^{1/2} \tag{11}$$

Substituting the VDW-EOS for the reactants into Eq. (11), gives:

$$c_1^2 = \gamma_1 P_1 v_1 \left(\frac{1+\varepsilon_1}{1-b_1^*} + 2\varepsilon_1 \right) \tag{12}$$

where $b_1^* = b_1 / v_1$ is the non-dimensional covolume of reactants. Replacing Eq. (12) into Eq. (10) yields:

$$\mu = \gamma_1 \left(\frac{1+\varepsilon_1}{1-b_1^*} + 2\varepsilon_1 \right) M_1^2 \tag{13}$$

where

$$M_1^2 = \frac{1}{\gamma_1} \left(\frac{1+\varepsilon_1}{1-b_1^*} + 2\varepsilon_1 \right)^{-1} \left(\frac{p-1}{1-v} \right)$$

and the specific heat ratio of reactants is

$$\gamma_1 = \frac{\gamma_1^0 \left(1+\varepsilon_1\right) - 2\varepsilon_1 \left(1-b_1^*\right)}{1-\varepsilon_1 + 2\varepsilon_1 b_1^*}.$$

Rankine-Hugoniot Relations for VDW Gases

Substituting Eq. (8) into Eq. (6), the energy equation becomes:

$$c_{P1}^0 T_1 + \left(1+\varepsilon_1\right) b_1 P_1 - 2\varepsilon_1 P_1 v_1 + \frac{u_1^2}{2} + q = c_{P2}^0 T_2 + \left(1+\varepsilon_2\right) b_2 P_2 - 2\varepsilon_2 P_2 v_2 + \frac{u_2^2}{2} \tag{14}$$

where $q = h_{f,2} - h_{f,1}$ is the heat of reaction.

Then, replacing $u_i^2 = m^2 v_i^2$, $T_i = P_i \left(1+\varepsilon_i\right)\left(v_i - b_i\right) / R_i$ and the Rayleigh line equation into Eq. (14), after algebraic manipulations, it follows that:

$$pv-\left[\frac{\gamma_2^0+2b_2^*-1+2\varepsilon_2b_2^*}{\gamma_2^0+1+2\left(2-\gamma_2^0\right)\varepsilon_2}\right]p+\left[\frac{\gamma_2^0-1}{\gamma_2^0+1+2\left(2-\gamma_2^0\right)\varepsilon_2}\right]v= \\ =\left[\frac{\gamma_2^0-1}{\gamma_2^0+1+2\left(2-\gamma_2^0\right)\varepsilon_2}\right]\left\{2\alpha+\left[\frac{\gamma_1^0-2b_1^*+1+2\left(2-\gamma_1^0-b_1^*\right)\varepsilon_1}{\gamma_1^0-1}\right]\right\} \tag{15}$$

where $\alpha = q/(P_1v_1)$ is the non-dimensional heat release, $b_i^* = b_i/v_1$ are the non-dimensional covolumes, with $c_{P,i}^0 / R_i = \gamma_i^0 / (\gamma_i^0 - 1)$, where γ_i^0 are the specific heat ratios at low pressure (perfect gas equivalent).

The canonical form of a hyperbola is $(p - p_0)(v - v_0) = K^2$, where *K* is a constant and $p = p_0$ and $v = v_0$ are horizontal and vertical asymptotes, respectively. This equation can be rewritten as:

$$pv - v_0p - p_0v + p_0v_0 = K^2 \tag{16}$$

Comparing this equation (16) with Eq. (15) yields:

$$v_0=\frac{\gamma_2^0+2b_2^*-1+2\varepsilon_2b_2^*}{\gamma_2^0+1+2\left(2-\gamma_2^0\right)\varepsilon_2};\ p_0=-\frac{\gamma_2^0-1}{\gamma_2^0+1+2\left(2-\gamma_2^0\right)\varepsilon_2} \tag{17}$$

Thus, after algebraic manipulation of Eq. (15), the Hugoniot curve for a VDW gas is obtained:

$$\left[p+\frac{\gamma_2^0-1}{\gamma_2^0+1+2\left(2-\gamma_2^0\right)\varepsilon_2}\right]\left[v-\frac{\gamma_2^0+2b_2^*-1+2\varepsilon_2b_2^*}{\gamma_2^0+1+2\left(2-\gamma_2^0\right)\varepsilon_2}\right]=\left[\frac{\gamma_2^0-1}{\gamma_2^0+1+2\left(2-\gamma_2^0\right)\varepsilon_2}\right] \\ \times\left[2\alpha+\frac{\gamma_1^0-2b_1^*+1+2\left(2-\gamma_1^0-b_1^*\right)\varepsilon_1}{\gamma_1^0-1}-\frac{\left(\gamma_2^0+2b_2^*-1+2\varepsilon_2b_2^*\right)}{\gamma_2^0+1+2\left(2-\gamma_2^0\right)\varepsilon_2}\right] \tag{18}$$

If reactants and products have equal and constant properties, $\gamma_1^0 = \gamma_2^0 = \gamma^0$, $b_1^* = b_2^* = b^*$ and $\varepsilon_1 = \varepsilon_2 = \varepsilon$, then Eq. (18) describes a hyperbola. Initial values for ε_2, b_2 e γ_2^0 can be estimated from the perfect gas solution and, then, more accurate values can be obtained by iteration of the VDW solution.

Figure 2 shows the Hugoniot curves for VDW gases, using different specific heat ratios, $\gamma_1^0 = \gamma_2^0 = \gamma^0$, non-dimensional heat release values, *α*, non-dimensional covolumes, $b_1^* = b_2^* = b^*$, and attraction parameters, $\varepsilon_1 = \varepsilon_2 = \varepsilon$.

When the Rayleigh line is tangent to the Hugoniot curve, the upper and lower CJ points are determined. It can be shown that $M_2 = 1$ in the CJ points in the case of VDW gases (Gonzales, 2009). The Rayleigh line, Eq. (10), for a VDW gas, can be written as:

Figure 2. Hugoniot curves for VDW gases

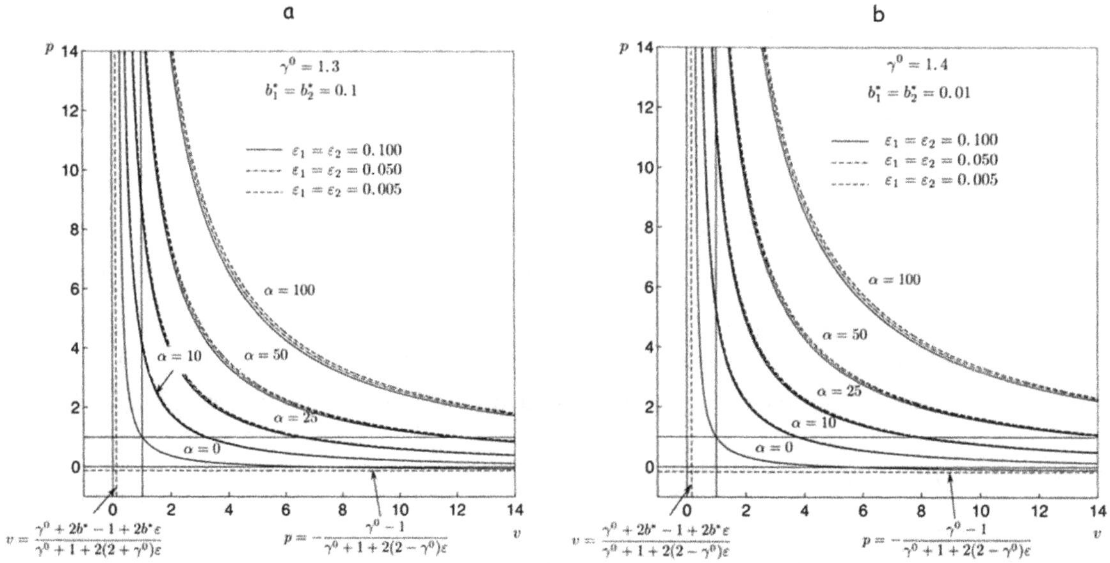

$$\frac{dp}{dv} = -\mu = \frac{p-1}{v-1} \tag{19}$$

Differentiating Eq. (18) yields:

$$\frac{dp}{dv} = -\frac{p\left[\gamma_2^0+1+2\left(2-\gamma_2^0\right)\varepsilon_2\right]+\left(\gamma_2^0-1\right)}{v\left[\gamma_2^0+1+2\left(2-\gamma_2^0\right)\varepsilon_2\right]-\left[\gamma_2^0+2b_2^*-1+2\varepsilon_2 b_2^*\right]} \tag{20}$$

Figure 2a

$$\gamma_1^0 = \gamma_2^0 = \gamma^0 = 1.3\,.$$

Figure 2b

$$\gamma_1^0 = \gamma_2^0 = \gamma^0 = 1.4\,.$$

Equating the derivatives (19) and (20), an expression for p in terms of v can be obtained and replaced into Eq. (18), then providing the following equations for p and v:

$$p_{\pm} = \frac{B}{2A}\left[1 \pm \left(1-\frac{4AC}{B^2}\right)^{1/2}\right] \tag{21}$$

$$v_{\mp} = \frac{B'}{2A'}\left[1 \pm \left(1 - \frac{4A'C'}{B'^2}\right)^{1/2}\right] \tag{22}$$

where

$$A = \left(\gamma_2^0 + 1\right)\left[\left(1 - \gamma_2^0\varepsilon_2\right) - \left(1 - \varepsilon_2\right)b_2^*\right] + 6\varepsilon_2\left(1 - b_2^*\right)$$

$$B = -\left(\gamma_2^0 - 1\right)\left[\gamma_2^0 + 1 + 2\left(2 - \gamma_2^0\right)\varepsilon_2\right]\left(2\alpha + X_1 - 1\right)$$

$$C = \left(\gamma_2^0 - 1\right)\left\{\left[1 + \left(2 - \gamma_2^0\right)\varepsilon_2\right]\left(2\alpha + X_1\right) - \left(1 + \varepsilon_2\right)b_2^*\right\}$$

$$A' = \left(\gamma_2^0 + 1\right)\gamma_2^0\left(1 - 3\varepsilon_2\right) + 2\left(4\gamma_2^0 + 1\right)\varepsilon_2$$

$$B' = -2b_2^*\left(\gamma_2^0 + 1\right)\left(1 - \varepsilon_2\right) - 12b_2^*\varepsilon_2 + \left[\left(\gamma_2^0 + 1\right)\left(1 - 2\varepsilon_2\right) + 6\varepsilon_2\right]\left(\gamma_2^0 - 1\right)\left(2\alpha + X_1 + 1\right)$$

$$C' = \left(\gamma_2^0 - 1\right)\left(2\alpha + X_1 + 1\right)\left[\left(\gamma_2^0 + b_2^*\right)\left(1 - \varepsilon_2\right) + 2\left(b_2^* + 1\right)\varepsilon_2\right] - \left[\gamma_2^0\left(\gamma_2^0 - 1\right)\left(1 - \varepsilon_2\right) + 2\left(\gamma_2^0 - 1\right)\varepsilon_2\right]$$

$$X_1 = \left[\gamma_1^0 - 2b_1^* + 1 + 2\left(2 - \gamma_1^0 - b_1^*\right)\varepsilon_1\right] / \left(\gamma_1^0 - 1\right).$$

In equations (21) and (22), the plus signs correspond to a CJ detonation and the minus signs correspond to a CJ deflagration. Dividing the VDW-EOS for reactants and products gives the temperature ratio across the combustion wave:

$$T_{\pm} = \frac{R_1}{R_2}\frac{\left(1 + \varepsilon_2\right)\left(1 - b_2^*\right)}{\left(1 + \varepsilon_1\right)\left(1 - b_1^*\right)} p_{\pm} v_{\pm} \tag{23}$$

The equation for M_1^2 as a function of *p* and *v* can be rewritten:

$$M_{1\pm}^2 = \frac{1 - \varepsilon_1 + 2\varepsilon_1 b_1^*}{\gamma_1^0\left(1 + \varepsilon_1\right) - 2\varepsilon_1\left(1 - b_1^*\right)}\left(\frac{1 - b_1^*}{1 + 3\varepsilon_1 - 2\varepsilon_1 b_1^*}\right)\left(\frac{p_{\pm} - 1}{1 - v_{\pm}}\right) \tag{24}$$

Deflagration and detonation velocities can be determined from $u_{1\pm} = \left(\frac{p_{\pm} - 1}{1 - v_{\pm}} P_1 v_1\right)^{1/2}$.

Figures 3 and 4 show the dependence of properties of the CJ combustion waves upon the dimensionless heat-release parameter α in VDW gases.

Figure 3. Properties of CJ combustion waves in VDW gases for $b^* = 0.1$, $\varepsilon_i = \text{constant}$ and $\gamma = 1.25$

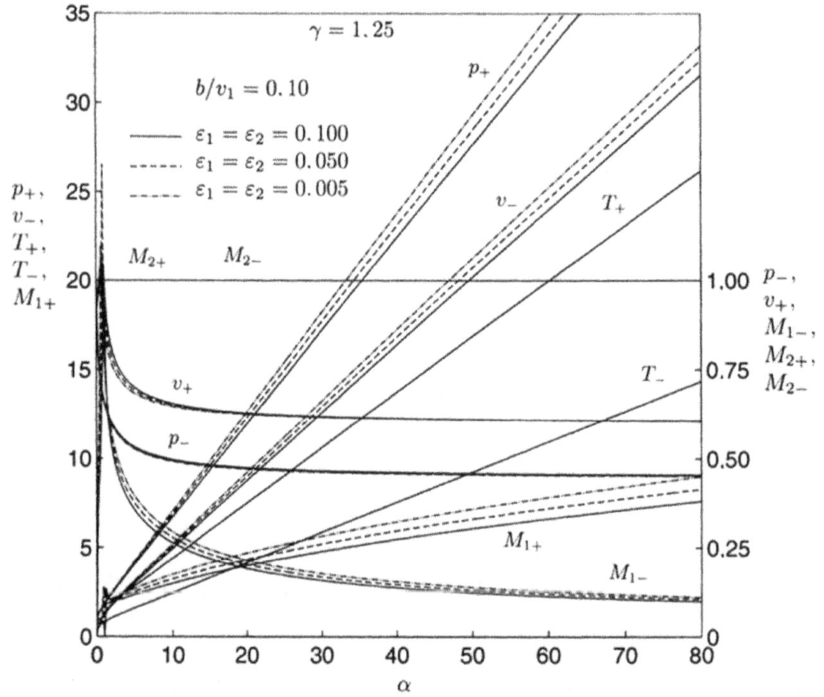

Figure 4. Properties of CJ combustion waves in VDW gases for $b^* = 0.1$, $\varepsilon_i = \text{constant}$ and $\gamma = 1.3$

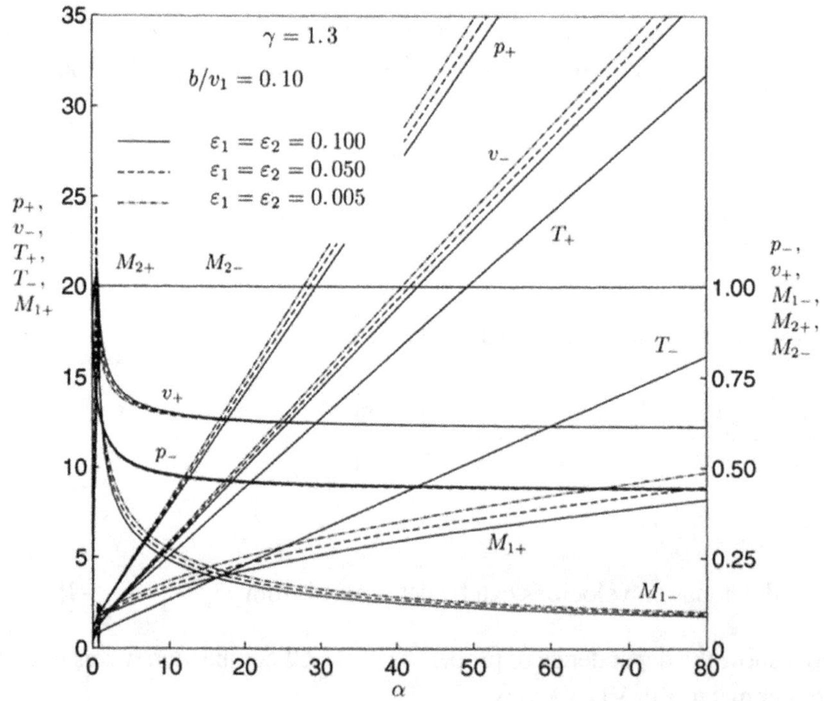

Stagnation Properties of VDW Gases

Stagnation conditions are attained when the flow velocity is reduced to zero by an isentropic process. Therefore, the energy equation can be written as

$$c_v T_0 + P_0 v_0 \left(1-\varepsilon_0\right) = c_v T + Pv\left(1-\varepsilon\right) + \frac{u^2}{2} \tag{25}$$

where $\varepsilon_0 = a/(P_0 v_0^2)$, c_v is the specific heat at constant volume, assumed as a constant, and subscript 0 indicates value at the stagnation condition. Consequently, the stagnation temperatures are obtained from:

$$\frac{T_0}{T} = \frac{1+\dfrac{\left(\gamma^0-1\right)}{\left(1+\varepsilon\right)\left(1-b^*\right)}\left[1-\varepsilon+\dfrac{\gamma^0\left(1+\varepsilon\right)-2\varepsilon\left(1-b^*\right)}{2\left(1-\varepsilon+2\varepsilon b^*\right)}\left(\dfrac{1+\varepsilon}{1-b^*}+2\varepsilon\right)M^2\right]}{1+\left(\gamma^0-1\right)\dfrac{\left(1-\varepsilon_0\right)}{\left(1+\varepsilon_0\right)\left(1-b_0^*\right)}} \tag{26}$$

Considering that $b_0^* = b/v_0 = b^* v/v_0$ and using isentropic relations for the VDW-EOS, it follows that

$$b_0^* = \left[1+\frac{1-b^*}{b^*}\left(\frac{T_0}{T}\right)^{-\frac{c_v}{R}}\right]^{-1} \tag{27}$$

and

$$\frac{1+\varepsilon_0^2}{\varepsilon_0} = \frac{1+\varepsilon^2}{\varepsilon}\left(\frac{T_0}{T}\right)^{\frac{R+c_v}{R}}\left[b^*+\left(1-b^*\right)\left(\frac{T_0}{T}\right)^{-\frac{c_v}{R}}\right]^2. \tag{28}$$

Then, the stagnation pressure P_0 can be obtained from:

$$\frac{P_0}{P} = \left(\frac{1+\varepsilon^2}{1+\varepsilon_0^2}\right)\left(\frac{T_0}{T}\right)^{\frac{R+c_v}{R}} \tag{29}$$

Indices for reactants and products were omitted in equations (25-29).
Determination of the stagnation conditions follows the sequence:

1. Input data: ε, b^*, γ^0, M
2. Estimate T_0/T for a perfect gas and calculate ε_0 and b_0^* using Eqs. (27) and (28);
3. With ε_0 and b_0^* calculated in ii) solve Eq. (26) and determine T_0/T;
4. With T_0/T calculate again ε_0 and b_0^* using Eqs. (27) and (28);
5. Return to step iii) until T_0/T vary less than 0.1%;
6. Calculate P_0/P using Eq. (29).

Once known the stagnation temperatures, their ratio can be calculated. Given the parameters a_i, b_i, γ_i, R_i and α of reactants and products, the pressure, specific volume and temperature ratios, Mach numbers and stagnation conditions in the CJ deflagrations and detonations in VDW gases can be calculated.

CHAPMAN-JOUGUET SOLUTIONS FOR NA GASES

The CJ solutions for NA gases can be obtained directly from the CJ solutions of VDW gases, making $\varepsilon = 0$ in equations (21-24, 26 and 29.

Therefore, for NA gases, the Rayleigh line equation is

$$\frac{p-1}{v-1} = -\mu \tag{30}$$

where

$$\mu = \frac{\gamma_1}{1-b_1^*} M_1^2$$

and

$$M_1^2 = \frac{1-b_1^*}{\gamma_1}\left(\frac{p-1}{1-v}\right)$$

and the Hugoniot curve becomes:

$$\begin{aligned}\left[p+\frac{\gamma_2-1}{\gamma_2+1}\right]\left[v-\left(\frac{\gamma_2+2b_2^*-1}{\gamma_2+1}\right)\right] &= \\ =\left(\frac{\gamma_2-1}{\gamma_2+1}\right)&\left[2\alpha+\left(\frac{\gamma_1-2b_1^*+1}{\gamma_1-1}\right)-\left(\frac{\gamma_2+2b_2^*-1}{\gamma_2+1}\right)\right]\end{aligned} \tag{31}$$

with asymptotes $v_0 = \frac{\gamma_2 + 2\left(b_2/v_1\right) - 1}{\gamma_2 + 1}$ and $p_0 = -\frac{\gamma_2 - 1}{\gamma_2 + 1}$.

Figures 5 and 6 show the Hugoniot curves of NA gases, for $\gamma = 1.3$ and $\gamma = 1.4$, respectively, considering $b_1^* = b_2^* = b^* = \text{constant}$.

The pressure, specific volume and temperature ratios along combustion waves in NA gases are given, respectively, by:

$$p_{\pm} = \frac{\left(\gamma_2 - 1\right)\left(2\alpha + X - 1\right)}{2\left(1 - b_2^*\right)}\left\{1 \pm \left[1 - \frac{4\left(1 - b_2^*\right)\left(2\alpha + X - b_2^*\right)}{\left(\gamma_2^2 - 1\right)\left(2\alpha + X - 1\right)^2}\right]^{\frac{1}{2}}\right\} \tag{32}$$

Figure 5. Hugoniot curves of NA gases for b = const and γ = 1.3*

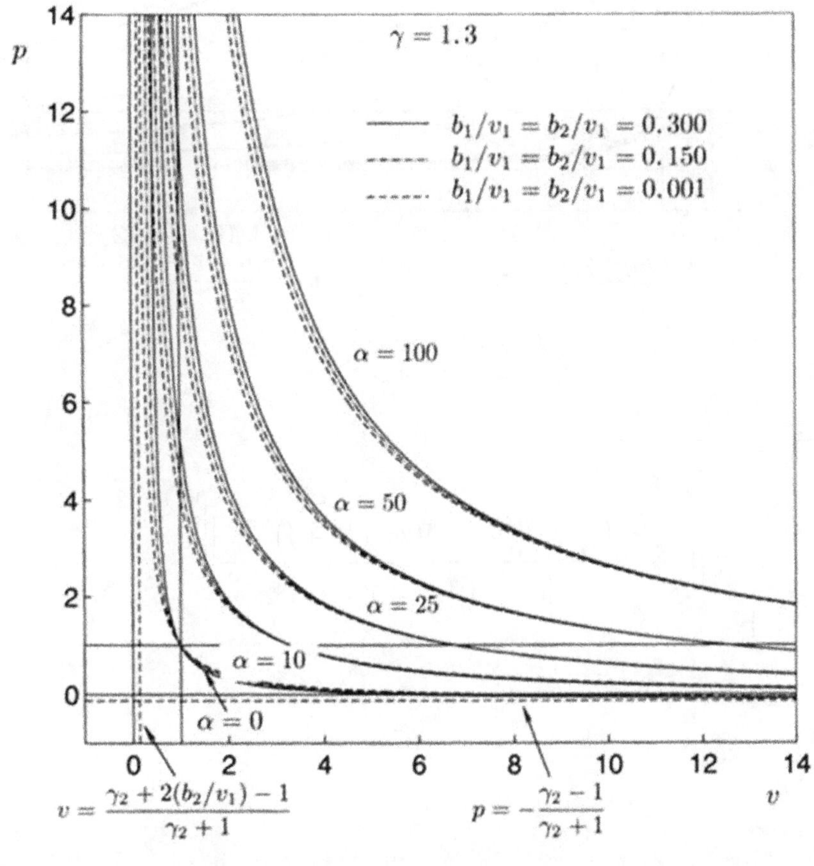

Figure 6. Hugoniot curves of NA gases for b^* = constant and γ = 1.4

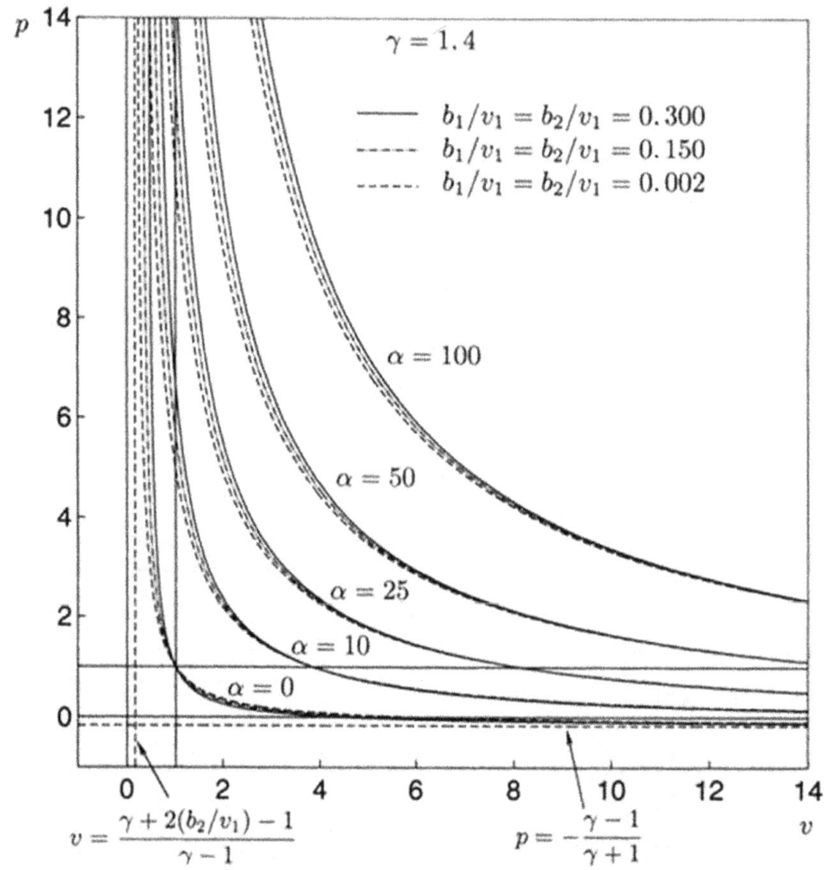

$$v_{\pm} = \frac{\left[\left(\gamma_2 - 1\right)\left(2\alpha + X + 1\right) + 2b_2^*\right]}{2\gamma_2} \cdot \left\{1 \mp \left[1 - \frac{4\gamma_2\left(\gamma_2 - 1\right)\left[\left(\gamma_2 + b\right)\left(2\alpha + X + 1\right) - \gamma_2\right]}{\left(\gamma_2 + 1\right)\left[\left(\gamma_2 - 1\right)\left(2\alpha + X + 1\right) + 2b_2^*\right]^2}\right]^{\frac{1}{2}}\right\} \tag{33}$$

$$T_{\pm} = \frac{R_1}{R_2}\frac{\left(v_{\pm} - b_2^*\right)}{\left(1 - b_1^*\right)}p_{\pm} \tag{34}$$

where

$$X = \left(\gamma_1 - 2b_1^* + 1\right)/\left(\gamma_1 - 1\right)$$

and

$$\frac{R_1}{R_2} \cong \frac{b_2}{b_1}.$$

The Mach numbers of the reactants for CJ detonations (plus sign) and CJ deflagrations (minus sign) in NA gases are given by

$$M_{1\pm} = \left[\frac{\left(1-b_1^*\right)}{\gamma_1}\frac{\left(p_\pm - 1\right)}{\left(1-v_\pm\right)}\right]^{\frac{1}{2}} \tag{35}$$

Figures 7 and 8 present the properties of combustion waves in NA gases as a function of the heat release α, with $b_i^* = b_i / v_1 = \text{constant}$ and $\gamma_i = \gamma = 1.3$ and 1.4, respectively.

Stagnation Properties of NA Gases

Stagnation temperature and pressure ratios along combustion waves in a NA gas are obtained, respectively, from:

$$\left(\frac{T_0}{T}\right) + \frac{\left(\gamma-1\right)b^*}{\gamma\left(1-b^*\right)}\left(\frac{T_0}{T}\right)^{\frac{\gamma-1}{\gamma}} = \frac{\gamma - b^*}{\gamma\left(1-b^*\right)} + \frac{\left(\gamma-1\right)}{2\left(1-b^*\right)^2}M^2 \tag{36}$$

$$\frac{P_0}{P} = \left(\frac{T_0}{T}\right)^{\frac{\gamma-1}{\gamma}} \tag{37}$$

Figure 9 shows the stagnation properties of NA gases in terms of Mach number. As seen in fig. 9, the stagnation temperatures and pressures of NA gases are lower than for ideal gases, with the same values of *M*, *T* and *P*.

DISCUSSION

Effects of Covolume

1. Case $b_1 = b_2 = b = \text{constant} \neq 0$ and $\gamma_1 = \gamma_2 = \gamma = \text{constant}'$.

Figure 7. Properties of CJ combustion waves in NA gases for $b_i^* = b_i / v_1 = \text{constant}$ *and* $\gamma_i = \gamma = 1.3$

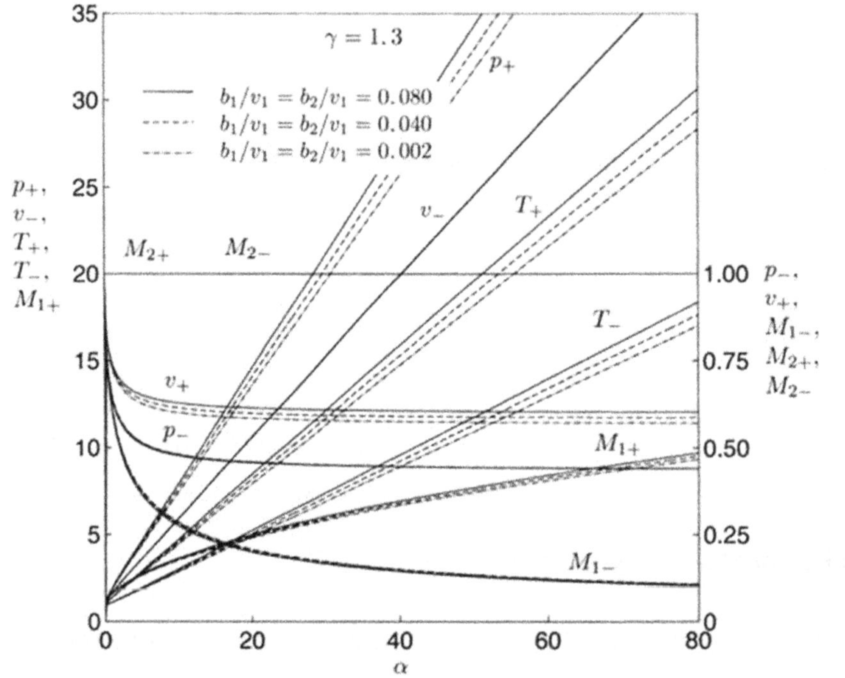

Figure 8. Properties of CJ combustion waves in NA gases, for $b_i^* = b_i / v_1 = \text{constant}$ *and* $\gamma_i = \gamma = 1.4$

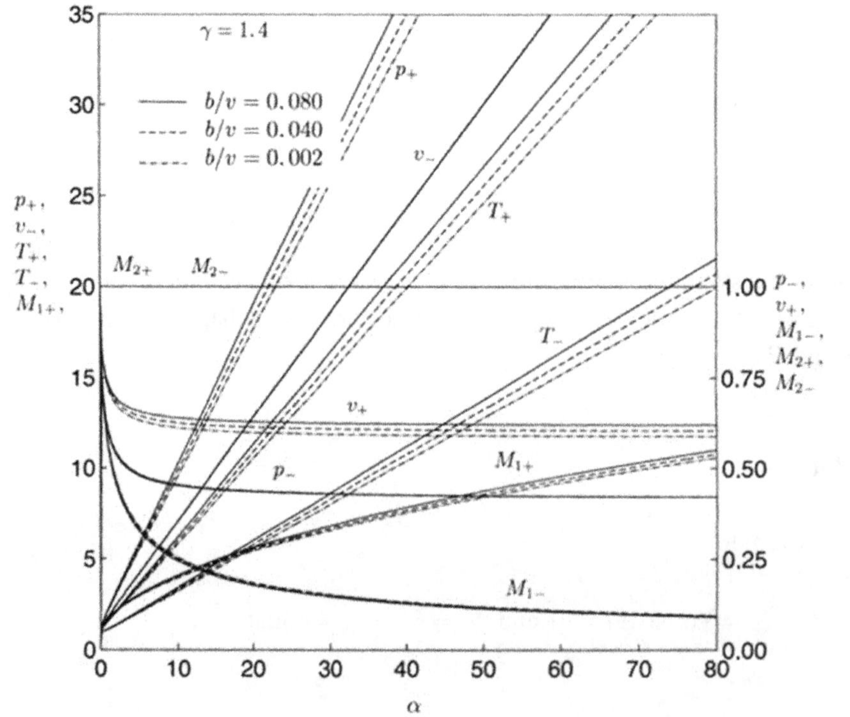

Figure 9. Stagnation properties of NA gases, for γ = 1.4

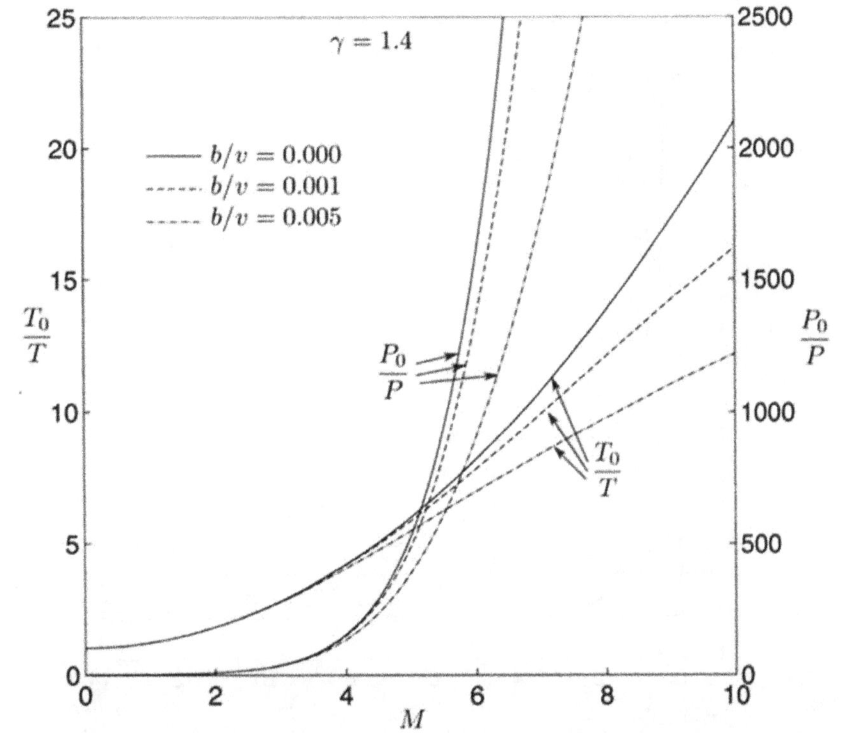

When $b_i^* = b / v_1 \neq 0$, for a given heat release α, the Hugoniot hyperbolas with different covolumes present a crossing point. Figure 10 depicts that, for increasing b values, the upper branches ($p > 1$) of the hyperbolas move to the left whereas their lower branches ($p < 1$) move slightly upwards. The vertex of the Hugoniot curve moves a little closer to its center when the covolume increases. The hyperbolas present a horizontal asymptote at $p=-(\gamma-1)/(\gamma+1)$, identical to the ideal-gas asymptote, and a vertical asymptote located at $v = (\gamma + b^* - 1)/(\gamma + 1)$ that depends also on the covolume.

The speed of sound in a NA gas is larger than the speed of sound of an ideal gas at the same thermodynamic conditions. When the covolume increases, the sound speed also increases, since there is less free space among the molecules. On the other hand, as seen in Figure 9, when the covolume is larger, the stagnation properties, T_0 and P_0, are lower, for a given Mach number and same static temperature and pressure, T and P.

In the case of CJ detonations, Figure 11 shows that, for a given heat release α, all properties (pressure ratio p_+, temperature ratio T_+, specific volume ratio v_+ and detonation Mach number M_{1+}) increase when b increases. Gases with larger b values, have larger molecules which can store more energy, consequently reach higher temperatures and pressures. In the case of CJ deflagrations there is an increase of temperature ratio T_- with increasing b, however there is no significant variation in pressure ratio p_-, in reactants Mach number M_{1-} and in specific volume ratio v_- since the expanded volume is much larger than the covolume.

Figure 10. Hugoniot curves for $b_i = b$ = constant and $\gamma_i = \gamma = 1.25$

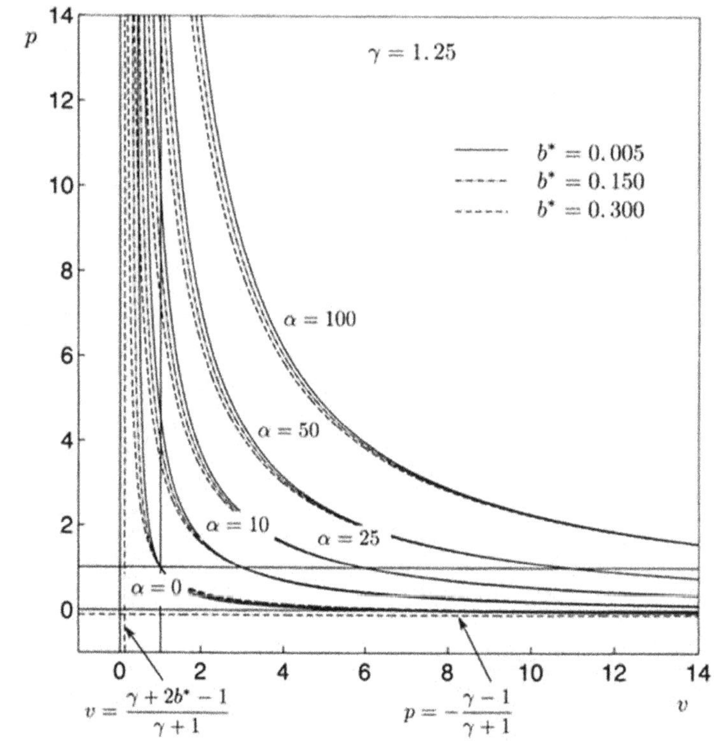

Figure 11. Properties of CJ combustion waves for $b_i = b$ = constant and $\gamma_i = \gamma = 1.25$

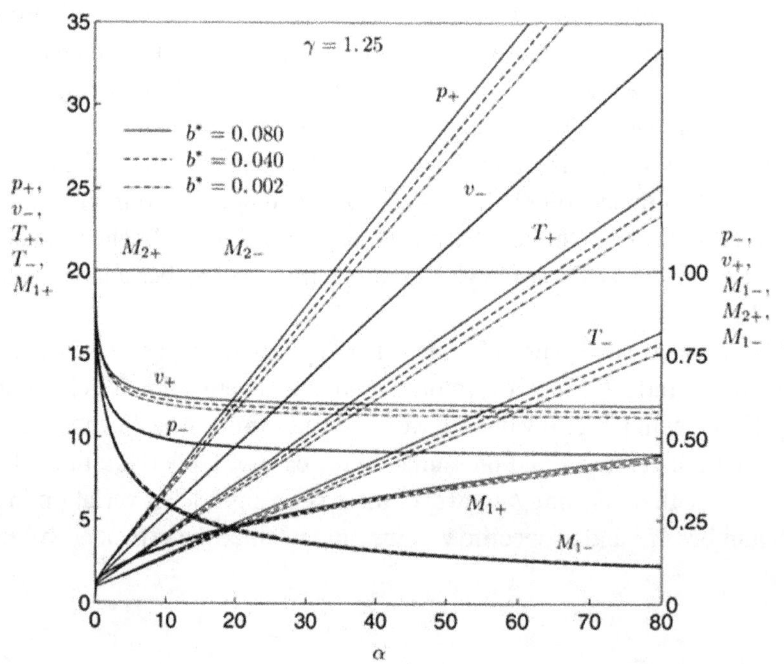

2. Case $b_2 < b_1$ and $\gamma = \text{constant}$.

The Hugoniot curves for $b_2 < b_1$ and $\gamma = 1.25$ are shown in Figure 12. For a given heat release α and covolume b_1, increasing values of the covolume b_2 move the upper branches ($p > 1$) of the Hugoniot curves slightly to the right, and, therefore, there is a smaller gas compression ($v_+ < 1$) in CJ detonations. The lower branches ($p < 1$) of the Hugoniot curves do not show significant variation, since in CJ deflagrations there occurs a large volume expansion, and the influence of the products' covolume on the specific volume becomes small.

A Hugoniot curve has two asymptotes: a horizontal asymptote that depends only on the specific heat ratio, which is constant, and a vertical asymptote that depends on the specific heat ratio and products covolume, thus moving slightly to the right for increasing b_2 covolumes.

In the case of CJ detonations, Figure 13 shows that, for a given heat release α, when the covolume b_2 increases, the specific volume ratio, pressure ratio and detonation Mach number, M_{1+}, increase, whereas the temperature ratio has no significant variation since a large covolume b_2 allows a smaller gas compression and higher pressures. Once b_1 is larger than b_2, in general, the gas constant R_1 will be lower than R_2, compensating for the variations in specific volume and pressure ratios, as seen in Eq. (34). In the case of CJ deflagrations, the covolume b_2 has no significant influence in any property, except in the pressure ratio with small heat release.

Figure 12. Hugoniot curves for $b_2 < b_1$ and $\gamma_i = \gamma = 1.25$

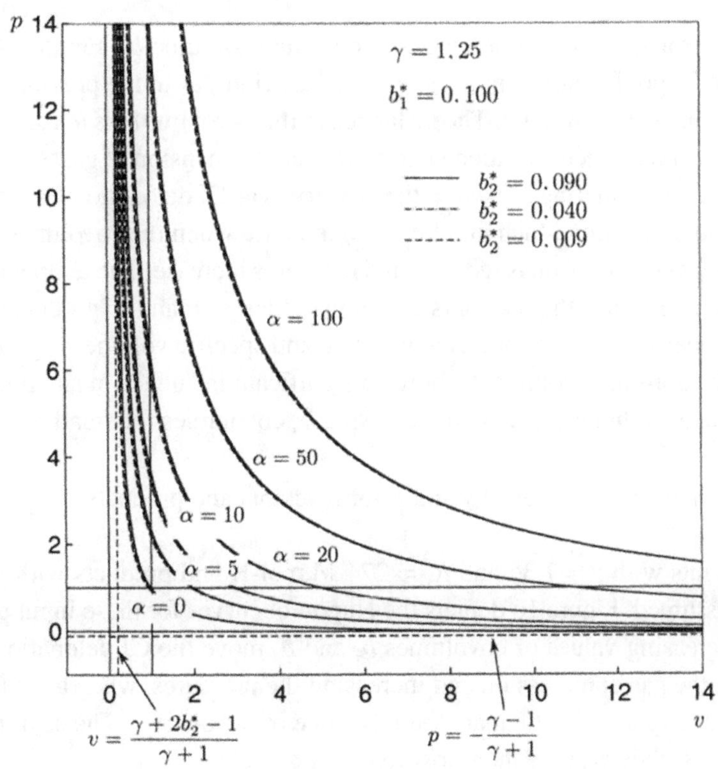

Figure 13. Properties of CJ combustion waves for $b_2 < b_1$ *and* $\gamma_i = \gamma = 1.25$

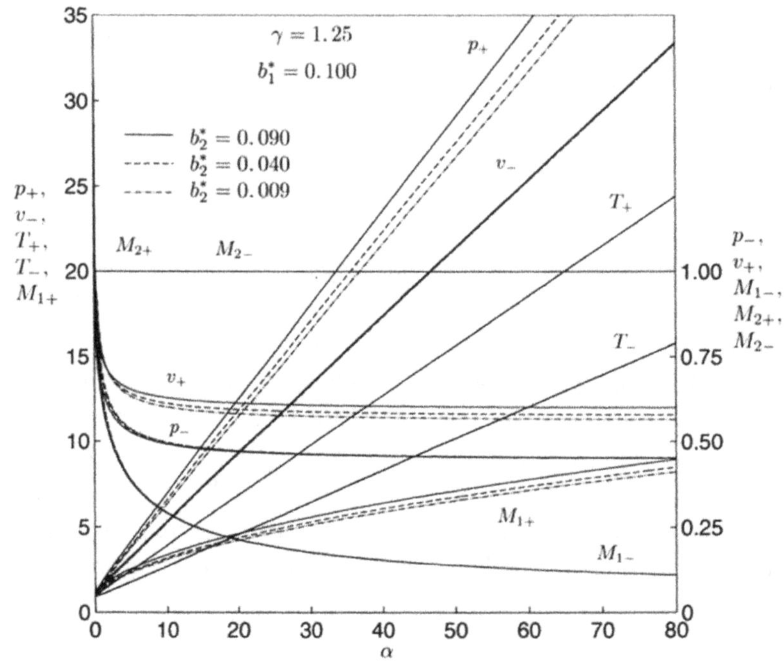

3. Case $b_2 > b_1$ and $\gamma = \text{constant}$

When b_2 increases, for given heat release α and reactant covolume b_1, Figure 14 indicates that the upper branches of the Hugoniot curves move slightly to the right, as in the previous case, reducing the gas compression and raising the pressure. The influence of the covolume b_2 is not significant in the lower branches of Hugoniot curves, since it is attenuated by the large expansion of gases, as mentioned before.

Figure 15 shows that, considering a fixed α, the properties of CJ detonation waves (pressure ratio p_+, specific volume v_+ and detonation Mach number M_{1+}) increase when the covolume b_2 is larger, similar to the previous case. Gas compression is reduced and pressures increase. The temperature ratio T_+ shows no significant variation, because the reactants covolume is larger than the products covolume and the ratio $R_1/R_2 \sim b_2/b_1$ compensates for variations in pressure and specific volume, as seen in Eq. (34). In CJ deflagrations, the variations in covolume b_2 show no significant influence in any property, except pressure ratios that present a slight decrease with increasing b_2 covolumes for small heat releases ($\alpha < 10$).

4. Case $b_1 < b_2$ with different values of γ and R for reactants and products.

In this case, reactants with $\gamma_1 = 1.35$ and $R_1 = 274$ J/kmol-K and products with $\gamma_2 = 1.25$ and $R_2 = 290$ J/kmol-K were assumed. Figure 16 depicts the Hugoniot curves for these input parameters and different covolumes. Increasing values of covolumes b_1 and b_2 move the CJ detonation branches slightly to the right, reducing the gas compression and increasing the pressures, whereas in the CJ deflagration branches the larger gas expansions attenuate the influence of covolumes. The asymptotes depend only on products' parameters, thus varying their position with b_2 and γ_2.

Figure 14. Hugoniot curves for $b_2 > b_1$ and $\gamma_i = \gamma = 1.25$

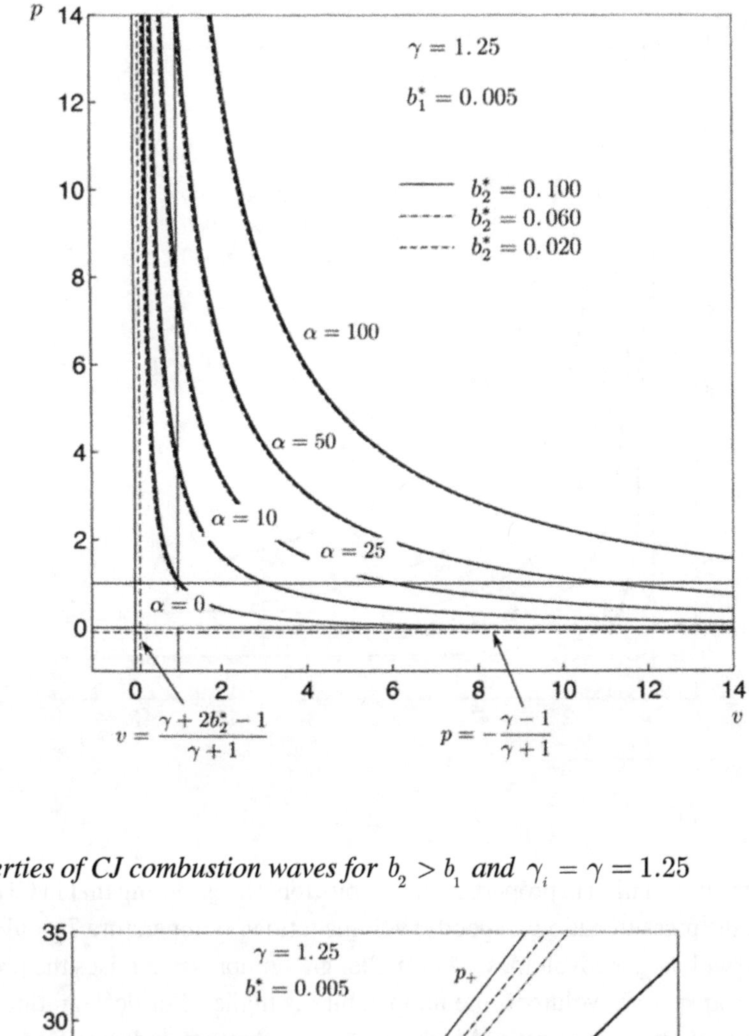

Figure 15. Properties of CJ combustion waves for $b_2 > b_1$ and $\gamma_i = \gamma = 1.25$

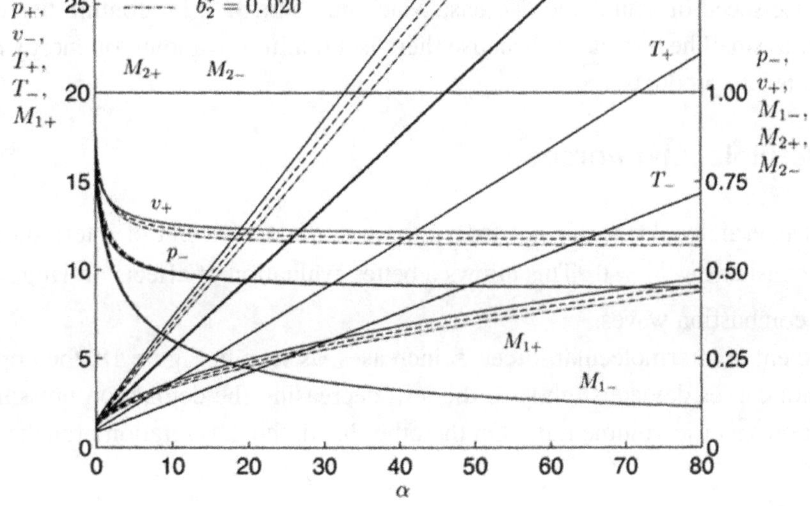

Figure 16. Hugoniot curves for $b_1 \neq b_2$ *and* $\gamma_1 \neq \gamma_2$

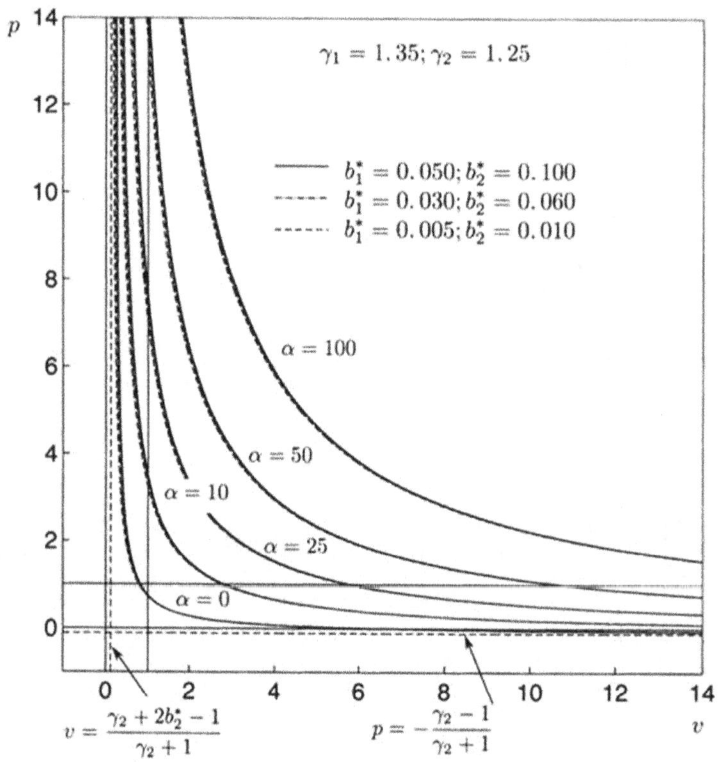

Figure 17 presents the variation in properties of combustion waves, noting that in CJ detonations, for a given heat release α, the pressure ratio p_+, specific volume ratio v_+, temperature T_+ and detonation Mach number M_{1+} increase with larger covolumes. The smaller gas compression raises the pressure and due to the larger pressures and specific volumes, the temperature is higher. For deflagrations, the covolumes have no significant effect on pressure ratio, Mach number of reactants and specific volumes. Temperature ratios T_- increase for larger covolumes, since larger molecules can store more energy. Increasing covolumes increase speed of sound and decrease Mach numbers M_{1-}. Discontinuities or jumps in the properties appear at small heat releases, because there is a minimum amount of energy α_{min} required to transform reactants into products.

Effects of Intermolecular Forces

The influence of intermolecular forces is verified through the coefficient of intermolecular attraction $\varepsilon = a/Pv^2$, and considering $b_i^* = 0$. This allows a better evaluation of effects on Hugoniot curves and properties of CJ combustion waves.

As the coefficient of intermolecular forces ε increases, as seen in Figure 18, the upper branch ($p > 1$) of the Hugoniot curves deviate slightly to the left, decreasing the detonation pressure, but with no significant effect on specific volume ratio. On the other hand, the deflagration branches ($p < 1$) of the

Figure 17. Properties of CJ combustion waves for $b_1 \neq b_2$ and $\gamma_1 \neq \gamma_2$

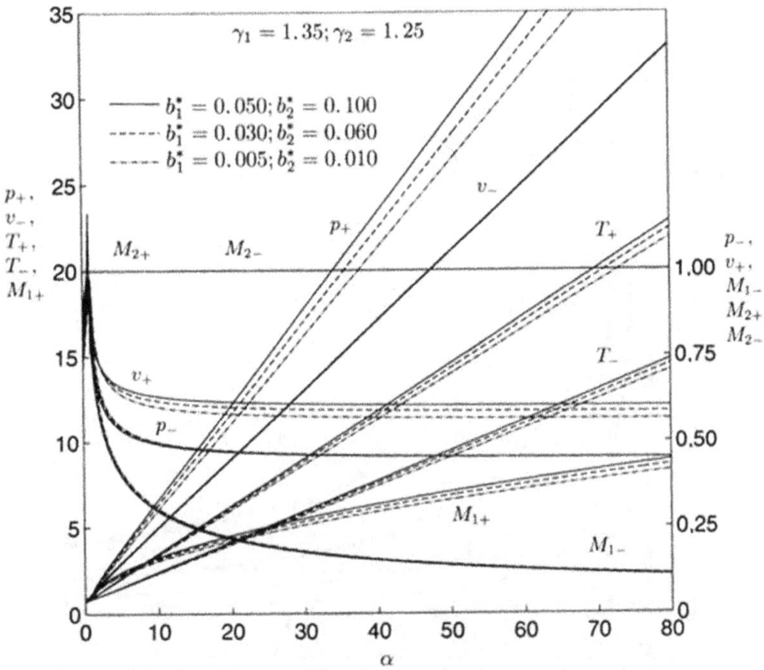

Hugoniot curves deviate very slightly down with higher intermolecular forces and, consequently, there is a larger gas expansion, with no significant influence on pressure.

The effects on properties of CJ combustion waves due to variations in the intermolecular force coefficients are shown in Figure 19. For CJ detonation waves, with a given heat release α, and for increasing coefficients of intermolecular forces, the detonation pressure ratio p_+ and the detonation Mach number M_{1+} decrease, whereas the specific volume ratio v_+ and temperature ratio T_+ show no significant variation. In CJ deflagrations, there are no significant effects of ε on pressure ratio p_-, temperature ratio T_- and Mach number M_{1-}, whereas there is a significant decrease of specific volume ratios v_- with larger ε.

APPLICATIONS

CJ Solutions for Stoichiometric Mixtures of *n*-Octane and Air

The burning characteristics of CJ combustion waves propagating in a stoichiometric mixture of *n*-octane and air, assuming $P_1 = 101325$ Pa and $T_1 = 300$ K, were calculated using the previous equations, considering ideal gas, NA and VDW equations of state. The following reaction was considered:

$$C_8H_{18} + 12{,}5\left(O_2 + 3{,}76N_2\right) \Rightarrow 8CO_2 + 9H_2O + 47N_2$$

Figure 18. Hugoniot curves for $\varepsilon_i = \varepsilon =$ constant *and* $\gamma_i = \gamma =$ constant

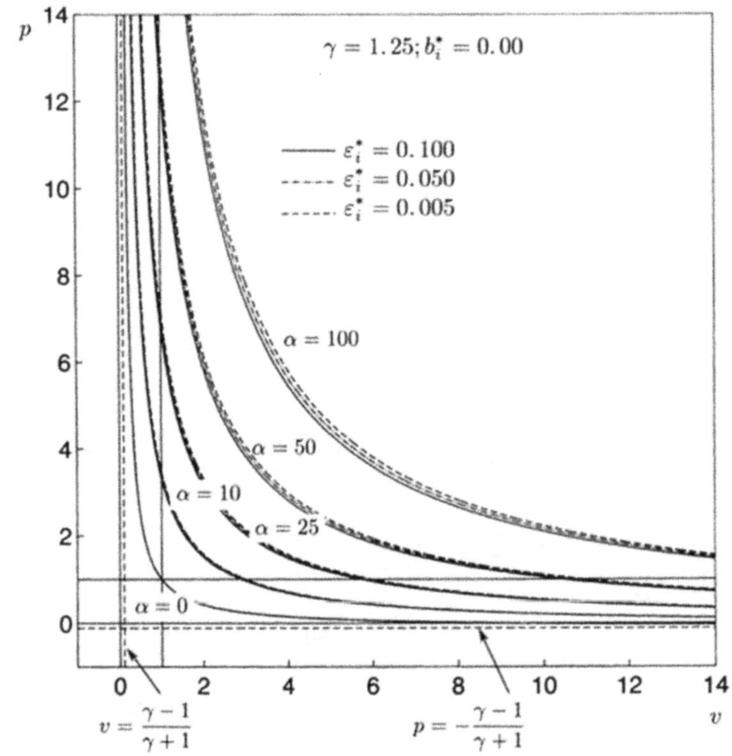

Figure 19. Properties of CJ combustion waves for $\varepsilon_i = \varepsilon =$ constant *and* $\gamma_i = \gamma =$ constant

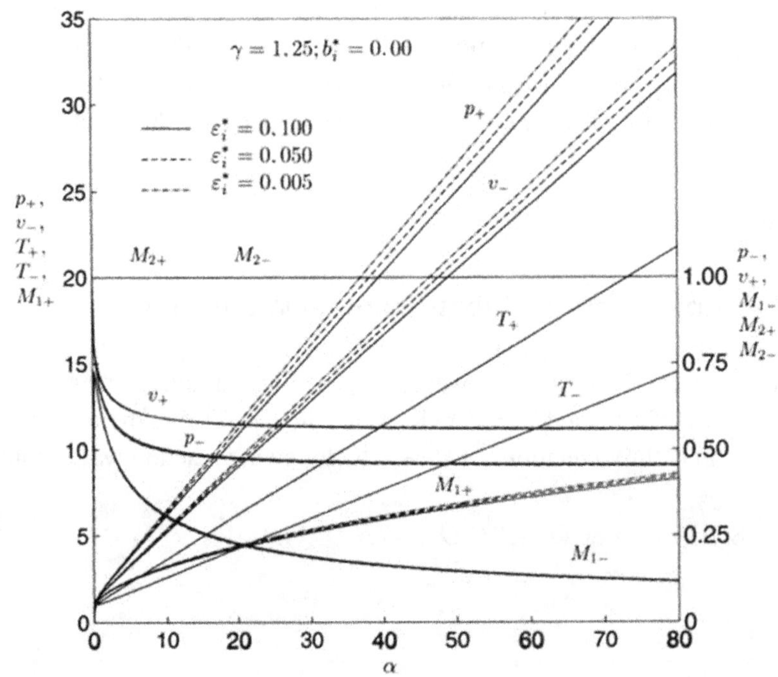

Table 1 shows the input data of reactants and products used in the calculations.

The constant and equal parameters used in the calculations were obtained considering three cases: *i)* γ, c_p, R, a, b of reactants; *ii)* γ, c_p, R, a, b of products; and *iii)* γ, c_p, R, a, b average of reactants and products.

$$C_8H_{18} + 12,5\left(O_2 + 3,76N_2\right) \Rightarrow 8CO_2 + 9H_2O + 47N_2$$

Results are depicted on Table 2 and are also compared to data of detonation simulations made with the NASA CEA 2004 code. This code calculates CJ detonations of ideal gases assuming chemical equilibrium, dissociation of products and different parameters of reactants and products.

Table 1. Properties of reactants and products at 300 K.

Species	#moles	Molar Mass kg/kmol	R kJ/kg-K	$c_{p,0}$ kJ/kg-K	$c_{v,0}$ kJ/kg-K	g°	a Pa.m^6/kg^2	b m^3/kg
C_8H_{18}	1	114.23	0.0729	17.113	16.385	1.044	290.3	0.002078
O_2	12.5	31.999	0.2598	0.918	0.658	1.395	3.115	0.043189
N_2	47	28.013	0.2968	1.039	0.743	1.400	4.931	0.048906
CO_2	8	44.010	0.1889	0.846	0.657	1.289	2.215	0.083117
H_2O	9	18.015	0.4615	18.723	14.108	1.327	9.398	0.307355

Table 2. Properties of CJ combustion waves in stoichiometric mixtures of n-octane and air, using ideal-gas, NA and VDW equations of state

Table 2a.

CJ PROPERTY	IDEAL GASES			CEA NASA
	Reactants	Products	Reac. & Prod.	
P_2 (deton.) (MPa)	25.675	17.772	17.27	18.900
v_2 (deton.) (m^3/kg)	0.47227	0.45839	0.45903	0.55184
T_2 (deton.) (K)	4413.5	2965.2	2727.6	2830.72
P_2 (deflag.) (MPa)	0.044439	0.04712	0.043445	
v_2 (deflag.) (m^3/kg)	15.364	11.5345	11.256	
T_2 (deflag.) (K)	2485.1	1992.5	1682.5	

Table 2b.

CJ PROPERTY	NOBLE ABEL GASES			VAN DER WAALS GASES		
	Reactants	Products	Reac. & Prod.	Reactants	Products	Reac. & Prod.
P_2 (deton.) (MPa)	25.675	18.65	18.066	25.756	18.837	18.460
v_2 (deton.) (m^3/kg)	0.51913	0.52265	0.523	0.5192	0.53197	0.52273
T_2 (deton.) (K)	4412.7	3017.4	3008.9	4600.75	3388.6	2977.6
P_2 (deflag.) (MPa)	0.0444	0.04704	0.047095	0.04441	0.04697	0.04696
v_2 (deflag.) (m^3/kg)	15.411	11.635	11.277	15.445	117.254	114.746
T_2 (deflag.) (K)	2485.1	2086.8	2024.5	2360	1894.87	1660

Effects of Initial Pressure on Detonations of Mixtures of Propane and Diluted Air

A study of the effects of initial pressure on the detonation velocity of a mixture of propane and diluted air was made, using the derived equations. The following reaction was considered:

$$C_3H_8 + 4.3O_2 + 8.7N_2 \rightarrow products$$

Figure 20 compares experimental data to theoretical results for VDW and NA gases using parameters of the products.

The detonation velocities calculated using the VDW-EOS followed closely the experimental detonation velocities for initial pressures up to 40 bar. Calculated values for detonation velocities using the perfect gas equation are lower than experimental data, whereas results for Noble-Abel gases are larger than experimental data, for initial pressures up to 40 bar. Therefore, the effects of intermolecular forces and covolumes should be taken into account for initial pressures above about 3 bar.

Figure 20. Influence of initial pressure on detonation velocities of a mixture of propane and diluted air (Source of experimental data: Schmitt and Butler, 2008)

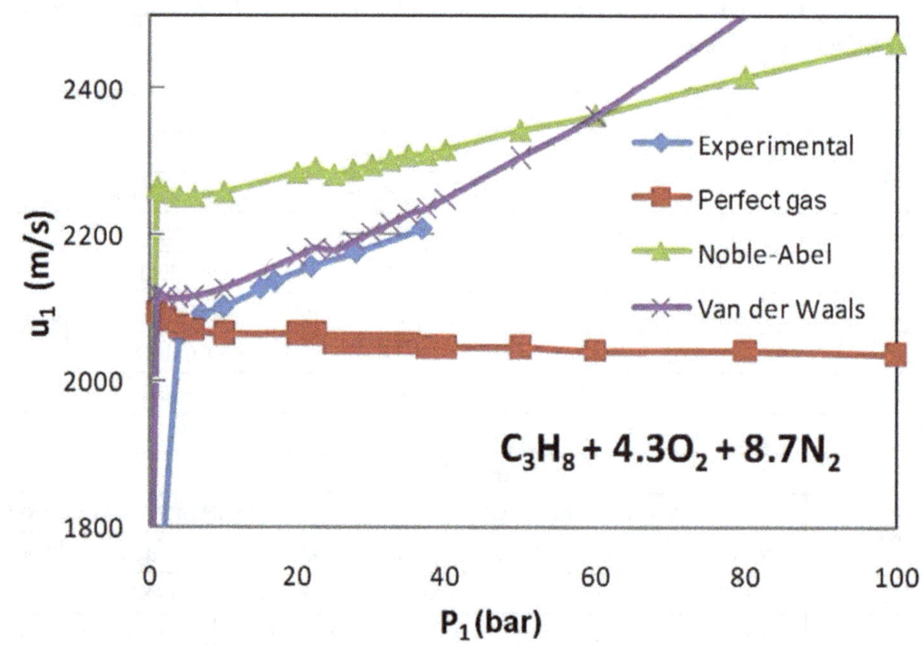

CONCLUSION

The CJ theory of combustion waves was extended to VDW and NA gases. Analytical expressions for the jump conditions along combustion waves were derived in terms of the covolumes and intermolecular force parameters. In general, the effects of covolumes were more significant than the effects of the attraction forces on Hugoniot curves and on properties of combustion waves. However, theoretical results using the Van der Waals equation of state matched more closely the experimental results for detonations of mixtures of propane and diluted air at high initial pressures.

REFERENCES

Astapov, N. S., Nikolaev, Y. A., & Ul'yanitskii, V. Y. (1984). Detonation Parameters of Hydrogen-Oxygen and Hydrogen-Air Mixtures at High Initial Density. *Fizika Gorenia i Vzryva*, *20*(1), 98–105.

Bauer, P., Dabora, E. K., & Manson, N. (1991). Chronology of the Early Research on Detonation Wave. In B. A. L. Kuhl et al. (Eds.), *Dynamics of Detonations and Explosions: Deflagrations, Progress in Astronautics and Aeronautics* (Vol. 133, pp. 3–18). Washington, DC: AIAA.

Bauer, P., Krishnan, S., & Brochet, C. (1979). *"Detonation Characteristics of Gaseous Ethylene, Oxygen, and Nitrogen at High Initial Pressures", Progress in Astronautics and Aeronautics* (Vol. 75, pp. 408–421). AIAA.

Dremin, A. N. (1999). *Toward Detonation Theory*. New York: Springer. doi:10.1007/978-1-4612-0563-0

Fickett, W., & Davis, W. C. (2000). *Detonation Theory and Experiment*. New York: Dover Publications.

Gealer, R. L., & Churchill, S. W. (1960). Detonation Characteristics of Hydrogen-Oxygen Mixtures at High Initial Pressures, A.1. *Ch.E. Journal*, *6*(3), 501–505. doi:10.1002/aic.690060330

Glassman, I. (1996). *Combustion* (3rd ed.). California: Academic Press.

Gonzáles, C. A. Q. (2010). Propagation of Combustion Waves in Real Gases [Master Dissertation] (In Portuguese). INPE, São José dos Campos, Brazil.

Kuo, K. K. (2005). *Principles of combustion* (2nd ed.). New Jersey: John Wiley & Sons.

Lee, J. H. S. (2008). The Detonation Phenomenon (illustrated ed.). Cambridge: University Press – UK.

Manson, N., & Dabora, E. K. (1993). Chronology of Research on Detonation Waves: 1920-1950. In B. A. L. Kuhl et al. (Eds.), *Dynamic Aspects of Detonations, Progress in Astronautics and Aeronautics* (Vol. 153, pp. 3–42). Washington, DC: AIAA.

Marchionni, M., Aggarwal, S. K., Puri, I. K., & Lentini, D. (2007). The Influence of Real-gas Thermodynamics on Simulations of Freely Propagating Flames in Methane/Oxygen/ Inert Mixtures. *Combustion Science and Technology*, *179*(9), 1777–1795. doi:10.1080/00102200701259999

Nettleton, M. A. (1987). *Gaseous Detonations: their Nature, Effects and Control*. London: Chapman and Hall Ltd. doi:10.1007/978-94-009-3149-7

Oates, G. C. (1984). *Aerothermodynamics of Gas Turbine and Rocket Propulsion*. New York: American Institute of Aeronautics and Astronautics.

Schmitt, R. G., & Butler, P. B. (1995a). Detonation Properties of Gases at Elevated Initial Pressures. *Combustion Science and Technology*, *106*(1-3), 167–193. doi:10.1080/00102209508907773

Schmitt, R. G., & Butler, P. B. (1995b). Detonation Wave Structure of Gases at Elevated Initial Pressures. *Combustion Science and Technology*, *107*(4-6), 355–386. doi:10.1080/00102209508907811

Williams, F.A. (1985). Combustion Theory (2nd ed.). MA: The Benjamin/Cummings Publishing Co.

Wingerden, K. V., Bjerketved, T. D., & Bakke, J. R. (1999). *Detonation in Pipes and in the Open*. Porsgrunn-Norway, Christian Michelsen Research and GexCon.

Chapter 4
Laser Ignition of Metalized Solid Propellants

Igor G. Assovskiy
Semenov Institute of Chemical Physics of RAS, Russia

ABSTRACT

This Chapter presents a theoretical analysis of radiation interaction with a semi-transparent metalized energetic material. Main regularities of the laser pulse interaction with metalized compositions are considered within the framework of non-resonant interaction of radiation with matter. The large variety of metalized composite propellants with different properties of the components, their ratio and dispersion can be divided into two classes, depending on the ratio of the laser irradiation's characteristic time (tr) and the thermal relaxation time of the propellant characteristic cell containing one metal particle (tm). Analysis of the role of metallic particles shape shows that in the case of spherical metal particles, duration of the laser pulse corresponds to the optimal size of particles, heated to a maximum temperature. In the case of flat metallic particles and constant pulse duration, the critical radiation flux and the critical density of ignition energy significantly decrease with decreasing thickness of the particle.

INTRODUCTION

Experimental and theoretical study of interaction of thermal radiation with energetic materials have a great history starting from the pioneering works of P. Pokhil and A. Kovalskiy with co-workers in the fifties in ICP (Pokhil, 1982) (Koval'skiy, Khlevnoy, & Mikheev, 1967) (Pokhil, Maltsev, & Zaitsev, 1969), and Beyer, Fishman, Ohlemiller, and Summerfield in the USA (Beyer & Fishman, 1960) (Ohlemiller & Summerfield, 1968) at the same time.

Main purpose of first investigations was mechanism of processes in gas and condensed phases of solid propellants during their ignition and combustion. One of main findings of those investigations was ability of the double-based (DB) propellants to burn in deep vacuum (Pokhil, Maltsev, & Zaitsev, 1969). This means that chemical destruction of condensed phase of DB-propellant is exothermal enough to support self-destruction of the propellant (in contrast to theoretical models of solid propellants ignition and combustion existed at that time (see the review (Librovich, 1963)). This result led to the development

DOI: 10.4018/978-1-5225-2903-3.ch004

of the solid-phase models of ignition and combustion of solid propellants (see the reviews (Merzhanov & Averson, 1971; Vilyunov & Zarko, 1989; Ohlemiller & Summerfield, 1968).

The new period of investigations of ignition and initiation of energetic materials by thermal radiation has been started, since creation of first lasers, from pioneering work of A.A. Brish with employees (Brish, Galeev, & Zaitsev, 1969). The effect of laser radiation on energetic materials (rocket propellants, explosives, pyrotechnics, etc.) is nowadays the subject of large number of publications. The main purposes of the majority of these studies are the mechanisms and regularities of the emergence of wave modes of exothermic reactions propagation. In spite of long history of the study of the laser ignition (LI) of metalized solid propellants (MSP) there is no commonly accepted opinion on mechanism of the laser initiation. It is due to wide variety of energetic materials and sources of laser irradiation. Meanwhile, the examined characteristics of the process, which are specific and sometimes inconsistent for various materials, are related by a number of quite general regularities. The main of them are the thermal mechanism of ignition and the spatial inhomogeneity of chemical conversions in the materials. A chemical reaction is localized at optically dense particles and at other inhomogeneities related to specific features of the structure, irradiation, and transformation of the material (see the reviews, e.g., in (Kondrikov, Ohlemiller, & Summerfield, 1970; DeLuca, Ohlemiller, Caveny, & Summerfield, 1976; DeLuca, Ohlemiller, Caveny, & Summerfield, 1976; Karabanov & Bobolev, 1981; Assovskiy, 1994; Tarzhanov, 1998; Assovskiy, 2000; Aluker et al., 2008). These facts show on effective methods to control laser ignition of composite propellants.

In this chapter, we consider main regularities of the laser pulse interaction with MSP within the framework of non-resonant (thermal) interaction of radiation with MSP (Assovskiy, 2000). Certain attention is given to features of a local initiation of metallized energetic materials by short laser pulses of small energy. A role of shape and size of optical discontinuities in the pulse intensity dependence on pulse duration is considered in detail.

Ignition of Quasi-Homogeneous Propellant by Laser Pulse

For the laser ignition of MSP, the principal difference between the metal particles from other particles constituting the composite propellant is their high value of the coefficient of light absorption in the particles material and high value of their thermal diffusivity. The first leads to elevated temperature of the metal particles compared to the ambient temperature. The second circumstance causes a rapid equalization of temperature inside the metal particles and a relatively slow relaxation of the temperature of their environment. For example, the thermal diffusivity of the aluminum particles is equal approximately to 1 cm^2/s, and the thermal diffusivity of the other particles is equal about to 10^{-3} cm^2/s. A characteristic time

t^* of temperature equalization of metal particles and the environment is equal to r^2/κ, where r is a characteristic distance between metallic particles. If we take a particle size equal to 10^{-3} cm and the distance between the particles $r=10^{-2}$ cm, the thermal inertia of the aluminum particles is equal to 10^{-6} s, and the thermal inertia of the surrounding particle volume - 10^{-1} s. Hence, we may conclude that when MSP is exposed to a light pulse with a characteristic duration of more than 10^{-1}s we can neglect the difference between the temperature of the metal particles and the temperature of their surrounding layers and consider MSP as a quasi-homogeneous medium with averaged thermal characteristics. In other case when the characteristic duration of exposure is less than t^* we must consider the difference between the temperature of the metal particles and the temperature of their surrounding layers.

Figure 1. Sketch of laser-matter interaction and temperature distribution in depth of semi-transparent substance

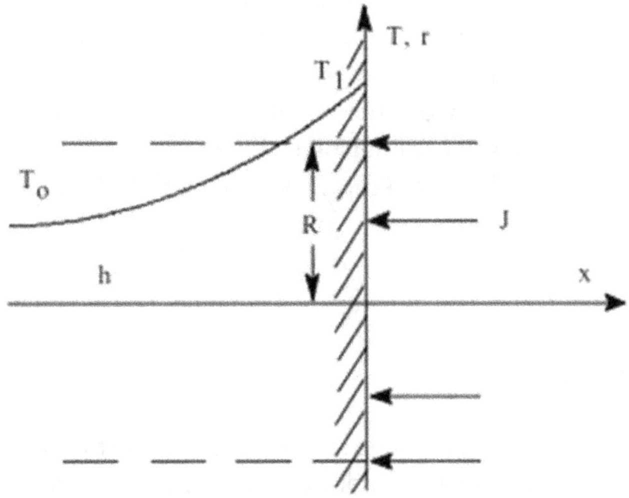

In this section of the chapter we will analyze the critical phenomena in the LI of MSP when the exposure time t_e is longer than the thermal relaxation time t^* of disturbances caused by the presence of metallic particles in the propellant. We also assume that the laser-beam dispersion in the MSP can be neglected. In such a case the problem of thermal regime in irradiated cylindrical volume ($x \leq 0, 0 \leq r \leq R$), Figure 1, may be formulated using the one-dimensional equation of thermal conductivity (1). Where $T(x, t)$ is the temperature averaged over the cross section of the cylinder. The boundary condition on the front surface ($x=0$) is Eq. (2), and a homogeneous initial distribution is Eq. (3):

$$\rho c \partial T/\partial t = \lambda \partial^2 T/\partial x^2 + \rho q_0 F_0(T) + J(t)e^{x/h}/h - 4Q/D \quad (1)$$

$$(\partial T/\partial x)_O = \beta(T_2 - T_1) + q_s \Phi_s(T_1),\ T_1 = T(0,t) \quad (2)$$

$$T(x,0) = T_O = T(x \rightarrow -\infty, t) \quad (3)$$

Here ρ, c and λ are respectively the density, heat capacity and thermal conductivity of the composite substance; q_0F_0 and q_sF_s are the heat release rates of bulk and heterogeneous reactions; J is the density of absorbed light flux; h is the characteristic absorption's depth; Q is the density of heat-flux going through the side of the irradiated cylindrical volume; β and T_2 represent the coefficient of heat exchange and the temperature of the surrounding medium at the front section of the cylinder; D and R are the diameter and radius of laser beam.

An asymptotic analysis of the temperature distribution outside of irradiated volume (under condition h >> D) shows that the intermediate asymptote for the heatflux Q (on times $t \sim D^2\rho c/\lambda$) has the form (Assovskiy, 1994):

$$Q \approx 2\lambda(T(x, t) - T_0)/D \qquad (4)$$

The activation energy E o of bulk reaction is assumed to be large enough; therefore the reaction takes place mainly in the thin layer (thickness $l_0 << h$) located near the position of maximum temperature (Frank-Kamenetskiy, 1969). Resulting rate of heat release in the reaction zone is positive if T>T*, where T* is the temperature of non-stable heat equilibrium of the cylinder (homogeneously heated) with environment (Assovskiy, Interaction of Laser Radiation with Reagents. Critical Diameter of Laser Beam, 1994), (Assovskiy & Leipunskiy, Theory of Propellant Ignition by Pulse Radiation, 1980):

$$\rho q_0 F_0(T^*) = 8\lambda(T^* - T_0)/D^2 \qquad (5)$$

Homogeneous Heating

When T > T* throughout the cylinder, it results in self-heating (thermal explosion) of substance. Whether T > T* only in a small part of the irradiated volume, possibility of the thermal explosion depends on parameters of the temperature distribution, on characteristics of the bulk reaction, and on the heat-release due to the radiation absorption.

We assume further the reaction rates are varying with temperature according to the Arrhenius law:

$$F_0 = k_0 \exp(-E_0/R_g T);\ F_s = k_s \exp(-E_s/R_g T) \qquad (6)$$

Using (5), (6) and Frank-Kamenetskiy approximation (Frank-Kamenetskiy, 1969), a relationship between T* and D may be presented in the form:

$$D^2/l_0^2(T_0) = 8x\ \exp(-x),\ l_0^2(T) = \lambda R_g T^2/E_0 \rho q_0 F_0(T) \qquad (7)$$

where l_0 is the characteristic thickness of the bulk reaction zone, and x is a relative temperature of irradiated volume.

$$x = (T^*-T_0)E_0/R_g T_0^2 \qquad (8)$$

The plot of function (7) is presented on Figure 2. This figure shows the existence of a critical diameter D* of laser beam, which corresponds to the maximum of function (7) (at x = 1).

$$D^* = 2(2/e)^{1/2} l_0(T_0) \qquad (9)$$

According to (9), D* has an order of magnitude equal to $l_0(T_0)$. Figure 3 shows plots of l_0 as a function of temperature T_0 for some exothermically reacting substances which thermochemical characteristics are presented in Tab. 1.

There are two states of the heat equilibrium for each D < D*(low-temperature x_1 and high temperature x_2), Figure 2. The value x_1 corresponds to a small deviation of the irradiated volume's temperature T_1 from the environment's temperature T_0:

Figure 2. Dependence of the relative radius R/l_o on the relative temperature x of irradiated cylindrical volume

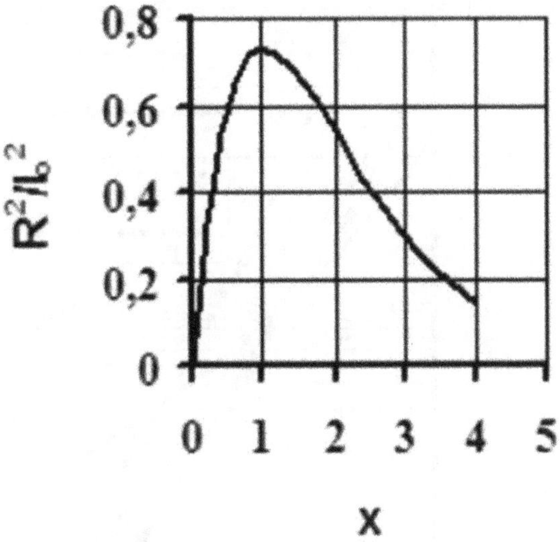

Table 1. Thermochemical characteristics of some exothermically reacting substances.

Substances	$lg(\rho q_o k_o)$	E, kJ/mol	λ, 10^{-3} W/cmK
Lead azide (LA)	16.83	152	1.76
Nitro cellulose (NC)	22.56	201	2.35
Hexogen (RDX)	19.01	172	1.67
Octogen (HMX)	23.26	220	2.90
Model gas-mixture (MG)	10.00	100	0.50

$T_1\text{-}T_o < R_gT_o^2/E_o$

The high temperature x_2 defines the upper limit of homogeneous heating of the irradiated volume, not exciting the thermal explosion:

$$T^*(D)=T_o+x_2R_gT_o^2/E, (x_2>1) \tag{10}$$

The value of critical temperature, Eq. (10), increases rapidly with decreasing the beam diameter (due to increase of x_2, Figure 2). So, it may considerably exceed the minimal temperature corresponding to $D=D^*$: $T^*(D^*) = T_o + R_gT_o^2/E_o$, that is usually considered as the pre-explosion temperature (Frank-Kamenetskiy, 1969). Typical values of the critical temperature T* for some exothermically reacting substances are presented in Tab. 2.

Figure 3. Scale of reaction zone thickness lo vs. reaction zone temperature T.

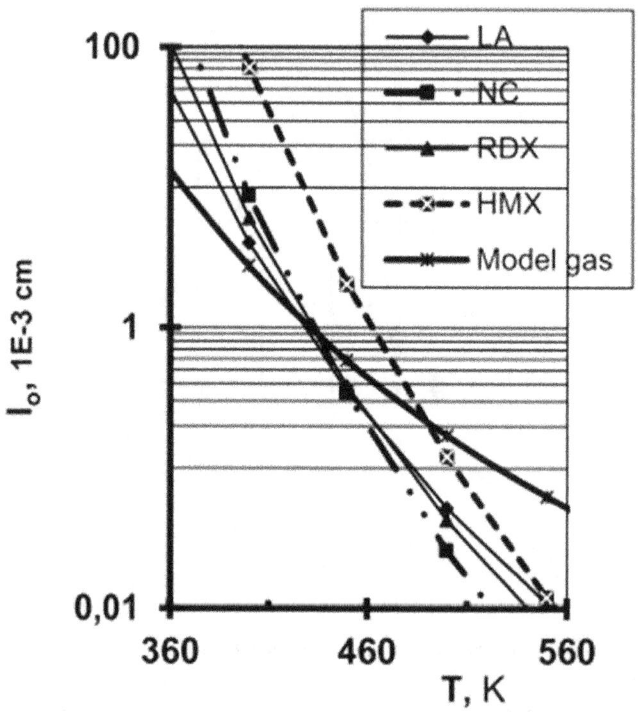

Table 2. Critical temperatures T for the thermal explosion of uniformly irradiated volume, °C*

Substance	HMX	RDX	NC
R = 2.0 mm	250	220	200
R = 5.0 mm	200	185	180

Non-Homogeneous Heating

The reaction zone temperature may far exceed the value (10), under nonhomogeneous heating of the irradiated volume. The asymptotic analysis of the system (1) - (4) (made in condition: $l_o << h$) shows possibility of the quasi-steady interaction with the reaction zone temperature T_1 determined by equation (Assovskiy, Heat Transfer in Laser Pulse Interaction with Reactive Substances, 2000), (Assovskiy & Leipunskiy, Theory of Propellant Ignition by Pulse Radiation, 1980):

$$Q_1(T_1) = 2\sqrt{2}\,\lambda(T_1-T_o)/D,\ T_1=To+JD^2/8(h + D/2\sqrt{2}) \qquad (11)$$

Here Q_1 is the heat flux from the reaction zone to colder layers of the irradiated volume:

$$Q_l^2(T) = (\beta(T_2-T)+q_sF_s(T))^2 + 2\lambda R_gT^2E^{-1}(\rho q_oF_o(T)-8\lambda(T-T_o)/D^2) \quad (12)$$

The maximum diameter of the radiation beam, such that the quasi-steady thermal regime is yet possible, can be defined like the Semenov's theory of thermal-explosion (Frank-Kamenetskiy, 1969). The critical condition correlates with the contact of curves corresponding to the left (Eq. (12)) and right parts of Eq. (11), see Figure 4.

Ignition by Polychromatic Pulse-Irradiation

In many cases, the laser pulse consists of multiple modes of different wavelengths. The problem of thermal regime of chemical reaction exposed by polychromatic pulse-radiation can be solved using results of previous consideration of monochromatic irradiation. We assume that the absorption of each mode, having a wavelength m, does not depend on other one. Taking into account Eq. (1), the heat-conductivity equation can be presented in the form:

$$\rho c\partial T/\partial t = \lambda\partial^2T/\partial x^2 + \rho q_oF_o(T) + \int_0^\infty j(t,m)h^{-1}\exp(x/h)\,dm - 4Q/D \quad (13)$$

A parametric analysis of Eqs. (13) with (2) and (3) shows that the thermal regime of interaction depends significantly on relationship between following timescales (designation corresponds to (Assovskiy & Leipunskiy, 1980)): the pulse duration tJ, the characteristic scales of thermal relaxation of irradiated volume $t_1=h^2_m/\chi$ and $t_2=R^2/\chi$, and the time-scale of volume's self-heating due to reaction $t_3 = c(T_1-T_o)/q_oF_o$. Here h_m is a characteristic depth of absorption.

Figure 4. Relation between heat removal Q_l (lines: 1,2, and 3) and heat flux from the reaction zone Q_l (curves: 4,5, and 6) for exothermic (4), endothermic (5), and neutral (6) reactions at different values of the laser beam diameter: (1) D < D, (2) D = D*, (3) D > D*.*

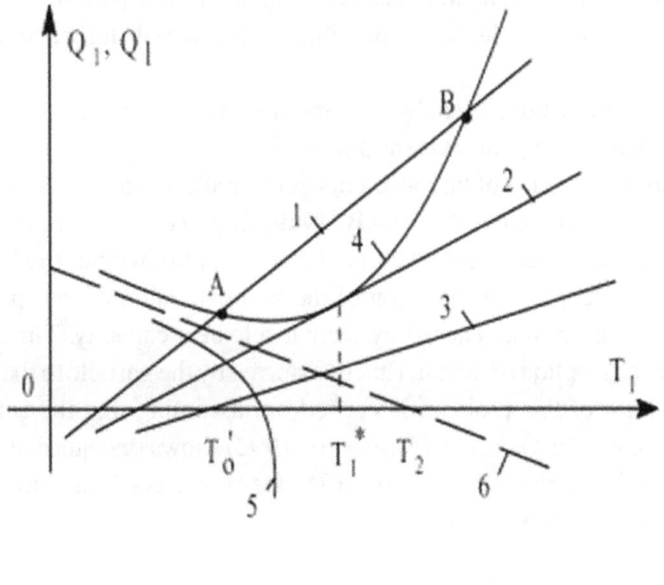

If the pulse duration t_J is much smaller than other time-scales, the volume temperature distribution depends (during irradiation) only on spectrum of absorbed radiation and on function h(*m*):

$$T(x,t)=T_o+\int_0^t\int_0^\infty j\ (t,m)\ h^{-1}\exp(x/h)\ dm\ dt$$

Transformation of this distribution after irradiation depends on a ratio of timescales t_1, t_2, t_3. In limit cases of such a ratio, it is possible to find the solutions in analytical form.

For example, the critical relationship between characteristics of igniting pulse, under assumption $t_J << t_1, t_2$, has a form:

$$a = Q_l(T_1)h_m/\lambda(T_1\text{-}T_o) > a^* \quad (14)$$

Here the criterion **a** is a ratio of the total heat-release in reaction-zone $Q_l(T_1)$, Eq. (12), in the end of irradiation, to the heat-flux from the front-surface to colder layers of irradiated cylindrical volume.

The critical value a* is a function of ratio h_m/D. If $D >> h_m$, then $a^* < 1$, and there exists a region of parameters $1 > a > a^*$, where the system self-ignition takes place later a time-delay after irradiation. If $D < h$, then $a^* > 1$, and there exists a region $1 < a < a^*$, where the initial self-heating of system, after irradiation, changes then to the cooling due to the heat-losses.

INFLUENCE OF OPTICAL DISCONTINUITY ON IGNITION PROCESS

Optical discontinuity such as metallic particles being present in the reacting substance may essentially influence not only optical characteristics but the macrokinetics of chemical process as well. That is why the phenomenon of anomalously low threshold for laser-pulse initiation of some explosives is usually explained by the presence of absorbing microparticles (see, for example, (Karabanov & Bobolev, 1981)).

Common physical considerations make clear that large sooty or metallic particles heated by irradiation may considerably intensify the reaction in adjacent matter and, in particular, initiate a local thermal explosion of exothermic reaction. Evaluation of possible temperature disturbances is necessary to analyze this phenomenon.

This section presents an asymptotic analysis of the thermal regime of an optically dense spheroid particle located in transparent energetic reagent, Figure 5.

It is useful to compare formulation of this problem with formulation of well-known problem of thermal explosion of spherical reactor (Frank-Kamenetskiy, 1969). In the case under consideration the reaction volume is out of sphere, and the reactor wall temperature is not known but the law of heat exchange is given. It is necessary to underline that initiation of thermal explosion of transparent reactive medium under continuous irradiation does not caused by the particle heat capacity. The reason for the thermal explosion is the effective absorption of radiant flux transferred by the particle to the surrounding medium.

The asymptotic analysis of this problem is carried out assuming that the activation energy of the bulk reaction is large enough. This analysis (Assovskiy, 1992) shows that quasi-steady temperature field around the particle can exist if the particle diameter D_1 does not exceed the critical one. The maximum temperature T_1 of adjacent layers is defined by

Figure 5. Diagram of irradiation of a particle enclosed in a scattering medium

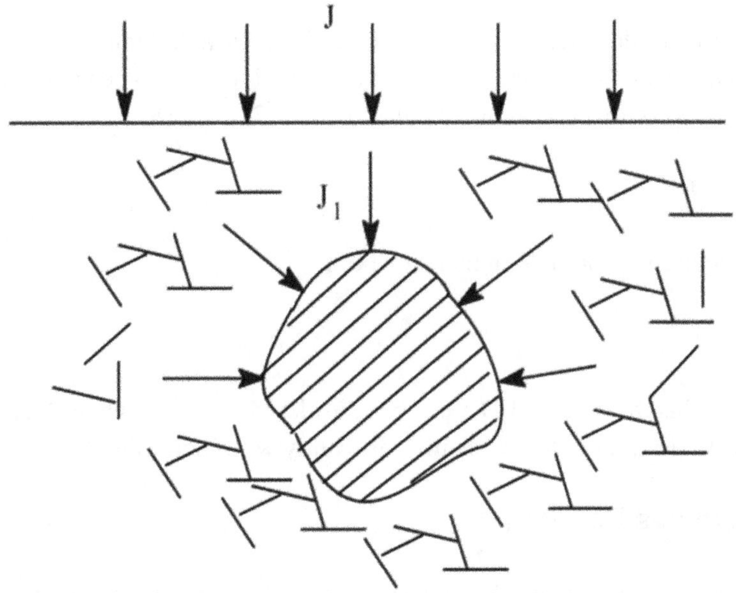

$$2R_gT_1^2E_o^{-1}q_oF_o(T_1)+\beta_1(T_2-T_1)^2 = 4\lambda^2(T_1-T_o)/D_1^2 \tag{15}$$

where $T_2=T_o + z\,J_1/\beta_1$ is an effective temperature; J_1 is the density of radiative flux absorbed by particle; z is relative size of lighted part of particle surface, depending on light scattering in the substance ($1/4 \leq z \leq 1$); and β_1 is the coefficient of heat exchange.

The equation (15) has a solution only in case $D_1 \leq D_1^*$. Functional relationship between critical values of particle diameter D_1, particle temperature T_1, and the radiation flux J_1 may be presented in parametric form (the asterisk indicating the critical values is omitted):

$$D_1 = 2\sqrt{2}\,l_o(T)/(e - 2\beta_1 l_o(T)/\lambda) \tag{16}$$

$$J_1 = ez\,\lambda\,(T - T_o)/\sqrt{2}\,l_o(T),\ T_1 = T + 2R_gT^2/E,\ e=2.72 \tag{17}$$

Here, T is an independent parameter, corresponding in the physical sense to the temperature of the particle (with diameter D1*) included in an inert substance (q_o=0). The width of reaction zone at this temperature is the characteristic scale of the critical particle diameter.

Relationship between the critical diameter D_1^* and the characteristic width of the reaction zone l_o is a function of the particle re-emission intensity. In a case of rather high reemission ($\beta_1 \geq e\lambda/\sqrt{2}\,l_o$), particles of any diameter do not induce the thermal explosion. If re-emission of the particle is negligibly small ($\beta_1 << \lambda/l_o$), the ratio of the critical diameter to the width lo is constant.

$$D_1^*/l_o = 2\sqrt{2}/e \cong 1.04 \tag{18}$$

In this case, the critical diameter D_1^* as a function of the particle temperature is similar to lo(T1) (Figure 3), and the relationship is mutually unique. The explicit ratio between the critical particle diameter and intensity of laser irradiation can be obtained from (17) and (18)

$$zJ_1 = 2\lambda(T_1^o(D_1) - T_o)/D_1^* \tag{19}$$

where T_1^o is the temperature of particle in inert medium:

$$T_1^o = T_o + zJ_1D_1/(2\lambda+\beta_1D_1) \tag{20}$$

It follows from (19) that the product $J_1D_1^*$ can be considered constant (when D_1 changes by one order of magnitude), since $T_1^o(D_1)$, (20), decreases insignificantly with increasing D_1.

On the Small Particles Paradox

The heat transfer from particle to environment during irradiation is frequently neglected in the literature on short laser pulse interaction with substances. In such a case, the particle temperature in result of irradiation is proportional to the specific surface area of the particle and to the pulse duration (Figure 6):

$$T_1^o - T_o = 6zJ\,t_J/D_1\rho_1c_1 \tag{21}$$

where c_1 is the heat capacity of the particle.

A paradoxical result follows from (21) considering laser ignition of metallized solid propellants: the smaller the particle, the more dangerous it is with respect to the explosives sensitivity to irradiation. In framework of this result some authors believed that the particles which are smaller than 10^{-5} cm cause the thermal explosion of lead azide (Karabanov & Bobolev, 1981). Additional factors, for example, the impossibility of combustion propagation beyond the boundary of a small particle, are invoked in the literature to explain the existence of the low threshold. However, the paradox can be resolved in framework of the thermal explosion theory.

Actually, duration of a short laser-pulse in the experiments is usually equal to $(3/5).10^{-8}$ s. It is comparable with or greater than characteristic time of formation of the quasi-stationary temperature distribution around a small particle (with diameter $D_1 < 10^{-5}$ cm) located in substance having the thermal diffusivity $k{\sim}10^{-2}$ cm^2/s.

The Eq. (21) is thus not applicable for such a small particle. It is more accurate to use Eq. (20) for estimation of the particle temperature. According to Eq. (20) the increase in the particle temperature is proportional to the particle diameter. At the same time, the smaller the particle the faster is increase of its temperature (see

Fig. (6). For this reason, the extremely small chemically inert particles cannot induce the thermal explosion of environment.

In the case of rather big particles the problem of metallized propellant ignition by short pulse irradiation requires to verify the condition of quasi-stationary temperature distribution around the particle. For quite big particles the thermal relaxation time ($\sim D_1^2/k$) can be comparable with the pulse duration tj. In such a case the thermal inertia of particle must be taken unto account.

Analysis of the non-steady temperature field around the particle (Assovskiy, Zemskih, Lejpunskij, & Marshakov, 1981), (Assovskiy, 2005) is illustrated by Figures 6 and 7. The particle temperature in adiabatic case can significantly exceed that one in non-adiabatic case, even during short pulse-irradiation, as it is shown on Figure 6.

For each duration of the radiation pulse t_j there is the maximum temperature disturbance caused by particles of an appropriate medium size (Figure 7):

$$D_1 = 2(3t_j\, \lambda/\rho_1 c_1)^{1/2}$$

Only the particles of this size act as photo-sensitizers for MSP irradiated during period t_j. Appropriate temperature disturbance induced by such a particle is proportional to intensity of radiation flux and can be estimated using asymptotic formula [17, 26: T_1-T_o = zJ $(3t_j/\, \lambda\rho_1 c_1)^{1/2}$.

The critical condition of local explosion corresponds to a situation when appropriate characteristic thickness of reaction zone $l_0(T_1)$ is comparable with $D_1(\rho_1 c_1/\, \rho c)^{1/2}$.

In conclusion, we underline important role of the characteristic thickness of reaction zone $l_0(T)$ (Figure 3), in non-isothermal and non-uniform processes in MSP. This thickness defines a scale for different critical sizes in the problem of laser pulse interaction with MSP.

REDUCTION OF ENERGY OF LASER IGNITION OF ENERGETIC MATERIALS

Recently developed semiconductor lasers combine small size and low energy consumption with relatively high efficiency of energy conversion to coherent radiation over a wide range of possible radiation wave lengths and, therefore, show strong promise for using laser radiation in many technical devices using

Figure 6. Temperature of inert particle under irradiation in adiabatic and non-adiabatic cases

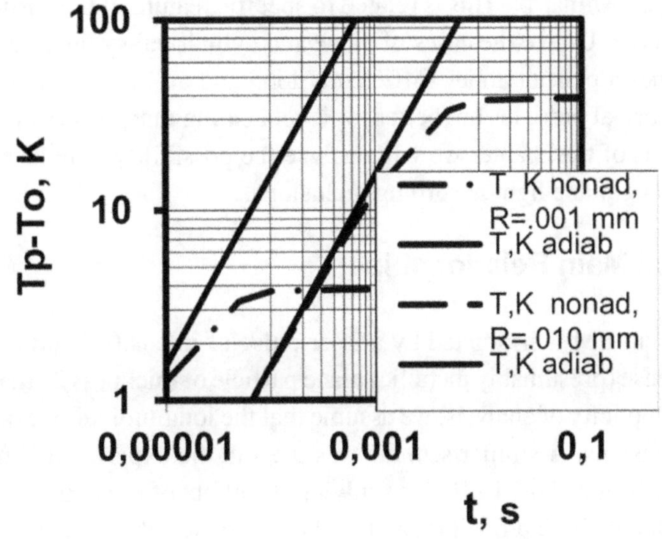

Figure 7. The envelope $T_{max}(t_J)$ of the curves-family $T_1(t)$

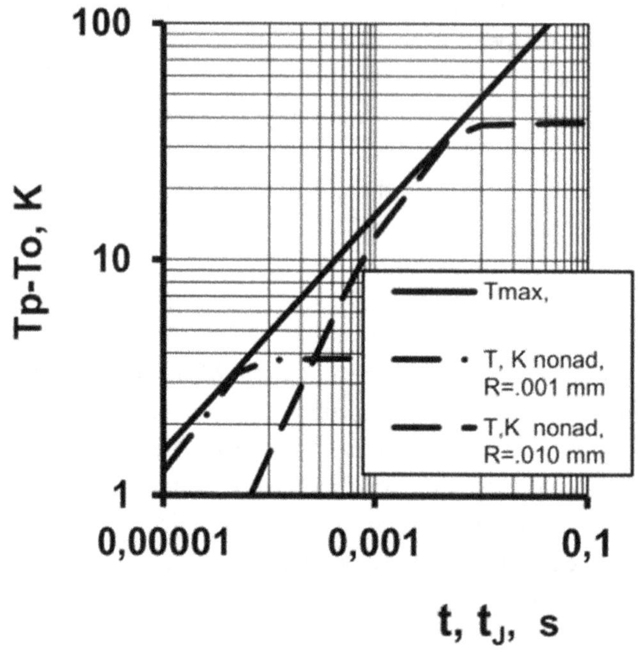

energetic materials (EM). Obviously, the field of application of such lasers for EM initiation is the wider, the lower are the critical energies of EM initiation. Taking into account the available data on the ability of optical inhomogeneity to reduce the critical energies of EM initiation, we may assume that the success of using low energy lasers for this purpose depends on creation of optically dense inhomogeneities in EM.

The optical inhomogeneities as spherical metal or carbon particles are among the best studied, especially for initiating explosives, such as metal azides. At the same time, as was shown in the part 2, the optimal size of a spherical inhomogeneity at the maximal temperature and the temperature itself essentially depend on the laser pulse duration. The temperature of suboptimal size particles differs slightly from the initial temperature of the EM matrix. This is related to specific feature of the particle heat transfer in the spherically symmetric case. Unlike the cases of plane and cylindrical symmetries, the temperature of a spherical particle reaches a plateau under sufficiently long and arbitrarily intense irradiation. For this reason, plane and cylindrical particles are better for local heating than spherical ones.

Therefore, in this part of the chapter we will analyze the possibility to reduce the critical energy of laser ignition of EM using plane light absorbing inclusions.

Process Model and Main Relationships

A sample of a transparent EM is irradiated by a laser pulse of intensity J and duration ti. The sample contains an optically dense (presumably metallic) plane particle of thickness 2h that is normal to the laser beam (Figure 3). For simplicity of analysis, we assume that the longitudinal size of the particle is similar to the laser beam dimensions. A similar scheme was used in the experimental procedure (Assovskiy, Zemskih, Lejpunskij, & Marshakov, 1981) for studying the effect of the temperature distribution in the condensed phase of a double base propellant on the characteristics of unsteady combustion, extinction, and reignition. In this procedure, a plate (nichrome foil $2h = 20$ μm thick with a 4×6 mm working zone) was heated by an electric current pulse.

Because of the plane symmetry, the heat problem is equivalent to the problem of laser pulse heating of a plate of thickness h one side of which contacts EM, and the other is adiabatically insulated. The heat conduction equation for an EM ($x > h$) with the consideration of the heat of a chemical reaction and the light absorption has the form:

$$\rho c \partial T/\partial t = \lambda \partial^2 T/\partial x^2 + \rho q_0 F(T) + \sigma J(t)\exp(-\sigma x) \qquad (22)$$

$$T(x,0) = T(\infty,t) = T_0$$

Here, T is temperature, K; ρ is the density, kg/m^3; c is the specific heat, J/(kg K); λ is the thermal conductivity, W/(m K); t is time, s; x is the spatial variable, m; $\sigma = h^{-1}$ is the coefficient of light absorption by EM, m^{-1}; J is the laser radiation flux density, W/m^2; q0 is the heat of EM decomposition, J/kg; and F is the reaction rate, which depends on temperature and the material conversion. The heat conductivity equation for an inert metallic plate ($0 < x < h$) under the assumption of its negligible transparency ($\sigma 1 \to \infty$) has the form:

$$\rho_1 c_1 \partial T/\partial t = \lambda_1 \partial^2 T/\partial x^2 \qquad (23)$$

The absorbed light flux is taken into account in the boundary condition on the surface of contact of the plate and EM ($x = h$):

$$\lambda_1(\partial T/\partial x)_{h-0} = \lambda(\partial T/\partial x)_{h+0} + J,\ \lambda_1(\partial T/\partial x)_0 = 0$$

$$T(x = h_{-0},t) = T(x = h_{+0},t) = Ts(t) \qquad (24)$$

Figure 8. Arrangement of a metallic particle 1 in the material 2 under irradiation

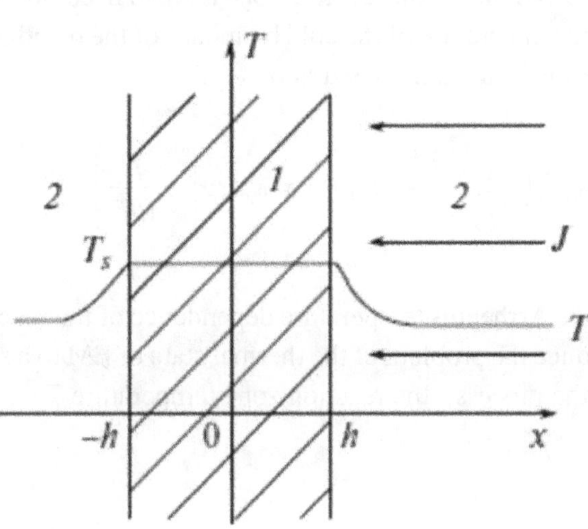

J = const at $0 < t < t_i$, J = 0 at $t > t_i$

In Eqs. (22)–(24) and below, the quantities without subscripts refer to EM; with the subscript 1 - to the plate; with the subscript I - to the moment of ignition or to the pulse characteristics; and with the subscript s, to the surface of contact of the plate and EM.

As is seen from Eqs. (22)–(24), the problem has several spatial and time scales. First of all, these are the pulse duration ti and its corresponding characteristic depths $l = \sqrt{\chi t_i}$ and $l_1 = \sqrt{\chi_1 t_i}$, of heating of the substance and the plate, respectively, where ($\chi = \lambda / \rho c$). Assuming that the plate thickness is sufficiently small $h << \sqrt{\chi_1 t_i}$, we can consider that the plate at each moment of time is uniformly heated to the temperature Ts($_t$).

Integration of the heat conduction equation with respect to the plate thickness under the boundary and initial conditions gives the equation for Ts(t):

$$\rho_1 c_1 h\, \partial T_s / \partial t = q_{s1} = J - q_s \tag{25}$$

where q_{s1} is the heat flux entering the plate through the surface of contact with EM, which is equal to the absorbed light flux J, minus the flux q_s consumed for EM heating. If h is small enough, we can neglect the heat consumed for the plate heating in comparison with the heat consumed for EM heating and can consider q_s = J. However, in the general case, the heat accumulated by the plate can be comparable or exceed the heat absorbed by the heated EM layer.

Let us estimate the flux q_s in a general case. If the activation energy is high enough, the reaction occurs mainly within a thin layer in the neighborhood of the maximal temperature. The maximal temperature of EM is reached near the contact surface and is equal to T_s. The flux q_s forms the heat flux q propagating to cold EM layers, which is determined by the relation (26):

$$q^2 = q_s^2 + q_r^2,\ q_r^2 = 2\lambda \rho q0 \int_T^{T_s} F\, d_T \tag{26}$$

Because the temperature difference in the reaction zone is small in comparison with the difference ΔT =Ts – T0, we can consider the temperature of the cold boundary of the reaction zone to be approximately equal to Ts and can estimate the heat fluxes q and qr a_s:

$$q \cong \lambda \frac{\left(T_s - T_0\right)}{\sqrt{\chi t_i}},\ q_r^2 = 2\lambda \rho q_0 F\left(T_s\right).\frac{RT_s^2}{E},\ F\left(T_s\right) = k_0 exp\left(-\frac{E}{RT_s}\right) \tag{27}$$

In Eq. (27) and further, the Arrhenius temperature dependence of the reaction rate is assumed. Thus, Eqs. (25)–(27) help us to reduce the problem of the thermal state of EM to a single equation for the most important characteristic of the process - the reaction zone temperature T_s:

$$\rho_1 c_1 h\, \partial T_s/\partial t = J - (\lambda\rho c(T_s\text{-}T_0)^2/t_i - 2\lambda\rho q0F(_Ts)\bullet \frac{RT_s^2}{E})_{1/}2 \tag{28}$$

The EM ignition corresponds to an unlimitedly fast increase in the temperature Ts af$_t$er termination of the laser pulse. Therefore, for determining the critical condition for ignition, it is most important to know the temperature Ts an$_d$ the ratio between the heat fluxes at the moment t = *ti.*

$_D$uring the laser pulse, the heat of chemical reaction can be neglected in comparison with the flux J. If, in this case, the plate thickness is small, then Eq. (28) implies that q = J and at the moment t = *ti,* $_t$he reaction zone temperature is:

$$Ts(_ti) = T0 +_J (ti\ \lambda\rho c)1/^2 \tag{29}$$

The correction to Ts$_($*ti*$_)$ for the heat content of the plate can be estimated using the enthalpy conservation law:

$$JSti = {}_m 1c1\Delta T + {}_m c\Delta T \tag{30}$$

where m1 is the $_p$late mass, m is the heated layer mass of EM, and S is the cross sectional area of the laser beam.

With the consideration of the thickness of the heated layer of EM, we obtain from Eq. (30) that

$$Ts(ti) - {}_{T0} = \frac{J\sqrt{t_i}}{\sqrt{»Ác}} \cdot \left[1 + \frac{\rho_1 h c_1}{\sqrt{»Ác t_i}}\right]^{-1} \tag{31}$$

$_A$s follows from Eq. (10), the effect of the plate thickness on the temperature Ts(ti) is d$_e$sc$_r$ibed by the criterion

$$K1 = \frac{\rho_1 h c_1}{\sqrt{»Ác t_i}} \tag{32}$$

At K1 << 1, the temperature Ts is de$_t$ermined by Eq. (29), but at K1 >> 1, the heat content of the plate makes a major contribution to Ts(ti):

$$Ts_(ti) - {}_T0 = \frac{Jt_1}{\rho_1 h c_1} \tag{33}$$

After termination of the irradiation, the chemical reaction becomes the only heat source. The critical condition for ignition of the substance is the thermal state in which the heat release of chemical reaction becomes equal to the heat loss from the reaction zone to the cold layers of the material: qr ≥ q. W*i*$_t$h *the* consideration of dependence (27) of qr on Ts, $_t$he cr$_i$tical condition for ignition takes the form

$$q_r = \left[2\lambda\rho q_0 k_0 exp\left(-\frac{E}{RT_s}\right)\cdot\frac{RT_s^2}{E}\right]^{1/2} \geq q = \lambda\frac{\left(T_s - T_0\right)}{\sqrt{\chi t_i}} \qquad (34)$$

where the dependence of Ts(ti) $_0$n $_t$he laser pulse intensity is given by Eq. (31).

Dependence of Critical Ignition Energy on Pulse Duration and Initial Temperature

Let us apply the derived criterion to evaluating the critical conditions for ignition and their dependence on the laser pulse parameters and the characteristics of EM. For this purpose, it is convenient to consider the system of Eqs. (10), (13) as a parametric dependence of, e.g., the critical laser flux J on the pulse duration ti, in which the independently varying parameter is the temperature Ts(ti$_)$. $_A$s examples, we consider three typical EM: the secondary explosives HMX and RDX and the double-based propellant N, which contain a light absorbing aluminum plate. The table presents the thermophysical and kinetic characteristics of these materials. References to the literature sources of the given values are available in (Assovskiy, 2000), (Assovskiy, 2005).

Figure 9 demonstrates the effect of the plate thickness on the dependence of the critical light flux J (a) and the critical ignition energy density E* = Jti $_($b) on the laser pulse duration for HMX. With increasing plate thickness at constant pulse duration, the critical light flux and the critical ignition energy density significantly increase.

Within the considered t_i range, the dependence of the critical light flux on the laser pulse duration does not change (Figure 9a), whereas the dependence of the critical ignition energy density on the pulse duration changes fundamentally. At h ≤ 100 nm (Figure 9b, curves 3–5), the critical ignition energy density increases with increasing t_i, which is typical for ignition of homogeneous EM by a heat pulse (see, e.g., (Pokhil, Maltsev, & Zaitsev, Methods of Experimental Study of Combustion and Detonation, 1969), (Ohlemiller & Summerfield, A critical Analysis of arc ignition of solid propellants, 1968), (Kondrikov, Ohlemiller, & Summerfield, 1970)). However, at h ≥ 500 nm (curves 1, 2), the critical ignition energy density decreases with increasing t_i. This is caused by high heat loss consumed for plate heating at small time t_i, which corresponds to K_1>>1. With increasing t_i, the EM heating depth increases and $T_s(t_i)$ decreases, which leads to a decrease in the heat loss consumed for plate heating. With increasing t_i, the difference between the ignition energies at different h decreases and vanishes at K_1<< 1, when the heat loss consumed for plate heating become negligible in comparison with the heat content of the heated layer of EM. In this case, the critical ignition energy density increases with increasing t_i as $l(t_i)$, i.e., proportionally to $t_i^{1/2}$.

Table 3. Thermophysical and kinetic characteristics of some EM

Substances	ρ, g/cm^3	c, J/(g K)	λ,W/(cm K)	q_0k_0, J/(g s)	E, kcal/mol
HMX	1.75	1.25	2.9x10^{-3}	10^{23}	52.4
RDX	1.7	1.0	1.67x10^{-3}	6.0 x10^{18}	41.0
Powder N	1.6	1.3	1.25x10^{-3}	2.3 x10^{22}	48.0
Aluminium	2.69	0.9	2.37	--	--

Figure 9. Dependence of (a) the critical light flux J and (b) the critical ignition energy density E on the laser pulse duration t_i for HMX at T_0 = 300 K at various plate thicknesses of h = (1) 1000, (2) 500, (3) 100, (4) 10, and (5) 0 nm*

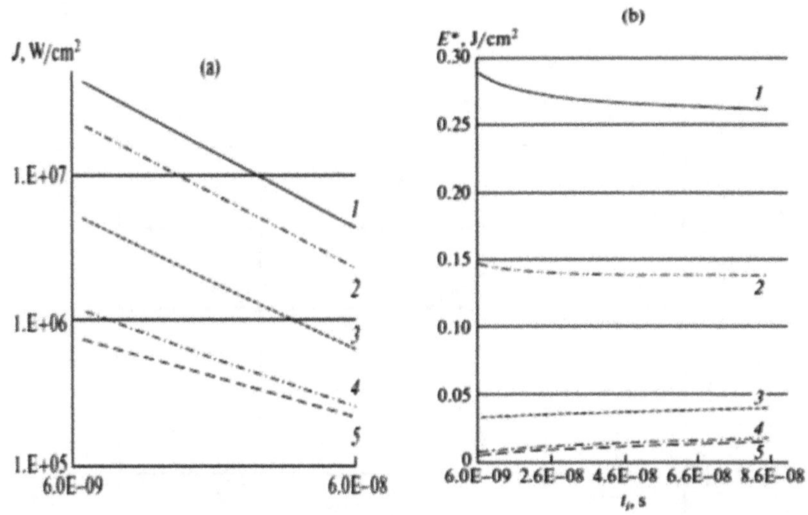

In conclusion, we consider the effect of the initial temperature of EM on the critical conditions for EM ignition. This issue is important in the context of both the effect of the operating conditions of the system, and the evaluation of the possibility of reducing the critical initiation energy by varying the initial temperature of the material. Figure 10 illustrates the effect of the nature of EM on the dependence of the critical ignition energy density E* on the initial temperature T_0 of the material. In the calculations, we assumed the constancy of the light flux density (J = 200 W/cm^2); therefore, the energy density changed only because of a change in the pulse duration. The presented results show, first of all, that, with increasing initial temperature, the critical ignition energy density of EM can significantly decrease, which is consistent with available experimental data.

Comparison of the presented curves also implies that the activation energy is not the main factor for evaluating the dependence E*(T_0). For example, the ignition energy of powder N, the kinetic constants of which are similar to those of HMX, is less dependent on the initial temperature than the HMX initiation energy. At the same time, the initiation energy of RDX, the activation energy of which is much lower than that of HMX, decreases with increasing T_0 approximately so as the HMX activation energy. This suggests that the temperature dependence of the sensitivity of the energy of ignition by a laser pulse is a complex factor, which depends on both kinetic, and thermophysical characteristics of the material.

Figure 10. Effect of the nature of energetic material on the dependence of the critical density of ignition energy E on the initial temperature T_o at J = 200 W/cm² and h = 0 for (1) HMX, (2) RDX, and (3) Double-based propellant N*

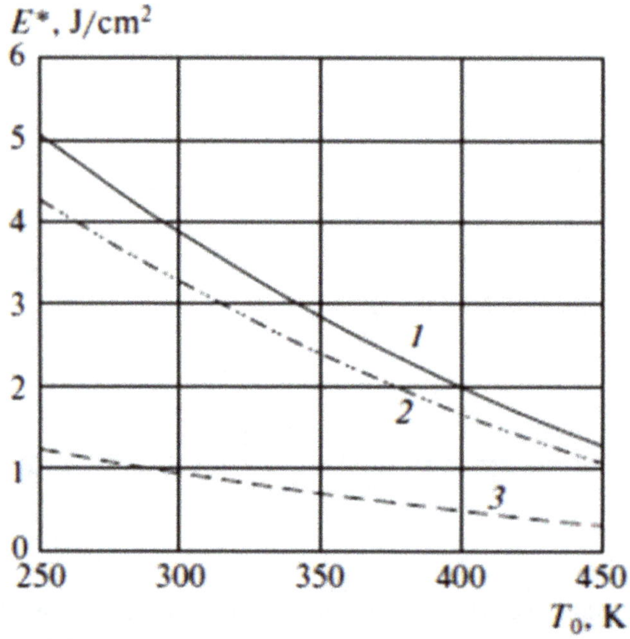

CONCLUSION AND FUTURE WORK

The main results of the conducted analysis are the following:

1. The large variety of metalized composite propellants with different chemical and physical properties of the components, their ratio and dispersion can be divided into two classes of materials, depending on the ratio of the characteristic time of the laser irradiation (t_r) and the thermal relaxation time of the propellant characteristic cell containing one metal particle (t_m).

For a relatively long laser irradiation when $t_r >> t_m$, the composite propellant can be considered as thermally quasi-homogeneous material with appropriate averaged thermal characteristics. For this case of laser ignition of MSP could be applicable corresponding experimental and theoretical data on the laser ignition of homogeneous propellants, such as single-, double-, and triple-based propellants (Ohlemiller & Summerfield, A critical Analysis of arc ignition of solid propellants, 1968) and MSP with metallic nanoparticles (Arkhipov, Korotkikh, Kuznetsov, Razdobreev, & Evseenko, 2011) (DeLuca et al., 2005).

In the case of relatively short laser pulse duration, when $t_r << t_m$, the metalized propellant must be considered as heterogeneous material structure, taking into account geometric shape of metal particles and heat transfer characteristics of the surrounding material layers of the matrix.

2. In the case of spherical metal particles, each duration of the laser pulse corresponds to the optimal size of the particles, which are heated to a maximum temperature. This finding explains why in some experiments an increasing of dispersion of the metal particles in the propellant does not lead to reduction in the laser ignition time delay (see the review in (Arkhipov, Korotkikh, Kuznetsov, Razdobreev, & Evseenko, 2011)).
3. In the case of the flat metal particles and the constant pulse duration, decreasing particle thickness leads to significant decrease of the critical radiation flux and the critical ignition energy density. In the limit when the heat loss consumed for the particle heating become negligible in comparison with the heat content of the heated layer of environment, the critical ignition energy density increases with increasing t_i proportionally to $t_i^{1/2}$.

Presented analysis is based on assumptions of the non-resonant (thermal) interaction of radiation flux with energetic material, assumption of local thermodynamic equilibrium in the system, and assumption on independence of physical and chemical properties of MSP from radiation flux. The tendency in development of laser application for problems of energetic materials application shows advantages of the pulse-lasers having low energy of radiation pulse, but very short pulse duration (much less than 10^{-9} s). In this case, the assumptions mention above need significant development, associated, for example, with the electrodynamic nature of laser fluxes. This development has been started by author with coworkers in (Assovskiy, Melik-Gaikazov, & Kuznetsov, Direct laser initiation of open secondary explosives, 2015), (Melik-Gaykazov, Kuznetsov, & Assovskiy, 2016) and plan to continue in not so far future.

ACKNOWLEDGMENT

This work has been supported by the Russian Foundation of Basic Research and by the Program of Presidium of the Russian Academy of Sciences.

REFERENCES

Aluker, E. D., Krechetov, A. G., Loboiko, B. G., Nurmukhametov, D. R., Filin, V. P., & Kazakova, E. A. (2008). Effect of temperature on the laser initiation of pentaerythritol tetranitrate (PETN). *Russian Journal of Chemical Physics*, *27*(5), 53–55.

Arkhipov, V. A., Korotkikh, A. G., Kuznetsov, V. T., Razdobreev, A. A., & Evseenko, I. A. (2011). Influence of the Dispersity of Aluminum Powder on the Ignition Characteristics of Composite Formulations by Laser Radiation. *Russian Journal of Physical Chemistry*, *5*(4), 616–624. doi:10.1134/S1990793111040026

Assovskiy, I. G. (1992). Effect of Absorbing Microinclusions on the Interaction of Light with a Reacting Substance. *Dokladi Physical Chemistry (Proc. Rus. Acad. Scien.)*, *324*(1-3), 205-210.

Assovskiy, I. G. (1994). Interaction of Laser Radiation with Reagents. Critical Diameter of Laser Beam. *Doklady Physical. Chemistry, Proc. RAS*, *337*(6), 752-756.

Assovskiy, I. G. (2000). Heat Transfer in Laser Pulse Interaction with Reactive Substances. In A. E. Bergles, & I. Golobic (Eds.), Thermal Sciences 2000 (pp. 107-112). Ljubljana, Slovenia. doi:10.1615/ICHMT.2000.TherSieProcVol2TherSieProcVol1.130

Assovskiy, I. G. (2005). *Physics of Combustion and Interior Ballistics*. Moscow: Nauka.

Assovskiy, I. G., & Leipunskiy, O. I. (1980). Theory of Propellant Ignition by Pulse Radiation. *Combustion, Explosion, and Shock Waves*, *16*(1), 310.

Assovskiy, I. G., Melik-Gaikazov, G. V., & Kuznetsov, G. P. (2015). Direct laser initiation of open secondary explosives. *Journal of Physics: Conference Series*. doi:10.1088/1742-6596/653/1/012014

Assovskiy, I. G., Zemskih, V. I., Lejpunskij, O. I., & Marshakov, V. N. (1981). *Experimental Method for Determination of Ignition Characteristics of Solid Propellants* (Report of ICP AS USSR No. 75063377).

Beyer, R. B., & Fishman, N. (1960). Solid Propellant Ignition Studies with High Flux Radiant Energy as a Thermal Source. In *Progress in Astronautics and Rocketry: Solid Propellant Rocket Research* (Vol. *1*, pp. 673–692).

Brish, A. A., Galeev, I. A., & Zaitsev, B. N. (1969). The initiation mechanism in condensed explosives by optical quantum generator radiation. *Combustion, Explosion, and Shock Waves*, *5*(4), 475. doi:10.1007/BF00742068

DeLuca, L., Ohlemiller, T. J., Caveny, L. H., & Summerfield, M. (1976). Radiative Ignition of Double Base Propellants: II. Pre-ignition Events and Source Effects. *AIAA Journal*, *14*(8), 1111–1117. doi:10.2514/3.7193

DeLuca, L., Ohlemiller, T. J., Caveny, L. H., & Summerfield, M. (1976). Radiative Ignition of Double-Base Propellants. *AIAA Journal*, *14*(7), 940–946. doi:10.2514/3.7167

DeLuca, L. T., Galfetti, L., Severini, F., Meda, L., Marra, G., Vorozhtsov, A. B., & Babuk, V. A. et al. (2005). Burning of Nano-Aluminized Composite Rocket Propellants. *Combustion, Explosion, and Shock Waves*, *41*(6), 680–692. doi:10.1007/s10573-005-0080-5

Frank-Kamenetskiy, D. A. (1969). *Diffusion and Heat Transfer in Chemical Kinetics*. New York: Plenum.

Goldshleger, U. I., Barzykin, V. V., & Mezhanov, A. G. (1971). On mechanism and regularities of solid propellant ignition by disperse flow. *Combustion, Explosion, and Shock Waves*, (3), 319–332.

Karabanov, Y. F., & Bobolev, V. K. (1981). Experimental Study of Laser Pulse Ignition of Explosives. In *Proceedings of the U.S.S.R. Academy of Sciences. Atmospheric and Oceanic Physics* (Vol. *256*, pp. 1152–1154).

Kondrikov, B. N., Ohlemiller, T. J., & Summerfield, M. (1970). Ignition and Gasification of a Double-Base Propellant Induced by CO2 Laser Radiation. In *Proceedings of the Thirteenth International Symposium on Combustion*. Salt Lake City, Utah.

Koval'skiy, A. A., Khlevnoy, S. S., & Mikheev, V. F. (1967). On ignition of doublebased propellant. *Combustion, Explosion, and Shock Waves*, *3*(4), 527–541.

Librovich, V. B. (1963). Ignition of Powders and Explosives. *Zhurnal Prikladnoi Mekhaniki i Technicheskoi Fiziki*, *6*, 74.

Melik-Gaykazov, G. V., Kuznetsov, G. P., & Assovskiy, I. G. (2016). On mechanism of laser initiation of secondary explosives. In *Space Technologies, Materials, and Devices* (pp. 255-259).

Merzhanov, A. G., & Averson, A. E. (1971). The Present State of Thermal Ignition Theory: An Invited Review. *Combustion and Flame*, *16*(1), 89–124. doi:10.1016/S0010-2180(71)80015-9

Ohlemiller, T. J., & Summerfield, M. A. (1968). A critical Analysis of arc ignition of solid propellants. *AIAA Journal*, *6*(5), 878–886. doi:10.2514/3.4613

Pokhil, P. F. (1982). On combustion mechanism of double based propellants. *J. Physics of Combustion and Explosion*, 1982, 117-140.

Pokhil, P. F., Maltsev, V. M., & Zaitsev, V. M. (1969). *Methods of Experimental Study of Combustion and Detonation*. Moscow: Nauka.

Tarzhanov, V. I. (1998). *Fast Initiation of Explosives*. Special Regimes of Detonation.

Vilyunov, V. N., & Zarko, V. E. (1989). *Ignition of Solids*. New York: Elsevier Science Publishers.

Chapter 5
Dynamics of the Internal Flow in Swirl Atomizers by CFD Simulations

Roman Ivanovitch Savonov
National Institute for Space Research, Brazil

ABSTRACT

This work presents the simulation of the internal flow in a swirl atomizer. The geometry of the atomizer is calculated by analytical equations used in engineering. The numerical simulation of the two-phase flow is performed by using two equations k-ε turbulence model. The fluids are presented as two-fluid homogeneous model. The interface between two phases is calculated by free surface model. The distribution fields of the axial and tangential velocities, pressures and air core are obtained. The aim of this work is to compare the results obtained by numerical simulation with ones obtained analytically. Also, to study the internal fluids flow inside the atomizer.

INTRODUCTION

The bi-propellant core injectors inject the components of the propellant into the combustion chamber in such a way as to optimize the combustion process by generating greater thrust of the engine depending on the bi-propellant torque employed. The proper mixing ratio between fuel and oxidant is done by means of thermodynamic calculations. During the operation of the engine the combustion process takes place in distinct regions of the chamber and the central part of the chamber is the one with the highest temperature. The peripheral injectors are regularly distributed close to the wall of the combustion chamber and have the function of thermally insulating the wall by the column of liquid represented by the fuel.

In these (peripheral) injectors only fuel is injected in such a way that in the region near the walls the richest fuel mixture forms. The combustion of this mixture occurs with reduced temperatures when compared to the burning in the central region of the chamber. Therefore, the temperature developed in the central region of the chamber is approximately 3000 - 3200 K and in the periphery around 1000 - 1200 K. The constructive material of these walls of the chamber must withstand such temperatures and are usually

DOI: 10.4018/978-1-5225-2903-3.ch005

Figure 1. Schematic of injection of rocket propellant liquid engine developing 400 N of thrust

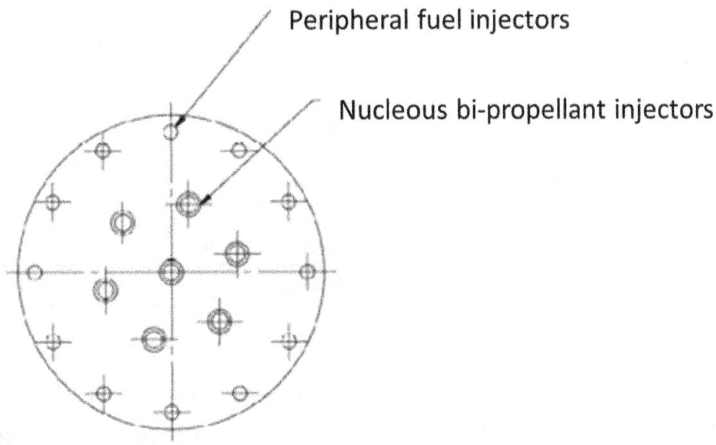

inconel and niobium. In some cases of bi-propellant pairs, when the thermal properties of the oxidant are higher than that of the fuel, the oxidant can be used to protect the walls of the combustion chamber. But, generally for this purpose is used the fuel because the oxidant is the active component, for this reason, to avoid the oxidation of the walls, the application of the fuel is justified. In most liquid propellant-based propellant designs, the inner shell of the combustion chamber walls is formed by jet-type injectors, also known as "impinging jet". This is due to the fact that this type of injector has a smaller spray angle than the core, for example, and therefore the injection of the propellant can be directed directly into the walls of the combustion chamber, which provides a better contact of the fuel with the wall.

In this case the heat exchange between the fuel and the wall of the chamber is favored, making the cooling of the walls more efficient.

INJECTOR TYPES

The main functions of the injectors are described below:

1. Ensure continuous injection of the propellant into the combustion chamber without interruption.
2. Inject exactly the determined mass quantity of propellant to ensure the mixing ratio between the fuel and oxidant.
3. Ensure the immediate defragmentation of the liquid sheet.
4. Properly atomize the bi-propellant pair ensuring a droplet size dispersion leading to an efficient combustion process.
5. Ensure the homogenous distribution of the propellant in the combustion chamber.

Injectors can be classified in several ways, but according to (Bayvel & Orzechowski, 1993), the best way to classify them is by the type of energy that is used in the atomization of the liquid. Table 1 shows this classification of injectors.

Table 1. Classification of injectors by the type of energy used in the atomization of the liquid (Bayvel & Orzechowski, 1993).

Energy Type	Injector Type
Energy from Liquid	Jet Injectors
	Centrifugal Infectors
	Mixed Injectors (centrifugal – jet)
Energy from Gas	Pneumatic Injectors
Mechanical Energy	Rotational Injectors
Vibrational, Electrical and other Energies	Acoustic, Ultra-sonic, Electrostatic and other Injectors

Jet injectors are less complex in their manufacture. The liquid enters the injector at high pressure, between 0.5 and 10 MPa, and leaves the device through the outlet orifice forming a jet of liquid as shown in Figure 2. Due to the high speed of the liquid, the jet, running a distance after the exit, defragments forming a spray.

This type of injector has wide application in injection systems in internal combustion engines with Diesel cycle. Jet injectors are also used in rocket engines as peripheral injectors due to the characteristics of the spray they form. The major disadvantage of these injectors for application in rocket engines is that they develop a small spray angle (maximum 15-20 °) and produce large diameter droplets. According to (Alemasov, 1980) the average diameter of the droplets is dm = 200 - 500 μm, while for centrifugal injectors it is dm = 25 - 250 μm. The spray angle influences the design of the engine combustion chamber considered for its application. The lower the spray angle, the greater the axial velocity of the liquid, so

Figure 2. Spray formation by jet type injector

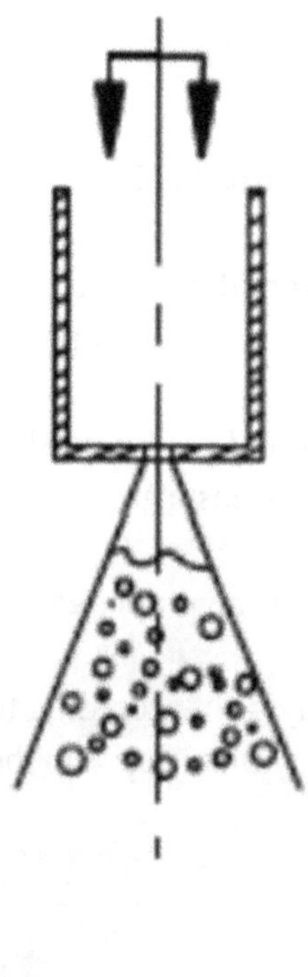

to satisfy the complete burning of the propellant it is necessary to have a larger combustion chamber, which consequently increases the mass of the engine. The size of the droplets influences the stability of the combustion process. The smaller the droplet size the more stable the combustion process.

In Figure 3 the atomization model of a liquid through the centrifugal injector is presented. The injector has this designation due to the rotation of the liquid within the device's torsion chamber. This type of injector presents an adequate disintegration of the liquid sheet from low operational pressures. Due to this, allied to the simplicity of manufacture, the centrifugal injectors are very employed. The liquid enters the torsion chamber through the tangential holes. Thus, the fluid gains axial velocity through the torsion chamber to the outlet orifice. At the exit of the injector a liquid sheet cone is formed. The disintegration of this sheet, compared to that found in jet injectors, starts at the smallest outlet distance, which increases the residence time of the propellant in the combustion chamber and decreases the size thereof. Centrifugal injectors are interesting because, for the same mass flow, the angle of the spray cone can be adjusted by changing the dimensions of the inlet, outlet and torsion chamber.

This property is very useful in the design of rocket engines to liquid propellant. This is because the designer has freedom in the design of these injectors and can position them in the easiest and most suitable way in the injector head. Injector flow rates are determined by thermodynamic calculations and should not be modified because this would cause a change in the mixing ratio of the bi-propellant pair and consequently influence the thrust developed by the engine. Thus, the designer has, for the same flow, the possibility of varying the dimensions of the injector and choosing which ones are the most appropriate, this facilitates the manufacturing, reducing the design cost.

Figure 3. Formation of spray by centrifugal type injector (swirl)

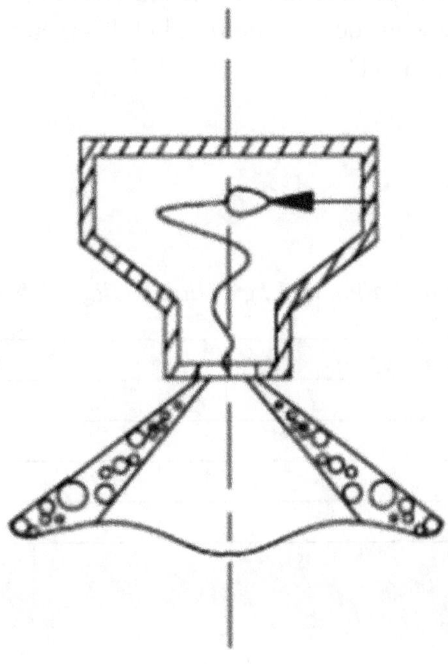

Pneumatic injectors use the kinetic energy of the gases to disintegrate the liquid sheet. This type of injector is complex when compared to the two types previously mentioned, but is often used in several areas. The most important use of these injectors is the atomization of oily fuels with high viscosities.

Rotational injectors are usually driven by electric motors. The axis of rotation can be horizontal or vertical. The rotational energy is transmitted to the liquid, which causes it to leave the injector with high kinetic energy coming to defragment itself. These injectors are used for the atomization of aqueous solutions. Injectors of this type ensure good atomization at low pressures.

In the aerospace area jet and centrifugal injectors receive special attention because they are considerably simpler and have greater efficiency compared to others. Table 1.5 shows the energy demand required to atomize 1 kg of liquid depending on the type of injector. Centrifugal injectors have the same efficiency as jet injectors and have the advantages discussed above.

Swirl atomizers are widely used in rocket propulsion systems. Its present a good quality of the atomization with relatively small length of the liquid spray in comparison with jet atomizers, and more simple design then others. Applications of swirl atomizers in rocket motors permit reduce dimensions of the combustor chamber and, consequently, weight of the engines. Due of these properties, swirl atomizers were intensely investigated from 1940 up to present time.

The first approximate analytical solution of the ideal fluid flow inside the injector was present by (Abramovich, 1944), six decades ago. This solution admits calculus of principal characteristics of the injector and it`s geometry. Later, several researchers investigated flow processes inside the swirl atomizer with the aim to perform the Abromovitch solution. (Kliachko, 1962) proposed calculate the coefficient of discard by instead the geometrical characteristic A by equivalent characteristic, it is calculated with consideration friction losses inside the injector. All calculus in this method is based on calculus by Abramovitch theory. Kliachko considerate that all hydraulic losses are due the friction, but its take only small part of total losses and depend on several factors. (Doble & Halton, 1947) applied the theory of the cyclone to calculate the flow inside the atomizer, but this theory is based on the supposition that rotation of liquid is calculated by Eq. (1)

$$V_{\phi}^{2}r = const \tag{1}$$

Table 2. Energy demand for disintegration of 1 kg of liquid (Bayvel & Orzechowski, 1993).

Injector Type	Output (W kg^{-1})
Jet Injectors	2-4
Centrifugal Injectors	2-4
Pneumatic Injectors	50-60
Rotary Injectors	15

This equation can be used only for half of exit radius. There are many works about the investigation of the swirl atomizer, but all of it used a same process to obtain the parameters of injector. First is calculated the size of the air core, and then is obtained the consumption through atomizer. None of these works and proposed methods has any advantage over Abramovitch's theory. Therefore, analytical calculus of injector's parameters in this paper is based on the Abramovitch theory.

The computational equipment developed very much during past decade. It stimulated to apply more sophisticated numerical methods and solution to investigation the flow dynamics of the atomizer.

(Alajbegovic & Meister, 2002) presented a three-phase flow simulation of the flow in the high-pressure swirl injector, with application in direct gasoline injection engines. The model was based on the multiphase extension of the two-fluid model. The authors obtained most important characteristics of the flow. A thin conical sheet with the air core was predicted. Cavitation, which occurs in the pressure depression located in the air core, was predicted. The analysis presented very promising approach for the simulation of flows in DGI injectors. (Shanwu, Yang, Hsiao, Hsieh, & Mongia, 2007) presented an investigation on confined swirling flows in an operational gas-turbine injector. The study was performed by means of large-eddy simulation. The co- and counter-rotation configurations were considered, and the effects of swirl direction on flow characteristics were examined. Good agreement was obtained between the measured and calculated mean velocity fields and turbulence properties. This work demonstrated the feasibility of using LES to study complex flow fields. (Hansen & Madsen, 2001) presented detailed investigation of the two-phase internal flow in a large-scale pressure-swirl atomizer. The study was performed by practice measuring and CFD modeling using three approaches: volume of fluid model (VOF) with assumption a laminar flow, VOF using LES turbulence modeling, and two-fluid Euler/Euler method using a laminar flow assumption. All simulations presented a similar result and produce air-core that matches those observed in the experiments. The tangential and axial velocity profiles in the conical swirl chamber and static wall pressure obtained by simulation demonstrated a good agreement with measure data. For the two flow rates, the VOF and Two-Fluid approaches assuming laminar flow appear to give the best agreement.

ANALISYS OF THE SWIRL ATOMIZER

The aim of this work is to investigate the internal flow in the injector. The internal flow characteristics, such as dimensions of the air core, distribution of the pressure inside of the atomizer, field of tangential and axial velocities, take a key role in the formation of the liquid cone and atomization process. Therefore, it is very important to know full information about the behavior of a fluid inside the injector to predict more completely processes, which occur outside the atomizer, such as formation of the liquid cone, pulverization angel, thickness of the liquid sheet, defragmentation, atomization and others. The first pass of this work is to calculate the injector geometry.

Abramovitch Theory: Calculus of the Dimensions of Atomizer

Abramovitch, in his theory, considerate, that friction losses in the injector are fully eliminated, and the angular momentum in the swirl chamber is constant Eq. (2).

$$V_e R = V_r r \tag{2}$$

where

V_e is the velocity of the liquid in the enter of the atomizer;
V_t is the tangential velocity of the fluid;
R is the radius of the swirl chamber;
r is the rotation radius of the liquid particle.

The liquid is considerate to be inviscid fluid. The pressure in any point of the injector is prescribed by Bernoulli equation. The radial velocity on the interface between the liquid and gas phase is considerate to be zero. The determination of the size of air core is based on the maximum consumption principle. Full description of the Abramovitch theory is discussed in (Abramovich, 1944) or (Vasiliev, Kudriavtcev, & Kuznetcsov, 1993). Here is presented the equations to calculate principal characteristics of the swirl atomizer, such as φ – coefficient of the nozzle filing Eq. (3), A, B – geometrical characteristics Eqs. (4, 5, μ - discharge coefficient.

$$\phi = 1 - \frac{r_v}{r_n} \tag{3}$$

where r_v and r_n are radiuses of air core and nozzle respectively.

Figure 4. Swirl atomizer

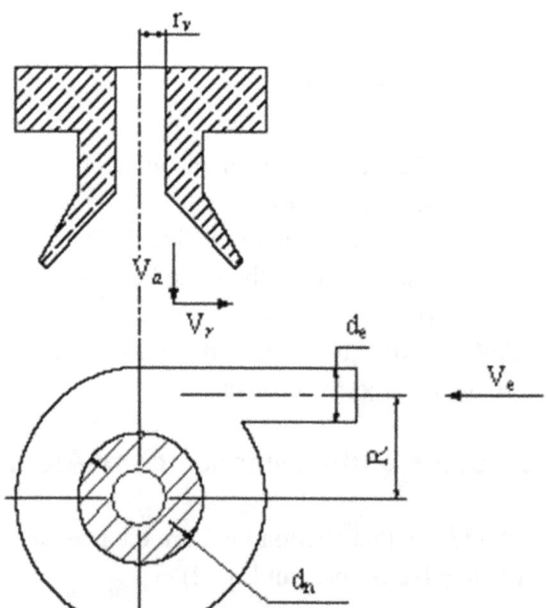

$$A = \frac{Rr_n}{nr_e^2} \tag{4}$$

where n is number of the tangential orifice;r_e is the radius of tangential orifice.

$$B = \frac{R}{r_e} \tag{5}$$

The relation between φ, μ, *A* and *tan(α)* is prescribed by (Abramovich, 1944) and (Vasiliev, Kudriavtcev, & Kuznetcsov, 1993) as:

$$A = \frac{1-\phi}{\sqrt{\frac{\phi^3}{2}}} \tag{6}$$

The Eq. (4) is used to appraise the error of the theoretical calculus with practice data, and Eq. (6) in initial calculus of the injector's geometry.

$$\mu = \sqrt{\frac{\phi^3}{2-\phi}} \tag{7}$$

$$\tan\alpha = \frac{(1-\phi)\sqrt{8}}{(1+\sqrt{1-\phi})\sqrt{\phi}} \tag{8}$$

where α is a pulverization angle.

The Abramovitch theory is applied to an inviscid fluid. It can be applied to viscous fluid too. According to (Vasiliev, Kudriavtcev, & Kuznetcsov, 1993),if

$$\frac{B^2}{n} - A \leq 6.2 \tag{9}$$

then friction losses can be considerate equal zero and Abramovitch theory give good precision of calculus. The consumption of liquid through atomizer is

$$\dot{m} = \mu F\sqrt{2\rho\Delta P} \tag{10}$$

where

F is area of the injector's nozzle;
ρ is density of the liquid;
ΔP pressure difference;

Using equations above are calculated the dimensions and geometrical characteristics of the injector for nominal regime of water, for: $\Delta P = 10$atm, consumption 50 g/s, its are presented in Table 3.

To confirm and proof the precision of the calculus based on Abramovitch theory, the pressure was varied from 10 to 1 atm, and were calculated analytically the consumption, values of the axial and tangential velocities on the nozzle exit at itch pressure level for the same atomizer, and then were compared with result obtained by numerical simulation using ANSYS CFX. To calculate these values were used equations presented by (Vasiliev, Kudriavtcev, & Kuznetcsov, 1993) or (Khavkin, 2004).

$$V_{eq} = \mu \sqrt{\frac{2\Delta P}{\rho}} \tag{11}$$

where V_{eq} is the equivalent velocity

$$V_a = \frac{V_{eq}}{\phi} \tag{12}$$

where V_a is axial velocity in the exit of the nozzle.

$$V_t = V_{eq} \frac{Rr_n^2}{rr_e^2} \tag{13}$$

$$\dot{m} = V_{eq} r_n^2 \pi \rho \tag{14}$$

The results of the calculus are presented in Table 4.

Table 3. Dimensions and geometrical characteristics of atomizer.

Input Data				Parameters of the Atomizer							
Pressure, atm	Consumption, g/s	Density, kg/m³	Number of orifice	φ	μ	A	B	α,°	D, mm	d_e, mm	d_n, mm
10	50	998	2	0.345	0.158	4.569	3.658	105	4.39	1.2	0.3

Table 4. Results of the analytical calculus.

ΔP, atm	m, kg/s	V_a, m/s	V_t, m/s	V_{eq}, m/s
10	0.049992467	20.57623	34,72147	7.100855
9	0.047427019	19.52032	32,93968	6.736463
8	0.044714622	18.40394	31,05583	6.351198
7	0.041826699	17.21531	29,05007	5.941002
6	0.038723999	15.93828	26,89514	5.500299
5	0.035350013	14.54959	24,55179	5.021063
4	0.031618013	13.01355	21,95979	4.490975
3	0.027382002	11.27006	19,01773	3.889299
2	0.022357311	9.201968	15,52792	3.175599
1	0.015809006	6.506774	10,97989	2.245488

NUMERICAL SIMULATION

In recent decades, the computer industry and development of microprocessors have made great progress. This made it possible to introduce numerical analysis in industry and research in full. If in the 1970s, and up to 90, the numerical analysis took weeks now can be accomplished in just a few hours. Nowadays, numerical modeling is one of the main steps of any project. The main techniques of numerical analysis are: the finite element method in the area of structures; And finite volumes in the fluid area. Several commercial software developed to achieve and simplify the application of numerical analysis are available. Among them is "ANSYS" which has modules to perform structural analysis, electrical, fluid dynamics, thermal, iteration of structures with fluid besides others such as: ABAQUS, PATRAN, NASTRAN, FLUENT, FEMAP. In this study of the fluid flow inside the centrifugal injector, the software "ANSYS CFX" was used.

Any numerical modeling can be divided into two steps: selection and generation of mesh (discretization) and formation of the physical problem (application of boundary conditions, choice of fluid type, type of turbulence, among others).

Mesh Generation

The generation of the mesh is one of the main steps of the numerical analysis. There are many physical domain spatial discretization techniques. Publications discussing this subject can be easily obtained. There is an enormous amount of "computational fluid dynamics" books that address the issue of mesh generation detailing and deepening the subject. ((Blazek, 2001), (Cebeci, Rshao, Kafyeke, & Laurendeau, 2005), (Chung, 2002). The spatial discretization of the Navier-Stokes equations involves a large number of schemes. These schemes can be divided into three main categories: finite differences, finite volumes and finite elements. All of these methods depend on one type of mesh.

There are two types of grid generation: structured and unstructured.

- **Structured:** Each knit point is uniquely identified by the indexes i, j, k and corresponds to the Cartesian coordinates xi, j, k, yi, j, k, zi, j, k. The mesh cells are quadrilateral for the bi-dimensional and hexahedral cases for 3-D.
- **Unstructured:** Both mesh cells and nodes have no particular ordering, so cells (elements) or nodes cannot be directly identified by their indexes. The mesh elements are triangles in the two-dimensional case and tetrahedral in 3-D. The unstructured mesh consists of a mixture of quadrilateral and triangular elements, for the two-dimensional case; And hexahedral, tetrahedral, prisms and pyramids for the 3-D case to solve the boundary layer problem.

The main advantage of the use of the structured mesh follows from its property, that the indexes i, j, k have a linear address space called computational space.

In this way, they correspond directly to the way the flow variables are stored in the computer memory. This property allows access to neighboring points with ease and agility. In this way, the evaluation of gradients, flows and also the application of boundary conditions is very simplified.

The main disadvantage is the generation of the structured mesh for complex geometries. The solution to this problem is the division of physical space into several topologically simple parts, called blocks. These blocks can be divided into the structured mesh with less complexity. The multi-block methodology is interesting to be employed by the implementation of the problem to be solved in parallel using several microprocessors through domain decomposition. Structured mesh generation in complex geometries requires a lot of processing time.

Unstructured mesh offers great flexibility when applied to complex geometries. Higher advantage of unstructured mesh is based on the fact that the mesh composed of triangles (2-D) or tetrahedra (3-D) can be generated automatically regardless of the domain complexity. In practice, it is necessary to apply only a few parameters to obtain a good quality mesh. In order to solve the boundary layer problem, it is recommended to introduce the rectangular elements (2-D) or prismatic elements and hexahedrons (3-D) in the approach of the walls. Another benefit of this mixture of elements is the reduction of number of elements, faces, edges and sometimes knots. The non-structured mesh generation for cases of complex geometry is not a trivial task, but, in any case, it is much simpler and faster compared to the generation of the structured mesh using the multi-block method. A further advantage of unstructured mesh is that, depending on the need for solution, the refinement and thickening of the mesh can be performed in a simple manner. One of the major disadvantages of using unstructured mesh is the need to use sophisticated data structures within the solver. These structures work with indirect addressing, which leads to a reduction in computational efficiency. Also, this type of mesh requires higher computational memory compared to structured meshes.

The geometry of the atomizer is not very complex, and was possible to create the structured grid. In beginning, a simplification of the geometry was improved. The geometry of the tubular channels was substituted by orifice formed by intersection of these channels with cylindrical part of vortex chamber, as shown at Figure 5 (red region).

The multiblocks technique is used to generate grid. The physical domain was divided by 69 blocks as shown on the Figure 6.

O'Grid function was used to obtain a good grid quality on cylindrical regions. It is available in ICEM CFD software. Initially, homogeneous regular mesh without refinement regions was generated. The numerical simulation with this mesh is not presented good solution. The air core was formed only on the nozzle region, and velocities' field near the walls had poor quality. But, this first analysis helped to

Figure 5. Atomizer`s geometry

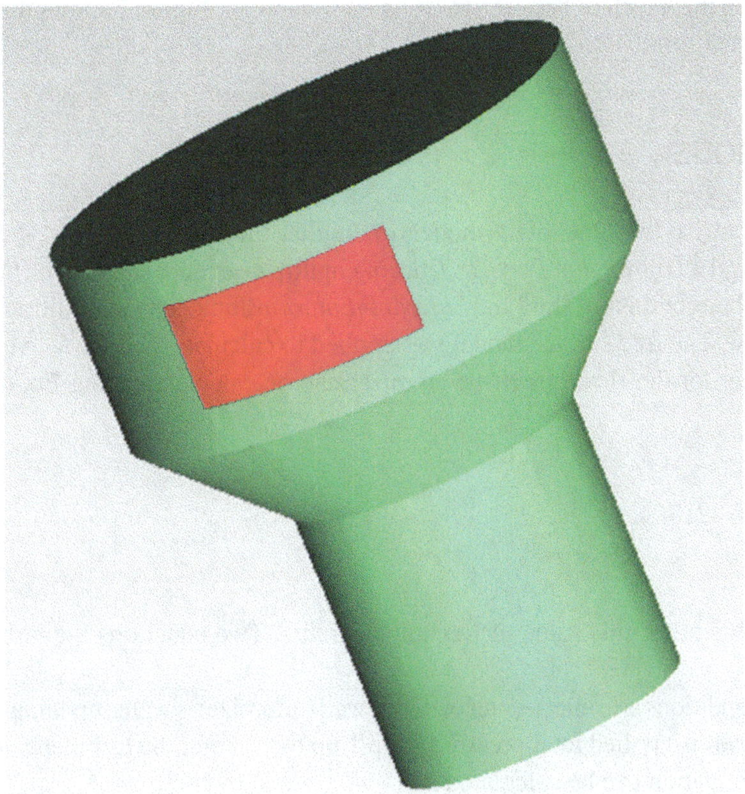

Figure 6. Block structure for grid generation

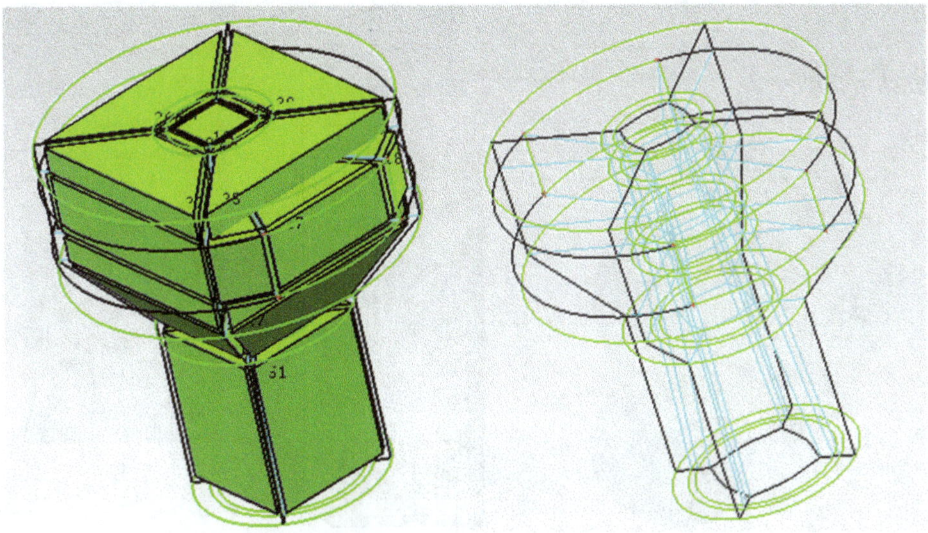

determine the regions of higher gradient values. The block structure is adopted. The mesh was refined near the walls and on the interface between two phases water – air. Figure 7 shows the final mesh, which was used in numerical simulation.

NUMERICAL MODEL

ANSYS CFX software has two types of the models to simulate the multiphase flow, an *Eulerian–Eulerian* multiphase model and a *Lagrangian Particle Tracking* multiphase model. The two fluids - water and air, were considerate to be a continuous fluid and *Eulerian–Eulerian* homogeneous multiphase model was used to simulate this flow. The free surface model was applied to calculate the interface between two fluids.

Reynolds number for the flow inside the atomizer can be calculated using Eq. (15), Vasiliev et al. (1993).

$$\mathrm{Re} = \frac{4\dot{m}}{(\pi \rho \nu \sqrt{n} d)} = 4215 \qquad (15)$$

The flow is turbulent for this value of Reynolds number. Two equations k-ε turbulence model was used in simulation.

The boundary conditions are: inlet – enter orifices, wall – atomizer`s walls, opening – the exit of nozzle.

At inlet boundaries is applied total pressure from 1 up to 10 atm., the turbulence kinetic energy and turbulence eddy dissipation can be calculated as:

$$k_{inl} = \frac{3}{2}\left(T_i \sqrt{u^2 + v^2 + w^2}\right) \qquad (16)$$

Figure 7. Mesh refinement

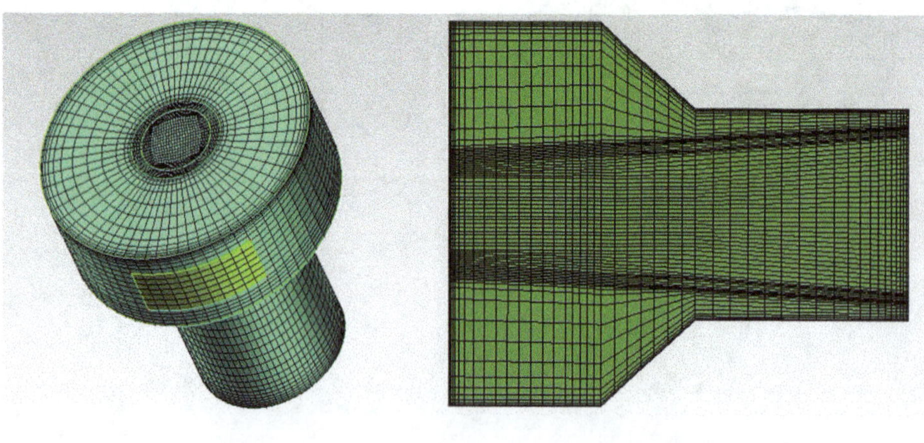

$$\varepsilon = \frac{k^{\frac{3}{2}}}{0.3l} \tag{17}$$

where

u, v, w – velocity component;
l – hydraulic diameter;
T_i – turbulence intensity.

At the opening boundary condition was applied a static pressure, and turbulence option is zero gradient.

Initial condition is static pressure equal ambient pressure, water volume fraction is zero and air volume fraction is 1.

RESULTS

Turbulence Models Evaluation

Figure 8 (a) (κ-ε) and (b) (RNG κ-ε) shows the volumetric fraction of water and air inside the injector and the formation of air core.

As shown in Figure 8, the air core formation is identical for the two cases. This means that the coefficient of rotation in this injector is less than 0.1, and numerical modeling can be done using the turbulence model κ-ε which is simpler and faster in the numerical solution. The distribution of the turbulent viscosity inside the injector for the two models of turbulence presents different results. Figure 9 shows the turbulent viscosity distribution diagram of air inside the injector for two cases.

Figure 8. Volumetric fraction of water and air inside the injector according to the fashion model

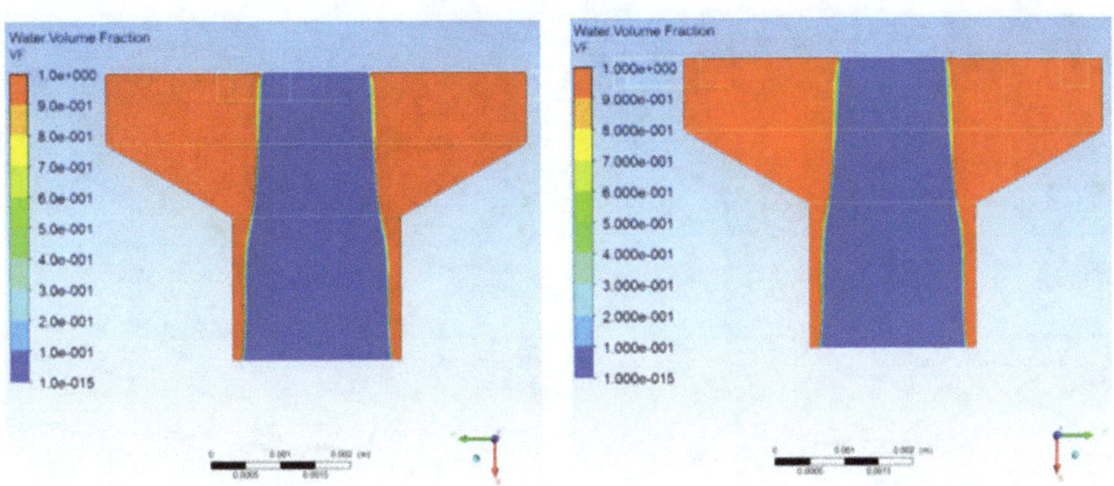

(a) κ-ε Turbulence model

(b) RNG κ-ε Turbulence model

From Figure 9 it can be observed that the maximum value of air viscosity in the case of model application is almost twice as high as in the RNG κ-ε model. The higher values of the air viscosity are concentrated in the central part of the injector where the air core is formed (Figure 9 (a)). The formation of the higher values of turbulent viscosity starts from the exit to the top of the injector, while in the case of application of the RNG κ-ε model the higher values are formed only at the top of the injector. The turbulent viscosity of the fluid has a great influence on the flow. High viscosity values decrease fluid flow velocity and increase energy dissipation. The site of formation of these values (model κ-ε) for the flow in the centrifugal injector is critical because it is located exactly in the central part where the air core is formed. In the case of this first injector analyzed, the number of rotation (S) is small, so the viscosity values in the case of κ-ε model even being larger than in the case of model RNG κ-ε do not yet affect the formation of the vortex chamber.

Figure 10 shows the total pressure distribution within the injector in the two simulation cases.

The total pressure distribution is the same for two turbulence models. In Abramovich's theory it is assumed that flow energy is defined by the Bernoulli equation. By Bernoulli's equation, the energy of the flow is equal to the total pressure and is constant along the flow. According to this theory, it is considered that the fluid has no viscosity. In numerical simulation, the viscosities of air and water were included in the calculations. For this reason, the total pressure does not remain constant. As shown in Figure 10, the total pressure is distributed across the layers. The highest values correspond to the larger radii, near the inputs. With the radius decrease the total pressure drops gradually. This pressure drop is related to the loss of energy due to the friction between the layers of the fluid.

In Figure 11 the static pressure distribution diagrams are presented. The static pressure distribution for the two cases of turbulence models is also the same. The total pressure is 1.5 bar higher than the static pressure. This difference of 1.5 bar between total and static pressure is called dynamic pressure. The angular momentum of the flow tends to be constant. Therefore, with a decrease in radius, the tangential flow velocity increases, consequently the dynamic pressure increases. The sum of the static and dynamic pressure is constant except for small losses due to friction. Therefore, the static pressure with

Figure 9. Distribution of turbulent viscosity

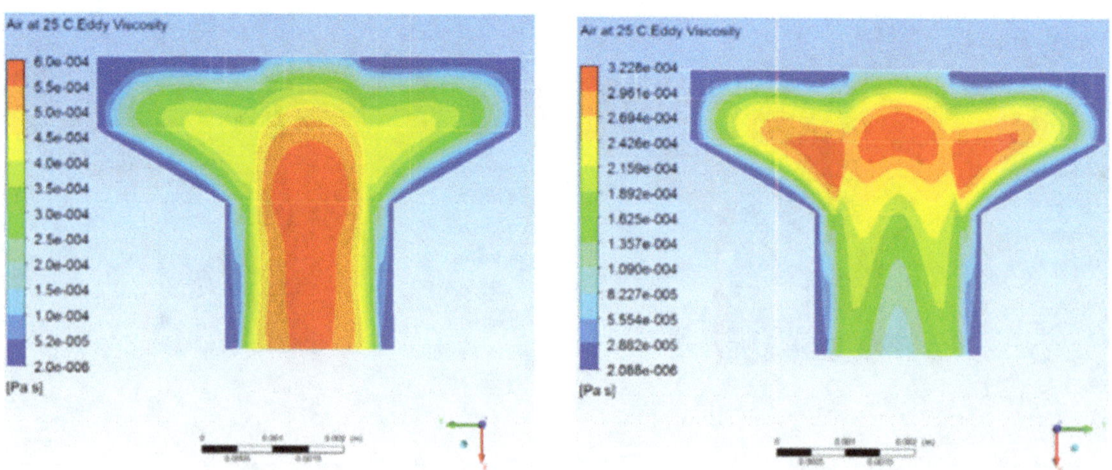

(a) κ-ε Turbulence model

(b) RNG κ-ε Turbulence model

Figure 10. Distribution of total pressure inside the injector

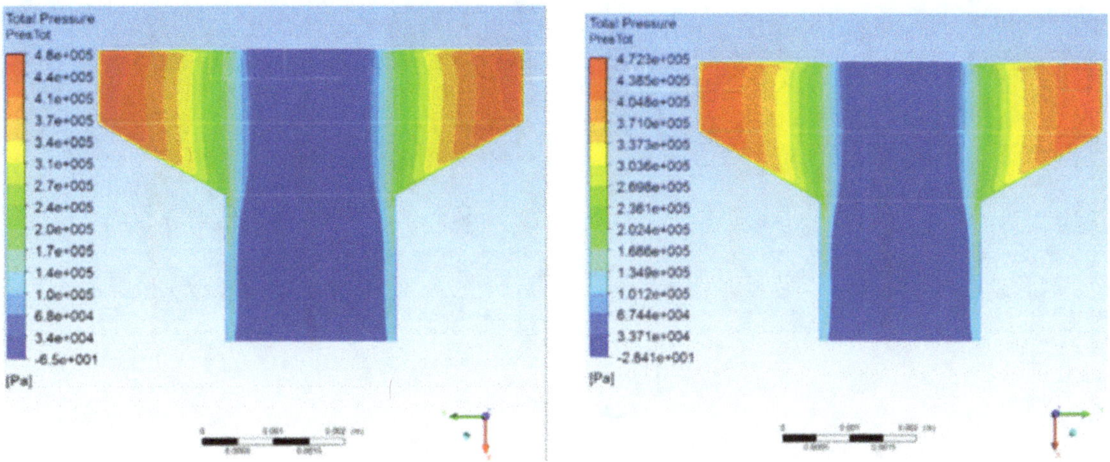

(a) κ-ε Turbulence model (b) RNG κ-ε Turbulence model

Figure 11. Static pressure distribution inside the injector

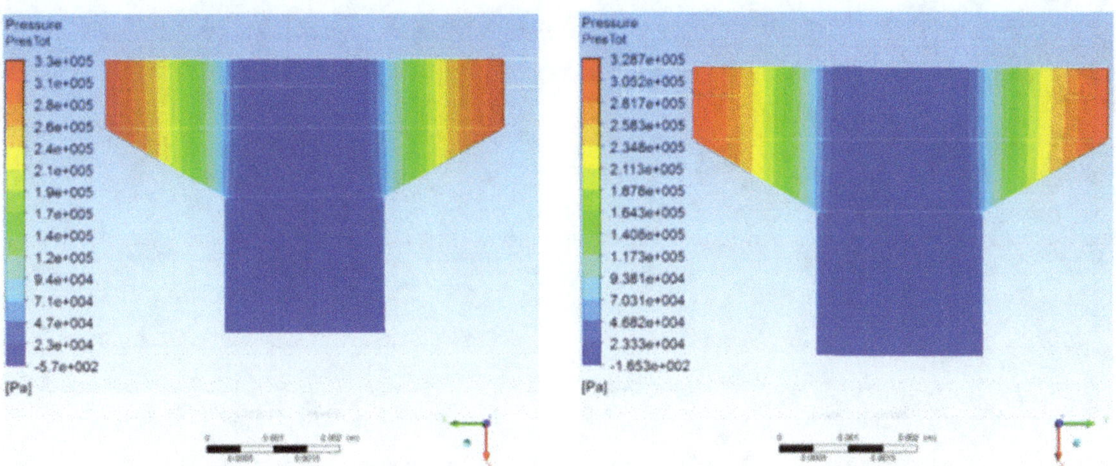

(a) κ-ε Turbulence model (b) RNG κ-ε Turbulence model

the radius decrease decreases. As shown in Figure 11, the static pressure distribution is well defined by the distance to the axis of rotation.

Figure 12 shows the axial velocity distribution diagram for the two cases of turbulence model. The axial velocity has low values in the rotating chamber (upper and wider injector). The increase of the axial velocity of the fluid occurs in the approximation with the air core. In order to better analyze the velocity distribution, in Figure 13 the velocity vector field is observed. Observing Figure 12 and Figure 13, it can be seen that the main axial flow occurs at the interface between air and water.

Figure 12. Axial velocity distribution within the injector

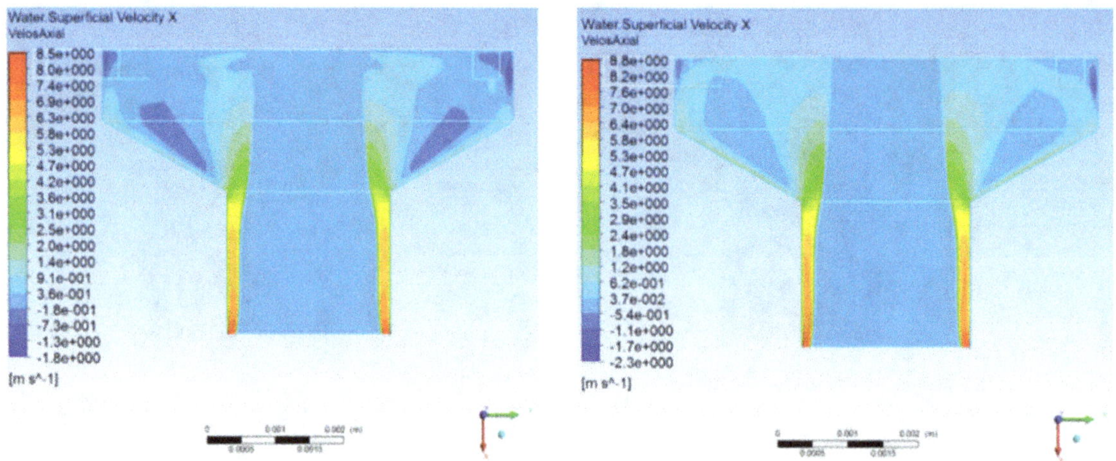

(a) κ-ε Turbulence model (b) RNG κ-ε Turbulence model

Figure 13. Vector field of speeds

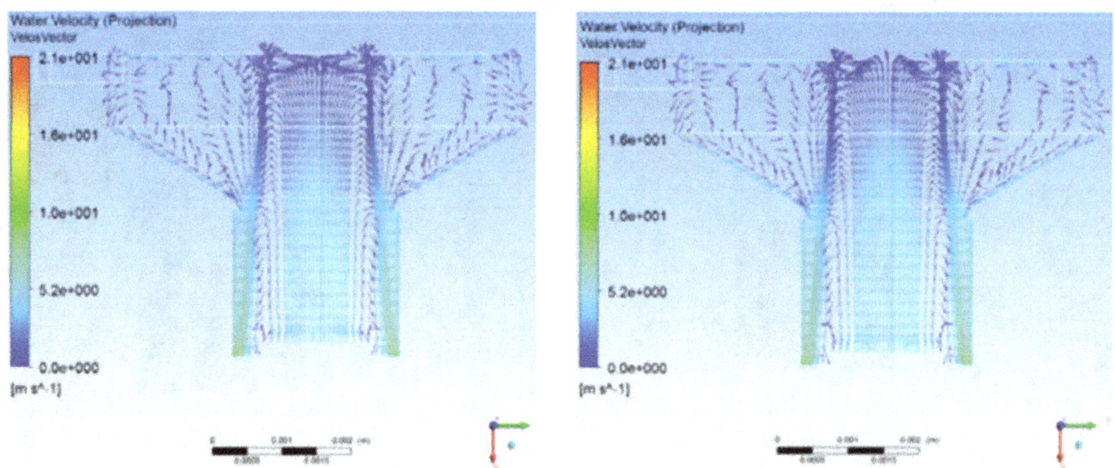

(a) κ-ε Turbulence model (b) RNG κ-ε Turbulence model

Also, in this region a recirculation zone is formed. In the upper part, the air meets the upper wall of the injector and forms a further recirculation zone; In this case, take the reverse direction drawn by the water towards the injector outlet.

In the rotating chamber, at the locations farthest from the water / air interface, the axial velocity component is low where the tangential component prevails. In the convergent part of the injector there is a small layer of fluid that goes towards the outlet through the wall. The main axial velocity increase starts exactly in the convergent zone of the injector, in the smaller diameter layers. This increase in axial velocity can be explained by the law of continuity. The rotating chamber has the largest radius, hence the cross-sectional area through which the water flows is larger. In the converging zone, this area has its section reduced due to the radius reduction of the injector in the direction of exit. In this way, the axial velocity increases so that the mass flow rate is the same. In the nozzle, the axial velocity reaches a maximum value, because the radius of the nozzle is of smaller section.

Figure 14 shows the radial velocity distribution diagrams. The main movement of the fluid inside the centrifugal injector is rotating; Because of this, in most of the injector the radial velocity component is close to zero (yellow and light green). The radial component has its value increased in the recirculation zones and in the places where the air nucleus presents greater radius. From the comparison of Figure 14 to Figure 8 these locations are easily identified. This increase in the radius of the air core occurs at the beginning of the nozzle and at the exit of the injector.

Figure 15 shows tangential velocity distribution diagrams. According to Abramovich's theory it was adopted that the angular momentum is constant inside the injector. According to this hypothesis, the tangential velocity corresponding to the same radius e must have the same value. This hypothesis is close to the numerical solution found for viscous fluid. In the rotating chamber, and in the convergent zone, the tangential velocities corresponding to the same radii are close and do not change their value along the axis of the injector. In the zone of the nozzle of the injector the value of the tangential velocity of the same level begins to fall until the exit of the device. This decrease in tangential velocity, and hence angular momentum, occurs due to friction-related losses.

In the case of a turbulence model (κ-ε), the tangential velocity is close to zero (blue color), and is larger than in the (κ-ε RNG) model. This is because the air in the central region enters due to the difference of pressures inside the injector and the environment, and has only axial velocity component. The tangential velocity component is transmitted to the air by its interaction with the fluid. In the case of the turbulence model (κ-ε) the turbulent viscosity is twice as large as in the model (RNG κ-ε). Because of this, the amount of layers to which the velocity is transmitted is also larger.

In the rotation chamber (Figure 15) the tangential velocity value hardly changes. According to the law of conservation of angular momentum, the tangential velocity should increase with the decrease of the radius. This does not occur because all the tangential velocity that flow should gain due to radius

Figure 14. Radial velocity distribution within the injector

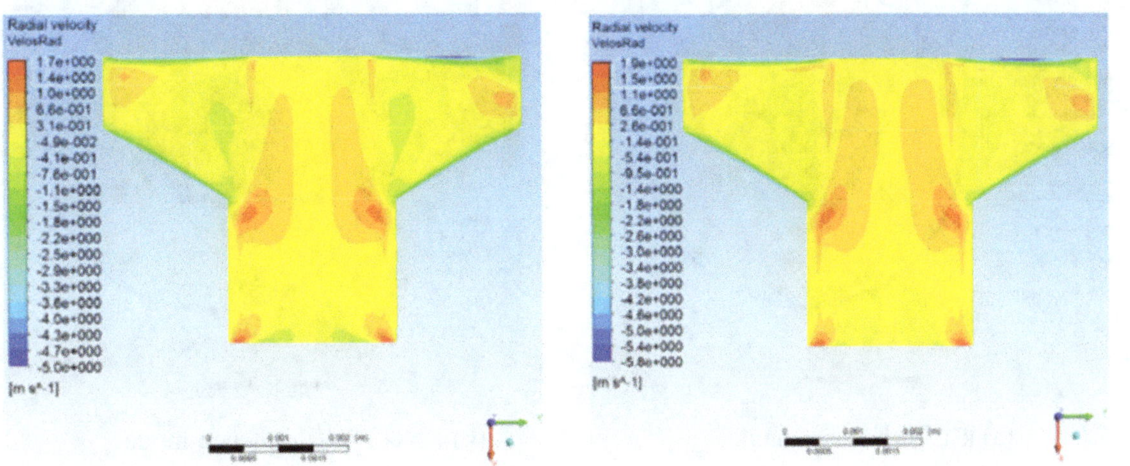

(a) κ-ε Turbulence model

(b) RNG κ-ε Turbulence model

Figure 15. Distribution of the tangential velocity inside the injector

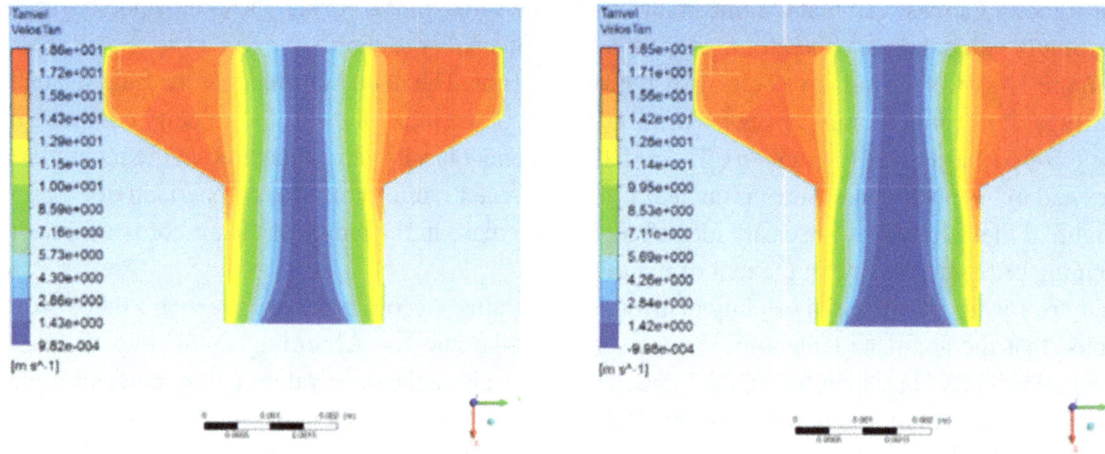

(a) κ-ε Turbulence model

(b) RNG κ-ε Turbulence model

reduction is lost by frictional forces. However, the tangential velocity distribution in the cross-section of the rotating chamber in a narrow speed range (15-18 m/s) is shown in Figure 16. In this figure, we can see a small increase of tangential velocity with approach of the center, until the region of interface of the water with the air. The tangential velocity of the water begins to fall again near this zone region to the interaction with the air.

Figure 16 shows the tangential velocity distribution in the cross-section of the rotation chamber.

Figure 16. Distribution of the tangential velocity in the rotational chamber cross section according to the turbulence model considered

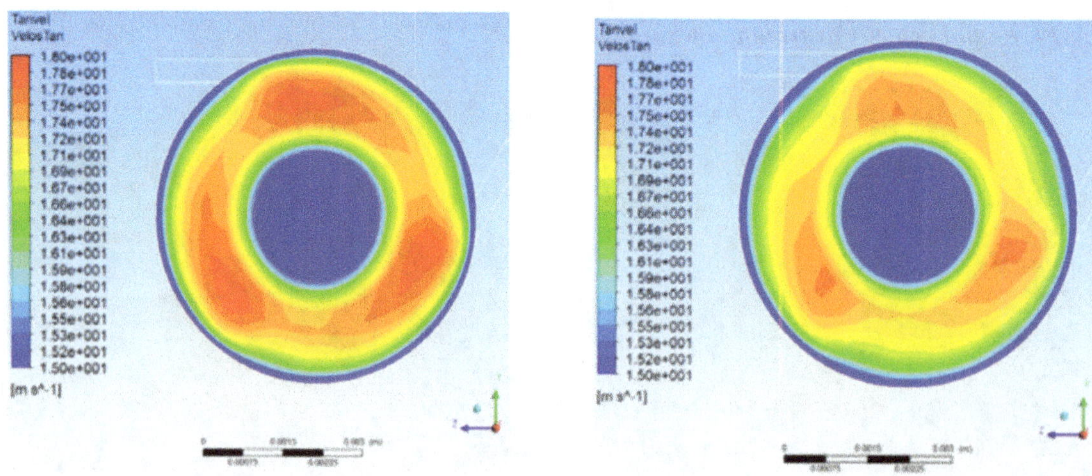

(a) κ-ε Turbulence model

(b) RNG κ-ε Turbulence model

The rated operating pressure of the injector is 5 Atm. However, the numerical simulation was performed for various operational pressures; In the range of 0 to 5 Atm. Figure 17 shows the axial velocity distribution graph in the injector outlet section as a function of radial coordinate for various pressures.

With increasing pressure, the axial velocity increases. At higher speeds, the level of turbulence increases and the losses near the next sections of the injector walls and in the zone of interaction of the water with the air also grow. The central segment of the graph, where velocity is equal to zero, corresponds to the core location of air, where the volumetric fraction of water is zero.

The tangential velocity distribution in the injector outlet section is similar to the axial velocity and is shown in Figure 18.

Figure 19 shows the comparison of the tangential velocity dependence versus pressure curves. The results obtained by the theoretical calculations present values reasonably higher than those presented by the numerical solutions.

This is because both theories (fluid without viscosity, viscous fluid) cannot fully describe the losses of the angular momentum inside the injector.

In Abramovich's theory fluid viscosity was not considered in the calculations, so friction losses do not exist in this model and the tangential velocity values obtained by this theory are larger. The second theory (Kliachko, 1962) involves correction of calculations due to fluid viscosity. The values of the tangential velocity are smaller in comparison with the ones obtained using Abramovich's theory, because angular momentum losses are included in the calculations.

Figure 20 shows the axial velocity dependence curves as a function of pressure. The values of the axial velocities obtained theoretically are also higher than the values presented by the numerical solution. The difference in results for the axial velocity is smaller than in the case of the tangential velocities. In both theories, it is considered that the total pressure inside the injector is constant, but as discussed

Figure 17. Axial velocity distribution in the outlet section of the injector

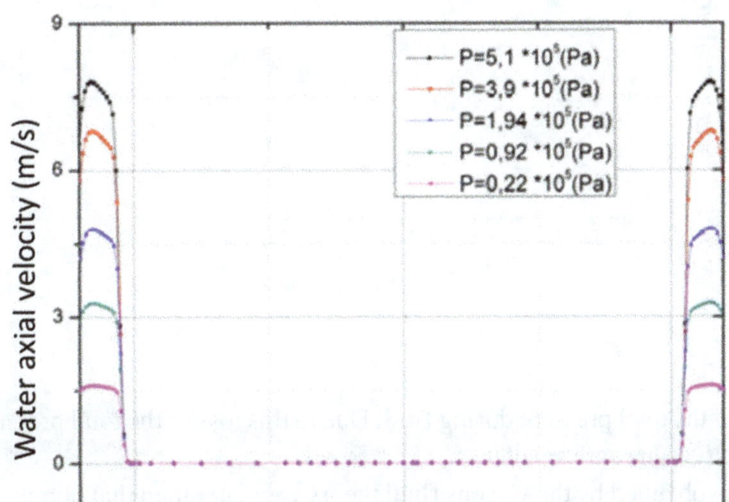

Figure 18. Tangential velocity distribution in the injector outlet section

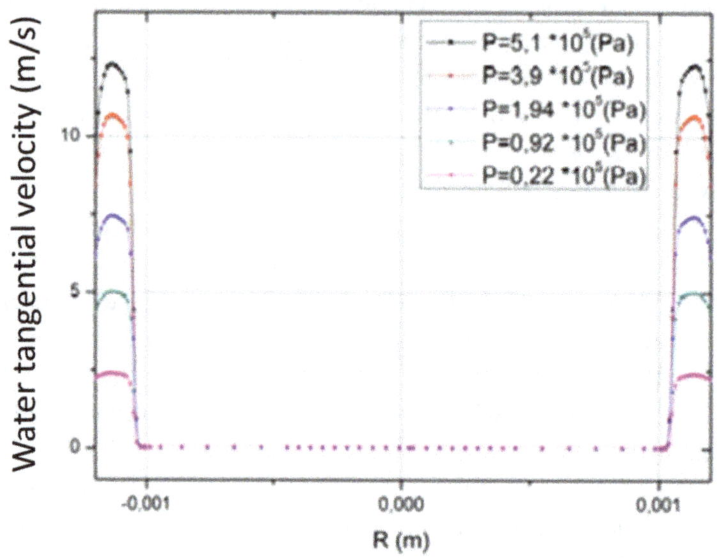

Figure 19. Solutions for the tangential velocity curves as a function of pressure

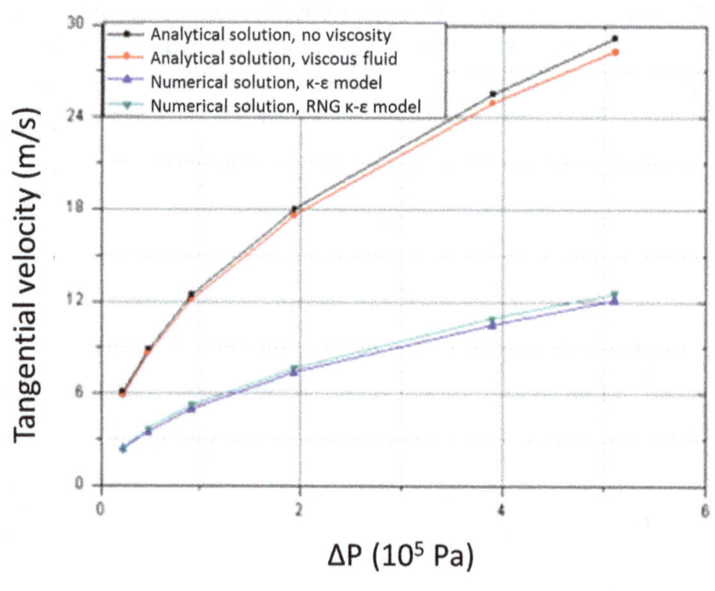

earlier there is loss of the total pressure during flow. Due to this loss of the total pressure the numerically obtained axial velocity values are smaller.

The axial velocity obtained by the viscous fluid theory is greater than that obtained by Abramovich's theory. Due to the viscosity of the fluid there is loss of angular momentum. Also in the viscous fluid model, it is adopted that the total pressure is constant. Considering the angular momentum losses, the

Figure 20. Solutions for axial velocity curves as a function of pressure

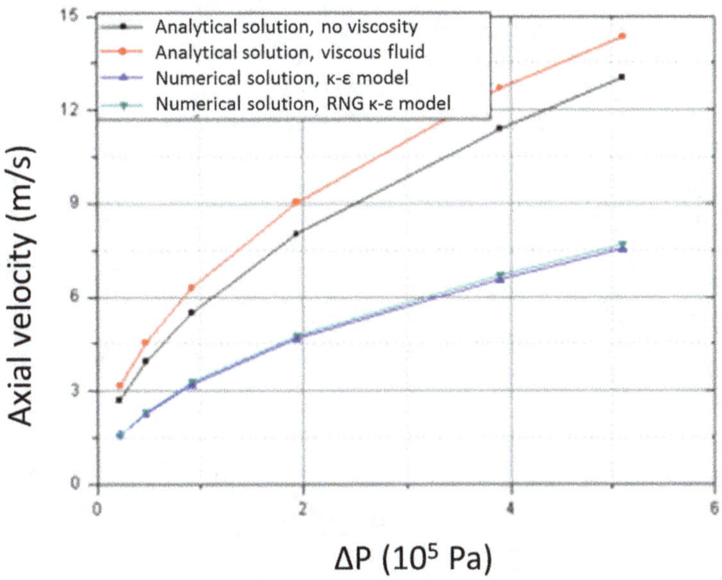

tangential velocity decreases and for the total pressure to be constant, the axial component must grow as tangential velocity decreases.

The difference between numerical and theoretical results increases with increasing pressure. This behavior of the curves shows that the influence of the viscosity, and the working regime of the injector, is more relevant at high pressures, when the flow rates are higher and the flow rate is turbulent.

Figure 21 shows the dependence curves of the spray angle as a function of pressure. The black curve corresponds to Abramovich's theory. In Abramovich's theory, it is considered that the fluid has no viscosity; For this reason, all the geometric characteristics of the injector are constant values and do not depend on the flow regime. The red curve shows the spray angle values obtained by the viscous fluid theory; For this reason, there is variation of the angle due to the variation of the inlet pressure. In the device with increasing pressure, the angle increases. This behavior is in agreement with numerical solution (green and blue curve). As discussed earlier, when the viscosity of the fluid is greatest, the flow at the injector outlet has an axial component of velocity greater than the tangential component.

Consequently, the cone angle of liquid is lower in the case of the higher viscosity fluid; As shown in Figure 21. Also, as shown in Figure 21, there is a difference between the numerical solution results and the viscous fluid theory. This difference indicates that the viscous fluid theory does not include all the effects of viscosity on the flow inside the injector.

The fuel injector is the outside of the bi-propellant injector. Therefore, so that the flow inside this injector is not affected, it is necessary that the diameter of the air core is greater than external diameter the internal injector. To confirm this with therefore, a model of the external injector with internal body was made. The result of the numerical simulation is shown in Figure 22.

As shown in Figure 22 there is no interference from the walls of the internal injector with flow at the external injector outlet. This means that the flow area of the fluid is the same as for the case of model without internal body, so the mass flow and other flow parameters are not affected.

Figure 21. Solutions for the liquid cone angle curves as a function of pressure in the injector device

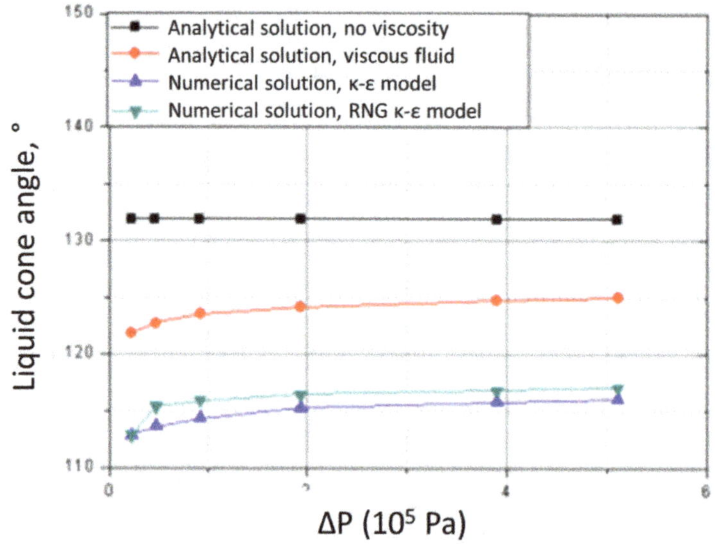

Figure 22. Distribution of water inside the injector with internal body

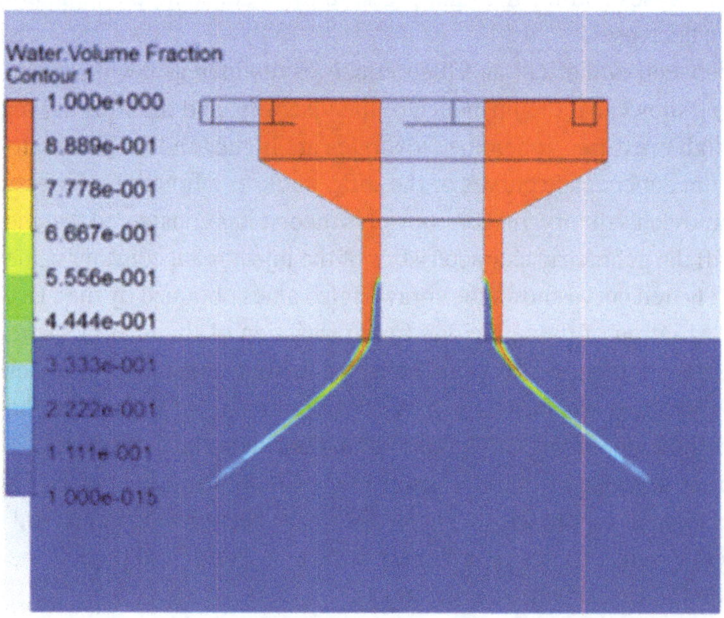

Figure 22 shows the formation of liquid cone outside the physical domain of the injector. The cone walls are not linear. The walls at the exit of the injector have a parabolic profile. This detail is important when it is necessary to profile the injector nozzle for the purpose of reducing the liquid spray angle. If a direct profile is drawn from the cylindrical part of the injector nozzle, there is a discontinuity in the flow that affects the angle and can also affect the liquid distribution homogeneity as shown in Figure 23.

Figure 23. Profile of the discontinuity of flow in the case of profiling of the injector nozzle

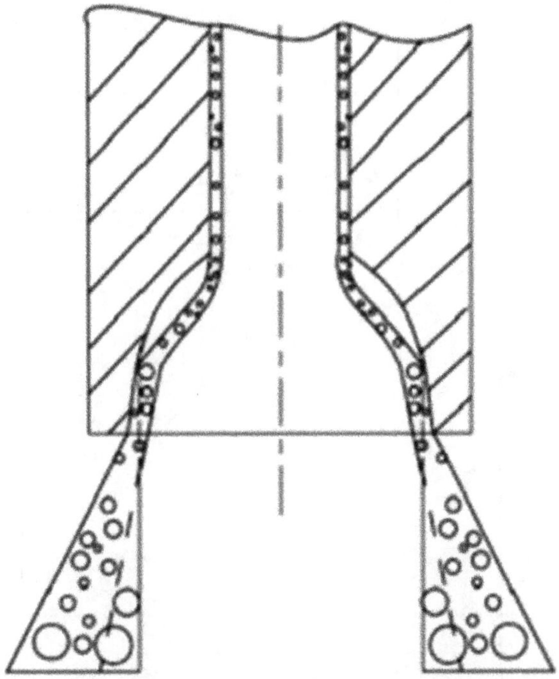

In order to avoid discontinuity of flow of the fluid through the injector, it is necessary that the profile of the injector nozzle follows the flow at the beginning. The profile start curve can be calculated. A schematic of the equation of the cone wall curve is shown in Figure 24.

The EA curve in Figure 24 is the liquid cone profile to be calculated. The curve EA is formed through the points of intersection of the straight-line BA with ADC. The curve BA is the trajectory of the water particles externally to the injector. The circumference with center at the D point represents the exit orifice of the injector. The particles of water, after leaving the injector have linear trajectory (BA). Accompanying the curve BA along the entrance orifice the water cone is formed. When cutting the cone with ADC plane, the cone wall will take the form of curve EA. In order for the curve EA to be formed the plane tangential to the radius or to the cylindrical surface of the injector nozzle must travel a quarter of the circumference of the exit orifice of the injector. In other words, the radius r must change its angular position φ from 0 to π/2. According to Figure 22, the DC segment is the x coordinate of the intersection point of line BA with plane ADC, or coordinate x of profile point A. So, segment CA is the y coordinate of this point. To deduce the equation of the cone profile it is necessary to find the functional dependence between the coordinate y and x. The angle α/2 is known as the spray angle calculated by the centrifugal injector theory or, otherwise, can be obtained by practical tests. Thus, from the triangle ΔDBC we have:

$$x = DC = \frac{r}{sen\varphi} \tag{18}$$

The coordinate y can be calculated from the triangle ΔACB.

Figure 24. Schematic drawing for equation deduction of the liquid cone wall curve

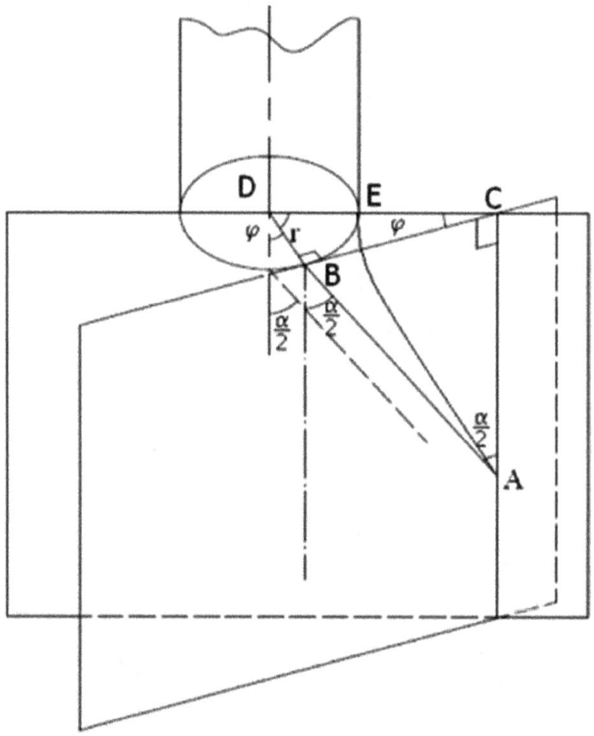

$$y = AC = -\frac{BC}{tg\frac{a}{2}} \tag{19}$$

From the triangle ΔDBC

$$BC = \frac{r}{tg\varphi} \tag{20}$$

Substituting the value of the BC of Eq. (20) in Eq. (19) yields:

$$y = -\frac{r}{tg\varphi tg\frac{a}{2}} \tag{21}$$

Extracting r from Eq. (18) and inserting into Eq. (21) yields:

$$y = -\frac{x \operatorname{sen}\varphi}{tg\varphi \tan\frac{a}{2}} = -\frac{x\cos\varphi}{tg\frac{a}{2}} \tag{22}$$

The angle φ can be expressed through the x-coordinate:

$$\varphi = arcsen\frac{r}{x} \tag{23}$$

By inserting Eq. (23) into Eq. (22) we obtain:

$$y = -\frac{x\cos\left(arcsen\frac{r}{x}\right)}{tg\frac{a}{2}} \tag{24}$$

The graph of the function is shown in Figure 25 compared to the numerical solution.

Flow Simulation

The simulation is done for ten values (1 – 10 atm.) of the total pressure at enter of the swirl atomizer. Figure 26 shows the air core formation corresponding to nominal pressure 10 atm.

Figure 27 shows the distribution of the total and static pressures. It can be observed, that total pressure have higher values due the contribution of the dynamic component.

Figure 28 presents the distribution of the water axial velocity inside the atomizer. Axial velocity has low values on the top of injector in the swirl chamber. In this region, the axial velocity increases with approximation to the air core. With passage to the nozzle the axial velocity has greater values and reaches the maxim value on the exit of the nozzle.

Figure 25. Profile of liquid cone wall obtained analytically (left) and numerical solution (right)

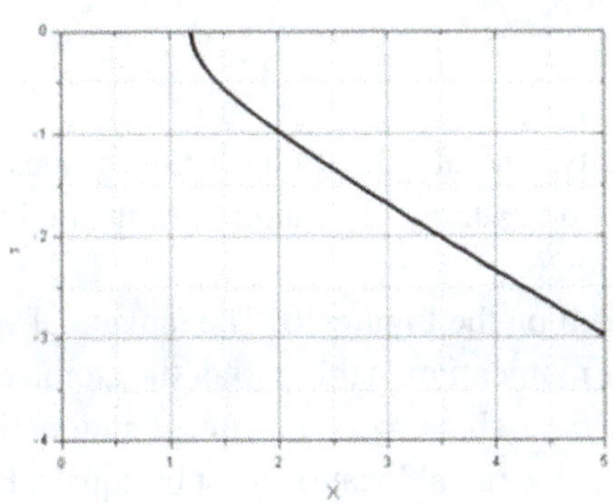

Figure 26. Water volume mass fraction (10 atm)

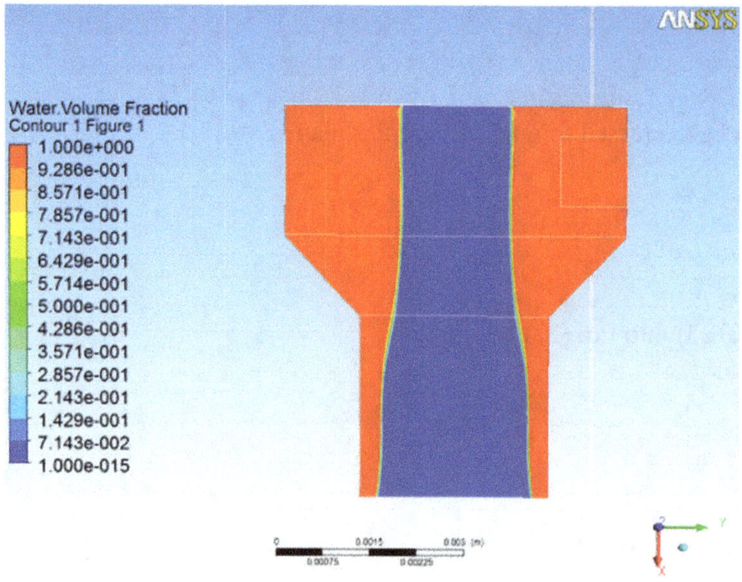

Figure 27. Distribution of static a), and total b) pressures

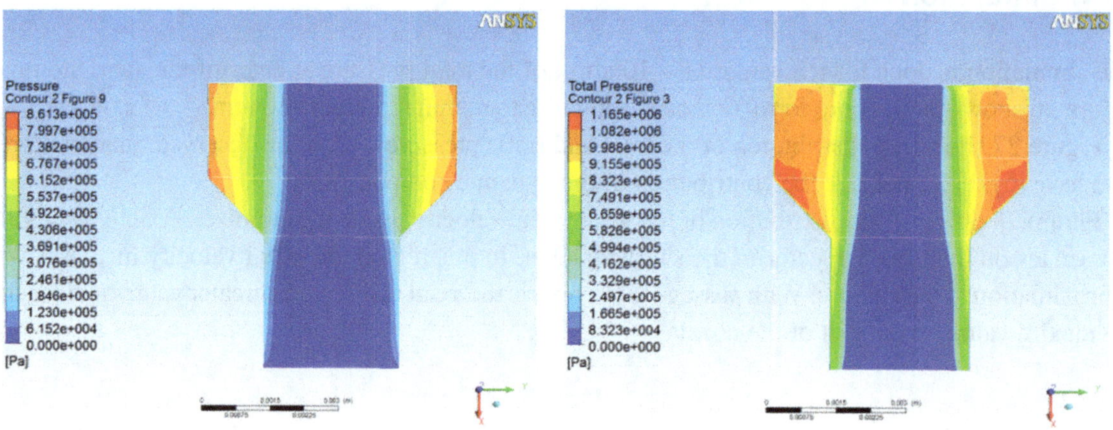

a) b)

Figure 29 shows the distribution of the radial velocity. Radial velocity has raised values on the two regions of the nozzle – on the top part of the nozzle and on exit, due the increase of the radius of the air core. It can be repaired comparing Figure 29 and Figure 26.

The distribution of the tangential velocity is presented on the Figure 30. The tangential velocity has greater values on the swirl chamber because the angular momentum in this part of the atomizer is greater then in the nozzle. It is opposed to the Abramovitch theory, which based on consideration, that angular momentum is constant, and tangential velocities on the same radial location mast be equals, Eq. 2. But it has clear physical explication. The Abramovitch considerate the inviscid fluid, but in simulation is used

Figure 28. Water axial velocity

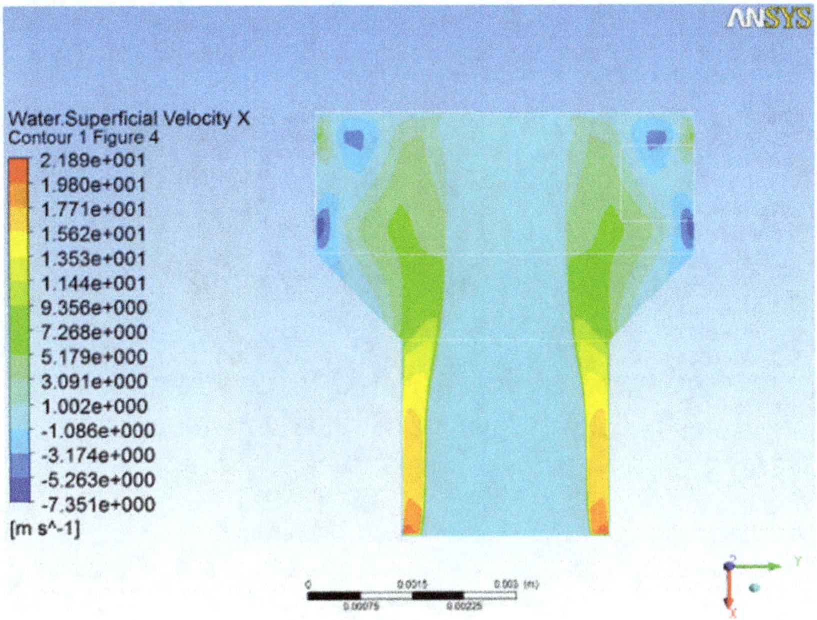

Figure 29. Water radial velocity

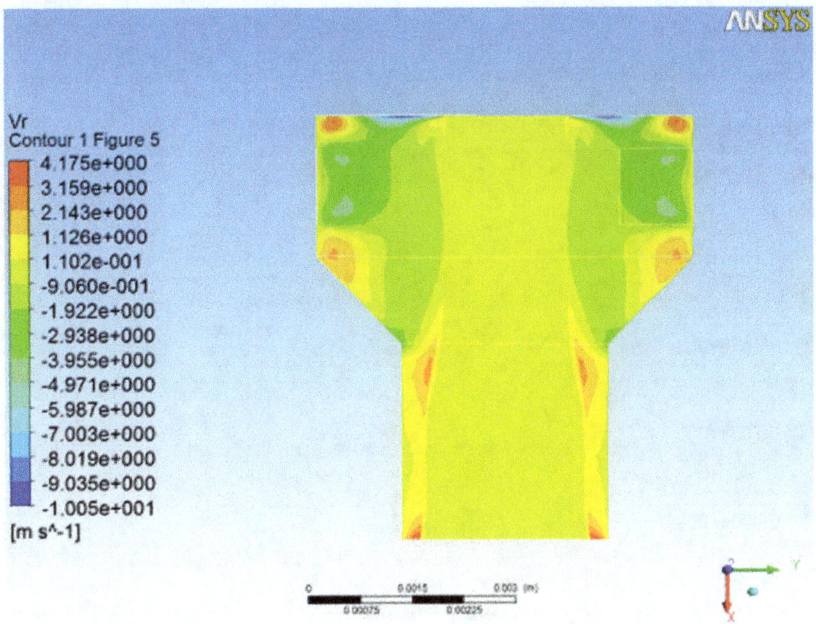

Figure 30. Water tangential velocity

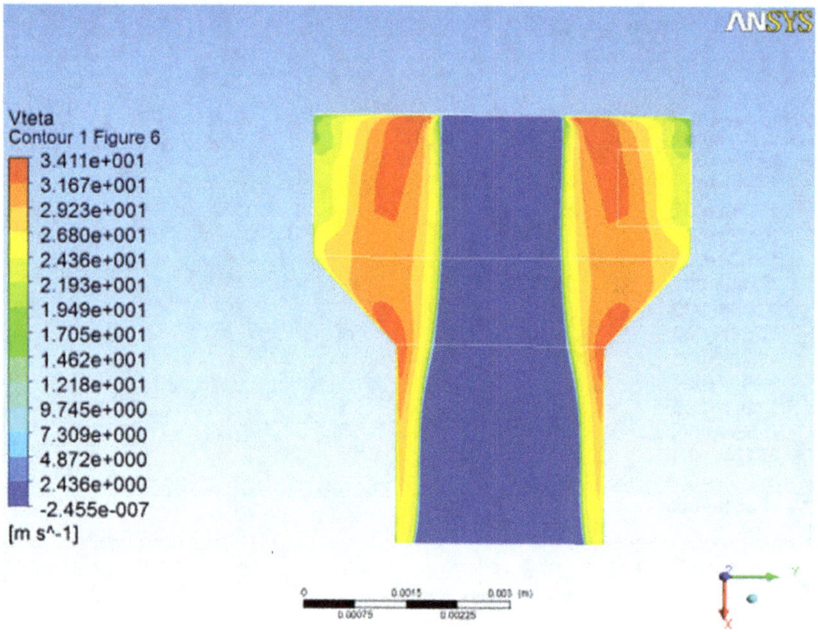

Figure 31. Vector plot of the water velocity

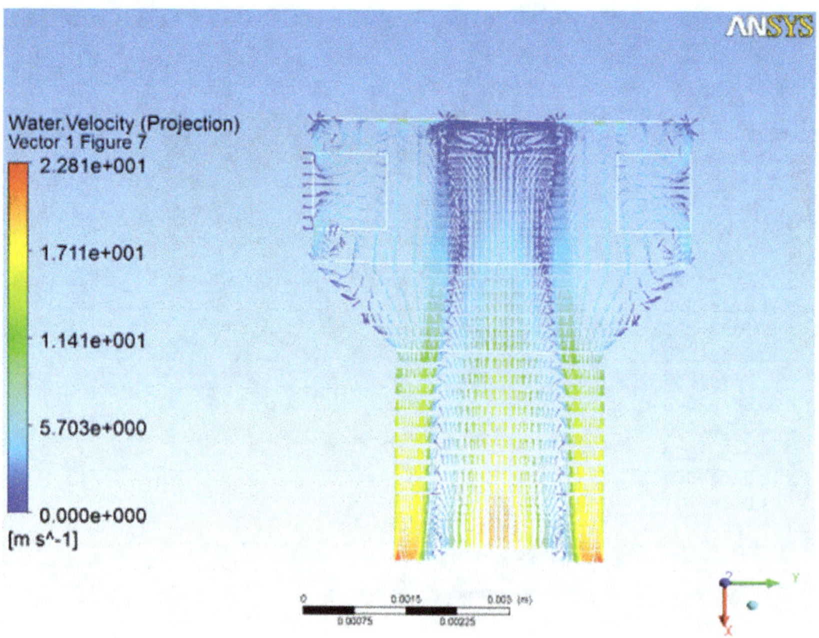

the viscous fluid – water. The tangential velocity in the nozzle is smaller because the momentum in the nozzle decries due of the friction losses. The behavior of the tangential velocity on the same section plane ZY is with accord to conservation of the angular momentum, Eq. 2. It increases when the radius reduces. Only on region of the interface with the air core it decreases due of the interaction between water and air.

Analyzing Figure 31 can be observed some recirculation zones. The recirculation zones are localized on the swirl chamber above and below the enter orifice. The fluid flow has hydraulic losses on these regions; therefore, it is good to reduce the height of the swirl chamber up to the height of the tangential orifice. Others recirculation zones is concentrated on the interface of two fluid due the opposed flow direction of the air and water. The recirculation is observed on top of atomizer too. It appears because the air enters in the nozzle and flows off the top of injector. On top, it changes the flow direction and flows off together water to the exit of the nozzle. The air core on the top has greater radius due of these recirculation zones.

Water mass flow in the transversal section located on the exit of the nozzle is presented on Figure 32. Two zones of the maximum mass flow can be observed here. This no-uniform distribution of the mass flow due of the number of the tangential enters of the injector. In this case the atomizer has two orifices of the fluid enter, consequently, its form two zones of the maxim mass flow. According to (Vasiliev, Kudriavtcev, & Kuznetcsov, 1993) no-uniformity reduces if increase the number of enters, and distribution of the mass flow is nearly uniform if number of the orifices is six. In case of the six orifices the zones of the maximum mass flow are closed together and distribution become uniform.

Figure 32. Water mass flow on the exit of the nozzle

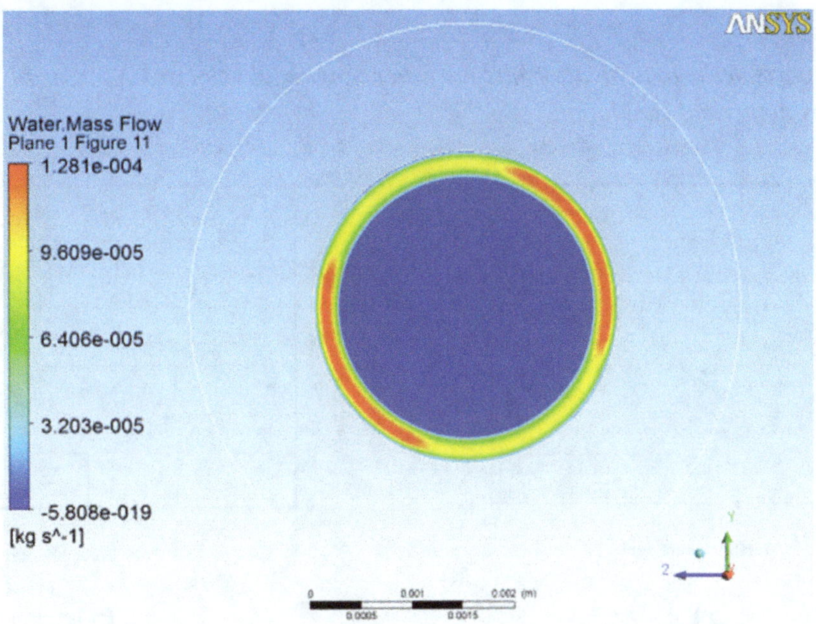

Figure 33. Comparison the analytical results for water mass flow with ones obtained by numerical simulation

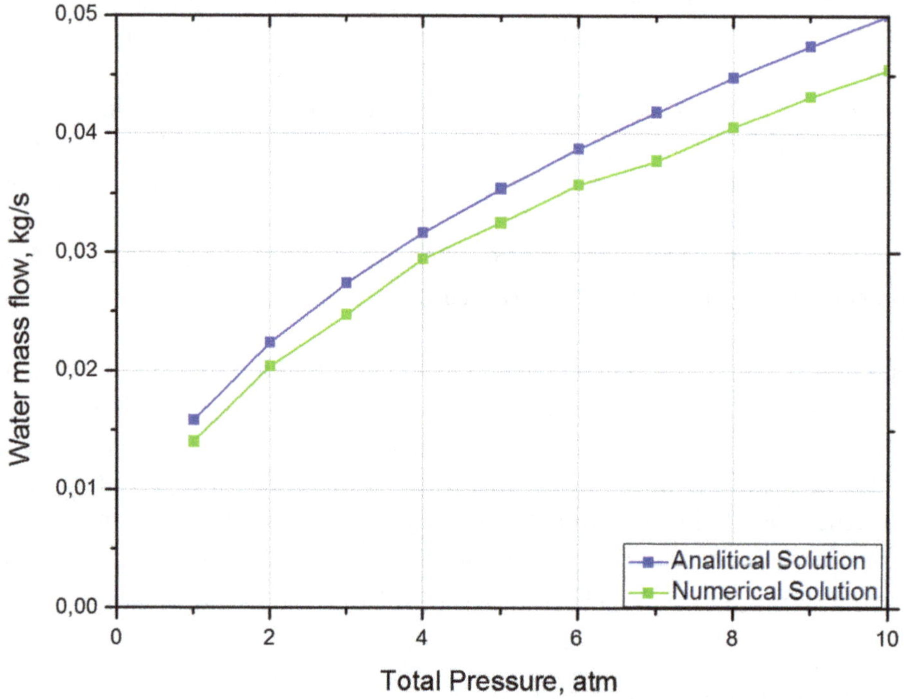

Figure 34. Comparison the analytical results for water tangential velocity a) and axial velocity b) with ones obtained by numerical simulation

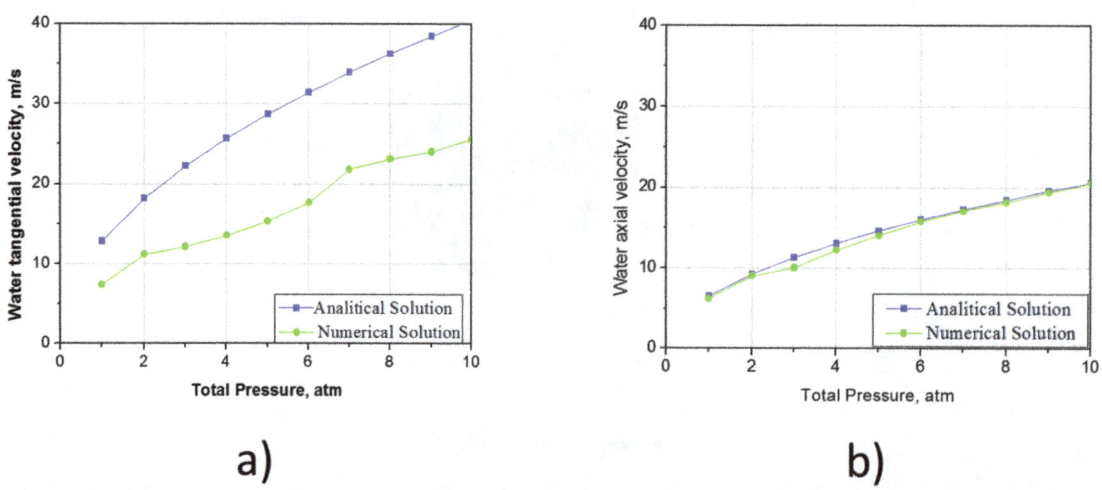

CONCLUSION

The numerical simulation of the two-phase flow was performed by *k-ε* turbulence model, after simulations and analysis of the *k-ε* and RNG *k-ε* models. The boundary layer on the walls and on interface between two fluids was possible to characterize due the good refinement of the mesh in these regions. The analytical solution for axial, tangential velocities and mass flow on the exit of nozzle were obtained. The numerical and analytical results for axial velocity and mass flow present a good concordance as shown on the Figure 33 and Figure 34. b). The analytical and numerical results for water tangential velocity presented on Figure 34. a) have significant deviation, because during analytical calculus the friction losses are disrespected. The angular momentum on the swirl chamber is greater than one on the nozzle. It affects the calculus for the pulverization angle. The angle of the liquid cone on the exit of the atomizer will have minor value than calculated analytically. The numerical results for the distribution of the mass flow on the transversal section localized on the exit of nozzle shown on Figure 32 confirm the practice results presented by (Vasiliev, Kudriavtcev, & Kuznetcsov, 1993), (Khavkin, 2004) and others.

ACKNOWLEDGMENT

The author wishes to thank the financial support of the National Research Committee – CNPq.

REFERENCES

Abramovich, G. N. (1944). The Teory of Swirl Atomizers. In *Industrial Aerodynamics* (pp. 114–121). BNT ZAGI.

Alajbegovic, A., Meister, G., Greif, D., & Basara, B. (2002). Three phase cavitationg flows in high-pressure swirl injectors. *Experimental Thermal and Fluid Science*, *26*(6-7), 677–681. doi:10.1016/S0894-1777(02)00179-6

Alemasov, V. E. (1980). *Teoria dos motores-foguete*. Machinostroenie.

Bayvel, L., & Orzechowski, Z. (1993). *Liquid Atomization*. Washington: Taylor & Francis.

Blazek, J. (2001). *Computational Fluid Dynamics*. Baden-Daettwil, Switzerland: Alstom Power Ltd.

Cebeci, T., Rshao, J., Kafyeke, F., & Laurendeau, E. (2005). *Computational Fluid Dynamics for Engineers*. Long Beach, California: Horizons Publishing Inc.

Chung, T. J. (2002). *Computational fluid dynamics*. Cambridge University Press. doi:10.1017/CBO9780511606205

Doble, S. M., & Halton, E. M. (1947). The Application of Cyclone Theory to Centrifugal Spray Nozzles. *Institute of Machenical Engineers*, *154*(1), 111–119. doi:10.1243/PIME_PROC_1947_157_015_02

Hansen, K. G., & Madsen, J. (2001). *A Computational and Experimental Study of the Internal Flow in a Scaled Pressure-Swirl Atomizer* [MSc Thesis]. Denmark: Aalborg University Esbjerg.

Khavkin, Y. I. (2004). *Theory and Practice of Swirl Atomizers*. New York: Taylor & Francis.

Kliachko, L. A. (1962). Theory of Centrifugal Injectors. *Teploenergetica*, 34-38.

Shanwu, W., Yang, V., Hsiao, G., Hsieh, S., & Mongia, C. (2007). Large-eddy simulations of gas-turbine swirl injector flow dynamics. *Journal of Fluid Mechanics*, 583, 99–122.

Vasiliev, A. P., Kudriavtcev, V. M., & Kuznetcsov, V. A. (1993). *Theory and calculos of liquid propellant rocket motors.*

Chapter 6
Disintegration of Liquid Sheet Produced by Swirl Injector

Leopoldo Rocco Jr.
Flowtest Engineering and Research Co, Brazil

ABSTRACT

The development of liquid sheets that emerges from nozzles is influenced mainly by their initial speed and by the physical properties of the liquid and the ambient gas. A minimum speed of the sheet is necessary for its enlargement against the superficial tension that tends to contract the surface. As this speed increases, the sheet expands until a main extremity is formed, where balance exists among the superficial tension and the inertial forces. The form and regularity of the sheet's disintegration process has influence in the size distribution of the produced drop and in the Sauter mean diameter (SMD). The initial thickness of the produced liquid sheet is important to determine the medium size of obtained drops. It was observed that thicker films produce thicker ligaments and larger drops. The medium drop diameter produced in conical sheets of pressurized swirl atomizers is calculated according to the thickness of the sheets and in the wavelength for the maximum growth tax.

INTRODUCTION

The development of a liquid sheet that emerges form a nozzle is influenced mainly by its initial speed and for the physical properties of the liquid and of the ambient gas. It is necessary a minimum speed of the sheet for its enlargement against the force of superficial tension, which tends to contract the surface. As larger this speed is, more the sheet expands and it is prolongated until that a main extremity is formed, where balance exists between the superficial tension and the inertial forces. Its disintegration may happen with or without perforations in the liquid sheet.

The generation of a wave movement in the sheet, through the one which areas corresponding to one or half wavelength of the complete oscillation, will be torn apart before the extremity is reached. These torn areas, when don't suffer disintegration for the action of the air or turbulence, they contract quickly under the action of the superficial tension forming a net of lines.

DOI: 10.4018/978-1-5225-2903-3.ch006

Fraser (Fraser, Dombrowski, & Routley, 1963) showed that the regularity of the disintegration process and the uniformity of production of lines has a big influence in the distribution of the size of the drop. Perforations that happen in the sheet at the same distance of the nozzle have a similar formation and the size of the drop is constant in the disintegration of perforated sheet. The disintegration of the wavy sheet is highly irregular, and consequently the drop sizes are much more varied.

Atomizers that unload the liquid in the form of a sheet are usually capable to exhibit all the three manners of sheet disintegration. Sometimes, two different manners happen simultaneously, and its relative importance can influence a lot, so much in the drop size, as in the size distribution.

Dombrowski and Fraser (Fraser, Dombrowski, & Routley, 1963) used an improved photographic technique and a source especially projected to combine a discharge and intense illumination with short duration, and established that the ligaments are caused, mainly, for perforations in the liquid sheet. If the perforations are caused by attrition with the air, the ligaments separate very quickly. However, if the same ones are created through other means, as turbulence in the nozzle, the ligaments break up more slowly. They concluded that (1) liquid sheets with high superficial tension and viscosity are more resistant to breaking and (2) the effect of the density of the liquid in the disintegration of the sheet is very small.

Dombrowski and Johns (Dombrowski & Johns, 1963) observed that the interaction between liquid and air produces unstable waves, which dissolve in fragments that contract in ligaments and these ligaments break up to generate minor drops due to the air velocity, because the ligaments are finer and are formed closer to the nozzle.

The initial thickness of the liquid sheet is directly proportional to the size of the nozzle, and is important to determine the medium size of the produced drop, as it was observed that thicker films produce thicker ligaments and larger drops.

The comparison among the magnitude of the disturbances wavelength that cause rupture of the liquid sheet with its thickness is used by Dumouchel (Dumouchel, Ledoux, Bloor, Dombrowski, & Ingham, 1990) to classify the liquid sheets in two groups. In the first case, when the magnitude of the wavelength of the disturbances in the liquid surface is in the same order of the sheet's thickness, the rupture process begins close to the exit of the nozzle and extends along the length of the sheet, and the ligaments, great structures and drops are unstable and break up in smaller drops.

In the second case, when the magnitude of the wavelength of the disturbances in the surface of the liquid is smaller than the thickness of the sheet, disturbances in the surface of the liquid are confined to a small area close to the surface and, as the disturbances grow, fine ligaments and small drops are removed from the surface. This disintegration process happens in areas far from the limits of the nozzle.

Plane Sheets

The mechanism for the disintegration of plane liquid sheets moving in the air was studied theoretical and experimentally by York et al. (York, Stubbs, & Tek, 1953). They concluded that the instability and wave formation in the interface among the continuous and discontinuous phases are the main factors in the rupture of a liquid sheet in drops. Forces of surface tension try to do the protuberances return for its original position, but the local static pressure of the air decreases (corresponding to the local increase in speed) and that tends to enlarge as becomes more distant to the external protuberance. This corresponds to the pattern of instability for induced wind, where the forces of surface tension are opposed to any movement of the interface of its initial plan and try to reestablish the balance, while a forced aerodynamics increases any divergence of the interface and promote, like this, the instability.

To formulate the problem, York et al. (York, Stubbs, & Tek, 1953) considered an infinite two-dimensional sheet of liquid of finite thickness with air on both sides, without viscous effects and non-rotational flow. Using the speeds obtained by calculations with the equation of Bernoulli, the pressure and certain displacements can be described. As in the case of a liquid jet, an exponential increase in the width of the wave happens under certain conditions.

The growth tax has a defined maximum for a certain number of Weber, especially to high numbers of Weber, and in the disturbance of that wavelength, it will dominate the interface and will quickly disintegrate the sheet.

Fan Sheets

Fraser (Fraser, Dombrowski, & Routley, 1963) has proposed an expression for drops sizes produced by the rupture of a sheet with low viscosity in form of a fan. His model assumes that the fastest wave growth (β_{max}) separates from the border in the form of a ribbon with the width of half wavelength ($\lambda_{opt}/2$). This ribbon immediately contracts in a filament with diameter D_L, which subsequently dissolves in drops of equal diameter.

In agreement with the analysis of Rayleigh, the collapse of the ligaments produces drops with the diameter D around 1,89 D_L.

The Sauter medium diameter of drop, SMD, is the diameter of the drop whose proportion of the volume for the superficial area is the same of the total spray.

For a real case, the liquid has finite viscosity and the thickness of the sheet decreases with the movement, as it increases the distance to the exit nozzle. Dombrowski and Johns (Dombrowski & Johns, 1963) related the measurements of the Sauter medium diameter of their works for a nozzle fan spray in fan, theoretically, for a flow with low viscosity, empirically correlating the thickness of the liquid sheet, the distance of the nozzle, the dynamic viscosity of the liquid, its pressure and the speed of the sheet.

Conical Sheets

The theory of rupture of conical sheets produced by pressurized swirl nozzles is not totally developed, but there are evidences that the curvature radius of the cone has a destabilized effect in the flotations.

York et al. (York, Stubbs, & Tek, 1953) were capable to do a rude estimate of the size of the drops produced by pressurized nozzles of the type swirl, creating a significative approach to the experimental results.

Eisenklam (Eisenklam, 1964) defined three distinct ways for the sheet disintegration, described as ring, wave and perforated sheet.

In the ring way, as illustrated in Figure 1, there is an unbalance in the extremity free from the liquid sheet, due to the unbalance among the surface tension and viscous forces, which tend to unite the liquid, and the forces that tend to separate it, forming a margin that breaks according to the mechanism that corresponds to the disintegration of a free jet.

The resulting liquid continues to move in the direction of the original flow, remaining fixed to the surface and creating thin lines (ligaments) that also quickly fracture in rows of drops. This disintegration way is very prominent where the viscosity and surface tension of the liquid are high and produce big drops, with numerous small drops around them.

Figure 1. Hollow conical liquid sheet disintegration in ring and ligaments

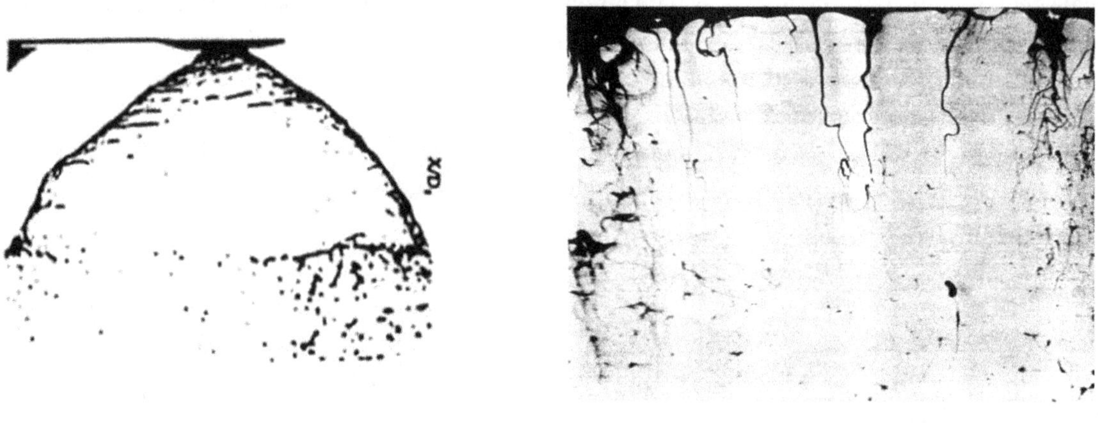

In the case of the waves, these are formed close to the nozzle and the wavelength for the maximum growth causes periodic thickening of the liquid sheet towards the normal flow. Rings break outside of the conical sheet, and the liquid volume contained in the rings can be equated to the volume of a ribbon (belt) cut out of the sheet, with the same thickness as the sheet in the rupture distance and a width similar to a wavelength. These cylindrical ligaments dissolve then in drops, in agreement with the mechanism of Rayleigh.

The formation mechanisms, stabilization and disintegration of conical liquid sheets were also studied by Bayvel and Orzechowski (Bayvel & Orzechowski, 1993). Figure 2 illustrates surface waves dominating the disintegration of a conical hollow liquid sheet under ideal conditions. The axial waves λ^* are responsible for the creation of toroidal ligaments of diameter d_1 that possesses circumferential waves λ_1 that drive to the formation of the drops.

In the disintegration of a perforated sheet, as shown in Figure 3, several holes appear in the sheet, which are delineated by formed margins of the liquid inside of the one that was formed initially. They grow quickly in size until reaching the margins of adjacent holes, and there is a consequent aggregation to produce ligaments in an irregular way, which then break up in drops of varied sizes.

These disintegration ways can happen in a same hollow conical liquid sheet produced by a pressurized swirl injector, under several conditions of injection as low or high pressure is applied or with or without the influences of the moving adjacent air.

Drops Diameters in Conical Sheets of a Swirl Pressurized Injector

The drop medium diameter produced in conical sheets of swirl pressurized atomizers is calculated according to the thickness of the sheet and to the wavelength for the maxim growth rate.

York et al. (York, Stubbs, & Tek, 1953) thought that the infinite plain sheet model, used in the theoretical analysis, didn't represent the conical spray of his experiences sufficiently.

For modulate the dilatation of the liquid conical sheet, the effect of changing the number of Weber in the length and in the time of rupture of the sheet is determined by the relative importance of the curvature radius in the main direction of the flow. For the sinuous modulation, the rupture of the film happens due to the non-linear way of joining dilatation and length. The time of rupture of the sheet due

Figure 2. Hollow conical liquid sheet disintegration in waves

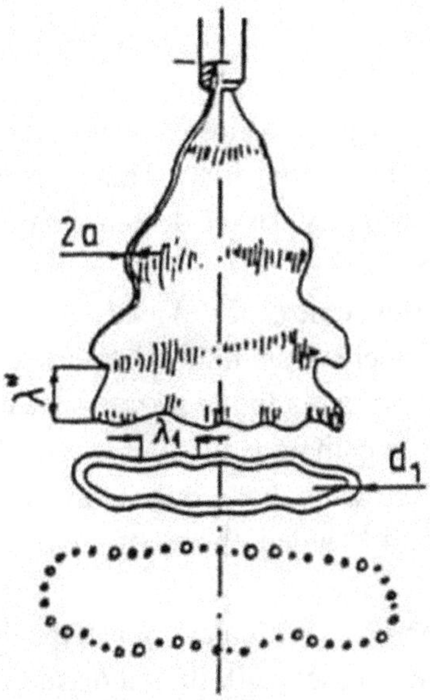

Figure 3. Hollow conical liquid sheet disintegration in perforated sheet (Bayvel & Orzechowski, 1993)

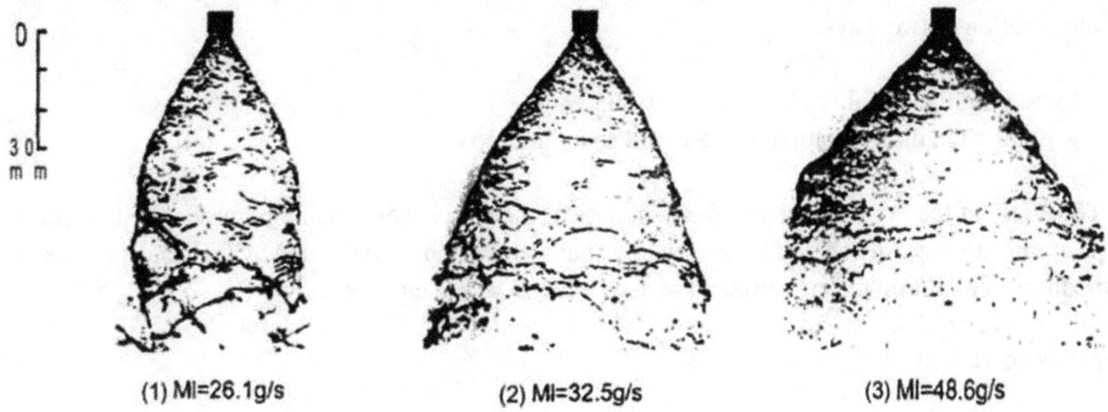

to changes in the number of Weber is also influenced by the non-linear way of joining for the behavior of linear and non-linear dilatations.

For conical films, increases in the deviations of the liquid film drive small drops to soon stand out of the fine ligaments and afterwards the film dissolves, in agreement with Rayleigh mechanisms of jet rupture (Bazarov, Yang, & Puri, 2004).

Couto et al. (Couto, Carvalho, & Netto, 1997) obtained an expression in which, starting with the expression of Dombrowski and Johns (Dombrowski & Johns, 1963) for plane liquid sheets in fan, they considered the conical format, the axial components and tangential of speed of a hollow conical spray produced by a swirl pressurized atomizer.

Thickness of the Liquid Sheet

In swirl type pressurized atomizers, the thickness of the liquid film in the end of the discharge hole is directly related to the area of the air nucleus.

Theory and practice indicate that, as higher the injection differential pressure, thinner is the produced liquid sheet and best is the quality of the atomization. This is usually attributed to an increase on the speed of discharge of the liquid, or also, partly, for the decrease on the thickness of the liquid film caused by the increase of the differential pressure.

An increase in the diameter of the final hole (nozzle) leads to a thicker film, as it reduces the discharge coefficient. The thickness of the film increases with the increase on the area of the tangential entrance holes of the liquid, because the flow increases in the nozzle, and it results in a thicker film. The effect of a reduction in the diameter of vortex camera is the increase of the thickness of the liquid film. That is attributed to the lowest action of the vortex that reduces the diameter of the nucleus of air inside the end of the discharge hole. The effects of hole length and of vortex camera length in the thickness of the film are quite small.

The thickness of the liquid film increases with increases in the dimensions of the nozzle, liquid flow rate and viscosity of the liquid, and decreases with the increase in the density of the liquid and in the injection pressure.

The surface tension represents an important paper in the subsequent collapse of the liquid sheet in ligaments and drops.

The viscosity of the liquid is very important in the atomization process, as the viscous forces oppose the atomization in two ways:

1. Increasing the initial thickness of the film;
2. Resisting the disintegration of the liquid sheet in drops.

The effect of the liquid density in the thickness of the film is very small and probably its influences in the atomization quality should be also small. That assumption is confirmed by the results of measures of medium size of drop for pressurized swirl-type atomizers (Lefebvre, 1989).

Experiment

A double and concentric swirl injector was manufactured in brass for cold tests with cut oil and kerosene oil, and for the hot tests with melted paraffin.

A milling cutter with digital marker was used, dividing plate and drilling to obtain the tangential holes of the swirl, besides a mechanical lathe with cut tools and drills.

To guarantee the close fitting of the vortex chambers, the injectors were projected to use nitrile rubber o-rings, derived from petroleum, ideal to use in work temperatures in the range of –50.0 to +120.0 °C, guaranteeing like this the gasket of the ensemble.

Figure 4 shows the internal injector and external injectors, and Figure 5 presents the adapter that unites the ensemble injector to the injector head.

The angles of the formed cones can be altered by several factors, like the format of the vortex chamber and injection pressure, among others.

Figure 4. Internal (longer) and external (shorter) injectors

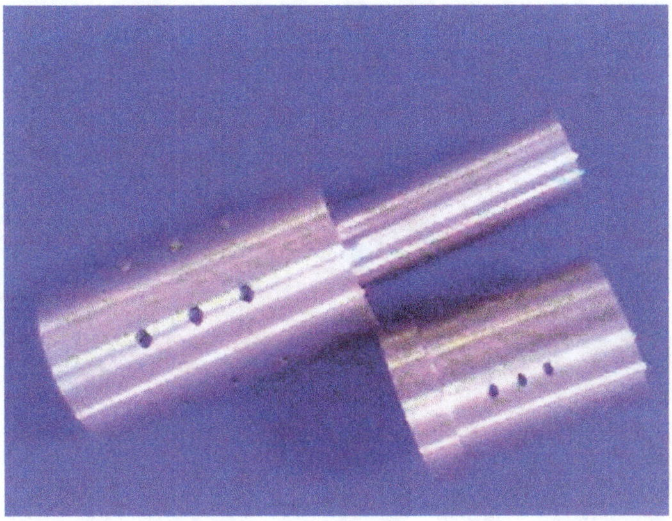

Figure 5. Adapter for uniting the ensemble injector to the injector head

Test Bench

The test bench, Figure 6, was used for the trials with cut oil for automatic lathes, commercial kerosene oil and melted paraffin.

Figures 7 and 8 show the external and internal injectors (not operating simultaneously), forming hollow conical liquid sheets of kerosene with pressure bellow 1 atm.

Figure 9 shows the external and internal injectors working simultaneously, forming hollow conical liquid sheets of kerosene with pressure bellow 1 atm.

Figure 6. Test bench

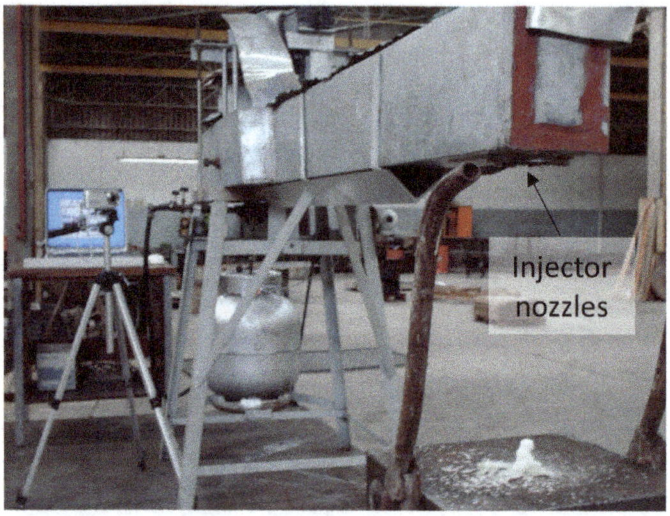

Figure 7. External injector conical liquid sheets

Figure 8. Internal injector conical liquid sheets

Figure 9. External and internal injectors simultaneously forming hollow conical liquid sheets

Figure 10 presents the external and internal injectors working simultaneously, showing an interaction between the flows, forming hollow conical liquid sheets of kerosene with pressure bellow 1 atm.

It can be observed that, very close to the exit of the nozzles, the sheets are still separated and are united soon afterwards.

The angle of the cone, formed by the external injector with melted paraffin injected at 5 atm, can be seen in Figure 11.

Figure 12 presents the angle of the cone formed by the internal injector, with melted paraffin injected at 5 atm.

CONCLUSION

The manufacturing process of of the injectors and injector head was considered efficient and satisfactory, as it allowed the production of all necessary pieces with good finishing, making possible all the tests, which were accomplished with good reliability pattern.

In several circumstances, when varying the injected fluids and working pressures, the cones were formed in the external and internal injectors according to the formation theories and disintegration of hollow conical liquid sheets, found in the literature.

Figure 10. External and internal injectors working simultaneously, with interaction of the sheets

Figure 11. Angle of the cone formed by the external injector with melted paraffin at 5 atm

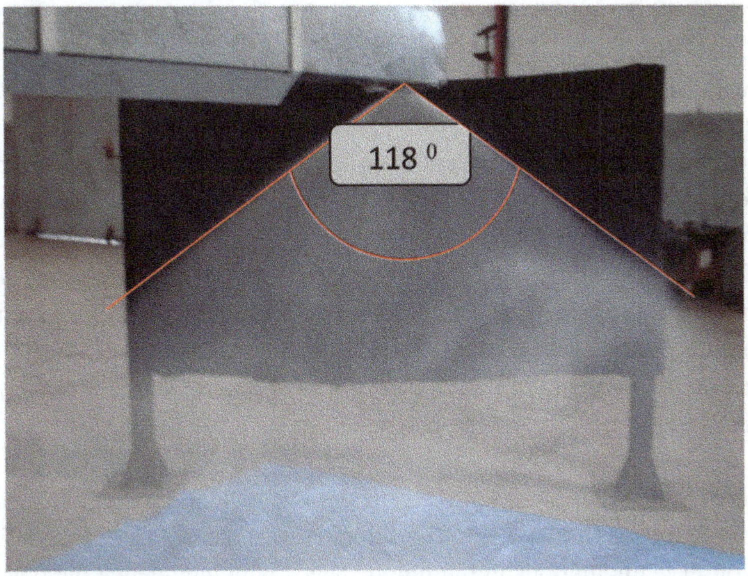

Figure 12. Angle of the cone formed by the internal injector with melted paraffin at 5 atm

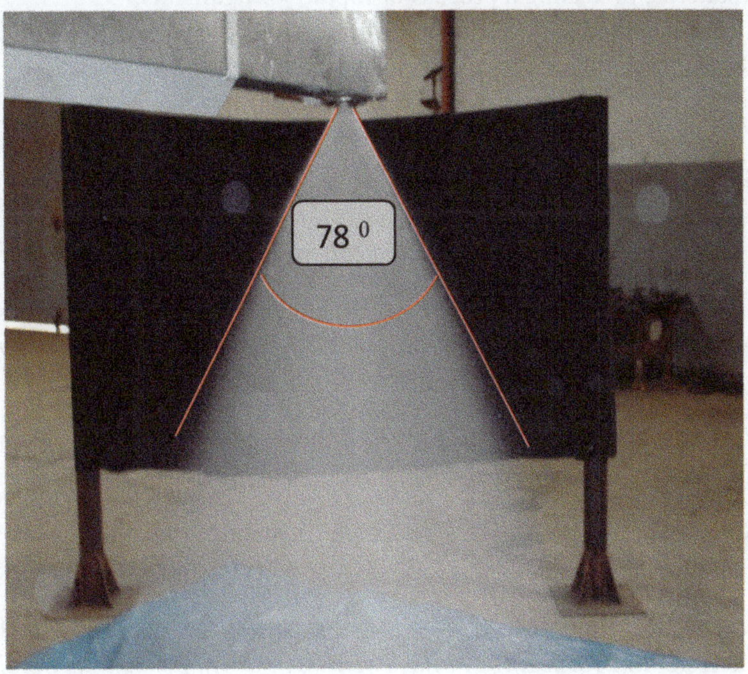

The initial observations of the injected melted paraffin (done in a profile projector) showed that, besides producing drops in the spherical format as would be expected, the same ones presented different diameters.

The rehearsals with wire cloth sieves made possible the calculations to determine the Sauter medium diameter of drops (SMD).

The formation of the liquid sheet is present even in low pressures and flows, both for the internal and external injectors.

Increasing the injection pressure does not have great effects in the angle of the spray, close to the exit of the nozzle, but it causes significant alteration in the curvature of the spray and in the length of the liquid sheet, which decreases as the pressure increases until not being visibly identified and probably in the size of the produced drops. Therefore, for high injection pressures, a fine fluid fog was formed.

Perforations are not observed in the liquid sheets in the area between the nozzle and strip of drop formation, not evidencing the existence of the perforated sheets scheme, proposed by Eisenklam (Eisenklam, 1964).

In the region where there is formation of the drops, it is not clear enough to affirm that there are no perforations and nor clear enough to see rings, but ligaments seem to exist.

It is possible to observe undulations in the liquid sheets, which evidence the development of dissociate structures of the same ones, that end up promoting their rupture, resulting in formation of drops.

In the area where the drops are already observed, those are arranged in a specific form, aligned in vertical or oblique patterns, indicating a process of rupture of ligaments.

The accomplished calculations revealed that the dimensions of the injector and mainly the angles of the produced cones are compatible with the literature and the project methodology.

REFERENCES

Bayvel, L., & Orzechowski, Z. (1993). *Liquid Atomization*. Philadelphia: Taylor & Francis.

Bazarov, V., Yang, V., & Puri, P. (2004). Design and Dynamics of Jet and Swirl Injectors. In V. Yang, M. Habiballah, J. Hulka, & M. Popp (Eds.), *Liquid Rocket Thrust Chambers: Aspects of modeling, analysis and design* (pp. 19–103). Massachusetts: Progress in Astronautics and Aeronautics.

Couto, H. S., Carvalho, J. A., & Netto, D. B. (1997). Theoretical formulation for sauter mean diameter of pressure-swirl atomizers. *Journal of Propulsion and Power*, *13*(5), 691–696. doi:10.2514/2.5221

Dombrowski, N., & Johns, W. R. (1963). The aerodynamic instability and disintegration of viscous liquid sheets. *Chemical Engineering Science*, *18*(3), 203–214. doi:10.1016/0009-2509(63)85005-8

Dumouchel, C., Ledoux, M., Bloor, M. I., Dombrowski, N., & Ingham, D. B. (1990). The Design of Pressure Swirl Atomizers. In *Proceedings of the Twenty-Third Symposium on Combustion* (pp. 1461-1467). The Combustion Institute.

Eisenklam, P. (1964). On ligament formation from spinning discs and cups. *Chemical Engineering Science*, *19*(9), 693–694. doi:10.1016/0009-2509(64)85056-9

Fraser, R. P., Dombrowski, N., & Routley, J. H. (1963). The atomization of a liquid sheet by an impinging air stream. *Chemical Engineering Science*, *18*(6), 339–353. doi:10.1016/0009-2509(63)80027-5

Fraser, R. P., Dombrowski, N., & Routley, J. H. (1963). The filming of liquids by spinning cups. *Chemical Engineering Science*, *18*(6), 323–337. doi:10.1016/0009-2509(63)80026-3

Lefebvre, A. H. (1989). *Atomization and Sprays*. West Lafayette: Purdue University.

York, J. L., Stubbs, H. E., & Tek, M. R. (1953). The mechanism of disintegration of liquid sheets. Trans. ASME, 1953, 1279-1286.

Chapter 7
Solid Rocket Motor Internal Ballistics Simulation Considering Complex 3D Propellant Grain Geometries

Guilherme Lourenço Mejia
Aeronautics Institute of Technology, Brazil

ABSTRACT

Solid rocket motors (SRM) are extensively employed in satellite launchers, missiles and gas generators. Design considers propulsive parameters with dimensional, manufacture, thermal and structural constraints. Solid propellant geometry and computation of its burning rate are essential for the calculation of pressure and thrust vs time curves. The propellant grain geometry changes during SRM burning are also important for structural integrity and analysis. A computational tool for tracking the propagation of tridimensional interfaces and shapes is then necessary. In this sense, the objective of this work is to present the developed computational tool (named RSIM) to simulate the burning surface regression during the combustion process of a solid propellant. The SRM internal ballistics simulation is based on 3D propagation, using the level set method approach. Geometrical and thermodynamic data are used as input for the computation, while simulation results of geometry and chamber pressure versus time are presented in test cases.

INTRODUCTION

Rockets based on chemical energy (with solid or liquid propellants) are currently the only means to access Space and research and development of this area has vital importance for Aerospace Engineering. The Brazilian Aerospace Program has a long history using Solid rocket motor (SRM) technology since the sounding rocket Sonda I launched in 1965 (AEB, 2017). In this context, the reduction of project costs with modern computational simulations is an important interest of the involved companies and institutes.

DOI: 10.4018/978-1-5225-2903-3.ch007

Solid rocket motors are employed in satellite launchers, missiles and gas generators. The design takes on account propulsive parameters such as thrust versus time, with dimensional, manufacture, thermal, structural constraints. Experiments using real scale rocket models are costly, therefore computational simulations are required to decrease the overall project cost.

The combustion mechanisms for solid rocket propellants are quite complex and dependent on many local fluid, chemical, and thermal phenomena. Many solid rocket propellant burning-rate models are greatly simplified because of limited computational power and understanding of the combustion process (Wiedemanna, et al., 2009). In the exhaust gas of typical SRM engines, a considerable amount of small aluminum oxide particles is included. Aluminum is used as an additive in the motor propellant in order to increase performance and to dampen burn instabilities. During the burn process, most of this aluminum is transformed into aluminum oxide. A large number of micron-sized dust particles are generated continuously during a burn. At the end of a burn, a second group of much larger fragments from a slag pool clustering inside the motor leaves the nozzle.

There are several quasi-steady formulations to predict the burning rate of an energetic solid material. One of them is the Vieille's or Saint Robert's law (Marmureanu, 2014), which is an empirical model that represents the pressure dependence on burn rate, as will be shown further on.

Solid propellant geometry and modeling of its burning are essential for calculation of pressure versus time and thrust versus time curves (NASA, 1971). In addition, propellant structural integrity is mandatory during the rocket burn. Structural analysis use propellant geometry as input and should consider the burning surface regression, i.e., the propellant grain geometry changes during rocket burn. A computational tool for tracking the propagation of tridimensional interfaces and shapes is necessary for this task. This developed tool for SRM burning simulation (named RSIM) handles complex grain geometry for versatility. Figure 1 presents the design flowchart.

Solid Rocket Motor History in Brazil

The Brazilian Aerospace Program has a long history using SRM technology since the sounding rocket Sonda I (1965). In this context, the reduction of project costs with modern computational simulations is an important interest of the involved companies and institutes.

Figure 1. Program design flowchart

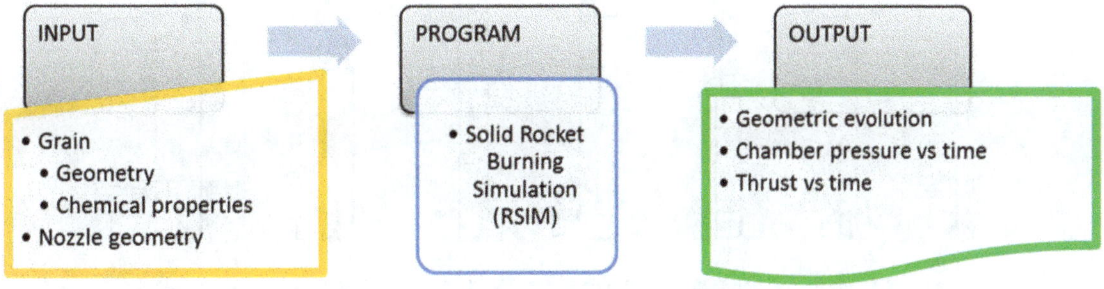

Figure 2. VLS launch trajectory (IAE, 2012)

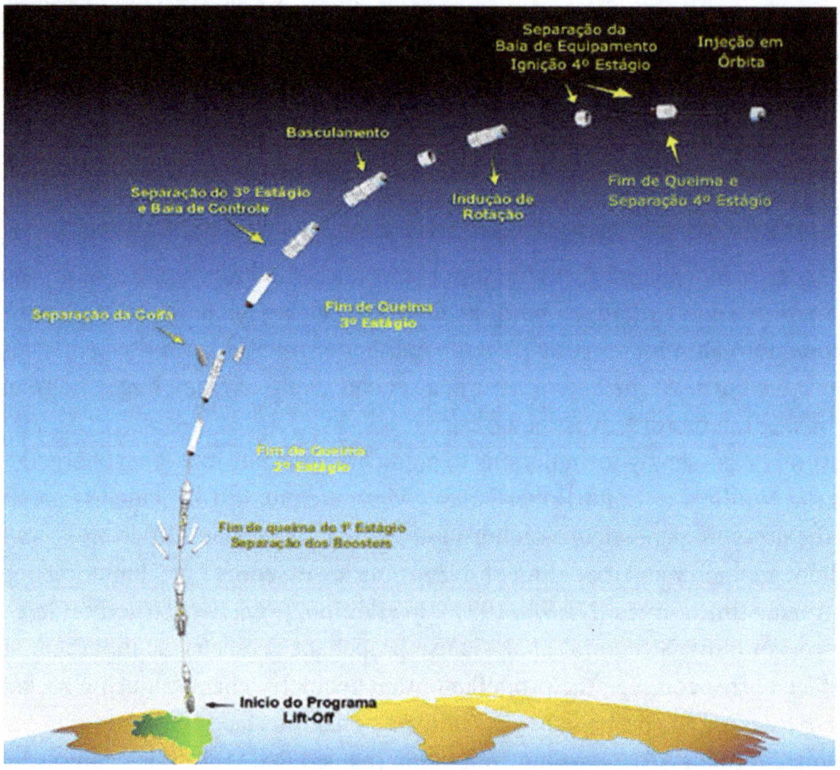

Figure 3. Brazilian sounding rockets, (AEB, 2017)

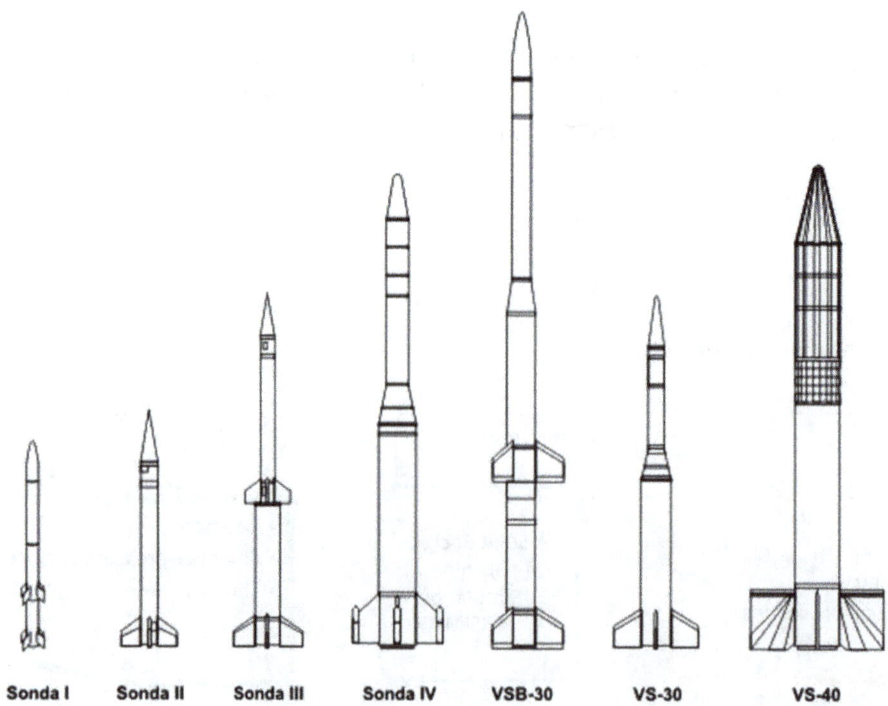

Technology Application

Solid rocket propellants are employed in complex systems, such as:

- Satellite launcher motors
- Missile motors
- Gas generators

The design takes into account: propulsive parameters (e.g. thrust versus time) dimensional, manufacture, thermal and structural constraints. Experiments using real scale rocket models are expensive, therefore computational simulations are required to decrease the overall project cost.

Structural analysis (e.g. FEA) uses propellant geometry as input. FEA should consider the burning surface regression, i.e., the propellant grain geometry changes during rocket burn. A computational tool for tracking the propagation of tridimensional interfaces and shapes is necessary for this task. The present developed tool RSIM handles complex grain geometry for versatility.

Computational power has exponentially increased over time, as illustrated in Figure 4. - this tendency is baptized Moore's law.

Figure 4. Computational capacity over time (computations/second per computer) (Koomey, Berard, Sanches, & Wong, 2011)

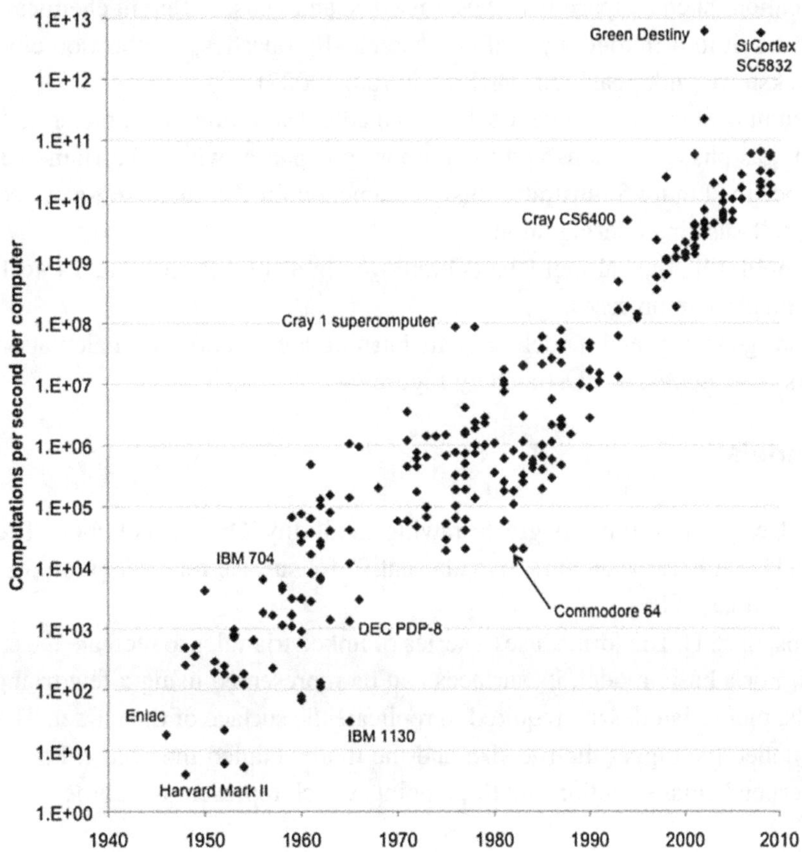

It is only natural to benefit from these technological advances. While hardware has progressed, so has software. The brain and mind of computaters.

The C programming language was devised in the early 1970s as a system implementation language for the nascent Unix operating system. Derived from the typeless language BCPL, it evolved a type structure; created on a tiny machine as a tool to improve a meager programming environment, it has become one of the dominant languages of today (Ritchie, 1993).

C++ was designed to provide Simula's facilities for program organization together with C's efficiency and flexibility for systems programming. It was intended to deliver that to real projects within half a year of the idea. It succeeded. While a modest amount of innovation did emerge over the years, efficiency and flexibility have been maintained without compromise. While the goals for C++ have been refined, elaborated, and made more explicit over the years, C++ as used today directly reflects its original aims (Stroustrup, 1994).

THEORETICAL STUDY

Thermodynamics Considerations

A solid propellant consists of several chemical ingredients such as oxidizer, fuel, binder, plasticizer, curing agent, stabilizer, and cross-linking agent. The specific chemical composition depends on the desired combustion characteristics for a particular mission. Solid propellants are often tailored and classified by specific applications such as space launches, missiles, and guns. Different chemical ingredients and their proportions result in different physical and chemical properties, combustion characteristics, and performance (Beckstead, Puduppakkam, Thakre, & Yang, 2007).

The phenomenon of a solid propellant self-sustained deflagration is characterized by decomposition, evaporation, gas-phase reactions and then a pure gas phase, where the flame surface is present and combustion occurs. Figure 5 illustrates this phenomenon for a combustion-wave structure of RDX monopropellant; self-sustained deflagration.

To better exemplify the typical ingredients employed in a solid propellant, Table 1 shows a list of components and their common cases.

Solid propellant geometry and modeling of its burning are essential for calculation of pressure vs time and thrust vs time curves, as illustrated by Figure 6.

Employed Models

Computer-Aided Design (CAD) offers great drawing flexibility. One of its famous file formats is stereolithography (STL). It became an attractive alternative for surface representation for rapid prototype industry (Rypl & Bittnar, 2006).

In simple terms, the STL file format uses a series of linked triangles to recreate the surface geometry of a solid model. For a basic model, its surfaces can be represented using a few triangles. For higher resolution models, more triangles are required to replicate the surface of the model. The more triangles that make up a model, the bigger the file size and the more detailed the object. Each triangle facet is described by a perpendicular direction and three points which represent the corners of the triangle. An

Figure 5. Schematic illustration of combustion-wave structure of RDX monopropellant; self-sustained deflagration (Liau & Yang, 1995)

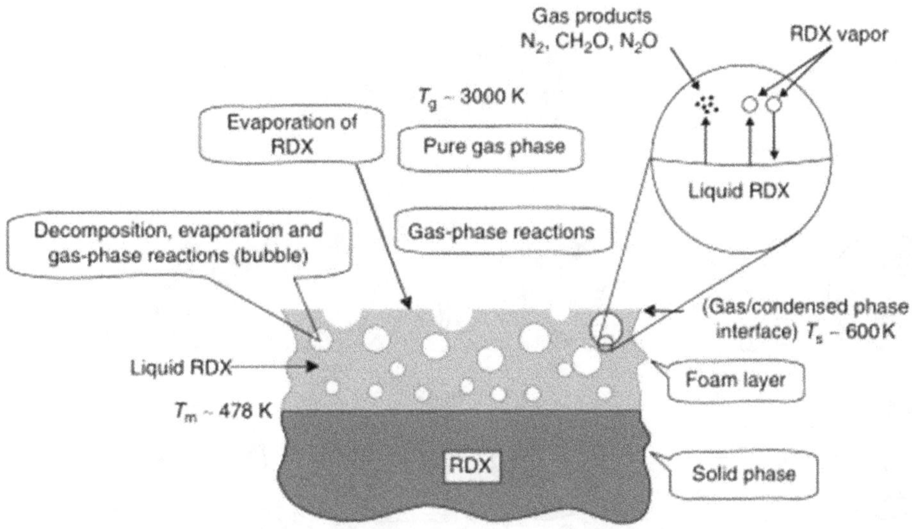

Table 1. Typical solid propellant chemical ingredients (Chaturvedi & Dave, 2015).

Component	Examples
Oxidizer	AP, AN, KN
Metal fuel	Aluminum
Binder	HTPB, CTPB, NC
Modifier or Catalyst	Iron oxide, lithium fluoride
Plasticizer	DOA, NG, GAP, DEP
Curing agent	HMDI, TDI, and IPDI
Additive	Opacifier, Bonding agents, Desensitizing agents

STL file provides a complete listing of the x, y and z coordinates of these corners and perpendiculars (All3DP, 2016).

The STL file format can define complex shapes (mathematically speaking, polyhedrons with any polygonal facets). In practice, it is mostly used to describe the layout of triangles in a virtual space. For example, Figure 7 shows different levels of a spherical surface representation.

METHODS

Level Set Method and Discretization

The level set method is a numerical technique for tracking the evolution of interfaces and shapes (Sethian, 1994). It allows performing computations involving curves and surfaces on a fixed grid without the need

Figure 6. Different SRM cross sections and thrust curves (Braeunig, 2014)

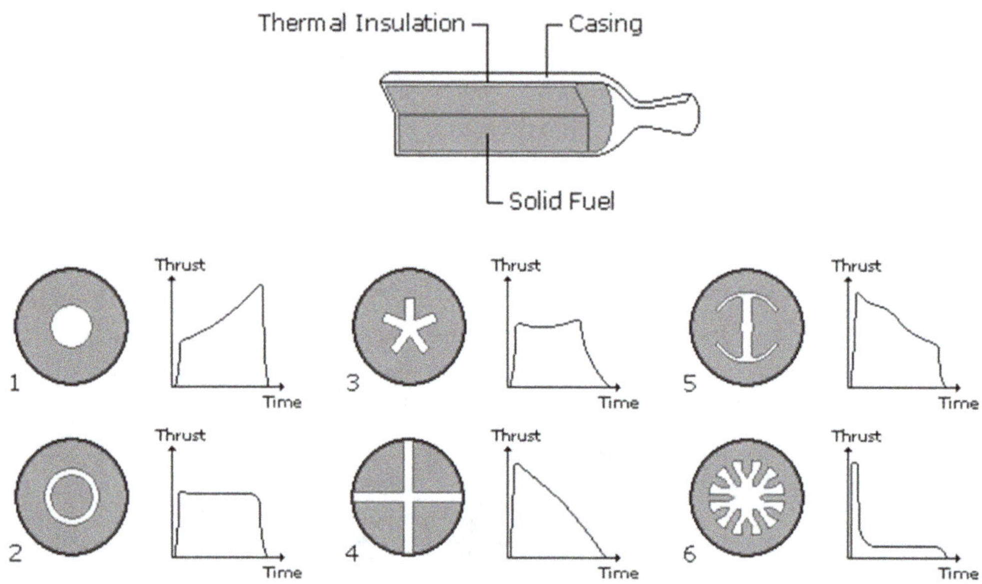

Figure 7. Spherical surface representation by STL (All3DP, 2016)

of parameterizing these objects, consisting on an Eulerian approach (Sethian, Curvature and the Evolution of Fronts, 1985), (Chorin, 1985), (Milos & Rasky, 1994). This method is employed in the geometry module (GEOM) of RSIM program, providing a time step iteration process over a discretized 3D space.

The problem of evolving interfaces in a 2D space is considered to illustrate the level set method. The perpendicular propagation of a boundary is determined by the time evolution of a function $\Phi(x,y)$. The propagation is given by a velocity V. The overall scheme is shown in Figure 8.

In the discretization, the scalar value Φ separates the inner and outer regions. Where $\Phi < 0$, the point is inside; where $\Phi = 0$, the point is on the border; and where $\Phi > 0$, the point is outside.

The LS equation, which describes the time evolving interfaces in a certain space, is the following:

Figure 8. Schematics of a propagating curve (Cavallini, Favini, Di Giacinto, & Serraglia, 2009)

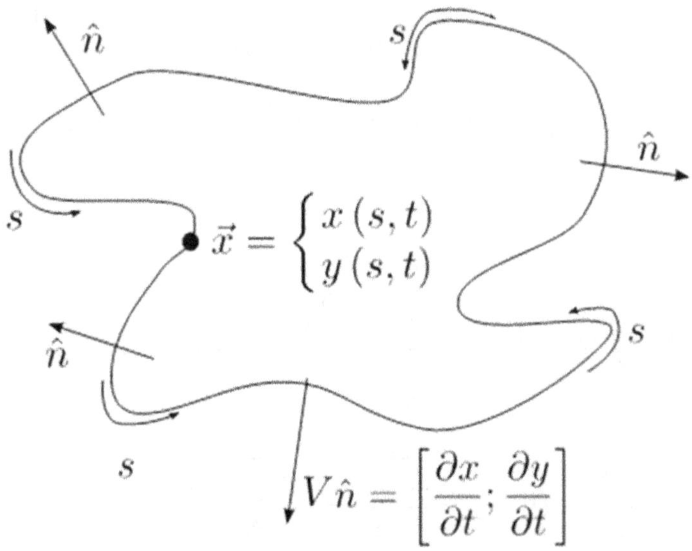

$$\frac{\partial \phi}{\partial t} + V\left|\vec{\nabla}\phi\right| = 0 \tag{1}$$

Since it's not always possible to express Φ as a function, a numerical solution for the previous equation is necessary. This level set equation can be expressed in the class of Hamilton-Jacobi equations (Osher & Sethian, 1988):

$$\phi_t + H\left(\phi_x, \phi_y, \phi_z\right) = 0 \tag{2}$$

Taking advantage of the cylindrical symmetry of most propellants, the LS equation can be expressed as follows:

$$\phi_t + \dot{r_b}\sqrt{\left(\phi_r\right)^2 + \left(\phi_\theta / r\right)^2 + \left(\phi_z\right)^2} = 0 \tag{3}$$

The space and time first order approximation gives the following numerical equation (Cavallini, Favini, Di Giacinto, & Serraglia, 2009), (Attili, 2008):

$$\phi_{ijk}^{n+1} = \phi_{ijk}^{n} - \Delta t^{n} \left(\dot{r}_b \Big|_{ijk}^{n} \left[\begin{array}{l} \max\left(\max\left(1/r_j D^{-\theta} \phi_{ijk}^{n}, 0 \right)^2, \min\left(1/r_j D^{+\theta} \phi_{ijk}^{n}, 0 \right)^2 \right) \\ + \max\left(\max\left(D^{-r} \phi_{ijk}^{n}, 0 \right)^2, \max\left(D^{+r} \phi_{ijk}^{n}, 0 \right)^2 \right) \\ + \max\left(\max\left(D^{-z} \phi_{ijk}^{n}, 0 \right)^2, \max\left(D^{+z} \phi_{ijk}^{n}, 0 \right)^2 \right) \end{array} \right]^{1/2} \right) \tag{4}$$

where:

$$\begin{array}{l} D^{-\theta}\phi_{ijk}^{n} = \dfrac{\Delta_{\theta}^{-}\phi_{ijk}^{n}}{\Delta\theta} = \dfrac{\phi_{ijk}^{n} - \phi_{i-1jk}^{n}}{\Delta\theta}; D^{+\theta}\phi_{ijk}^{n} = \dfrac{\Delta_{\theta}^{+}\phi_{ijk}^{n}}{\Delta\theta} = \dfrac{\phi_{i+1jk}^{n} - \phi_{ijk}^{n}}{\Delta\theta} \\ D^{-r}\phi_{ijk}^{n} = \dfrac{\Delta_{r}^{-}\phi_{ijk}^{n}}{\Delta r} = \dfrac{\phi_{ijk}^{n} - \phi_{ij-1k}^{n}}{\Delta r}; D^{+\theta}\phi_{ijk}^{n} = \dfrac{\Delta_{r}^{+}\phi_{ijk}^{n}}{\Delta r} = \dfrac{\phi_{ij+1k}^{n} - \phi_{ijk}^{n}}{\Delta r} \\ D^{-z}\phi_{ijk}^{n} = \dfrac{\Delta_{z}^{-}\phi_{ijk}^{n}}{\Delta z} = \dfrac{\phi_{ijk}^{n} - \phi_{ijk-1}^{n}}{\Delta z}; D^{+z}\phi_{ijk}^{n} = \dfrac{\Delta_{z}^{+}\phi_{ijk}^{n}}{\Delta z} = \dfrac{\phi_{ijk+1}^{n} - \phi_{ijk}^{n}}{\Delta z} \end{array} \tag{5}$$

Solid Propellants Burning Law

The solid propellant burn rate model employed in the burn module (BURN) of RSIM program is St. Robert's law, as follows (Beckstead, Overview of combustion mechanisms and flames structures for advanced solid propellant, 2000):

$$\dot{r}_b = aP_c^n \tag{6}$$

Considering stationary regime, the chamber pressure P_c is calculated by the following equation (Chorin, 1985):

$$P_c = \left[a \frac{S_b}{A_{th}} \rho_p c^* \right]^{\frac{1}{1-n}} \tag{7}$$

Thermochemical parameters (a, n), propellant density (ρ_p) and nozzle throat area (A_{th}) are inputs for RSIM.

Thrust Equation

The thrust equation is obtained considering an isentropic flow, conical nozzle, and a nozzle efficiency (Chankapoe, Winya, & Pittayaprasertkul, 2013): Figure 9 and conical nozzle schematics used in computation.

The following equations are related to conical nozzle model, including the efficiency parameter.

$$F = C_{F,act} A_{th} P_c \tag{8}$$

$$C_F^0 = \sqrt{\frac{2\gamma^2}{y-1}\left(\frac{2}{\gamma+1}\right)^{\frac{\gamma+1}{\gamma-1}}\left[1-\left(\frac{P_e}{P_c}\right)^{\frac{\gamma-1}{\gamma}}\right]} + \left(\frac{P_c - P_\infty}{P_c}\right)\varepsilon \tag{9}$$

$$\varepsilon = \frac{A_e}{A_{th}} \tag{10}$$

$$\lambda = \frac{1}{2}\left(1 + \cos\alpha\right) \tag{11}$$

Figure 9. Conical nozzle schematics

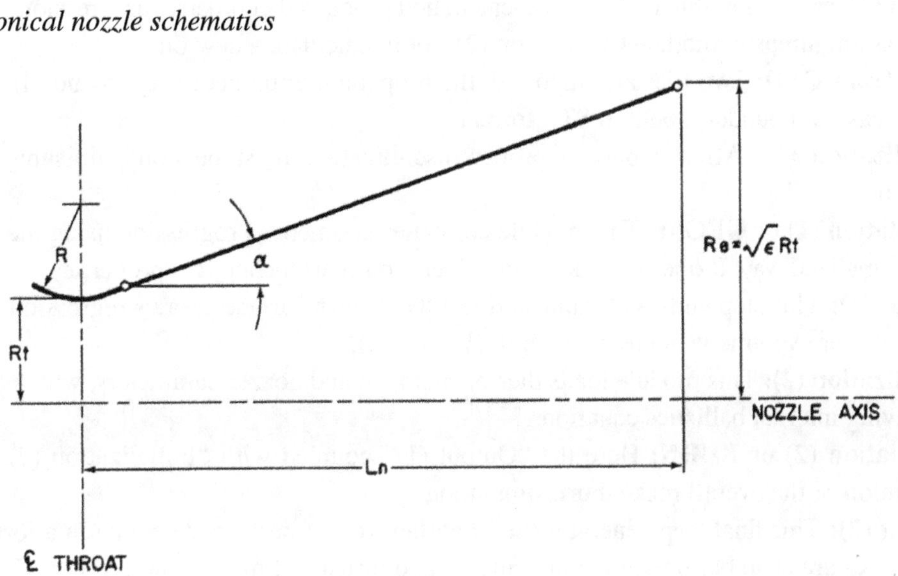

$$\eta_F = \frac{C_{F,act}}{\lambda C_F^0} \tag{12}$$

Solid Propellants Erosion Burning

Solid propellant erosion burning can be overlooked in small SRM. However, in large motors, this phenomenon, if not modeled, may result in loss of accuracy.

Nozzle Erosion

Nozzle erosion mainly manifests in large SRM, where the burn time is in the order of minutes. Aluminium, in the propellant formulation, is also an erosive agent, as this metal is expelled at high velocities, throat material is depleted.

SOFTWARE IMPLEMENTATION

Flowchart

The flowchart presented in Figure 10 details the overall steps of RSIM program since input choices until the output archives.

Description

Explaining each main module and its tasks (Figure 10):

- **Choice:** Here it is possible to load a pre-calculated processed geometric progression – in this case the program jumps to module Calculation (2) - or to calculate a new one.
- **Input from CAD:** Two files are imported, the propellant grain geometry (mandrel) and the propellant case or boundary, both in STL format.
- **Initialization (1):** After import, a proper discretization must be done, finishing with mesh generation.
- **Calculation (1) or GEOM:** This module computes geometric progression using the LS method, in a normalized way. It does not take in consideration thermochemical properties.
- **Output (1):** This step ends with tabulated results of burn surface area vs regression depth (S_b *vs web*) and bore volume vs regression depth (V_b *vs web*).
- **Initialization (2):** This module loads thermochemical and nozzle parameters, which are essential for solving internal ballistics equations.
- **Calculation (2) or BURN:** Here the "Output (1)" coupled with "Initialization (2)" allows the calculation of the overall rocket burn simulation.
- **Output (2):** This final step generates the important rocket performance data in a form of tables. The curves are chamber pressure, burn surface and thrust vs time – P_c *vs t*, S_b, *vs t*, *F vs t*

Figure 10. RSIM flowchart, including each main module

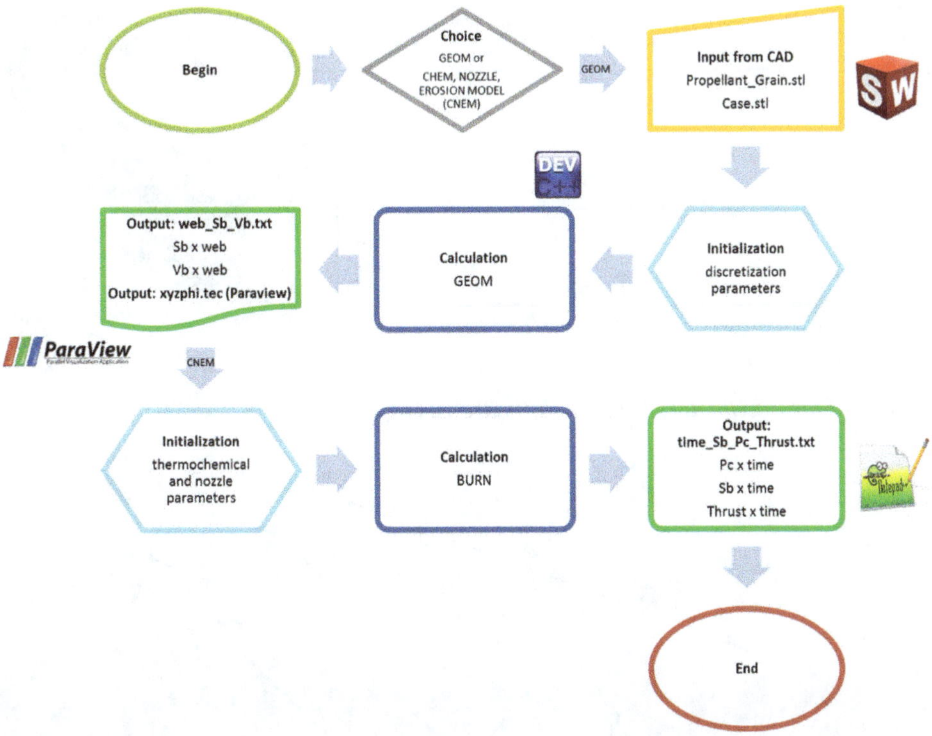

Mesh Refinement Impacts

One important challenge is to balance simulation accuracy and calculation time. Crude mesh is used when quick solutions are desired and medium errors are tolerated; while mesh refinements are employed when small errors are sought, this – of course – demands more computations and takes more time to finish the whole simulation.

RESULTS

Cases

RSIM program, written in C++ code, was employed for analyzing different cases of SRM: star; finocyl; multiperforated propellant grains; and also, NAWC SRM n. 6.

Case 1: Star Propellant Grain

The first case analyzed was a star shaped SRM. Figure 11 shows the propellant grain.

The propellant grain star has 9 points, dimensional parameters are presented in Figure 12. Note that the cylindrical and mirror symmetry are exploited in grid construction and level set computation. The geometry was created using a CAD tool and a proper curve discretization was done for RSIM input.

Figure 11. Case 1 - Propellant grain (Dimensions are in mm)

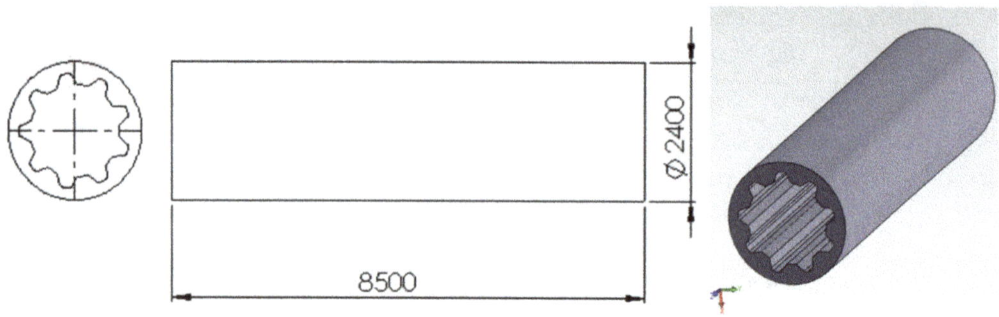

Figure 12. Case 1 - Star parameters (Dimensions are in cm)

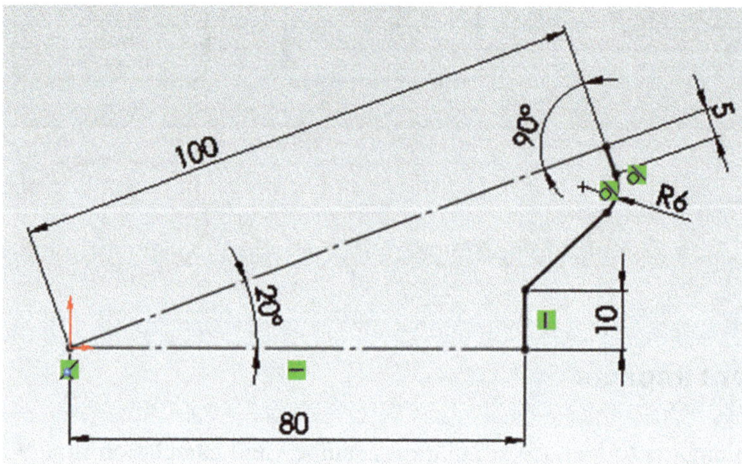

Results of the simulation consist in geometry and chamber pressure changes over time. The geometry progress is represented in Figure 13, with time step of 2 s between blue curves, while the red arrows represent the burn direction. Observe that the geometry changes are bounded by the outer diameter (2.4 m) of the propellant grain.

The chamber pressure versus time plot is presented in Figure 14. Note that the burn is progressive until 35 s and after that it becomes regressive as sliver appears. This result from RSIM is then compared with previous program, written in Fortran code, specifically designed for star or finocyl propellant grain geometry. This comparison is also shown in Figure 14. Observe that the two curves are very close during stationary burn.

Case 2: Finocyl Propellant Grain

Past works in the literature covered the finocyl SRM simulation (Willcox, Brewster, Tang, Steward, & Kuznetsov, 2007), (Qin, Guoqing, Peijin, & Jiang, 2006), (Yildirim & Aksel, 2005). In this trend, the second case considered is the S43 motor, employed in the Brazilian Satellite Launch Vehicle: VLS. The

Figure 13. Case 1 - Section view of the propellant burn propagation. Time step between curves is 2 s and dimensions are in m

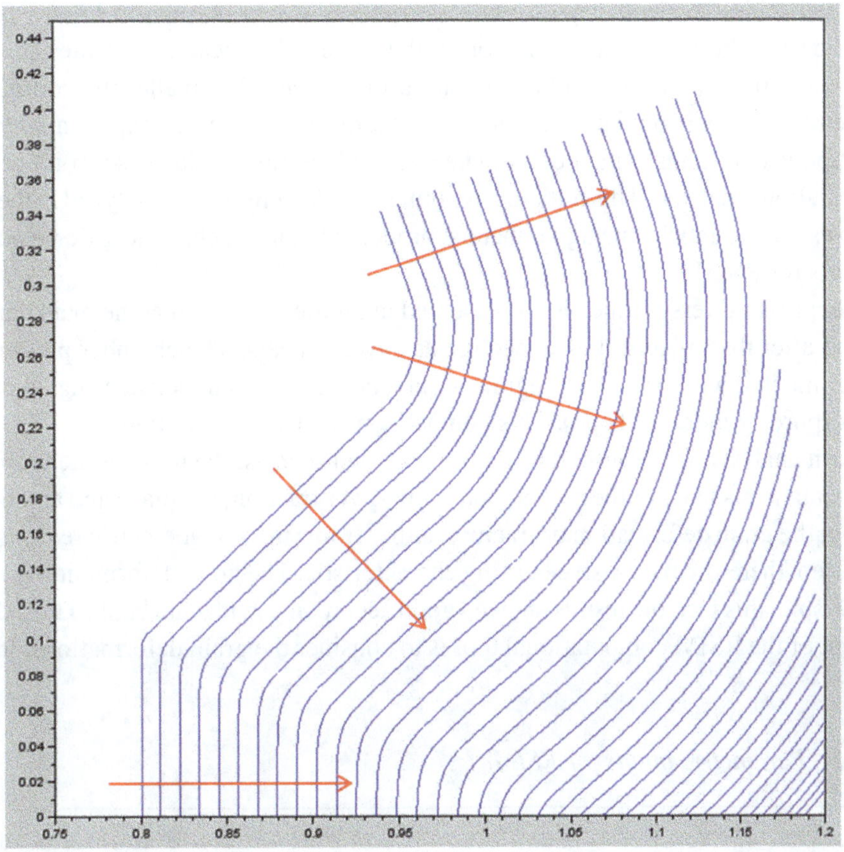

Figure 14. Case 1 - Chamber pressure (MPa) versus time (s) comparison

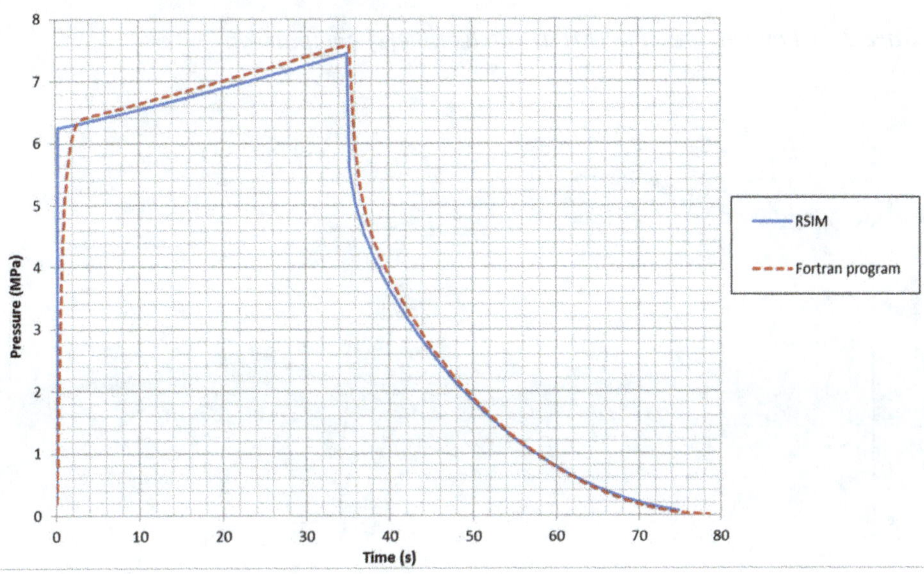

propellant grain is shown in Figure 16, a finocyl geometry with an outer boundary consisted of front dome, cylindrical region and aft dome.

The calculated grid after geometry discretization in RSIM can be seen in Figure 17. This grid is generated considering the proper discretization in the three cylindrical coordinates (θ, r, z). There is an important trade-off for precision and computation time spent: the smaller the coordinate step (Δθ, Δr, Δz), the better the precision, however the computation time increases exponentially. So, the most favorable solution is to establish small coordinate steps, where further decrease won't bring noticeable increase in simulation precision. This trade-off is applied for the three cases analyzed in the present work.

Table 2 shows the thermochemical parameters, propellant density, characteristic velocity and nozzle throat area inputs for RSIM.

The chamber pressure versus time plot is presented in Figure 15. Note that the main burn profile ends around 60 s and after that it has a drastic decline as sliver appears. The chamber pressure result from RSIM is then compared with previous program, data bank, and a result considering an outer boundary without domes (pure cylindrical shape). This comparison can be seen in Figure 15.

From the burn analysis, very useful data can be obtained such as: the mass of the propellant remaining and the instantaneous mass of the rocket motor, the sliver fraction; the place and amount of expected thermal loading; the grain deformation or fracture, either from stress or geometric reasons; the port area for every burn step, therefore the erosive burning characteristic. The grain deformation is not considered in the RSIM. Despite this fact, the results showed good accordance with data bank. Thus, the next step of the development of the RSIM computational tool is to consider the grain deformations. In order to treat

Table 2. Case 2 - Parametric inputs for RSIM.

Case 2	*a (m/s)*	*n*	$\rho_p(kg/m^3)$	$c^*(m/s)$	$A_{th}\ (m^2)$
RSIM	0.0018	0.275	1713	1550	0.030

Figure 15. Case 2 - Chamber pressure (MPa) versus time(s) comparison

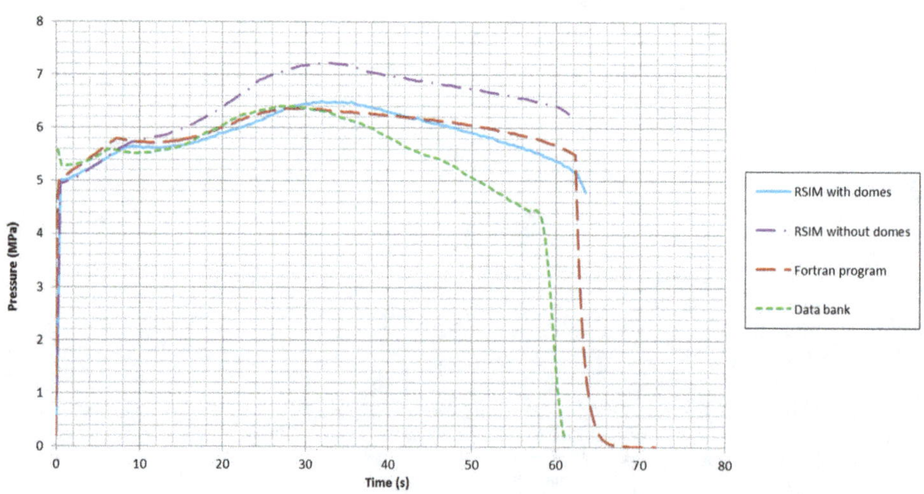

deformations of the computational domain, most codes make use of an Arbitrary Lagrangian-Eulerian (ALE) method (HIRT et al., 1997), but difficulties arise when the mesh undergoes dramatic deformations during the regression of star-shaped or finocyl grains (Gueyffier, Roux, Fabignon, & Chaineray, 2014). The most problematic deformations occur in region where the grain surface is initially convex toward the gas. In such regions, a geometric singularity (caustic) forms in finite time of the grain surface.

Observe that the "RSIM with domes" curve is close to "data bank" curve until 30 s and then exceeds the later until the end. This is mainly due to propellant and throat erosions, which were not modeled in RSIM and causes SRM performance losses. The erosion of rocket-nozzle materials during motor firing is one of the major limits in the advancement of solid-rocket propulsion. Prediction of carbon-carbon nozzle erosion behavior in solid rocket motors is based on numerical models, considering the solution of Reynolds averaged Navier-Stokes equations in the nozzle, heterogeneous chemical reactions at the nozzle surface, variable multi-component transport and thermodynamic properties, and heat conduction in the nozzle material (Bianchi, Nasuti, Onofri, & Martelli, 2009). Also, note that the "RSIM without domes" curve exceeds "RSIM with domes" curve during the entire burn time, because the first essentially considers more propellant to be burned than the later.

Table 3 shows the chamber pressure versus time integral comparison between RSIM and data bank. Naturally, this integral confirms the simulation overestimation of chamber pressure.

Case 3: Multiperforated Propellant Grain

Introducing a new type of simulation capability in addition of those present in references (Hertfield, Jenkins, Burkhalter, & Foster, 2004), (Dauch & Ribereau, 2002), the third case in this work is a multiperforated propellant grain, consisted of 36 holes in a cylinder. The calculated grid after geometry discretization in RSIM is presented in Figure 21.

The result of the burning surface area (S_b) vs web thickness (burn distance) is shown in Figure 22. Observe that S_b increases abruptly until web of 0.024 m, then decreases sharply and finally achieves a modest rise until web of 0.17 m. This behavior indicates that the SRM burn is initially progressive, then regressive and finally quasi-neutral. This can also be observed in the chamber pressure versus time plot, Figure 23.

Table 3. Case 2 - Pressure vs time integral comparison between RSIM and Data bank.

	Pressure vs Time Integral (MPa.s)
RSIM	368,1
Data bank	334,3
Difference	10,1%

Figure 16. Case 2 - Processed geometry input of VLS S-43 motor for RSIM program

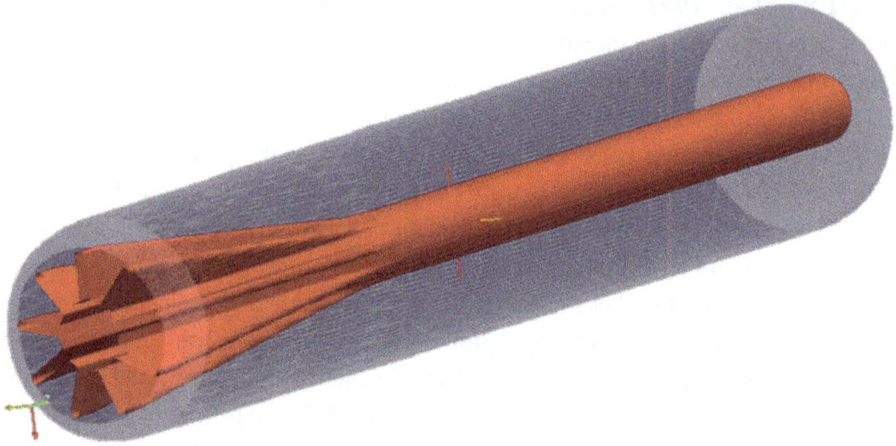

Figure 17. Case 2- Zoom of geometry discretization for VLS S-43 motor in RSIM

Figure 18. Case 2- Geometry progression results (perspective)

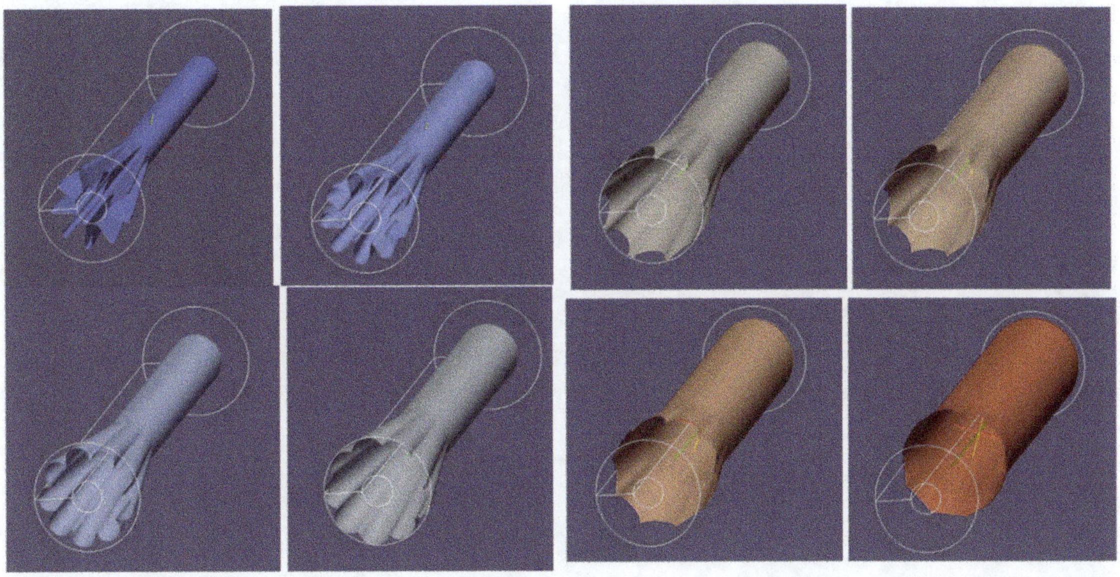

Figure 19. Case 2- Geometry progression results (sideview)

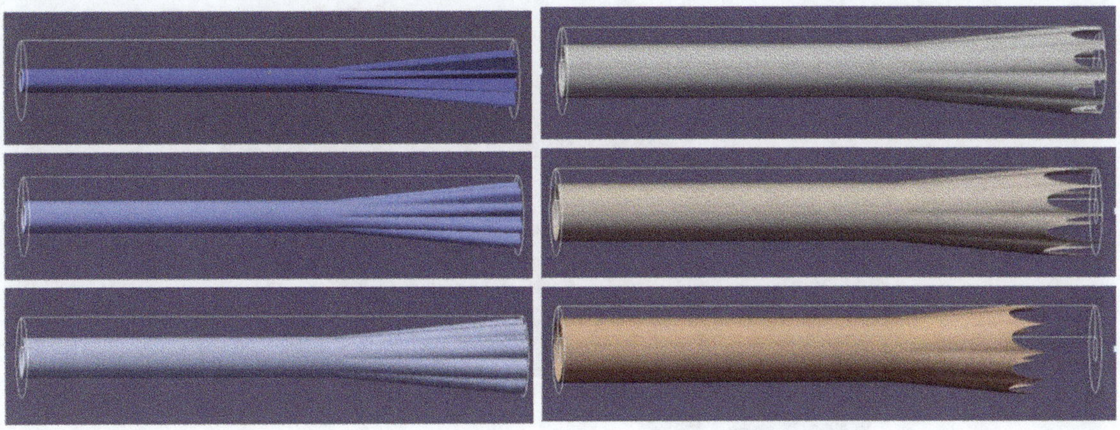

Figure 20. Case 3- Geometry for a multiperforated motor in RSIM.

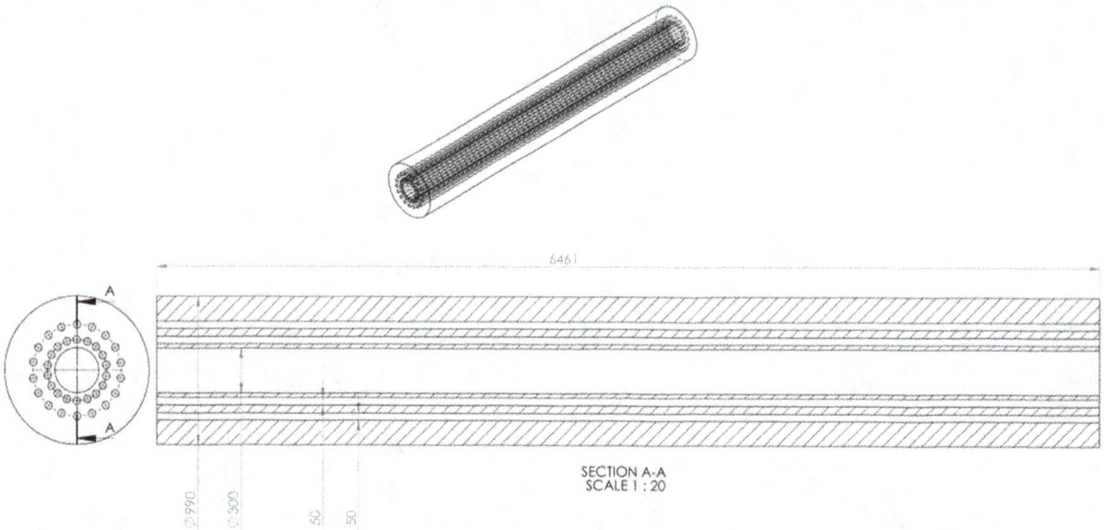

Figure 21. Case 3- Geometry discretization for a multiperforated motor in RSIM.

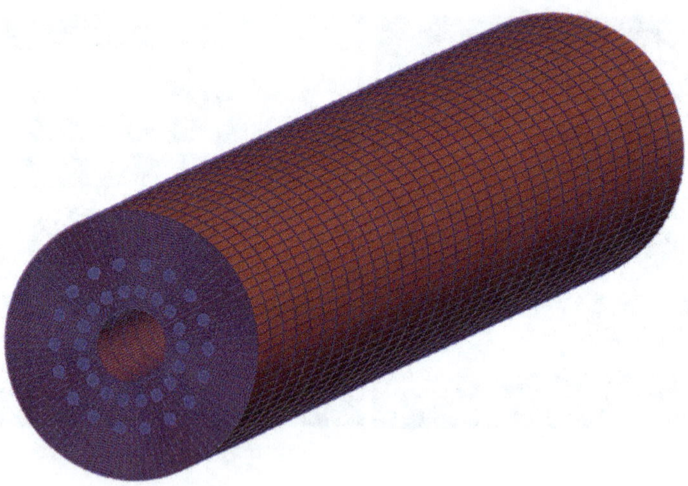

Figure 22. Case 3 - Burning surface area S_b (m^2) vs web thickness (m)

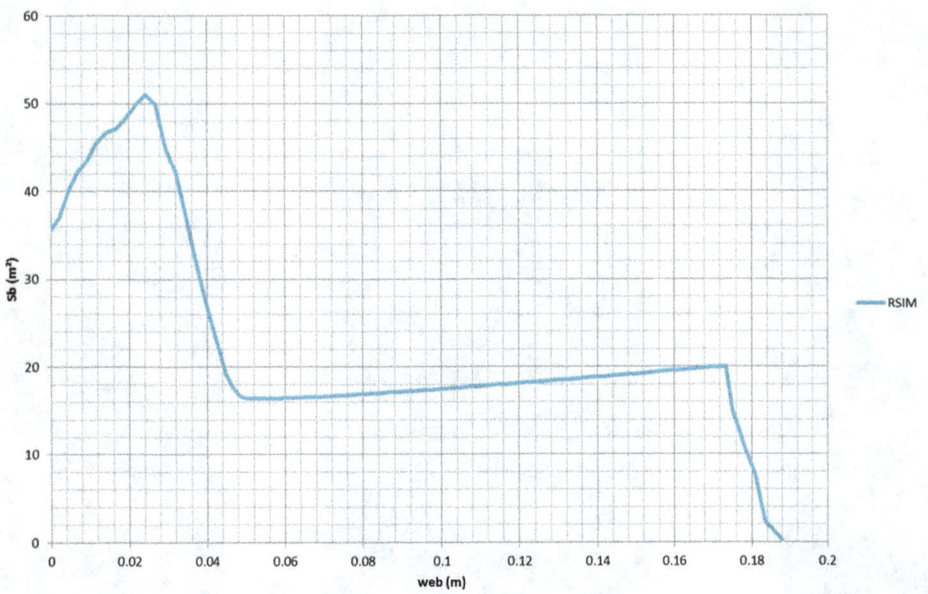

Figure 23. Case 3 - Chamber pressure (MPa) versus time (s)

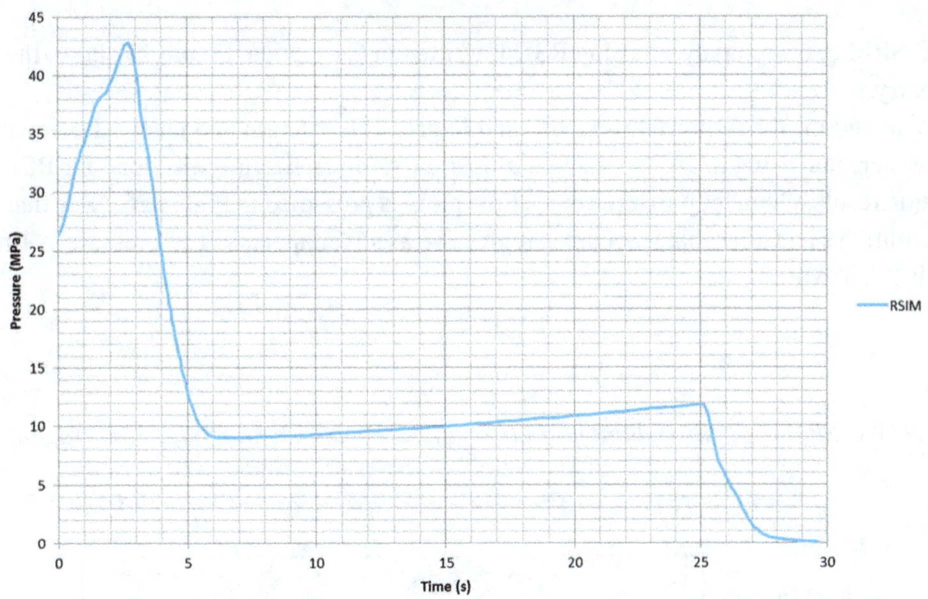

Figure 24. Case 3 - Geometry progression results (perspective)

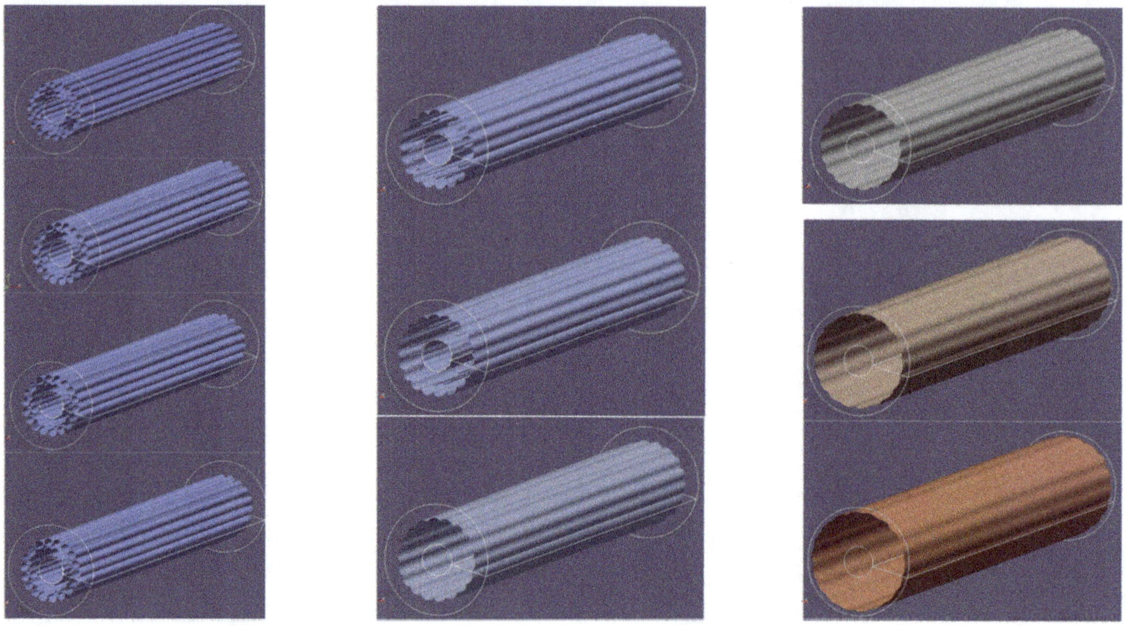

Case 4: NAWC SRM n. 6

The NAWC SRM n. 6 was analyzed using RSIM, written in C++ code. Figure 25 shows the propellant grain geometry.

Cylindrical and mirror symmetry are exploited in grid construction and level set computation. The geometry was created using a CAD tool and a proper curve discretization was done for RSIM input.

Simulation result - chamber pressure versus time plot - is presented in Figure 26. Note that the burn is progressive until 5 s and after that it becomes regressive as sliver appears. The pressure vs time integral calculated is 27.50 MPa.s.

Figure 25. NAWC SRM n. 6 - Propellant grain (Cavallini, Favini, Di Giacinto, & Serraglia, 2009)

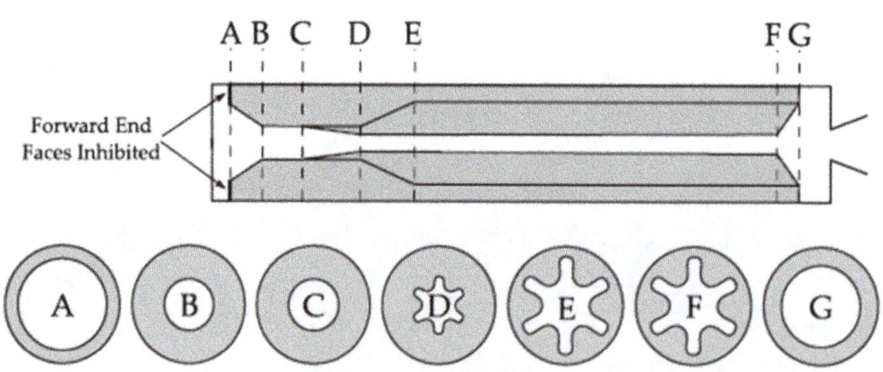

Figure 26. NAWC SRM n. 6 - Chamber pressure (MPa) versus time (s)

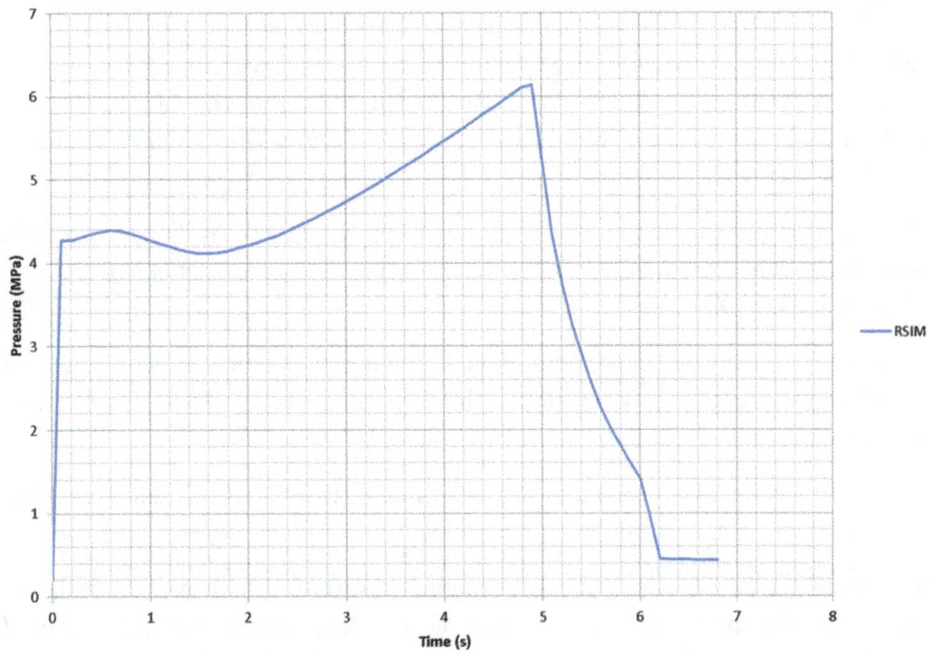

CONCLUSION

The solid rocket motor burn simulation with 3D propagation: RSIM, based on level set method, was verified with previous program using a simplified case of study. Results of chamber pressure versus time were equivalent considering the same geometrical and thermodynamic data input for the computation.

Other two cases were analyzed: finocyl and multiperforated propellant grains. This demonstrated that RSIM program is capable to handle complex grain and case geometries as input, useful for solid rocket motor designs. Nevertheless, it has potential to be further developed using thermochemical model, with transitory and erosion dynamics.

RSIM handles complex grain and case geometries as input. Simulation with 3D propagation: RSIM, based on level set method, was verified with using a simplified case. Pc vs t results were equivalent for the same geometrical and thermodynamic data input.

The RSIM program has potential to be further developed, adding these features: Transitory and erosion dynamics model; Thermochemical model; GUI (Graphical User Interface).

REFERENCES

AEB. (2017, April 03). *Veículos Lançadores, Brasília.* Retrieved from http://www.aeb.gov.br/programa-espacial/veiculos-lancadores/

All3DP. (2016, November 30). *STL File Format for 3D Printing - Explained in Simple Terms.* Retrieved from https://all3dp.com/what-is-stl-file-format-extension-3d-printing/

Attili, A. (2008). *Numerical simulation of internal ballistic of solid rocket motors* [PhD thesis]. Università degli Studi di Roma, Rome.

Beckstead, M. W. (2000). Overview of combustion mechanisms and flames structures for advanced solid propellant. In V. Yang, T. B. Brill, & W. Z. Ren (Eds.), Solid Propellant Chemistry, Combustion, and Motor Interior Ballistics (pp. 267-285). New York, USA: Progress in Astronautics and Aeronautics, AIAA. doi:10.2514/6.2000-3325

Beckstead, M. W., Puduppakkam, K., Thakre, P., & Yang, V. (2007). Modeling of combustion and ignition of solid-propellant ingredients. *Progress in Energy and Combustion Science*, *33*(6), 497–551. doi:10.1016/j.pecs.2007.02.003

Bianchi, D., Nasuti, F., Onofri, M., & Martelli, E. (2009). Thermochemical erosion analysis of Carbon-Carbon nozzles in solid-propellant rocket motors. In *Proceedings of the 45th AIAA/ASME/SAE/ASEE Joint Propulsion Conference & Exhibit*, Denver, Colorado, USA. AIAA.

Braeunig, R. A. (2014, June 14). *Rocket and Space Technology, 2012*. Retrieved from http://braeunig.us/space/index.htm

Cavallini, E., Favini, B., Di Giacinto, M., & Serraglia, F. (2009). SRM internal ballistic numerical simulation by SPINBALL model. In *Proceedings of the 45th AIAA/ASME/SAE/ASEE Joint Propulsion Conference & Exhibit*, Denver, Colorado, USA. AIAA. doi:10.2514/6.2009-5512

Chankapoe, S., Winya, N., & Pittayaprasertkul, N. (2013). Performance Investigation of Solid-Rocket Motor with Nozzle Throat Erosion. *International Journal of Mechanical, Aerospace, Industrial and Mechatronics Engineering, 7*(9).

Chaturvedi, S., & Dave, P. N. (2015). Solid propellants: AP/HTPB composite propellants. *Arabian Journal of Chemistry*.

Chorin, A. (1985). Curvature and Solidification. *Journal of Computational Physics*, *57*(3), 472–490. doi:10.1016/0021-9991(85)90191-3

Dauch, F., & Ribereau, D. (2002). A software for SRM grain design and internal ballistics evaluation, PIBAL. In *Proceedings of the 38th AIAA/ASME/SAE/ASEE Joint Propulsion Conference & Exhibit*, Indianapolis, Indiana, USA. AIAA. doi:10.2514/6.2002-4299

Gueyffier, D., Roux, F. X., Fabignon, Y., & Chaineray, G. (2014). High-order computation of burning propellant surface and simulation of fluid flow in solid rocket chamber. In *Proceedings of the 50th AIAA/ASME/SAE/ASEE Joint Propulsion Conference & Exhibit*, Cleveland, USA. AIAA. doi:10.2514/6.2014-3612

Hertfield, R., Jenkins, P., Burkhalter, J., & Foster, W. (2004). Analytical methods for predicting grain regression in tactical solid-rocket motors. *Journal of Spacecraft and Rockets*, *41*(4), 689–693. doi:10.2514/1.3177

Koomey, J., Berard, S., Sanches, M., & Wong, H. (2011). Implications of Historical Trends in the Electrical Efficiency of Computing. *IEEE Annals of the History of Computing*, *33*(3), 46–54. doi:10.1109/MAHC.2010.28

Liau, Y. C., & Yang, V. (1995). Analysis of RDX monopropellant combustion with two-phase subsurface reactions. *Journal of Propulsion and Power*, *11*(4), 729–739. doi:10.2514/3.23898

Marmureanu, M. I. (2014). Solid rocket motor internal ballistics simulation using different burning rate models. *U.P.B. Sci. Bull.*, *76*, 49–56.

Milos, F., & Rasky, D. (1994). Review of Numerical Procedures for Computational Surface Thermochemistry. *J. Thermophys. Heat Transfer*, *8*, 24–34.

NASA. (1971). *Solid rocket motor performances analysis and prediction (NASA SP-8039)*. Cleveland, Ohio, USA: NASA.

Osher, S., & Sethian, J. A. (1988). Fronts propagating with curvature-dependent speed: Algorithms based on Hamilton-Jacobi formulations. *Journal of Computational Physics*, *79*(1), 12–49. doi:10.1016/0021-9991(88)90002-2

Qin, F., Guoqing, H., Peijin, L., & Jiang, L. (2006). Algorithm study on burning surface calculation of solid rocket motor with complicated grain based on level set methods. In *Proceedings of the 42th IAA/ASME/SAE/SEE Joint Propulsion Conference & Exhibit*, Sacramento, CA. doi:10.2514/6.2006-4774

Ritchie, D. (1993). The Development of the C Language. In *Proceedings of the Second History of Programming Languages conference.* Cambridge, Mass.

Rypl, D., & Bittnar, Z. (2006). Generation of computational surface meshes of STL models. *Journal of Computational and Applied Mathematics*, *192*(1), 148–151. doi:10.1016/j.cam.2005.04.054

Sethian, J. A. (1985). Curvature and the Evolution of Fronts. *Communications in Mathematical Physics*, *101*(4), 487–499. doi:10.1007/BF01210742

Sethian, J. A. (1994). Level Set Techniques for Tracking Interfaces, Fast Algorithms, Multiple Regions, Grid Generation, and Shape/Character Recognition. In *Proceedings of the International Conference on Curvature Flows and Related Topics*, Trento, Italy (pp. 215-231).

Stroustrup, B. (1994). *A History of C++: 1979– 1991*. New Jersey: AT&T Bell Laboratories.

Wiedemanna, C., Homeistera, M., Oswalda, M., Stabrotha, S., Klinkradb, H., & Vörsmanna, P. (2009). Additional historical solid rocket motor burns. *Acta Astronautica*, *64*(11-12), 1276–1285. doi:10.1016/j.actaastro.2009.01.011

Willcox, M., Brewster, M., Tang, K., Steward, D., & Kuznetsov, I. (2007). Solid rocket motor internal ballistic simulation using three-dimensional grain burnback. *Journal of Propulsion and Power*, *23*(3), 575–584. doi:10.2514/1.22971

Yildirim, C., & Aksel, M. (2005). Numerical simulation of the grain burnback in solid propellant rocket motor. In *Proceedings of the 41th AIAA/ASME/SAE/SEE Joint Propulsion Conference & Exhibit,* Tucson, Arizona, USA. AIAA. doi:10.2514/6.2005-4160

Chapter 8
Green Propellants

Gilson da Silva
Instituto Nacional da Propriedade Industrial (INPI), Brazil

ABSTRACT

Friendly or green compositions are being sought in many kinds of applications, such as fertilization, building materials, energy generation, and so on. Thus, this classification (green) can be established after the subjection of the compound to a thorough toxicity study. The research for low toxicity and no damage to the environment has stimulated the development of specific investigation lines in many areas. Inevitably, the criteria for safe handling, sensitivity and, mainly, specific impulse (efficiencies) of propellant compositions is still superior compared to ecological appeals. Nowadays, however, the solid or liquid propulsion, as aerospace as military, has already compounds to efficiency and eco-friendly characteristics. Thus, the chapter proposed will show a review in the energetic materials area, aiming at the most promising materials to be used as oxidizers and combustibles in a green propulsion system.

INTRODUCTION

The research for low toxicity and no damage to the environment has stimulated the development of specific investigation lines in many areas. Inevitably, the criteria for safe handling, sensitivity and, above all, specific impulse (efficiencies) of propellant compositions are still superior compared to ecological appeals. Nowadays, however, the solid or liquid propulsion, as aerospace as military, already has compounds to efficiency and eco-friendly characteristics. Thus, the chapter proposed will show a review in the energetic materials area, looking for the most promising materials to be used as oxidizers and combustibles in a green propulsion system.

"Friendly" or "green" classifications are looked for in many kinds of applications, such as agrochemicals, home appliances (that can save energy or use no toxic gas), automotive engine, and so on, representing a very promising market strategy. In some areas, the "green" classification can be done only after the toxicity study of the compound/composition showing a reduced risk for the mankind and/or the environment.

The acronym "GEMs" is used for an energetic material that shows "green" behavior. In this case, it should show high oxygen balances and halogen and metal-free nature (Rahn, 2010).

DOI: 10.4018/978-1-5225-2903-3.ch008

In a common liquid propulsive system by using the mixture of liquid oxygen (LOX) and hydrogen (LH), the main product of the reaction is very friendly, water. On the other hand, the reactants are dangerous and unstable substances, in which storage (tanks, pressurized system and valves) and manipulation need an elevated level of complexity and risks.

There is different kind of chemical propulsion (monopropellant or bipropellant) used for distinct proposes (launch or positioning) where the compound is in varied stage (solid, liquid or gas).

Nowadays, Hydrazine is the only liquid monopropellant widely used for the generation of hot gases, despite its high toxicity, volatile and carcinogenic properties that require extreme handling precautions, increasing the operational complexity (Gronland et al., 2006).

The propulsive system LOX/LH can be used as an example of a bipropellant propulsion, in which oxidizer and fuel liquids are injected, atomized, and mixed at the first zone of the combustion chamber (Silva, Rufino, & Iha, 2013). Another example using hydrazine is a hypergolic bipropellant, where the droplets of hydrazine and nitrogen tetroxide in liquid phase react spontaneously (Hawkins, Schneider, Drake, Vaghjiani & Chambreau, 2011).

In a solid propellant rocket systems (heterogeneous propellant), with a binder Hydroxyl-Terminated Polybutadiene (HTPB) or Glycidyl Azide Polymer (GAP), aluminum (fuel) and ammonium perchlorate (AP – oxidizer), the toxicity does not stay only in the hydrogen chlorine, aluminum oxide and aluminum chloride, which affect the environment around rocket launch sites, burning the vegetation around them, but also in the isocyanate-based materials (Sciamareli, 2010), used to cure the binder to obtain a thermosetting hybrid fuel grain, which is a carcinogenic and has negative reproductive effects, as well as in the perchlorate ions which may act as a teratogen and has adverse effects on the function of the thyroid gland (Ilyushin, Tselinsky & Shugalei, 2012).

In spite of the liquid bipropellants (e.g. LOX/LH) which generally offer higher specific impulse than liquid monopropellants, the bipropellant systems tend to be more complicated because of the extra hardware components (pumps, valves, tanks, connections, etc.) needed to make sure the proper amount of fuel is mixed with that of the oxidizer.

Unlike from the liquid propellant, which is injected from the external tank into the combustion chamber at the time of ignition, solid propellants are placed directly on it. The solid propellants system requires, normally, pyrotechnic charges to initiate the combustion, but pyrotechnic charges are susceptible to initiation by electromagnetic radiation or accumulated electric charge, increasing the accidental initiation risks of the solid rocket motor, according to the Department of Defense Interface Standard (2010).

For some propellants systems, like a solid rocket motor, transmission lines and security systems for ignition and destruction can use quantities of explosives with high sensitivities to heat, shock, friction and flames, like lead azide, lead styphnate and mercury fulminate (primary explosives). But, mercury and its compounds are toxic towards biological systems; it can complex with proteins by inhibiting enzymes and it is also dangerous to the nervous system. Lead itself leads to blood-producing problems (Amenia), injuries to the central nervous system, it replaces calcium salts, depositing in bones, and disorganizes metabolic processes, among other problems (Ilyushin, Tselinsky & Shugalei, 2012).

Trying to mitigate this problem, Ilyushin, Tselinsky and Shugalei (2012) proposed the replacement of these primary metal explosives to energetic eco-friendly materials. The candidates must conform to the following primary "green" criteria: insensitivity to moisture and light; sensitivity to initiation but not too sensitive to handle and transport; thermally stable to 200°C at least; chemically stable for extended periods; devoid of toxic metals such as lead, mercury and some others and be free of perchlorate, which may act as a teratogen and has side effects on the function of the thyroid gland.

Organic compounds, inorganic salts, coordination complexes, metastable interstitial composites, in summary, any kind of primary explosive that follows the criteria above could be accepted as a candidate.

A nano aluminium and nano oxides of d-metals (Nano-scale thermite) were investigated, because they show combustion velocity in the same range of explosives. On the other hand, the oxidation of aluminum nano particles by the air and manufacturing hazard was the reason for their exclusion.

Tetracene hasn't the necessary stability required, but a reaction product of tetracene and sodium nitrite in aqueous acidic media, 5-tetrazolo-azo-1'-tetrazole-5'-amine, shows promising sensibility and thermal stability to be considered as a green initiation explosive.

Other compounds related by then (Ilyushin, Tselinsky & Shugalei, 2012) show different problems: 2-diazonio-4,6-dinitro-1-phenolate is not stable upon exposure to light and it actives an inappropriate immune system response that predisposes allergic syndromes; cyanuric triazide (1,3,5-triazido-2,4,6-triazine) which is a volatile organic compound and sublimates above 30°C.

Among the inorganic salts, potassium 4,6-dinitro-7-hydro-7-hydroxybenzfuroxanide showed low thermal stability, but a very similar salt 4,6-dinitro-7-hydro-7-hydroxybenzfuroxan has been recommended for USA weapons, for its thermal stability and safe handling characteristics.

Another technical problem to military application of the AP/aluminum solid rocket motors is the smoke production, due to the burn products in high air humidity amounts that can be seen (Langlet, Östmark & Wingborg, 1999).

The GEMs for defense and space applications have been increasing in number and importance and, among them, it is worth highlighting: Ammonium dinitramide (ADN) and hydrazinium nitroformate (HNF), in replacement of AP.

GREEN COMPOUNDS

Ammonium Dinitramide

According to Nagamachi, Oliveira, Kawamoto and Dutra (2009), there have been many routes for ADN synthesis since its origin in the USSR, but the only effective one for large-scale production has used sulfamate salts.

A method of preparing dinitramidic acid and its salts is described by Langlet et al. (1999). According to them, the acid and the salts, which the formula $M+^{n}(^{-}N(NO_2)_2)_n$, wherein M is a metal cation or a nitrogen-containing cation and n=1-3, is prepared by nitration of a compound selected from a group consisting of NH_2NO_2, $NH_4NH_2CO_2$; NH_2SO_3H, $NH(SO_3H)_2$, $N(SO_3H)_3$, and its salts with metal cations or organic cations, e.g. $NH(SO_3NH_4)_2$, and other products formed when ammonia is reacted with sulphuric trioxide, with a common nitrating acid such as nitric acid/sulphuric acid or nitric acid/acetic anhydride. They even taught a method of preparing dinitramide salt, comprising the neutralization of dinitramidic acid with a neutralizing agent (ammonia). The separation salt separation of the mixture can be gotten by passing the aqueous solution through a column containing an adsorbing agent, which adsorbs the salt by using activated carbon, silica gel and zeolites.

In an alternative method, the salt (ADN) is formed by the reaction of ammonia and a nitronium-containing compound, such as a nitronium salt or a covalently bonded compound containing a NO_2 group, at a temperature ranging from about -20°C to about -120°C. The nitronium-containing compound is preferably formed, dissolved, dispersed, or mixed in a liquid prior to its introduction into the reaction chamber, or formed in the chamber (Schmitt, Bottaro, Penwell & Bomberger, 1994).

In the preferred mode, the initial step is to form a solution, dispersion, or mixture of the nitronium-containing compound in a non-reactive aprotic liquid or solvent (methylene chloride (CH_2C_{12}), chloroform ($CHCI_3$), carbon tetrachloride (CCI_4), methyl ether ($(CH_3)_2O$), any freon gas (including fluorocarbons and chlorofluorocarbons), acetonitrile, ethyl acetate, ethyl ether, tetrahydrofuran, sulfolane, or mixtures of same) to facilitate the contact between the ammonia reactant dissolved in the aprotic solvent and the solid nitronium-containing reactant, the aprotic liquid does not enter into the reaction (Schmitt, Bottaro, Penwell & Bomberger, 1994).

At the end of the process, which should be carried out under anhydrous conditions, the salt produced may be extracted and separated from the other products by solvent extraction with acetone, isopropanol, t-butanol, ethyl acetate, tetrahydrofuran, and $CH_3(CH_2)_nOH$, where n = 0-7, or mixtures of the above solvents. The dinitramide salt can also be extracted by solutions of ammonia in chlorinated hydrocarbons such as CH_3Cl_2, $CHCI_3$, and CCI_4, as well as a mixture of ethyl acetate and n-butanol (Schmitt, Bottaro, Penwell & Bomberger, 1994).

Oliveira (2014) studied different routes to the ADN production, adapting the routes from the Langlet et al. (1999) and Vörde & Skifs (2005) teaching. The most promising route, according to Oliveira (2014) was the one in which ammonium sulfamate is nitrated by a mixture of nitric acid/ sulfuric acid. The product of the reaction is rich in ammonium nitrate and, to separate the ADN, needs to be purified with acetone, leaching the ADN. The study conducted by Oliveira (2014) showed yet the influence of the temperature and time in the routes analyzed, improving the separation process of the reaction products.

AND-Based Propellants

Anflo and Wingborg (2002) related a monopropellant composition based in ADN. According to them, the composition was suitable for rocket propulsion, vehicle propulsion or gas generation, more specifically, for space applications. The composition studied by them comprises a solution of ADN, a fuel and a stabilizing agent.

According to Anflo and Wingborg (2002), to prevent ion exchange with the dissolved ADN, in the propellant composition, a combustion stabilizing agent was employed. Among the bases suitable for the stabilization, hexamine and urea was highlighted by them, while methanol was elected the most preferred fuel in the propellant composition.

The composition related by them (Anflo and Wingborg, 2002) exhibits ranging from 15 to 55% by weight of the fuel in a solvent mixture (water + fuel), where a stabilizer is added in an amount of 0.1 to 5, based upon the total weight of the other components.

When methanol is the fuel, a composition consists of about 64.3% ADN, about 24.3% water and about 11.4% by weight of methanol, to which the stabilizer (i.e., urea) is added in the aforementioned amount.

The burn test carried out by Anflo and Wingborg (2002) shows that the chamber pressure was markedly stabilized, when compared to other tests without the stabilizing agent, increasing the combustion efficiency of the system.

In another study, Thormählen and Anflo (2012) related a deficiency in the propellant composition proposed by Anflo and Wingborg (2002). They discovered that the composition tends to undergo saturation with respect to the dissolved ADN, which forms solid crystals in the mixture of liquid solvents when the temperature of the monopropellant is being lowered to about -7°C. On the other hand, propellant for the purpose of orbit maneuvering and attitude control of satellites, and other space vehicles, in which case rocket engines or thrusters should keep a homogenous solution at temperatures below -7°C, and beyond that, at temperatures down to -30°C. Thereby, they related some modification in the propellant composition, which comprises 55-62% by weight of ADN; 10-18% by weight of methanol; 4-10% by weight of ammonia; and the balance water, which remains a homogeneous solution down to a temperature of -25°C, or below. In a practice realization, ammonia may be used as a 25% aqueous solution in the preparation of the monopropellant.

According to Thormählen and Anflo (2012), the monopropellant blend consisting of 62.0% of ADN, 13.1% of methanol, 6.2% of ammonia, and 18.7% of water has presented no crystallization, being able to storage down to -30°C at least.

In a recent publication, Anflo and Bergman (2014) taught a dual mode chemical rocket engine using storable low-hazard liquid propellants, which consist of a rich fuel monopropellant and hydrogen peroxide, respectively, used in a mode rocket engine comprises primary and secondary reaction chambers. The chemical rocket engine has a primary reaction chamber, for hydrogen peroxide, connected to a secondary reaction chamber, having means for injection therein of a fuel-rich monopropellant.

In the monopropellant mode operation, the engine uses hydrogen peroxide, which is catalytically decomposed in the primary reactor without pre-heating of the primary reactor.

In the bipropellant mode, the catalytic combustion of hydrogen peroxide, in the primary reactor, is used to provide an oxidizer and heat to initiate the thermal decomposition of a liquid ADN or hydroxylammonium nitrate (HAN) based monopropellant injected into the secondary reactor. The preferred fuel-rich propellant used is the one that had been taught by Thormählen and Anflo (2012). The Figure 1 shows the engine proposed by them.

The decomposition of the hydrogen peroxide happens in the reaction chamber 140, which comprises a catalyst bed for the decomposition. This first reaction chamber is connected to a second reaction chamber 150, which has means 125 for injection therein of a fuel-rich monopropellant. The fuel-rich monopropellant is injected from outside into the secondary reaction chamber of the engine.

The catalyst in the primary reaction chamber 140 would be the life limiting element of the thruster when exposed to the decomposition and combustion species and operated at high temperatures. While the temperature in the secondary chamber can be significantly increased, the temperature in the primary

Figure 1. Rocket engine with dual mode propulsion system

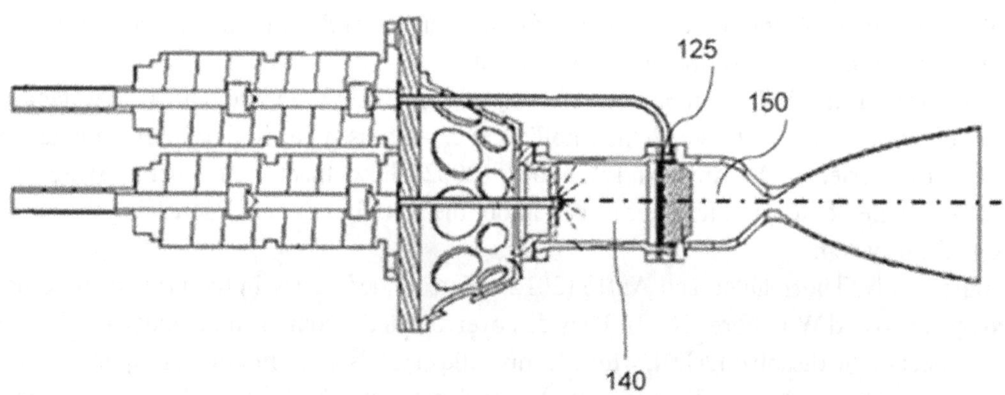

reactor can keep essentially unaffected. The rocket engine presented can also operate in the monopropellant mode (providing lower thrust and impulse bit) by injection of only the hydrogen peroxide.

Hydroxylammonium Nitrate

Hydroxylammonium nitrate (HAN) is an energetic oxidizer salt for use in liquid gun propellants, hybrid rocket motors, solution propellants and as pharmaceutical compound. The processes for preparing of HAN have some troubles, i.e., low profit, impurities and/or problems in accurately controlling the concentration of the desired salt. Some routes to the HAN process production use mercury, which has undesirable environmental issues.

In a process developed by Fuchs, Ritz, Thomas and Weiss (1988) the HAN can be produced by combining NO and H_2 gases in the presence of platinum. In another catalytic process, Fuchs et al. (1988) prepare HAN using the catalytic reduction of nitrogen monoxide with hydrogen at elevated temperature in a dilute aqueous solution of a mineral acid and in the presence of a particular supported platinum catalyst (suspension). However, the processes for producing HAN wherein nitrous oxide or nitric acid is hydrogenated in the presence of platinum (catalyst) which requires the use of hydrogen or nitrous oxide that are dangerous compounds with explosive or health risks.

Levinthal, Willer, Park and Bridges (1993) taught a process for producing an aqueous HAN solution which is alcohol-free, comprising the preparation of an alcohol-free hydroxylammonium sulfate in an aqueous medium, carrying out a simple neutralization reaction by the combination of the hydroxylammonium sulfate solution with an inorganic base (selected from the group consisting of an alkali metal oxide, an alkali metal hydroxide, an alkaline earth metal oxide, and an alkaline earth metal hydroxide) aiming to produce an aqueous hydroxylamine solution, distilling the aqueous hydroxylamine solution, under a reduced pressure at a temperature of less than about 65° C., and carrying out a neutralization reaction with an amount of nitric acid, distilling the HAN solution at a temperature of from above -50°C, but below a temperature at which the hydroxylamine moiety completely decomposes, the concentration of nitric acid up to about 70% by weight.

HAN-Based Propellants

A hybrid motor may be characterized as a cross between a solid and a liquid fuel motor, most of them utilize a liquid oxidizer and a solid fuel, but there are some hybrid motors which a combustible liquid fuel and a solid oxidizer are used. Since one of the fuels or the oxidizer should be inserted on the reaction chamber, the thrust may be adjusted or stopped by the flow governance of the liquid component. Loehr (2009) had revealed a hydroxylamine based staged combustion hybrid rocket motor.

In the system proposed by Loehr (2009), both liquid oxidizers and gas oxidizers may be used. The rocket motor, Figure 2, has at least two sections, a first end portion, next to the nozzle, and a second end portion, in the opposite side, which may be separated by a separator. The first end portion contains a first chamber which is a combustion chamber, that stores solid combustible materials for use in combustion, the solid combustible materials being magnesium/hydroxyl-terminated polybutadiene (Mg/HTPB) or others (like Plexiglas (polymethylmethacrylate (PMMA)), high density polyethylene (HDPE), double base, GAP, gas generator fuels and black powder). The second end portion contains a second chamber which may contain a bladder/tank for storing the oxidizers, which are pressurized liquids. A gas generator is placed next to the bladder, at the second end portion of the rocket motor housing, and may be

Figure 2. Loehr's hybrid rocket engine

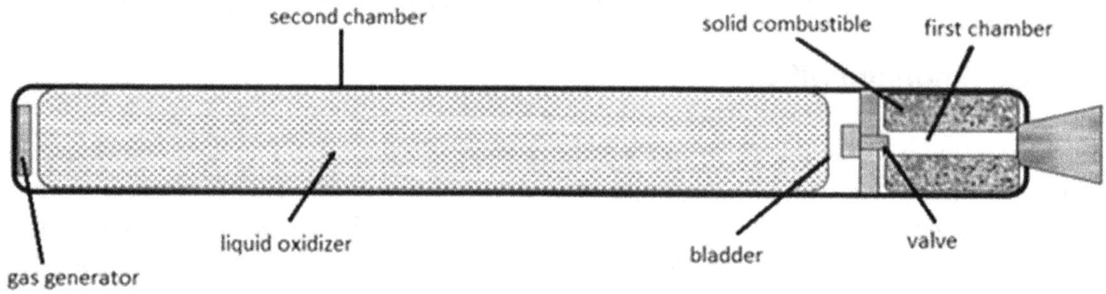

used to pressurize the bladder. The first and second chambers may be in fluid communication via a flow control means for coupling the first chamber and the second chamber, and for valves to the flow control of materials there between. An opening in the bladder, in the opposite side of the gas generator, allows for the flow of the oxidizer out of the bladder. The oxidizer that flows out of the bladder is forced to be in contact with a catalyst, which catalyzes the decomposition of the oxidizer.

The reaction products generated by the decomposition of the oxidizers are fed into the first chamber, through a valve flow. In case of HAN uses, the reaction products which contain hot gases (water steam, carbon dioxide and nitrogen). The oxygen rich gas generator input into the fuel grain chamber may exceed the auto ignition temperature of HTPB and may dispense an ignition grain heating charge, typically associated with HTPB hybrid rocket motors.

The system has several advantages, like the liquid oxidizer that is neither highly corrosive nor requires cryogenic conditions to exist in liquid state, it has a low hazard classification and is relatively stable for long-term storage.

According to Fujiwara et al. (2012), the reaction control of HAN-based propellants is difficult, for its high detonation resistance. Then, water added to HAN is used as a method for controlling the reaction. On the other hand, the addition of water lowers the specific impulse. Thereby, to solve this problem, Fujiwara et al. (2012) taught mixing HAN and ammonium nitrate (AN) drug, where the ratio of AN is less than 10% by weight of the HAH + AN total mass, water and methanol. The composition could have between 60 to 76% by weight of HAH+AN mixture; 16 to 32% by weight of methanol and 5 to 24% by weight of water. The composition is less toxic than hydrazine and the handling is easier.

The bipropellant mode rocket engine taught by Anflo and Bergman (2014), previously seen, could replace the ADN propellant by a HAN propellant, using the same arrangement.

Hydrazinium Nitroformate (HNF)

As reported by Silva et al. (2013), HNF is a desirable oxidizer to use in solid propellant formulations, by the energetic and performance aspects for it. Unfortunately, the most applied binder in solid propellants, HTPB, shows bad compatibility with HFN because the HNF attacks the HTPB backbone, breaking down the binder chain, then HNF cannot be used with this binder. When a binder containing double bonds is utilized, a typical shelf life with HNF will range from 2 to 15 days at a temperature around 20 to 30ºC, According to Low and Haury, 1972.

The synthesis of nitroform with hydrazines salts has been known since Thornton (1967), who taught a process that provides a nitroform salt stable over a long period of time and at temperatures above the decomposition temperature of nitroform. According to his process, a nitroform salt is provided which consists of nitroform and a member of the group consisting of hydrazine. The salts may be prepared by dissolving the nitroform component in a short chain alcohol (alcohols containing from 2 to 6 carbon atoms are adapted to use in the process) and then slowly adding hydrazine or a hydrazine derivative to precipitate the nitroform salt. Then, the nitroform-hydrazine salt or the nitroform-hydrazine derivative salt product may be recovered by separating it from the supernatant alcohol by filtration.

At a later period, Zee, Mul and Hordijk, 1994, improved the methodology of HNF's production. The method according to them is carried out in a solvent that is a dissolvent medium for the nitroform, but also a non-solvent for the HNF to be formed. Suitable solvents are organic solvents as dichloroethane (DCE) and dichloromethane (DCM), being the DCE the most appropriate. The ratio of nitroform to solvent is preferably between 3:1 and 1:9, and sufficient solvent should be present to dissipate the heat produced in the reaction. A small amount of water (between 0.5 - 20 vol.%) is required to the mixture of nitroform and solvent or Cl-C6 alcohol is then added. The mixture is then cooled and hydrazine is added dropwise, with the temperature of the reaction system should remain under 5°C. After stirring, the crude HNF is removed from the reaction medium by filtration (the supernatant liquid is decanted and the wet product is dried). The recrystallization can be done in a solution with short chain alcohol.

HNF-Based Propellants

Louwers and Heijden (2003) taught a method of HNF production, in a crystalline form, to be employed in a solid propellants composition. According to them, for the use of HNF in solid propellants, a high load of the solid material is required (around 80% or more). Then, HNF is often recrystallized after the synthesis to meet the requirements of purity and the desired average particle size. The sensitivity of the product to shock and/or friction is another important aspect of the use of HNF. It can be rapidly decomposed by shock and/or friction and the appropriate morphology of it is more suitable for processing into solid propellants with a high filling degree which may give rise to an increased sensitivity to shock and/or friction.

In the process taught by Louwers and Heijden (2003), there is the possibility of obtaining crystalline HNF having a length to diameter ratio of at most 2.5 and sensitivity to friction and shock above 20 N and 2 J, respectively. The crystalline HNF is obtainable by pressing crystals having a length to diameter ratio of at least 3 until a particular pressure, thereby obtaining a crystalline material that, after the mechanical treatment, is still free-flowing or can readily be made free-flowing. Thus, no solid cake of HNF is obtained but it seems that a loosely coherent whole is obtained, which has substantially retained its original diameter but has a considerably reduced length to diameter ratio. The product can be made with a multimodal fraction of particles having an average number of particle size ranging between 1 and 1,000 μm.

In another publication, Louwers, Heijden and Elands (2005) described a solid propellant combination that is prepared by blending solid oxidizers such as AP or HNF with a liquid precursor for the matrix material. The solid propellant is obtained after the binder cure. As HNF and HTPB mixtures were deemed unsuitable, the use of it is based on the use of amount substantially no hydrazine or nitroform in unreacted form, that is, the HNF should be obtained by means of the process taught by Zee et al. (1994), previously seen, which results in a purity of the recrystallized HNF between 98.8 and 100.3, based on H_3O^+ and a pH-value of a 10% weight aqueous solution of HNF of at least 4. In addition, the water content of the different propellant ingredients, especially the water content of the binder components influences the stability and accordingly a water amount of less than 0.01% weight in the binder is preferred. In addition to the aforementioned aspects, stabilizers may be added to further improve the shelf-life.

According to them (Louwers et al., 2005) a solid propellant can comprise 80 to 90% weight of HNF, in combination with 10 to 20% weight of HTPB and other standard binder components, such as curatives, plasticizers, crosslinking agents, chain extenders and anti-oxidants). Aluminum can be added in the ratio of 10 to 20% of the HNF.

Hydrogen Peroxide (H_2O_2)

According to Anflo and Bergman (2014), hydrogen peroxide is probably the most studied monopropellant worldwide. In despite of its relative low specific impulse (1,600-1,800 Ns/kg), that depends on the concentration of H_2O_2. On the other hand, it can also be used as an oxidizer in bi-propellant. Hydrogen peroxide is reactive and decomposes slowly over time when stored even in its most stabilized form. However, the toxicological and carcinogenic concerns of the current state-of-the-art propellants have led to renewed interest in hydrogen peroxide.

Schneider, Hawkins, Ahmed and Rosander (2014) describe the hydrogen peroxide with being the only known high performing storable oxidizer that can be considered environmentally benign. It's the most popular of the green propellants, which decomposes exothermically in the presence of a suitable catalyst, generating water vapors and oxygen. In a conventional thruster that utilizes H_2O_2 for propulsion (monopropellant thrusters) a silver catalyst to the H_2O_2 decomposition is required. But, silver-based catalysts rapidly deteriorate at high temperatures, making the conventional catalysts unsuitable for prolonged use at high concentrations of H_2O_2 (> 80%), According to Wynn and Musker (2015), who proposed a catalyst comprising an active layer on a carrier comprising gamma-phase alumina to increase the catalyst resistance.

In the catalyst taught by them, the carrier can further comprise aluminum and the gamma-phase alumina can be disposed on the surface of the aluminum as a passivation layer, where the active layer can comprise platinum.

In an embodiment proposed by them (Wynn and Musker, 2015), the catalyst can be used in a bipropellant thruster, that comprises a decomposition chamber with a fuel inlet arranged to receive fuel from a fuel tank, the fuel inlet being disposed so that the fuel combusts with the oxygen produced by the decomposition of the hydrogen peroxide. A hybrid thruster can be done too, by a solid fuel arranged to combust with oxygen produced by the decomposition of the H_2O_2.

In a hypergolic bipropellant, a combination of two propellant components that spontaneously ignite when they come into contact with each other is required. Then a hypergolic bipropellant consists of a fuel and an oxidizer that are incompatible (Mayer and Nardini, 2014).

Mayer and Nardini (2014) had developed a fuel composition that can be used with H_2O_2 to a hypergolic green bipropellant. The composition taught by them comprises an alcohol, a catalytic metal ion, an anion (nitrate) and an amide (N, N dimethyl formamide). In the system, upon contact with the fuel composition, the hydrogen peroxide is catalytically decomposed into oxygen and water by the metal ion that acts as a catalyst. The heat produced by this reaction initiates the interaction between N, N dimethyl formamide (DMF) and the anion, in which the interaction promotes the oxidation of alcohol by oxygen and/or H_2O_2.

The metal cations required for the fuel composition is a transition metal. Silver, iron, cobalt, copper and manganese showed good results. The alcohol is selected from the group consisting of methanol, ethanol, propanol, butanol and pentanol, with ethanol being the most appropriate.

Not only for rocket thrust, but also to explosive composition hydrogen peroxide is required. Compositions comprising H_2O_2 and a sensitizer (a compressible material and/or bubbles of gas) The sensitizers can be manganese salts, iodide salts, nitrogen-based compounds such as, for example, sodium nitrite, nitro-amines such as, for example, N, N'dinitrosopentamethyleletetramine; boron-based compounds such as, for example, sodium borohydride; carbonates such as, for example, sodium carbonate, microspheres of glass and resinous materials, etc. According to Araos (2013). On the other hand, mixtures of ethanol and hydrogen peroxide can be used to an emergence propulsion system in aircraft, as taught by Kwon, Lee and Jang, 2013.

As it can be seen in the literature related, hydrogen peroxide has been a very interesting material for mono e bipropellants systems until now, being a versatile material.

Nitrous Oxide

Nitrous oxide is much warmer than the typical cryogenic oxidizer (LOX), nitrous oxide is cooled at minus 10 to 37,8°C as compared to LOX's minus 147°C, and the pump's gas generator can be powered by decomposed monopropellant nitrous oxide, which when started, the decomposition of nitrous oxide is self-sustaining. Nitrous oxide is non-corrosive and may be used with common structural materials. It is stable and comparatively non-reactive at ordinary temperatures to such substances as ozone, hydrogen, halogens and alkali metals (Sarigul-Klijn and Sarigul-Klijn, 2007).

Cai et al. 2011 taught a nitrous oxide monopropellant. The thruster mainly comprises an electromagnetic valve, a propellant front guide pipe, a decomposition chamber and a spray pipe. The nitrous oxide was adopted by its following properties: non-toxic, pollution-free and corrosion-free. In addition, nitrous oxide has high stability at normal temperature and high saturated vapor pressure. When the nitrous oxide is used for propelling liquefied gas, the density of the nitrous oxide is higher than that of a cold gas propellant such as nitrogen; when the nitrous oxide is heated to a temperature higher than 520°C, the nitrous oxide releases heat and is decomposed into oxygen and nitrogen, and the oxygen and the nitrogen are close to the components of air, and the highest temperature of decomposition reaction is about 1,640°C. The hot gas generated through decomposition can serve as a heat source or a power source of an aircraft, the required power is low, and the hot gas is even used as a gas source required by life support system; moreover, a transport system of the nitrous oxide monopropellant thruster is simple in structure.

According to Kawaguchi, Tokudome, Habu, Yagishita and Hotta (2009), when nitrous oxide is used as a single propellant in a monopropellant thrust and a catalytic decomposition gas is obtained by the catalytically decomposing the nitrous oxide, the propellant used in the thruster can be made substantially nontoxic. In a bipropellant thruster, alcohols and LPG can be used in combination with nitrous oxide, serving as an oxidizer and the toxicity of the bipropellant can also be reduced. In this case, by taking advantage of a fact that nitrous oxide, unlike a toxic dinitrogen tetraoxide / hydrazine based propellant used in conventional bipropellant thrusters, does not have a characteristic of spontaneously igniting at normal temperature but is an energetic material (temperature of decomposition is about 1600°C), the heat obtained by catalytic decomposition of nitrous oxide can be used as ignition energy.

The freezing point of nitrous oxide is sufficiently low as -91°C, what is interesting to application in a deep space exploration mission in a low temperature environment of about -50°C. Then, there is no need to provide an anti-freezing heater in a propellant supply system in a thruster. In addition, in a bipropellant system with nitrous oxide and ethanol (freezing point: -114°C) can be used too without a heater even at a low temperature.

The saturated vapor pressure of nitrous oxide is relatively high (about 6.4 atm at -50°C), then it could be used as a propellant for itself (nitrous oxide gas). The same can be done to the fuel, when its saturated vapor pressure is not so high, supplying the fuel by the use of the vapor pressure of nitrous oxide gas can be done.

The nitrous oxide used in the thruster taught by Kawaguchi at al. (2009) is almost completely decomposed into oxygen and nitrogen by a decomposition catalyst. Hence, it is conceived that nitrous oxide can also be used as a source of life-support gaseous oxygen or thermal energy in a closed system such as a spaceship or space station. Furthermore, the oxygen obtained by the catalytic decomposition of nitrous oxide can be combined with fuel (hydrogen or methanol) and applied to fuel cells. The suitable catalyst has aluminum, magnesium and rhodium supported in a carrier or noble metal (rhodium, ruthenium and palladium) supported in a carrier of silica or silica-alumina, rhodium supported in alumina, cordierite or silicon carbide is the most appropriate.

In a monopropellant mode, the thruster can produce thrust by directly exhausting oxygen-nitrogen mixed gas of high temperature exceeding 1000°C, which is obtained by catalytically decomposing nitrous oxide to outside of the thruster. As silicon carbide has high oxidation resistance and heat resistance up to about 1600°C. It is desirable to use a silicon carbide structure as a carrier that also serves as a heater for a nitrous oxide decomposition catalyst.

Normally, liquid fueled rockets have a better specific impulse than solid rocket ones. The possibility of the control of the supply of fuel and oxidizer can shut down and restart the liquid thrusters.

The ignition process of a monopropellant can be done by an appropriate catalyst, high energy spark or raising a localized volume beyond the activation energy of the reaction. The chemical reaction (thermal decomposition and/or combustion) liberates energy. Commonly, the initiation mechanism is incorporated in the vicinity of a combustion chamber's propellant injector head. The design and manufacture of a propellant injector head used in a combustion chamber which is important to achieve effective and safe operation of the rocket thruster. If the design is not correct; flame can propagate back past the propellant injector head and into the propellant storage tank causing an explosion. Then, the configuration of the injector should provide a flame barrier. In the configuration proposed by Mungas, Fisher and Mungas, 2012, the flame barrier comprises fluid paths with a diameter of less than about 10 microns, and the electrode assembly comprises an interface sheath encompassing an insulating tube which encompasses and electrode; and a fuel inlet tube is disposed through the housing into the cooling chamber. The combustion system is provided wherein the interface sheath and the flame barrier comprise stainless steel alloys, pure nickel, nickel alloys, niobium, ruthenium, molybdenum, tungsten, tantalum, tantalum alloys, sintered ceramic or laminate structures.

Although of the lower specific impulse of hybrid motor using nitrous oxide compared to LOX, when nitrous oxide is used it can reach about 270 seconds, while LOX can reach 300 seconds. The problem with the specific impulse as a metric to compare rocket performance is that it does not consider propellant mass fraction, which is the mass of the propellant relative to the total vehicle mass. The most important, is the capability of the rocket propulsion system to accelerate the entire launch vehicle to the highest possible velocity.

Under this concept, Sarigul-Klijn and Sarigul-Klijn (2007) proposed a hybrid propulsion motor with nitrous oxide that can be comparable in performance to a solid motor or a liquid fueled engine. The main idea was to increase the performance of hybrid motors by increasing propellant mass fraction while maintaining specific impulse. The technologies are high propellant density, high oxidizer to fuel ratio, use of a propellant combination that have constant specific impulse over a wide oxidizer/fuel ratio, and a pump to feed the oxidizer into the motor case.

High propellant density decreases the size of the tank volume, and hence decreases the fraction of launch vehicle mass that must be devoted to tanks and motors. Dense propellants also reduce the size and weight of the flow passageways, pump size, and amount of the gas to pump the propellant.

While a bipropellant liquid-fueled engine requires two separate fluids for its gas generator, heavy high pressure helium tanks or a separate heat exchanger to vaporize fluids for tank pressurization (inert weights), nitrous oxide is an ideal oxidizer as a gas generator propellant as for the tank pressurization.

FINAL CONSIDERATIONS

The search for less hazardous and more efficient compounds for chemical propulsion have shown promising results, not only in terms of compounds that may be substituted for those less aggressive (green), but also changes in the settings of the equipment and systems, enabling the achievement of improvements in propulsion systems and supplanting the still present deficiencies in less aggressive compounds, but less energetic, admitted them to effective use.

REFERENCES

Anflo, K., & Bergman, G. (2014). Dual mode chemical rocket engine, and dual mode propulsion system comprising the rocket engine. *World Intellectual Property Organization, WO2014193300*, A1.

Anflo, K., & Wingborg, N. (2002). Ammonium dinitramide based liquid monopropellants exhibiting improved combustion stability and storage life. *World Intellectual Property Organization, WO02096832*, A1.

Araos, M. (2013). *Composition explosive amelioree.* CA: Office de la Propriété Intellectuelle du Canada.

Cai, G., Sun, W., Fang, J., Mao, L., Han, L., & Wu, J. (2011). Nitrous oxide monopropellant thruster and design method thereof. *State Intellectual Property Office of P.R.*

Department of Defense Interface Standard. (2010). *Eletromagnetic Environmental Effects requirements for Systems, MIL-STD-464*. Philadelphia: Author.

Fuchs, H., Ritz, J., Thomas, E., & Weiss, F. J. (1988). Verfahren zur Herstellung von Hydroxylammoniumsalzen. European Patent Office, EP0287952 A2.

Fujiwara, S., Shibamoto, H., Nakamura, T., Furukawa, K., Matsuo, T., Hori, K., . . . Katsumi, T. (2012). Propellant for thruster. Japan Patent Office, JP2012046390 A.

Gronland, T. A., Westerberg, B., Bergman, G., Anflo, K., Brandt, J., & Lyckfeldt, O. ... Wingborg, N. (2006). *Reactor for decomposition of ammonium dinitramide-based liquid monopropellant and process for the decomposition.* U.S. Patents 7,137,244 B2.

Hawkins, T. W., Schneider, S., Drake, G. W., Vaghjiani, G., & Chambreau, S. (2011). *Hypergolic fuels.* U.S. Patents 8,034,202 B1.

Ilyushin, M. A., Tselinsky, I. V., & Shugalei, I. V. (2012). Environmentally Friendly Energetic Materials for Initiation Devices. *Central European Journal of Energetic Materials*, *9*(4), 292–327.

Kartte, K., Fuchs, H., Jockers, K., Kurt, K., & Meier, H. (1972). *Procédé de preparation de nitrate d'hydroxylammonium* (p. 2121005). FR: Institut Nacional de la Propriété Industrielle.

Kawaguchi, J., Tokudome, S., Habu, H., Yagishita, T., & Hotta, M. (2009). *Thruster using nitrous oxide.* U. S. Patents, US20090007541 A1.

Kwon, S.J., Lee, J.S., Jang, D.W. (2013). *Environmental hydrogen peroxide monopropellant.* Korean Intellectual Property Office, KR20130064931.

Langlet, A., Östmark, H., & Wingborg, N. (1999). *Method of preparing dinitramide acid and salts thereof.* U.S. Patents 5,976,483 A.

Levinthal, M. L., Willer, R. L., Park, D. J., & Bridges, R. (1993). *Process for making high purity hydroxylammonium nitrate.* U.S. Patents, US5,266,290 A.

Loehr, R. D. (2009). *Hydroxyl amine based staged combustion hybrid rocket motor.* U.S. Patents, US20090211227 A1.

Louwers, J., & Heijden, A.E.D.M. (2003). *Hydrazinum nitroformate.* U. S. Patents, US6,572,717 B1.

Louwers, J., Heijden, A. E. D. M., & Elands, P. J. M. (2005). *Hydrazinum nitroformate based high performance solid propellants.* U.S. Patents, US6,916,388 B1.

Low, G. M., & Haury, V. E. (1972). *Hydrazinum nitroformate propellant stabilized with nitroguanidine.* U.S. Patents, US3,658,608.

Mayer, A. E. H. J., & Nardini, F. T. (2014). Fuel composition for hypergolic bipropellant. *World Intellectual Property Organization, WO2014193235*, A1.

Mungas, G. S., Fisher, D. J., & Mungas, C. (2012). *Nitrous oxide flame barrier.* U. S. Patents, US20120279197 A1.

Nagamachi, M. Y., Oliveira, J. I. S., Kawamoto, A. M., & Dutra, R. C. L. (2009). ADN – The new oxidizer around the corner for an environmentally friendly smokeless propellant. *Journal of AerospaceTechnology and Management*, *1*(2), 153–160. doi:10.5028/jatm.2009.0102153160

Oliveira, J. I. S. (2014). *Síntese e caracterização do oxidante ADN (dinitramida de amônio), usado em propelentes para aplicação aeroespacial.* Unpublished doctoral dissertation, Instituto Tecnológico de Aeronáutica, São José dos Campos, Brazil.

Rahn, M. (2010). *Green Propellants*. (Unpublished doctoral dissertation). Royal Institute of Technology, Stockholm, Sweden.

Sarigul-Klijn, M. M., & Sarigul-Klijn, N. (2007). *High propellant mass fraction hybrid rocket propulsion*. U. S. Patents, US20070144140 A1.

Schmitt, R. J., Bottaro, J. C., Penwell, P. E., & Bomberger, D. (1994). *Process for forming ammonium dinitramide salt by reaction between ammonia and a nitronium-containing compound*. U.S. Patents US5,316,749 A.

Schneider, S., Hawkins, T. W., Ahmed, Y., & Rosander, M. (2014). *Catalytic hypergolic bipropellants*. U. S. Patents, US8,758,531 B1.

Sciamareli, J. (2012). *Síntese, caracterização e aplicação do polímero metil azoteto de glicidila (GAP) no desenvolvimento de novos propelentes para o programa aeroespacial brasileiro*. (Unpublished doctoral dissertation). Instituto Tecnológico de Aeronáutica, São José dos Campos, Brazil.

Silva, G., Rufino, S. C., & Iha, K. (2013). Green Propellants: Oxidizers. *Journal of AerospaceTechnology and Management*, *5*(2), 139–144. doi:10.5028/jatm.v5i2.229

Thormählen, P., & Anflo, K. (2012). Low-temperature operational and storable ammonium dinitramide based liquid monopropellant blends. *World Intellectual Property Organization*, *WO2012166046*, A2.

Thornton, J. R. (1967). Salts of nitroform with hydrazines. U.S. Patents, US3,297,747.

Vörde, C., & Skifs, H. (2005). Method of producing salts of dinitramidic acid. *World Intellectual Property Organization*, *WO2005070823*, A1.

Wynn, J. T. W., & Musker, A. J. (2015). Hydrogen peroxide catalyst. *World Intellectual Property Organization*, *WO2015033165*, A1.

Zee, F. W. M., Mul, J. M., & Hordijk, A. C. (1994). *Method of preparing hydrazine nitroform*. World Intellectual Property Organization.

APPENDIX: ABBREVIATIONS

ADN: Ammonium Dinitramide
AN: Ammonium Nitrate
AP: Ammonium Perchlorate
DCE: Dichloroethane
DCM: Dichloromethane
DMF: N, N Dimethyl Formamide
GAP: Glycidyl Azide Polymer
GEMs: Energetic material which green behavior
HAN: Hydroxylammonium Nitrate
HDPE: High Density Polyethylene
HNF: Hydrazinium Nitroformate
HTPB: Hydroxyl-Terminated Polybutadiene
LGP: Liquefied Petroleum Gas
LH: Liquid Hydrogen
LOX: Liquid Oxygen
PMMA: Polymethylmethacrylate

Chapter 9
Design and Testing of a Lab-Scale Hybrid Rocket Motor to Evaluate Swirl Injection Characteristics

Susane R. Gomes
Aeronautics Institute of Technology, Brazil

Leopoldo J. Rocco
Flowtest Aerospace Research, Brazil

ABSTRACT

This research aims to provide a methodology for the project of labscale hybrid motors. This development began with the thermal analysis of the fuel grain using the Flynn, Wall and Ozawa method, generating simulation entry data to maximize the motor performance. The simulation was performed with the Chemical Equilibrium Specific Impulse Code. Based on the optimum oxidizer to fuel ratio, the literature was used to supply the mathematical background to calculate the motor geometrical parameters whose operating conditions were determined throughout the simulation. Finally, firing tests were conducted to verify the reliability of the project methodology. The firing tests were performed with three injectors: two swirling and one axial. The tests showed that the higher the operating pressure the more suitable is the project, meaning the methodology developed works best in hybrid rocket motors with high operating pressures. Additionally, it was shown that the swirling flow injectors produce higher efficiency.

INTRODUCTION

Hybrid rocket motors present increased safety and reduced development and recurrent costs in comparison to other chemical propulsion means, higher Isp and chemical simplicity than solids and higher fuel density and mechanical ease than liquids. Nevertheless, combustion inefficiencies and low regression rates prevent the broad use of this technology (Altman & Humble, 1995), (Marxman & Gilbert, 1963).

DOI: 10.4018/978-1-5225-2903-3.ch009

Hybrid combustion occurs as a macroscopic diffusion flame, and the flame zone is established within the boundary layer (Altman, Rocket Motors, Hybrid, 2001). The regression rate depends largely on the convective heat transfer from the flame to the fuel surface. The examination of techniques to increase the combustion efficiency drew the scientific community to the research of oxidizer injection methods, and it was verified that the entrance conditions influence severely the burn pattern and the whole combustion process.

Our previous research (Gomes, Rocco, Rocco, & Iha, 2010) deals with the comparison between the performances obtained with the use of a swirling type injector over an axial. This work will expand this comparison and show the methodology used in the project of the labscale motors.

The different aspects of the effects of the oxidizer injection on the combustion efficiency and fuel consumption are analyzed and compared. Moreover, it was verified that the swirling flow increases the regression rates in the whole grain and, particularly, in the impingement region, due to the convective heat flux augmentation.

METHODOLOGY

This section outlines the steps taken in the prototype project: selection of materials; thermal analysis of the grain; combustion simulation and geometry determination.

Selection of Materials

The selection of the oxidizer physical state can be done attempting to the complexity inserted by this decision. A swirling two phase flow is avoided when the oxidizer is gaseous, meaning that more accurate and resistant hydraulic valves, which are considered necessary to operate in cryogenic temperatures, are not needed; this choice decreases cost and engine weight and simplifies the systems design.

Gaseous oxygen was used for its good performance, ease of handling and cost. The desired fuel must present low burning rates, in such a way that small changes in regression rates can be easily determined, in other words, such attribute make comparisons between miscellaneous performance results readily feasible. Most polymers present high viscosity and high surface tension, which fulfills the features required. Polyethylene was found to be rather appealing, once it is cheap, easy to find in the market and to conform in the desired form.

Two branches of polyethylene were tested for fuel grains, Ultra High Molecular Weight (UHMW) and High-Density Polyethylene (HDPE). The first tests evidenced that UHMW has better combustion stability and did not produce excessive carbon black, see Figure 1a. Meanwhile, tests conducted with HDPE showed high concentration of carbon black deposited all over the grain surface, Figure 1b.

The first nozzle used was made of stainless steel. But it could not handle a 15 s test without substantial throat erosion, see Figure 2. Therefore, graphite (20 μm grain length) was used. Graphite could endure each test with a maximum throat expansion of 0.5 mm, Figure 3.

Long tests were performed to address nozzle endurance in high temperature conditions. As a result it was observed that the stainless steel around the nozzle suffered melting before problems could be noticed on the graphite nozzle, which confirms the high quality of the graphite used, Figure 3.

Figure 1. A: Ultra High Molecular Weight fuel grain after test. B: High Density Polyethylene fuel grain after testing

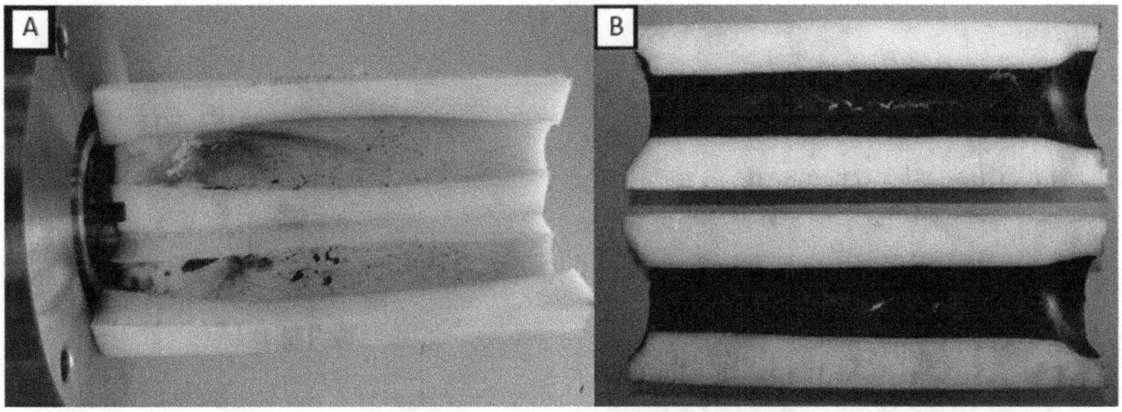

Figure 2. Stainless steel nozzle test (after 15s test)

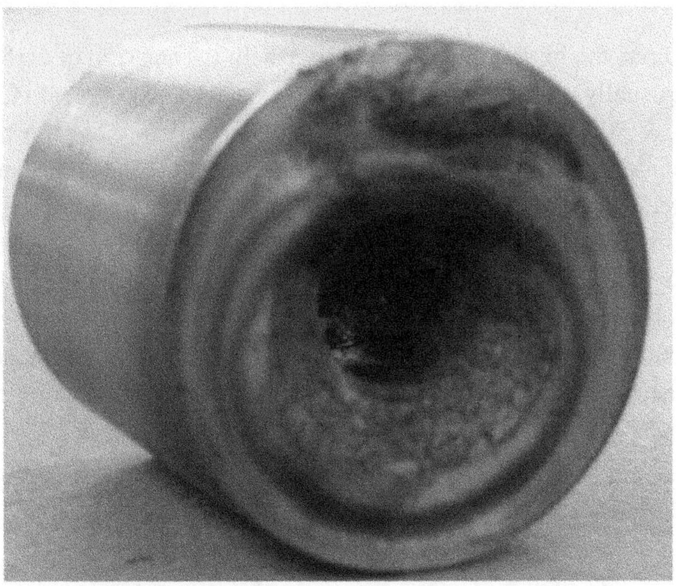

Thermoanalysis of the Grain

The thermal analysis provided the heat of formation of the ultra-high molecular weight polyethylene along with a general idea about its thermal decomposition. Some authors have proposed that the products of polyethylene decomposition include a wide variety of alkanes and alkenes (Grassie & Scott, 1985). In addition, the branching of polyethylene shows increased intramolecular hydrogen transfer, leading to lower thermal stability. The molecular change without vaporization is mostly occasioned by the weak links incorporated into the main chain as impurities (Grassie & Scott, 1985).

Figure 3. Graphite nozzles. A: Before test. B: After test, the nozzle is surrounded by melted stainless steel, 60s test

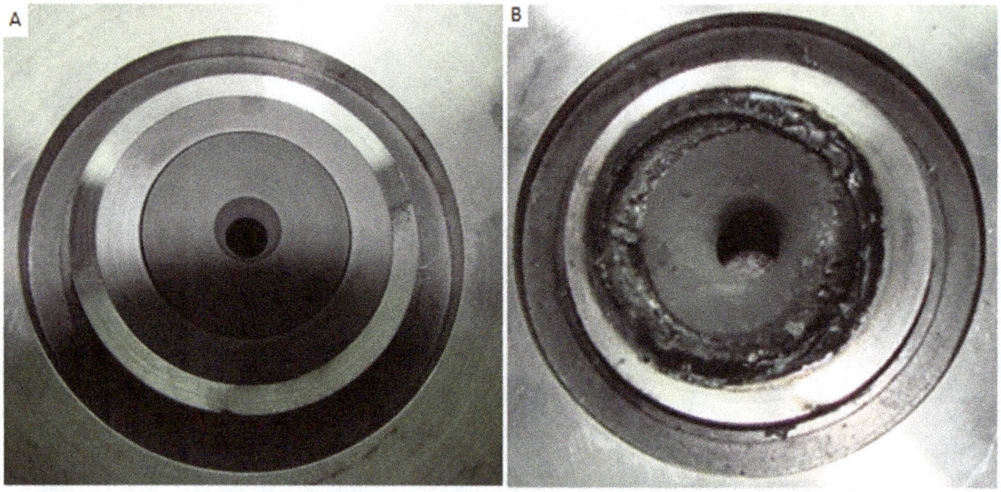

At higher temperatures, the initiation reactions occur with the scission of carbon bonds, usually tertiary, and products are usually propane, ethane, butane, hexane-1 and butane-1 (Grassie & Scott, 1985), (Peterson, Vyazovkin, & Wight, 2001), but propene and 1-hexene are the most abundant (Grassie & Scott, 1985), (Peterson, Vyazovkin, & Wight, 2001), (McGrattan, 1994). Grassie and Scott explained that the reaction between a radical with hydrogen on the fifth carbon should be more likely once the transition state is a six-membered ring (Grassie & Scott, 1985). Then, chain scission would produce two decomposition pathways: Route A producing propene and Route B, 1-hexene, see Figure 4.

DSC curves were obtained on a model DSC50 Shimadzu in the temperature range of 298-1000 K, under dynamic synthetic air atmosphere (ca. 50 mL/min). Sample masses were about 20 mg, and each sample was heated in hermetically sealed aluminum pans. Four different heating rates were used for the

Figure 4. Proposed mechanism of PE scission
McGrattan, 1994.

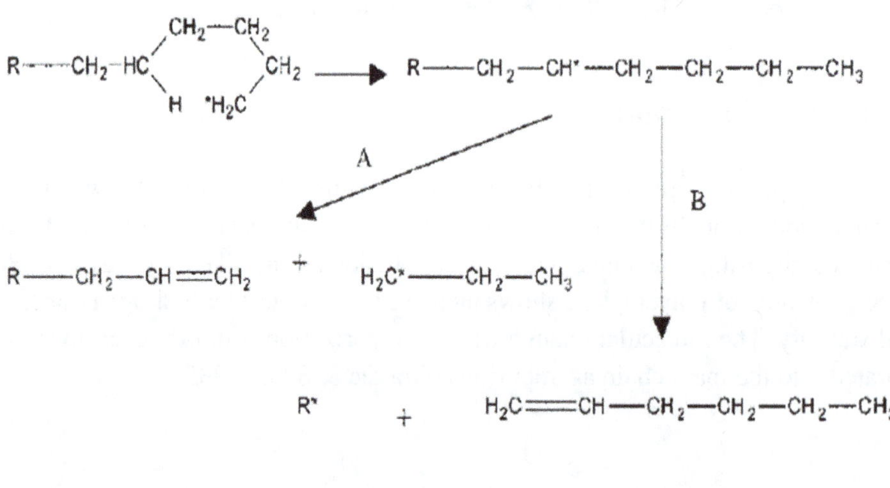

polyethylene samples: 20, 30, 35, 40 K/min. DSC system was calibrated with indium and zinc. TG curve was obtained with a heating rate of 20, 30, 40, 50, 60 K/min.

The activation energy of thermal decomposition of the fuel was approximately 200 KJ/mol and this process started at about 430 °C. The chemical properties of the fuel used in this work are shown in Table 1.

Combustion Simulation

The simulation is accomplished with the use of the American Air Force Chemical Equilibrium Specific Impulse Code – ISP code). This code performs the thermodynamic calculation necessary to predict the theoretical value of the specific impulse for a variety of pressure and oxidant to fuel ratios.

The assumptions of the program are one dimensional form of the continuity, energy and momentum equations; zero velocity at the combustion chamber inlet; complete and adiabatic combustion; isentropic expansion in the nozzle; homogeneous mixing; ideal-gas law; zero temperature and velocity lags between condensed and gaseous species.

The performance estimation encompasses rocket motor thrust and Isp evaluation through chemical equilibrium, isentropic expansion, and statistical calculations. Once the ratio of specific heats, temperature, and molecular weight of the exhaust gases are known, subroutines determine ideal nozzle performance in terms of thrust coefficient, characteristic exhaust velocity, and mass flow rate. These subroutines can utilize expressions for frozen one-dimensional, isentropic expansion.

The flows with null reaction rates are defined as frozen flows. This is the opposite of equilibrium flow, which implies infinitely fast reactions. Finally, a statistical sampling process corrects variations in performance due to real processes and provides indications of thrust and Isp uncertainty.

The flow through the nozzle can be calculated using two different methodologies, standard analysis assumes the flow is in stationary equilibrium, i.e., in any particular position down the nozzle the flow is in equilibrium, and along the expansion, the temperature and pressure alter. This method has the tendency to over evaluate the Isp result because, as expected, the flow is usually unable to reach a chemical equilibrium state due to its high velocity.

The other option is to use the frozen flow routine, which supposes a constant chemical composition along the expansion. The gas can be defined as frozen at the entry, throat or at the nozzle exit. The difference between the results is minor, but setting frozen at the nozzle requires the entry of the expansion ratio in the code.

The stoichiometric ratios were computed by balancing the fuel components with the exhaust products in the chemical formula for potential complete combustion.

$$C_2H_4 + 3O_2 \rightarrow 2CO_2 + 2H_2O \therefore O / F_{st,\ mass} = \frac{3 * 32}{28} = 3.43$$

Table 1. Chemical properties of the oxidizer and fuel

Fuel	GO2	UHMWPE
Molecular formula	O2	C2H4
Density (g/cm3)	0,0014	0,9
Enthalpy of Formation (cal/g)	0	-1270

Geometry Determination

The hypotheses assumed for the prototype design are the following:

1. The burning happens only in the cylindrical surface of the grain
2. The oxygen mass flow is constant and can be specified
3. The regression rate is constant over the whole burning
4. The throat erosion is considered negligible
5. The terms of mass accumulation are neglected.

Efficiencies from 0.93 to 0.97 are achieved in several experiments described in hybrid engines propellants1.In this work, it will be assumed a value of 0.93. The nozzle throat area was calculated assuming isentropic flow in the combustion chamber with Mach 0.2, which is a good approximation according to the literature (Altman, Rocket Motors, Hybrid, 2001).

Due to simplicity of calculation, conical nozzles will be used. An increased divergence angle raises the radial-velocity component, diminishing the thrust, but also shortens the nozzle for a specific expansion rate.

Usually, cone half-angles can range from 12° to 18°. The usual middle ground is a half angle of 15°. Bell nozzles present better performance than 15° half-angle conical nozzles of the same length. On the other hand, the development of bell nozzles involves complex calculations using computational fluid dynamics.

Nozzles are designed to be made from high-density graphite with a 30º half-angle convergent section and a 15º half-angle divergent section. Nozzle throat diameters vary with oxidizer flow rate while maintaining a constant expansion rate for each chamber pressure, as obtained in the theoretical analysis. In order to minimize the throat erosion, allowing the reutilization of the nozzle, this segment will have a 5 mm length.

With the assumed burn time of 20 seconds, the pre and post-combustion chambers are usually scaled with lengths ranging from half to one internal diameter1.The pre-chamber will have a length of one port diameter and the post-chamber, half that value, the inverse configurations may also be tested, just by altering the chamber position and turning the pre-chamber into the post-chamber.

A structural investigation is necessary to guarantee the safety during the engine operation. Axial tension and transverse efforts were taken into account considering the required precautions with a high safety margin.

THEORETICAL RESULTS

Simulation

The operating pressures range between 10 and 30 atm, and the external pressure is assumed to be 1 atm. The specific impulse and the vacuum specific impulse are plotted as function of the oxidizer to fuel ratio, see Fig. 5 and 6. From the best fit of the curves one can calculate the maximum values of Isp, thus obtaining the best O/F combination, used in the next design steps.

Figure 5. Comparison of specific impulse vs. oxidizer to fuel ratio for different pressure values

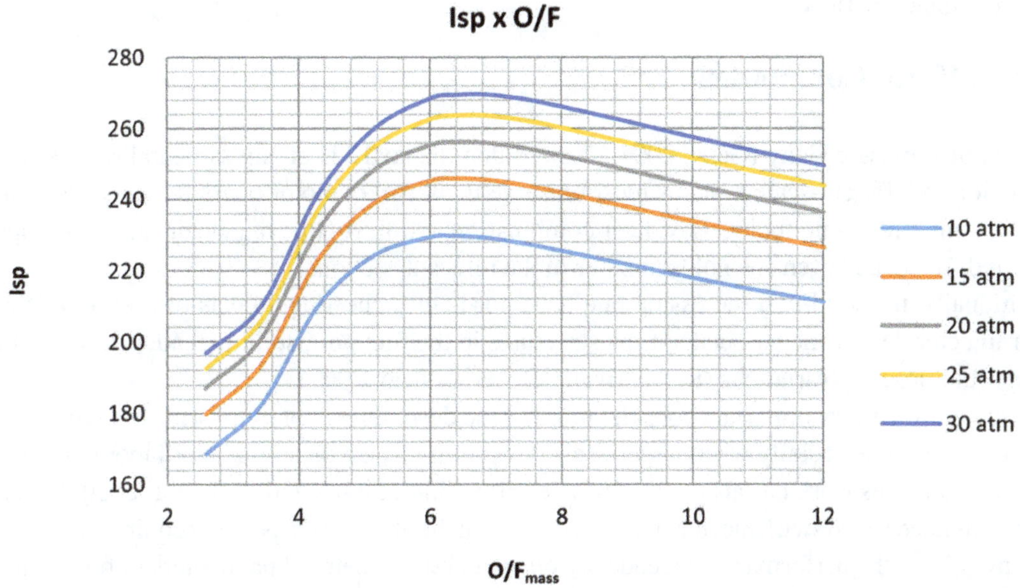

Figure 6. Characteristic velocity vs. oxidizer to fuel ratio for different pressure values

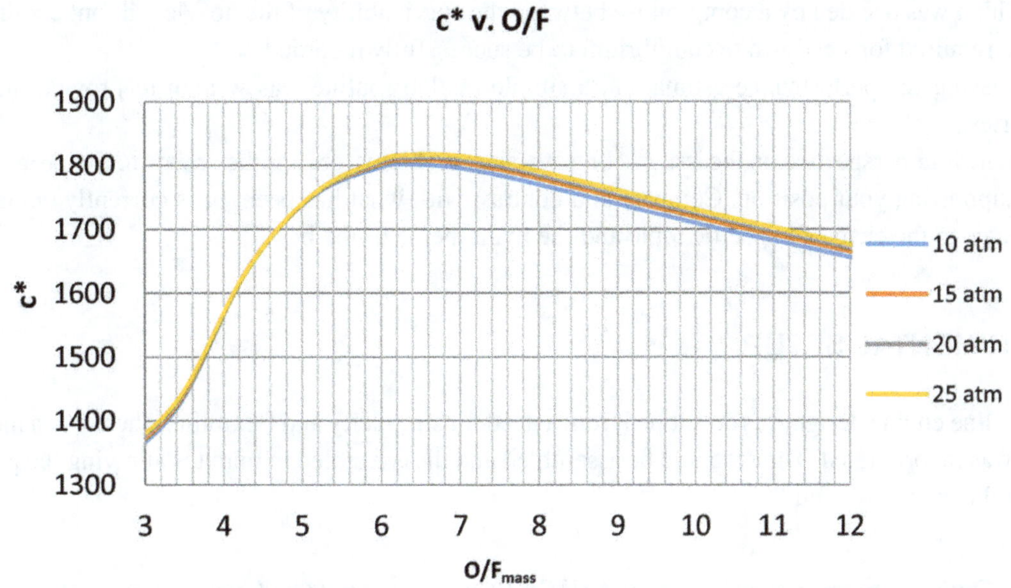

The best oxidizer-to-fuel ratios for each pressure were obtained. The optimum performance parameters are exhibited in Table 2.

Labscale Motor Geometries

The works of Professor Gany (Gany, 2007) showed that the analysis of experimental results and their extrapolation to different-scale systems must preserve main rocket motor operating conditions such as geometric similarity, same fuel and oxidizer combination, oxidizer flow rate determined in proportion to the length scale, and even Reynolds number in the relevant sections.

Additionally, the solid-fuel regression rate behavior is influenced by effects of fuel port geometry, oxidizer injector design and associated flow structures in the fuel port, solid fuel additives, and boundary layer fluid property variations (Risha, Evans, Boyer, & Kuo, 2007).

Size scale-up is a very important consideration for the development of large-scale operational hybrid propulsion systems, especially when larger-scale designs are based on subscale or laboratory-scale experimental data. This work has also the purpose of providing designs with different length to diameter ratios, port diameters and swirl injector characteristics with the ultimate purpose of acquiring experimental data to investigate the performance dependency on the cited geometrical parameters in future analyses.

Two auxiliary chambers were set up ahead and aft of the grain. The pre-combustion chamber provides an uniform entrance condition with the effort of increasing the overall regression rates, while the post-combustion chamber provides better mixing of the gasifying propellants (Altman & Humble, Hybrid rocket propulsion systems, 1995).

The burning duration time of the single cylindrical grain ports of UHMWPE was defined as 20 seconds, which was decided by a compromise between the survivability of the nozzle without cooling and the time required for steady-state equilibrium to be successfully recorded.

Possessing the performance estimation, a simple matlab routine was written to provide the final geometries.

Present your perspective on the issues, controversies, problems, etc., as they relate to theme and arguments supporting your position. Compare and contrast with what has been, or is currently being done as it relates to the chapter's specific topic and the main theme of the book.

EXPERIMENTAL SETUP

The baseline engine design was developed from a need for simplicity and flexibility; therefore a modular design was incorporated. The set could be assembled and disassembled in minutes allowing the practice of several tests per session.

Table 2. Optimum performance values for UHMW Polyethylene and GOX combustion

P (atm)	10	15	20	25	30
O/Fmass	6.34	6.54	6.58	6.60	6.61
Isp (s)	229	245	256	263	269
Isp vac (s)	269	280	287	293	297
c*(m/s)	1801	1807	1812	1815	1818

The case and the flanges were made from stainless steel. The case was machined to fit between the steel flanges. The nozzle was adapted in the aft flange, to prevent nozzle escape. A hydraulic system was settled to perform the thrust measurements. A schematic diagram of the test facility is exposed in Figure 7. A view of the motor attached to the test bench is shown in Figure 8. Compression fitting valves were used in the oxygen feed line. A pyrotechnic method was used to achieve ignition.

The chamber's length is 215 mm with an inner diameter of 68.3 mm. The grains are 195 mm long. The aft-chamber has a span of 15 mm and the pre-chamber is 5 mm long.

Injector geometries

Three injectors were built, one axial and two swirls. On behalf of clarification, the swirling types will be distinguished in reference to the size, the small opposed to the large one. The injectors can be used separately or combined. The basic characteristics of the systems are listed in Table 3.

The following pictures show different views of the injectors. The axial or small swirl can be used with the same setup, because there are separate hoses of oxygen supply connected to each injector as shown in Figure 9. The easy interchange of injectors in the head flange and effortless alteration in the oxygen feed line due to the compression fit valves connections contribute to the trouble-free experimental apparatus. Figures 10 and 11 display the injectors attached to the head flange of the test bench. Figure 12 shows the lateral view of the large swirl alone.

All injection exits and entrances are identical to guarantee homogeneous flow and regular burning along the port. The large injector outlet area works like an additional pre-chamber, enhancing the uniformity of the incoming flow conditions.

Figure 7. Diagram of the experimental apparatus

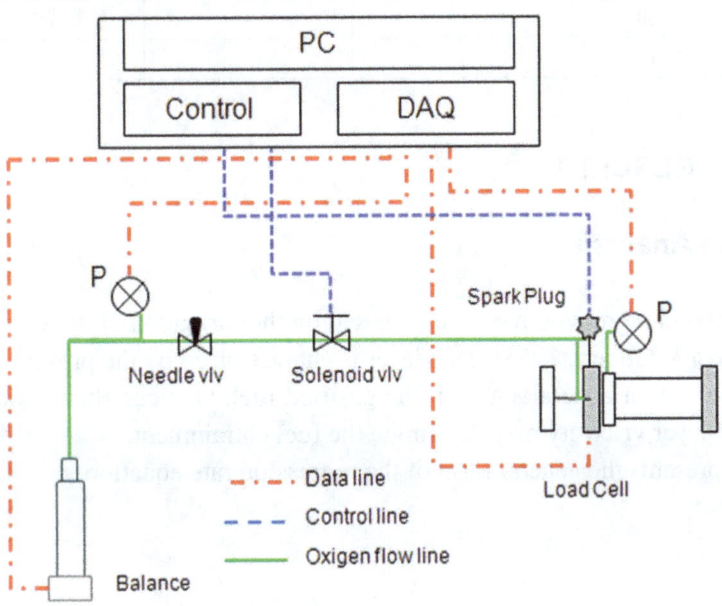

Figure 8. Picture of the hybrid rocket motor configuration

Table 3. Basic set configurations of single and combined injectors

Configuration	Injection Method	Number of Entrances	Exit Diameter (mm)	Number of Exits	Exit $A = \pi r2$ (mm2)
1	Axial	1	3	1	7.07
2	Small swirl	1	1.5	4	7.07
3	Large swirl	4	1.5	4	7.07
4	Axial + Small Swirl	2	-	5	14.14
5	Axial + Large swirl	5	-	5	14.14

EXPERIMENTAL RESULTS

Regression Rate Analysis

Some authors proposed a regression mechanism based on the formation of a melting layer at the burning surface (Marxman & Gilbert, 1963). This layer is characterized by the presence of droplets of fuel entrained in the combustion zone along with the gasified fuel. One can then deduce that the surface tension and the melt layer viscosity may determine the fuel entrainment.

Equation (14) represents the general form of the regression rate equation.

$$\dot{r} = aG^n x^m \tag{14}$$

Figure 9. Cross view of the injector with small swirl and axial exits with the two possible entrances and exits of oxidizer flux

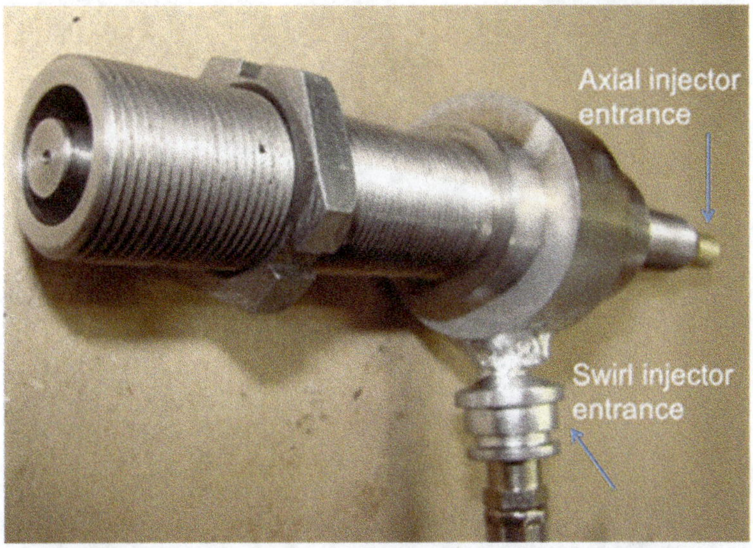

Figure 10. Injector mounted in the test bench

Figure 11. Front view of the large swirl with the four oxidizer supply entrances

Figure 12. Lateral view of the large view without the connectors and hoses

Along the port, two competing effects are the source of fuel consumption uniformity. As the boundary layer, denoted by x^{m}, increases, heat flux decreases, nevertheless at the same time the total mass flux increases due to the gasified fuel. Therefore, the regression rate may be expressed in terms of average values, as follows.

$$\dot{r}_{avg} = aG_{avg}^{n} L_{p}^{m} \tag{15}$$

Since the ballistic parameters depend on the mass flux, researchers found more appropriate, and with reasonable precision, to use the simplified burning expression, Eq. (3), for a given motor length and constant mass flow rate (Altman & Humble, Hybrid rocket propulsion systems, 1995), (Altman, Rocket Motors, Hybrid, 2001).

$$\dot{r}_{avg} = a_0 \bar{G}_{ox}^{n} \tag{16}$$

Data Reduction

The key purpose of the regression rate analysis is to determine the correlation between regression rates and easily controlled parameters along with experimentally derived constants. Correlation patterns were determined by making scatter plots and applying correlation analysis to the cluster of data points provided by the test firings.

The average regression rate was calculated as follows (Altman, Rocket Motors, Hybrid, 2001):

$$\bar{\dot{r}} = \frac{\overline{\dot{m}_f}}{\pi \left(\frac{D_i + D_f}{2} \right) L_p \rho_f} \tag{17}$$

The average oxygen mass flux equation considers the area associated to the medium diameter

$$\bar{G}_{ox} = \frac{4\overline{\dot{m}_{ox}}}{\pi \left(\frac{D_i + D_f}{2} \right)^2} \tag{18}$$

The medium propellant mass flow rate was calculated considering the mass of fuel and oxygen used.

$$\overline{\dot{m}} = \overline{\dot{m}_f} + \overline{\dot{m}_{ox}} \tag{19}$$

The I_{sp} values were obtained from the thrust data collected in the tests.

$$I_{sp} = \frac{F}{\dot{m}g_0} \tag{20}$$

Performance Parameters Comparison Between Different Injection Methods

The results of the three series of four tests are presented in Table 4. The plot is shown in Figure 13.

The diffusion-controlled model assumed that reaction rates are faster than the diffusion rates and that the boundary layer is much larger than the combustion zone. But it is observed that, at constant pressure, when mass flux decreases the regression is likely to deform and depend on the radiation environment, becoming larger than the extrapolated values (Altman & Humble, Hybrid rocket propulsion systems, 1995), (Altman, Rocket Motors, Hybrid, 2001). Differently, at high mass fluxes or low pressure, turbulence diffusion rates can dominate and reaction rates, which are influenced by pressure, become rate determining. Thus, the regression rates goes down the extrapolated values, see Figure 14.

The following graph, Figure 15, shows the data dispersion with smooth lines. The curve for the large swirl resembles the central line of Figure 14 associated with diffusion dominated. The values of the constant n obtained sustain this affirmation.

Table 4. Experimental regression rates and fluxes collected in the four firing tests of each injector

Injector Type	$\dot{r}$ (mm/s)	$\overline{G}_{ox}$ (kg/m²s)	$\dot{r}$ (mm/s)	$\overline{G}_{ox}$ (kg/m²s)	$\dot{r}$ (mm/s)	$\overline{G}_{ox}$ (kg/m²s)	$\dot{r}$ (mm/s)	$\overline{G}_{ox}$ (kg/m²s)
Axial	0,39	53,6	0,38	53	0,37	42,0	0,27	16,2
Large Swirl	0,55	40,3	0,52	34,1	0,32	11,4	0,21	4,5
Small Swirl	0,45	45,1	0,41	50,4	0,22	19,6	0,17	6,11

Figure 13. Regression rate vs. oxidizer mass flux for each injection method

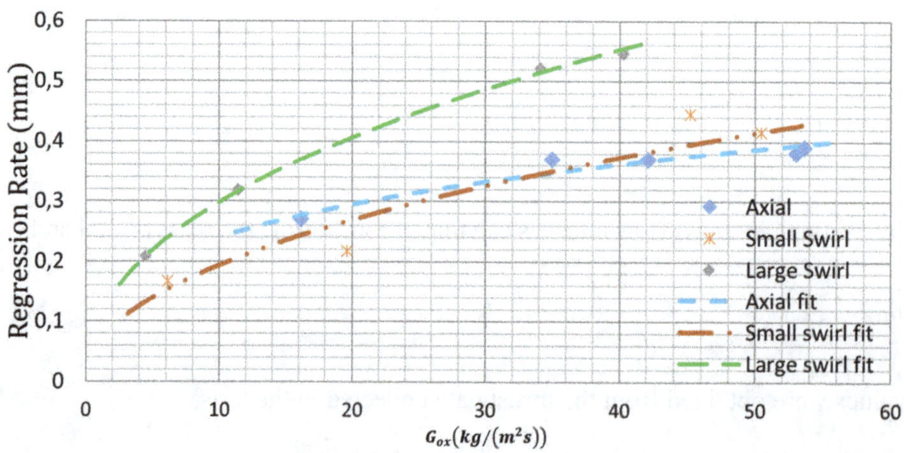

Figure 14. Diagram of the influence of radiation and kinetics effects on regression rates

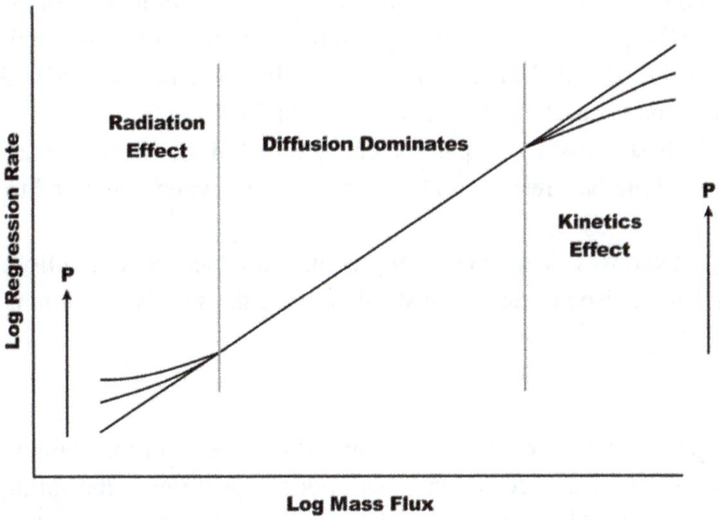

Figure 15. Scatted data with smooth lines

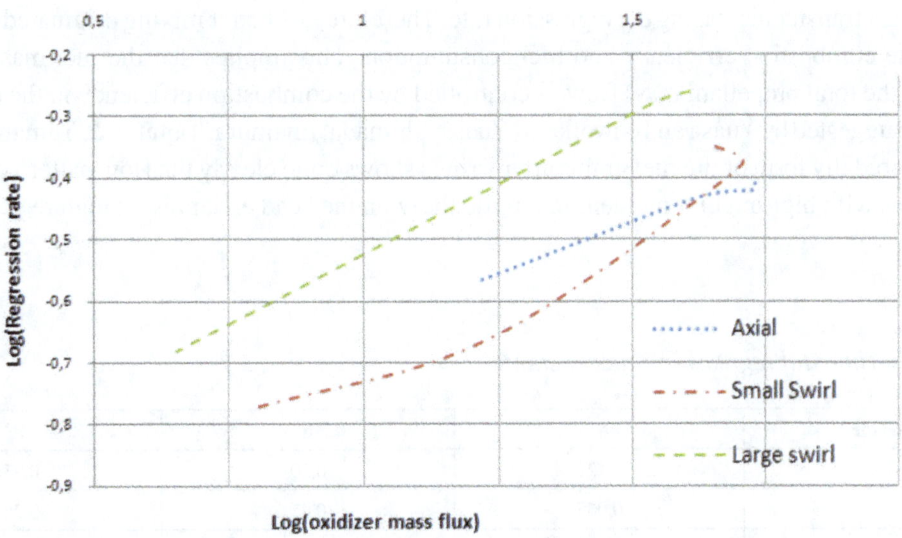

Figure 15 shows how this effect may change the combustion process and thus the regression rate equation. In the left tip of the small swirl curve, the values are higher than expected; this is probably due to the soot caused by the low oxidizer income flux. In this situation, the irradiative heat transfer becomes vital, and increases the overall efficiency. Differently, the axial injector curve shows a diminishment in the regression rate curve, to explain this some considerations must be done.

Tangential velocities reduce the boundary layer (Schlichting, 1995) and, therefore, flame zone. As the oxidizer mass flux grows the layer increases along until it reaches the point where the process becomes pressure dependent. This growth diminishes the reaction rates until they are smaller than the diffusion velocity, which conflicts with the diffusion dominant theory, and that's why the regression rate expression doesn't fit anymore and is smaller than what would be expected. Therefore, it points out that tangential velocities help increase the combustion efficiency and maintain the diffusion dominant, and this is also why the performance obtained with the axial injector is smaller than the obtained with the swirl injectors.

The regression equation for each configuration is shown in Table 5.

Table 6 presents the best performance parameters obtained in the firings for each injection method. The larger swirl provided the best results in all judgments even when compared to the combination of swirl methods.

From the data collected it was clear that higher pressures produce more stability and the tests become more repeatable. The thrust also appears to be steadier when the supply pressure is high.

Grain Analysis

Experimental investigations have shown that the controlling factors in the combustion process are the rate of heat transfer to the solid surface and the heat of decomposition of the solid-phase fuel (Altman, Rocket Motors, Hybrid, 2001). The mass flux, which is regulated by the rate of flow of the liquid phase, determines the rate of heat generated in the combustion zone and hence determines both the heat transfer and the thrust magnitude.

The oxidizer flow condition on the fuel surface causes the turbulent diffusion flame which in turn affects the heat transfer increasing the regression rate. Therefore, the better mixing originated by the swirl increases the combustion efficiency and fuel consumption. This implies that the fuel mass flow rate, and in turn, the total propellant mass flow, is controlled by the combustion efficiency in the combustion chamber as suggested by Yuasa and coworkers (Yuasa, Shimada, Imamura, Tamura, & Yamamoto, 1999).

The helicoidally form of the fuel grains after firings shows quite clearly the flow pattern over the surface. Regions with high angular momentum, particularly on the head end, present high regression rates.

Table. 5 Experimental regression rate constants

Injector	A_0	n	R^2
Axial	0,121	0,30	0,936
Small swirl	0,065	0,47	0,961
Large Swirl	0,107	0,45	0,999

Table 6. Performance comparison of the best parameters obtained for each injection configuration

Injector Type	$\dot{r}$ (mm/s)	O/F	Thrust (N)	Isp (s)	$\overline{G}_{ox}$ (kg/m²s)
Axial	0,39	4.63	55,1	134	53,6
Large Swirl	0,55	2.50	75,0	195	40,3
Small Swirl	0,40	3.60	73,9	187	45,1

The flow from the swirling type injector enters the combustion chamber with a centrifugal force; this condition is responsible for the high oxygen flow velocity, particularly the tangential component, and the augment of the density near the surface. These summing effects plus the amplified residence time account for the boundary layer shrink and, therefore, the enhancement of the burn rates.

The grains produced by each series of tests showed different consumption patterns. Both swirl injectors provided helicoids on the grain surfaces, but the areas of highest regression rates differ. The correspondent area in the larger swirl tests is on the head end, while in the small swirl this location is dislocated to the right.

The following picture, Figure 16, is a grain used with the axial injector. The primary remark is the irregular burn profile along the longitudinal axis, the regression rate presents a maximum at about 40% of the grains' length, and then the consumption keeps decreasing until the aft end.

The next observation is the flow pattern observed on the grain surface of the small swirl injector tests. The rotating entrance conditions can be perceived by the helicoidally disposed curves on the surface.

The regression rates achieved in the beginning of the grain are the highest values. This region is usually called impingement zone due to the strong influence of the incoming rotating flow. The high angular moment enhances the heat exchange which magnifies the velocity of fuel degradation or pyrolysis. However, the excessive carbon black in the cavity close to the head end means that there was a lack of oxidizer or the residence time, in this region, was insufficient for complete combustion.

The following analysis, Figures 19 and 20, regards the union of two different injector methods. The axial was used in parallel with the small swirl injector. It is shown that one side was more burned than the other; and the massive presence of carbon black on the surface probably indicates that the combustion was very unstable.

Subsequently, the grains obtained in the large swirl firings are shown in Figures 21 and 22.

Figure 16. Grain cross section after axial injector test

Figure 17. Helicoidally disposed curves on the grain surface consequent of the small swirling injector test

Figure 18. Grain superior view obtained from the small swirl injector test

Figure 19. Grain cross section after axial combined with small swirling injectors test

Figure 20. Grain superior view; helicoidally disposed imprints along the axis obtained after axial and small swirling injectors test

Figure 21. Grain cross sectional view after large swirl firings

Figure 22. Helicoidally consumption pattern after a typical large swirl test firing

The grain observation supports the superiority of the large swirl, which was already evidenced by the collected data, because the deposit of carbon black on the surface and on the nozzle entrance was by far the least and the regression rates were the largest.

In all tests, the lower end of the grain presented quite smaller regression rates than those from other cross sections, which is explained by the irradiative heat transfer phenomenon, dominant in this region due to the higher total mass flux.

A graphical explanation of the conclusion presented about the patterns evidenced in the highest regression rates is shown in Figure 23. To summarize, higher tangential velocities imply higher heat exchange, which induces higher regression rates.

The area obtained in the small swirl case is shifted to the right, see Figure 24. This effect is probably due to the smaller tangential velocities.

Injectors' Dimensions

The different tangential velocities are also largely influenced by the injectors' exit chamber dimension, see Figure 25. Exits with large diameters may be responsible for a wide consumption of the head end of the grain, while small chambers may resemble an axial injector and move the highest burn area towards the grain end.

Blocking Effect

All tests with axial injector harmed the injector material as shown in Figure 26, the integration of a long pre-chamber was imperative. The flow downstream the injector exit forms a recirculation zone which blocks heat transfer from the flame to the bulk head, therefore there is no need for insulation in the swirl case. The same barrier effect was obtained and documented by Jones and coworkers (Jones, Myre, & Cowart, 2009).

Figure 23. Region of highest regression rates in the grains tested with the large swirl

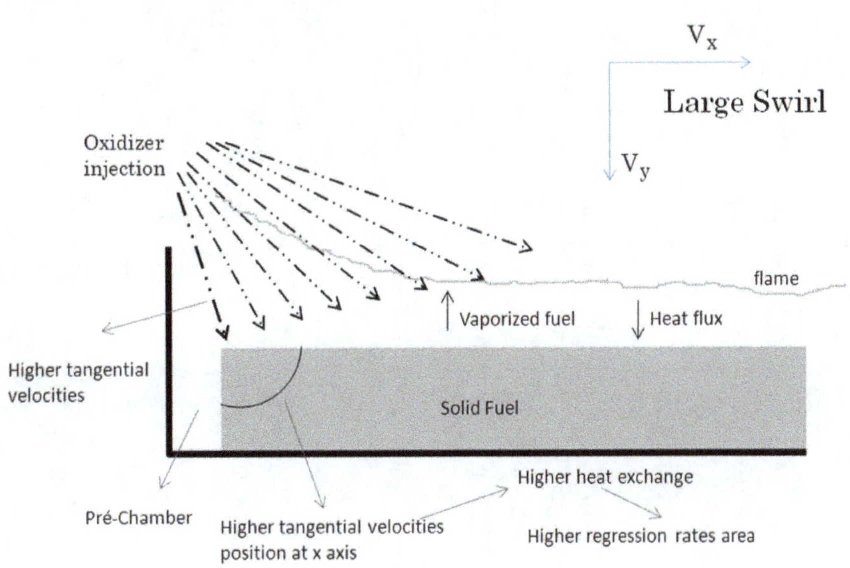

Figure 24. Region of highest regression rates in the grains tested with the small swirl

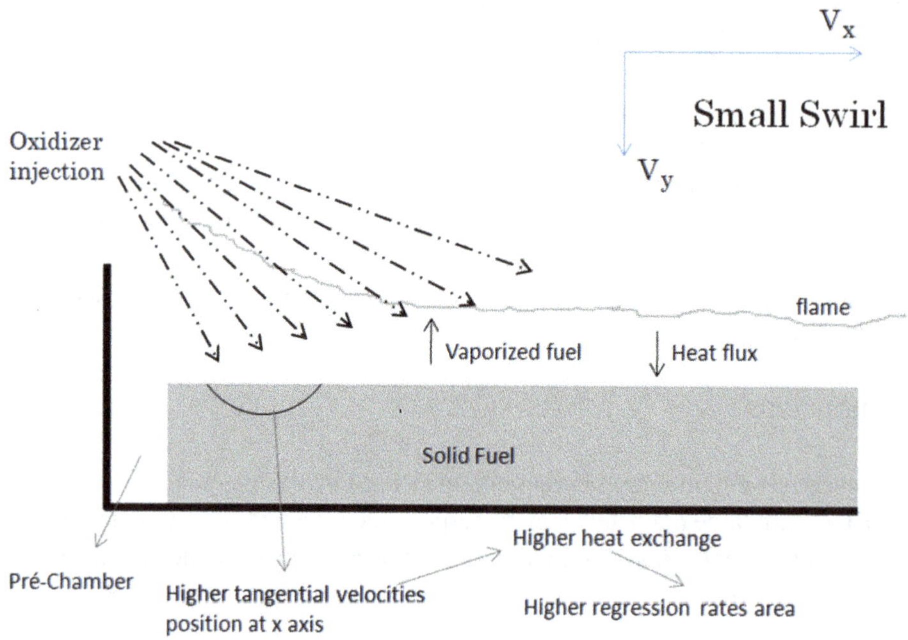

Figure 25. Relationship between tangential velocities of the incoming flow and the injectors' exit chamber dimension

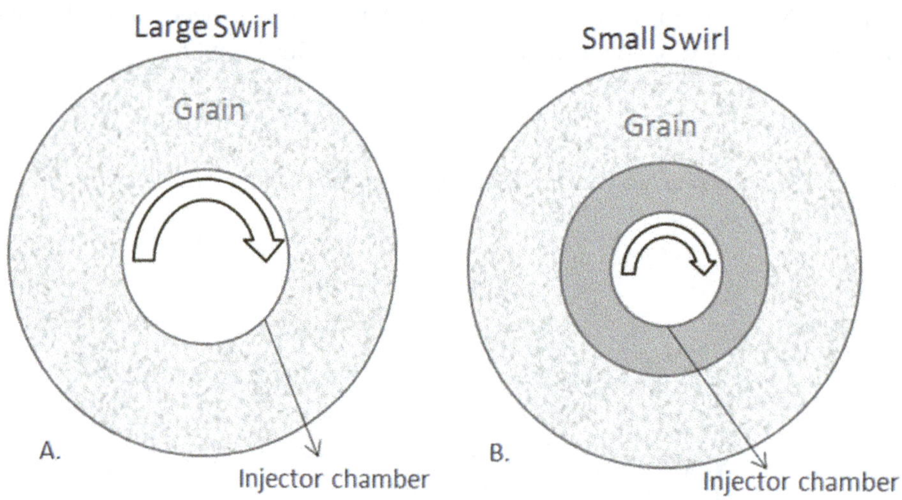

Figure 26. Comparison between injector cross-section after firings, axial test (left) and large swirl test (right)

COMPARISON BETWEEN THEORETICAL AND EXPERIMENTAL RESULTS

The pressures reached in the tests during the use of the axial injector didn't reach the operational pressures simulated, then, the reverse order was done, and simulations were produced for the pressures obtained in the experiments. The simulation inputs also include the oxidizer-to-fuel rations obtained. The results for the axial injector are shown in Table 7.

The comparison between the theoretical and the test data showed that the higher the pressure the stronger the convergence, but the limited amount of records makes it difficult to set a concrete conclusion.

The tests with the small and the large swirls achieved the simulated pressures. It is not clear why that has happened to the tests with the axial injector, one could guess that the reaction and regression rates were so much lower than the achieved in the other tests, that a sufficiently high pressure couldn't be obtained.

A comparison between the data for the swirl cases is shown in Tables 8 and 9.

CONCLUSION

A methodology for the design of labscale hybrid rocket motors was proposed. The theoretical results were compared to the experimental data, and a relationship between the accuracy of the model and the operational pressure was perceived.

An additional objective of this research was to evaluate and compare the regression rates of polyethylene fuel grains under three different oxidizer injection methods. Two swirl and one axial injector types were tested and the performance obtained was recorded and compared. The swirl injectors produced better conditions than the axial, which is in accordance with the literature data (Carmicino & Russo Sorge, 2003), (Knuth, Chiaverini, Gramer, & Sauer, 1998). The larger swirl provided an even better outcome in the regression rates, thrust and specific impulse.

Table 7. Comparison theoretical versus experimental for the axial injector

P (atm)	O/F	c^*_{exp} (m/s)	ISP_{exp} (s)	Isp_{theory} (s)	C^*_{theory} (m/s)	$Isp_{relative}$ $\frac{ISP_{exp}}{ISP_{theo}}$	$C^*_{relative}$ $\frac{c^*exp}{c^*theo}$
3,06	3,80	660,58	67,41	153,00	1643,79	0,44	0,40
4,96	4,84	946,87	96,62	173,22	1584,78	0,56	0,60
7,60	4,63	1311,08	133,78	194,30	1605,99	0,69	0,82

Figure 27. Comparison between theoretical and experimental results in relation to chamber pressure

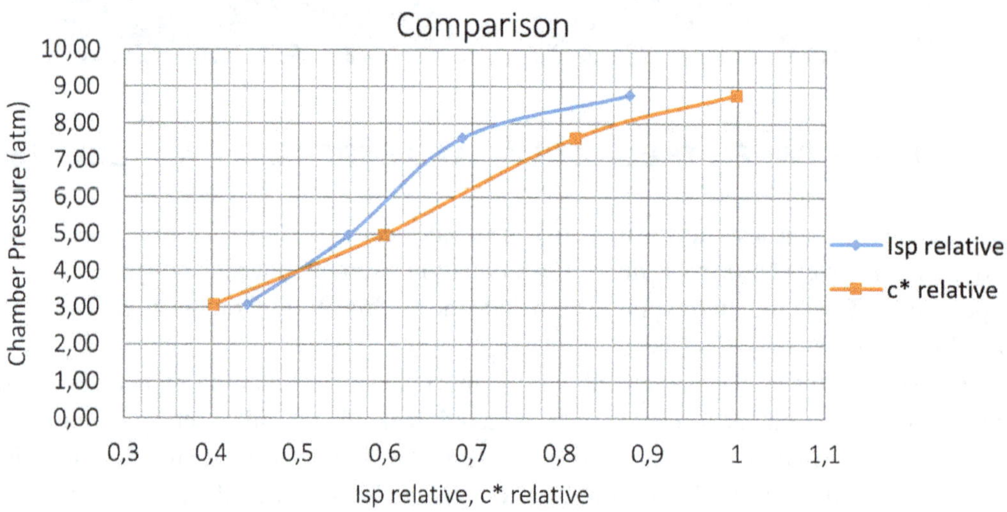

Table 8. Comparison theoretical vs. experimental for the large swirl injector

Pressure	O/F	Isp_{exp}	Isp_{theory}	$Isp_{relative}$
10.0	2.69	128	168	0.76
11.0	3.21	185	217	0.85

Table 9. Comparison theoretical vs. experimental for the small swirl injector

Pressure	O/F	Isp_{exp}	Isp_{theory}	$Isp_{relative}$
12.0	3.60	187	220	0.85
4.1	1.31	130	178	0.73

The non-uniform port diameter obtained in burned grains sustains that regression rate strongly depend on the incoming oxidizer flow pattern. Therefore, applying swirl to the oxygen jet improves low performance of the hybrid fuel combustion as proposed in the literature (Carmicino & Russo Sorge, 2003), (Knuth, Chiaverini, Gramer, & Sauer, 1998). However, future research should verify how the combustion pattern changes in different system scales, once this effect is not yet well quantified but is essential to the design of hybrid rocket motors based on labscale tests.

ACKNOWLEDGMENT

The authors gratefully acknowledge financial support from CNPq (National Council for Scientific and Technological Development).

REFERENCES

Altman, D. (2001). Rocket Motors, Hybrid. In D. Altman (Ed.), *The Encyclopedia of Physical Science and Technology*.

Altman, D., & Humble, R. (1995). Hybrid rocket propulsion systems. In R. Humble, G. Henry, & W. Larson (Eds.), *Space Propulsion Analysis and Design* (pp. 365–441). New York: McGraw-Hill.

Carmicino, C., & Russo Sorge, A. (2003). Investigation of the Fuel Regression Rate dependence on Oxidizer Injection and Chamber Pressure in a Hybrid Rocket. In *Proceedings of the 39th AIAA/ASME/SAE/ASEE Joint Propulsion Conference and Exhibit*, Huntsville, Alabama. doi:10.2514/6.2003-4591

Gany, A. (2007). Similarity and Scaling Effects in Hybrid Rocket Motors. In M. J. Chiaverini, & K. K. Kuo (Eds.), Progress in Astronautics and Aeronautics - Fundamentals of Hybrid Rocket Combustion and Propulsion (Vol. 218). AIAA.

Gomes, S. R., Rocco, L. J., Rocco, J. A., & Iha, K. (2010). Gaseous Oxygen Injection Effects in Hybrid Labscale Rocket Motor Operations. In *Proceedings of the 46th AIAA/ASME/SAE/ASEE Joint Propulsion Conference and Exhibit*, Nashville, Tennessee. doi:10.2514/6.2010-6545

Grassie, N., & Scott, G. (1985). *Polymer Degradation and Stabilization.* Cambridge: Cambridge University Press.

Jones, C. C., Myre, D. D., & Cowart, J. S. (2009). Performance and Analysis of Vortex Oxidizer Injection in a Hybrid Rocket Motor. In *Proceedings of the 45th AIAA/ASME/SAE/ASEE Joint Propulsion Conference and Exhibit*, Denver, Colorado. doi:10.2514/6.2009-4938

Knuth, W. H., Chiaverini, M., Gramer, D., & Sauer, J. (1998). Experimental investigation of a vortex-driven high-regression rate hybrid rocket engine. In *Proceedings of the 34th AIAA/ASME/SAE/ASEE Joint Propulsion Conference and Exhibit*, Cleveland, Ohio. doi:10.2514/6.1998-3348

Marxman, G. A., & Gilbert, G. M. (1963). Turbulent boundary layer combustion in the hybrid rocket. In *Proceedings of the Ninth Symposium (Intern.) on Combustion* (pp. 371-383). doi:10.1016/B978-1-4832-2759-7.50048-0

McGrattan, B. J. (1994). Decomposition of Ethylene—Vinyl Acetate Copolymers Examined by Combined Thermogravimetry, Gas Chromatography, and Infrared Spectroscopy. In T. Provder, M. W. Urban, & H. G. Barth (Eds.), Hyphenated Techniques in Polymer Characterization (pp. 103–115). Washington, D.C.: Americal Chemical Society. doi:10.1021/bk-1994-0581.ch008

Peterson, J. D., Vyazovkin, S., & Wight, C. A. (2001). Kinetics of the Thermal and Thermo-Oxidative Degradation of Polystyrene, Polyethylene and Poly(propylene). *Macromolecular Chemistry and Physics*, *202*(6), 775–784. doi:10.1002/1521-3935(20010301)202:6<775::AID-MACP775>3.0.CO;2-G

Risha, G. A., Evans, B. J., Boyer, E., & Kuo, K. K. (2007). Metals, Energetic Additives, and Special Binders Used in Solid Fuels for Hybrid Rockets. In M. J. Chiaverini, & K. K. Kuo (Eds.), Progress in Astronautics and Aeronautics - Fundamentals of Hybrid Rocket Combustion and Propulsion (pp. 423-442). AIAA.

Schlichting, H. (1995). *Boundary Layer Theory*. New York: McGraw-Hill Book Company Inc.

Yuasa, S., Shimada, O., Imamura, T., Tamura, T., & Yamamoto, K. (1999). A technique for Improving the performance of Hybrid Rocket Engines. In *Proceedings of the 35th AIAA/ASME/SAE/ASEE Joint Propulsion Conference and Exhibit*, Los Angeles, CA. doi:10.2514/6.1999-2322

Chapter 10

Thermal Decomposition Kinetics Studies of HTPB/Al/AP Solid Propellants Formulated With Iron Oxide Burning Rate Catalyst in Nano and Micro Scale

Luis Eduardo Nunes Almeida
Avibras Indústria Aeroespacial S.A., Brazil

Aureomar F. Martins
Avibras Indústria Aeroespacial S.A., Brazil

Susane R. Gomes
Aeronautics Institute of Technology, Brazil

Flavio A. L. Cunha
Avibras Indústria Aeroespacial S.A., Brazil

ABSTRACT

The thermal decomposition kinetics of ammonium perchlorate (AP)/hydroxyl-terminated-polybutadiene (HTPB) samples, with Iron Oxide catalyst at nano and micro scale were studied by thermal analysis techniques at different heating rates in dynamic nitrogen atmosphere. The exothermic reaction kinetics was studied by differential scanning calorimetry (DSC) in isothermal conditions. The Arrhenius kinetic parameters were obtained by Flynn-Wall and Ozawa Kissinger and Starink methods. The propellant samples thermal decomposition was studied simultaneously by TG-DTA. For this purpose, solid propellant grains containing nano and micro scale iron oxide were formulated. The effect of catalysts on

DOI: 10.4018/978-1-5225-2903-3.ch010

the propellant burning rate and the propellant initiation sensitivity were also evaluated by friction and impact. The effect of the catalyst in the propellant binder reaction was evaluated by viscosity and mechanical properties. SEM/EDS technique was used to evaluate the iron oxide morphology. Three bench firing tests were performed with rockets motor in order to know the ballistics parameters.

INTRODUCTION

Composite solid propellant is considered a heterogeneous mixture in which solid particles are embedded in a polymeric matrix (binder) (Kubota, 2007). There are many kinds of polymers used in binder composition and the most commonly used polymer is hydroxyl polybutadiene, which acts as a binder for the solid particles and also as a fuel during the propellant combustion. The solid particles of propellant are composed of an oxidizer, usually ammonium perchlorate (AP), and a metallic fuel, usually aluminum powder that is used to increase the temperature of the combustion products (Prajakta, Krishnamurthy, & Satyawati, 2006), (Ma & Li, 2006), (Li & Cheng, 2007), (Sciamareli, Takahashi, & Teixeira, 2002). In addition to the basic components, other ingredients like plasticizers, bonding agents and combustion catalysts are added in the propellant formulation. The function of the combustion catalysts is to increase the propellant burning rate (Prajakta, Krishnamurthy, & Satyawati, 2006), (Kubota, 2007), (Li & Cheng, 2007). The main catalysts used as accelerators for the thermal decomposition of AP based propellants were the transition metal oxides, as ferric oxide (III) (Fe_2O_3), cobalt oxide (III) (Co_2O_3), manganese oxide (MnO_2), chromium oxide (III) (Cr_2O_3) and copper chromite (II) ($CuCr_2O_4$) (Ma & Li, 2006). The characteristics of the metal oxides catalysts, such as particle size, surface area and defects in the crystalline structure may affect the burning behavior of ammonium perchlorate based propellants (Engen & Johannessen, 1990). Different mechanisms have been suggested to explain the thermal decomposition, but no model is completely satisfactory (Kishore, Verneker, & Sunitha, 1980), (Cavalheira, Gadiot, & Klerk, 1995). (Pekel, Pinardag, & Türkan, 1990) studied the effect of two iron oxides with different specific surface area on the burning rate of composite propellant. They observed that the burning rate increases proportionally with the specific surface area of the catalyst. The same conclusion was reached by Burnside (1975) and Engen and Johannessen (1990), on evaluating the effect of the specific surface area and particle size of different types of iron oxide on the composite propellants burning rate (Burnside, 1975) (Engen & Johannessen, 1990). The burning rate of propellant can be represented by analytical expression that defines burning rate as function of pressure at given temperature. The analytical expression most frequently used to describe the burning rate of composite propellants is the Saint Robert's and Vieille burning rate law.

$$r = aP_c^n$$

where:

r = propellant burning rate
a = coefficient of pressure or rate of burning constant
P_c = chamber pressure
n = pressure exponent

Most propellant systems are formulated to meet specific ballistic requirements of burning rate and performance. Binder systems are formulated and optimized to produce mechanical properties necessary for particular applications (Zyl & Keyser, 1996). In contrast, the pressure exponent is a parameter which normally receives little attention in this design and formulation effort. Various factors have an influence on the pressure exponent and a number of different methods and approaches have been used in controlling this property. This includes the effect of oxidizer, metal type, burning rate and ballistic additives, oxidizer coatings and binder effects. Normally the desire is to have a pressure exponent as low as possible, so that a more optimum motor design can be achieved. A low-pressure exponent result in low motor temperature sensitivity and therefore reduces variability of both pressure and thrust over operational temperature range. This in turn narrows the performance envelope and provides a more uniform performance under all conditions. Since the operating pressure range is reduced, lighter motor hardware can be used. This reduces inert weight, thus increasing missile performance by increasing the propellant mass fraction (Zyl & Keyser, 1996). Ammonium perchlorate (AP) is the most widely used crystalline oxidizer to formulate composite propellants for rockets. AP composite propellants are composed of fine AP particles as oxidizers with polymeric hydrocarbons as fuel components. Extensive experimental and theoretical studies have been carried out on the decomposition and combustion of AP composite propellants with a view to obtaining a wide spectrum of burning rates (Sutton & Biblarz, 2017). The combustion mode of AP composite propellants is controlled by the diffusion process of the gaseous decomposition products of the AP particles and the surrounding polymeric hydrocarbons at the burning surface. The AP particles decompose to produce perchloric acid, $HClO_4$, and the fuel binder decomposes to produce hydrocarbon fragments and hydrogen (Sutton & Biblarz, 2017). These gaseous decomposition products react to produce heat on and above the burning surface. A close-up view of the combustion process of the AP particles and the decomposing fuel binder (HTPB) is shown in Fig. 1.

An endothermic reaction involving pyrolysis of the binder to form gaseous fuel products, coupled with dissociative sublimation and/or decomposition of the AP particles to form ammonia and perchloric acid, occurs at the burning surface. The ammonia and perchloric acid molecules react exothermically to form premixed flamelets just above each AP particle. Since the reaction products of these premixed

Figure 1. A schematic view of the combustion process of AP particles and fuel binder

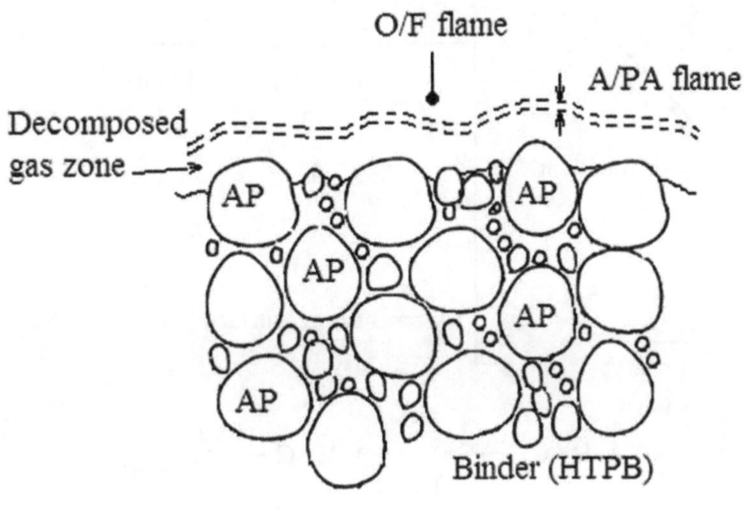

flamelets contain excess oxidizer fragments, these react with the fuel gases produced by the decomposition of the polymeric binder to produce diffusional flamelets. Thus, the combustion process of AP composite propellants consists of a two-step reaction (Sutton & Biblarz, 2017). The temperature in the condensed phase increases from the initial propellant temperature, T_0, to the burning surface temperature, Ts, through conductive heat feedback from the burning surface. Then, the temperature increases in the gas phase because of the exothermic reaction above the burning surface and reaches the final combustion temperature, Tg. Since the physical structure of AP composite propellants is highly heterogeneous, the temperature fluctuates from time to time and also from location to location. The temperature profile shown in Fig.2 thus illustrates a time-averaged profile. Thus, the burning rate of AP composite propellants depends largely on the particle size of AP, the mass fraction of AP, and the type of binder used.

The burning rate of propellants is one of the important parameters for rocket motor design. The diffusional mixing process between the gaseous decomposition products of the AP particles and of the polymeric binder used as a fuel component determines the heat flux feedback from the gas phase to the condensed phase at the burning surface. This process is a dominant factor in determining the burning rate, and it is for this reason that the burning rate is altered by changing the size of the AP particles. The heat flux feedback from the gas phase to the burning surface is also determined by the chemical reaction rate in the gas phase. The reaction rate in the gas phase is altered by the addition of catalysts. The catalysts act either on the decomposition reaction of the condensed phase or on the reaction in the gas phase of the gaseous decomposition products. The burning rates of AP composite propellants are increased by the addition of positive catalysts, which act to accelerate the decomposition reaction and/or the gas-phase reaction of the AP particles within the propellants (Sutton & Biblarz, 2017). Since catalysts act on the surface of the AP particles, the total surface area of the catalyst particles at a fixed concentration is an important factor for obtaining high catalyst efficiency. Though very finely divided iron oxides are used

Figure 2. An overall view of the reaction process and the temperature profile in the combustion wave of an AP composite propellant

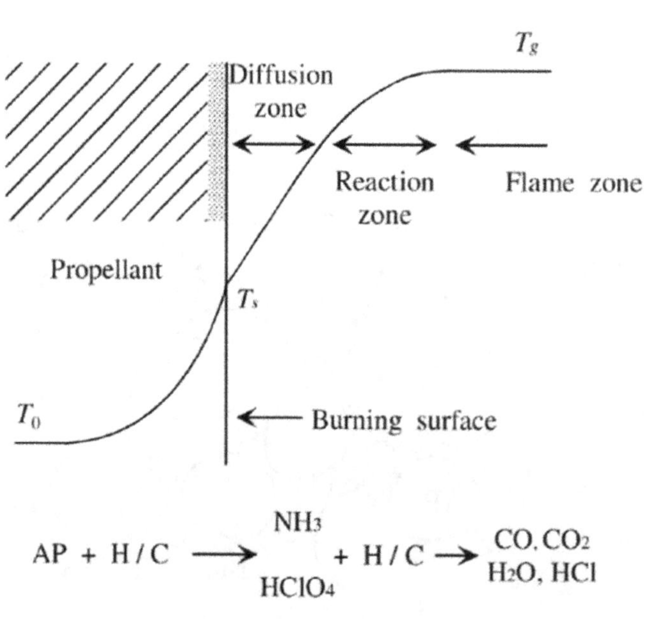

to increase the burning rate of AP composite propellants, organic iron compounds are more effective because of the formation of discrete iron oxide molecules during their decomposition. Different thermal analysis techniques such as Differential Scanning Calorimetry (DSC) and Thermogravimetric Analysis (TG) have been widely used in the study of thermal decomposition of energetic materials. Andrade et al. (2007) used DSC and TG in the study of thermal decomposition of composite solid smokeless propellant (Silva, Iha, Cardoso, Mattos, & Dutra, 2010). The knowledge of the thermal behavior of an energetic material is essential to guarantee safety during its production, storage and handling. In general, the kinetic models developed to study the kinetics of solid materials thermal decomposition are in principle based on consideration of the variation in the mass amount of the material in a time function, or the conversion rate of the sample (dα/dt), as it is also called, in a process of decomposition due to a thermal stimulation, which can be expressed by Equation 1 (Vyazovkin, et al., 2011). The majority of kinetic methods used in the area of thermal analysis consider the rate to be a function of only two variables, T and α:

$$\frac{d\alpha}{dt} = K\left(t\right)f\left(\alpha\right) \tag{1}$$

The dependence of the process rate on temperature is represented by the rate constant, k (T), and the dependence on the extent of conversion by the reaction model, f(α). One of the present recommendations is that it is crucial for reliable kinetic methods to be capable of detecting and treating the multi-step kinetics. Note that if a process is found to obey a singlestep equation (Eq. (1), one should not conclude that the process mechanism involves one single step. More likely, the mechanism involves several steps but one of them determines the overall kinetics. For instance, this would be the case of a mechanism of two consecutive reactions when the first reaction is significantly slower than the second. Then, the first process would determine the overall kinetics that would obey a single-step Eq. (1), whereas the mechanism involves two steps. The temperature dependence of the process rate is typically parameterized through the Arrhenius equation:

$$k\left(t\right) = A exp\left(\frac{-E}{RT}\right) \tag{2}$$

where A and E are kinetic parameters, the preexponential factor and the activation energy, respectively, and R is the universal gas constant. Combining Eqs. (1) and (2) yields:

$$\frac{d\alpha}{dt} = A exp\left(\frac{-E}{RT}\right)f\left(\alpha\right) \tag{3}$$

The resulting equation provides a basis for differential kinetic methods. In this form, the equation is applicable to any temperature program, be it isothermal or nonisothermal. It also allows for substitution of the actual sample temperature variation, T(t), for T that can be useful when the sample temperature deviates significantly from the reference temperature (i.e., temperature of the furnace). For constant heating rate nonisothermal conditions, Eq. (3) is frequently rearranged as:

$$\beta \frac{d\alpha}{dT} = A exp\left(\frac{-E}{RT}\right) f\left(\alpha\right) \tag{4}$$

Introduction of the explicit value of the heating rate reduces the applicability of Eq. (4) to processes in which the sample temperature does not deviate significantly from the reference temperature. Integration of Eq. (3) leads to:

$$g\left(\alpha\right) = \int_0^{\alpha} \frac{d\alpha}{f\left(\alpha\right)} = A \int_0^{t} exp\left(\frac{-E}{RT}\right) dt \tag{5}$$

where g(α) is the integral form of the reaction model. Eq. (5) lays a foundation for a large variety of integral methods. In this form, Eq. (5) is applicable to any temperature program that can be introduced by replacing T with T(t). This also means that one can use this equation to introduce in kinetic calculations the actual sample temperature variation, T(t), which is helpful in the situations when the sample temperature demonstrates significant deviation from the reference temperature. For constant heating rate conditions, the integral with respect to time is usually replaced with the integral with respect to temperature:

$$g\left(\alpha\right) = \frac{A}{\beta} \int_0^{T} exp\left(\frac{-E}{RT}\right) dT \tag{6}$$

This rearrangement introduces the explicit value of the heating rate in Eq. (6) that means that the application area of the equation is limited to the processes, in which the sample temperature does not deviate significantly from the reference temperature (Vyazovkin, et al., 2011).

All isoconversional methods take their origin in the isoconversional principle that states that the reaction rate at constant extent of conversion is only a function of temperature. This can be easily demonstrated by taking the logarithmic derivative of the reaction rate (Eq. (1) at α = const:

$$\left[\frac{\partial ln\left(d\alpha / dt\right)}{\partial T^{-1}}\right]_{\alpha} = \left[\frac{\partial ln\ k\left(T\right)}{\partial T^{-1}}\right]_{\alpha} + \left[\frac{\partial ln\ f\left(\alpha\right)}{\partial T^{-1}}\right]_{\alpha} \tag{7}$$

where the subscript α indicates isoconversional values, i.e., the values related to a given extent of conversion. Because at α = const, f (α) is also constant, and the second term in the right-hand side of Eq. (7) is zero. Thus:

$$\left[\frac{\partial ln\left(d\alpha / dt\right)}{\partial T^{-1}}\right]_{\alpha} = -\frac{E_{\alpha}}{R} \tag{8}$$

It follows from Eq. (8) that the temperature dependence of the isoconversional rate can be used to evaluate isoconversional values of the activation energy Eα without assuming or determining any particular form of the reaction model. For this reason, isoconversional methods are frequently called "model-free" methods. However, one should not take this term literally. Although the methods do not need to identify the reaction model, they do assume that the conversion dependence of the rate obeys some f(α) model. To obtain experimentally the temperature dependence of the isoconversional rate, one has to perform a series of runs with different temperature programs. This would typically be a series of 3–5 runs at different heating rates or a series of runs at different constant temperatures. It is recommended to determine the Eα values in a wide range of α = 0.05–0.95 with a step of not larger than 0.05 and to report the resulting dependencies of Eα vs. α. The Eα dependence is important for detecting and treating the multistep kinetics. A significant variation of Eα with α indicates that a process is kinetically complex, i.e., one cannot apply a single-step rate equation (Eq. (1) and/or (3)) to describe the kinetics of such a process throughout the whole range of experimental conversions and temperatures. Note that the occurrence of a multi-step process does not immediately invalidate the application of the isoconversional principle, although the latter holds strictly for a single-step process. The principle continues to work as a reasonable approximation, because isoconversional methods describe the process kinetics by using multiple single step kinetic equations, each of which is associated with a certain extent of conversion and a narrow temperature range (ΔT) related to this conversion (Fig. 3). As a matter of fact, the Eα dependencies evaluated by an isoconversional method allow for meaningful mechanistic and kinetic analyses as well as for reliable kinetic predictions. The isoconversional principle lays a foundation for a large number of isoconversional computational methods. They can generally be split in two categories: differential and integral.

Integral isoconversional methods originate from the application of the isoconversional principle to the integral equation (5). The integral in Eq. (5) does not have an analytical solution for an arbitrary temperature program. However, an analytical solution can be obtained for an isothermal temperature program:

$$g(\alpha) = A exp\left(\frac{-E}{RT}\right) t \tag{9}$$

Some simple rearrangement followed by the application of the isoconversional principle gives rise to Eq. (10):

$$lnt_{\alpha,i} = ln\left[\frac{g(\alpha)}{A_{\alpha}}\right] + \frac{E_{\alpha}}{RT_{i}} \tag{10}$$

where tα,i is the time to reach a given extent of conversion at different temperatures Ti. This is an equation for an integral isoconversional method for isothermal conditions. The value of Eα is determined from the slope of the plot ln tα,i vs. 1/Ti. For the commonly used constant heating rate program, Eq. (5) transforms into Eq. (6) that does not have an analytical solution. For this reason, there is a number of integral isoconversional methods that differ in approximations of the temperature integral in Eq. (6). Many of these approximations give rise to linear equations of the general form (Starink):

Figure 3. Each single-step rate equation is associated with a single value of α and a narrow temperature region ΔT related to it

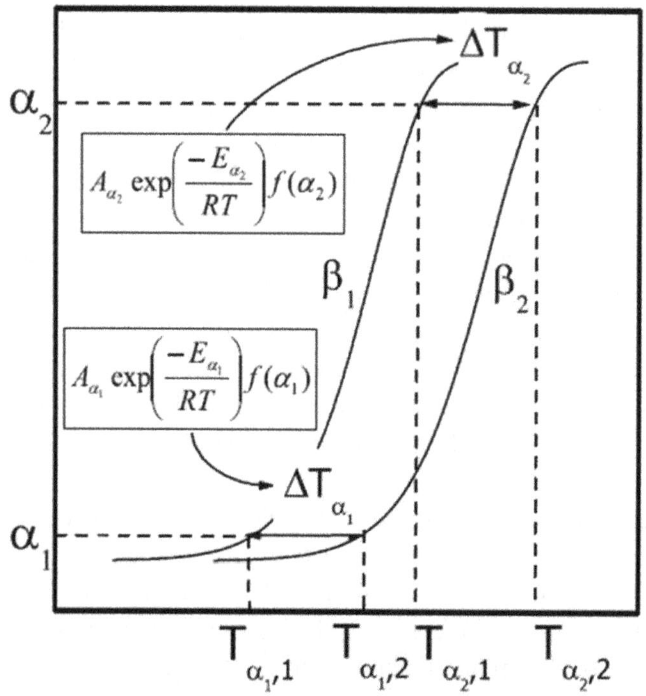

$$ln\left(\frac{\beta_i}{T_{\alpha,i}^B}\right) = Const - C\left(\frac{E_\alpha}{RT_\alpha}\right) \tag{11}$$

where B and C are the parameters determined by the type of the temperature integral approximation. For example, a very crude approximation by Doyle yields B = 0 and C = 1.052 so that Eq. (11) takes the form also known as the Ozawa, and/or Flynn and Wall equation:

$$\ln\left(\beta_i\right) = Const - 1{,}052\left(\frac{E_\alpha}{RT_\alpha}\right) \tag{12}$$

The crude temperature integral approximation results in inaccurate values of Eα. A more accurate approximation by Murray and White gives rise to B = 2 and C = 1 and leads to another popular equation that is frequently called the Kissinger–Akahira–Sunose equation:

$$ln\left(\frac{\beta_i}{T_{\alpha,i}^2}\right) = Const - \frac{E_\alpha}{RT_\alpha} \tag{13}$$

Compared to the Ozawa–Flynn–Wall method, the Kissinger–Akahira–Sunose method offers a significant improvement in the accuracy of the Eα values. As shown by Starink, somewhat more accurate estimates of Eα are accomplished when setting B = 1.92 and C = 1.0008 so that Eq. (11) turns into:

$$ln\left(\frac{\beta_i}{T_{\alpha,i}^{1,92}}\right) = Const - 1,0008\left(\frac{E_\alpha}{RT_\alpha}\right) \tag{14}$$

Since the aforementioned Eqs. (11)–(14) are equally easy to solve by applying linear regression analysis, it is recommended to use the more accurate equations such as (13) and (14). Eq. (12) is very inaccurate and should not be used without performing an iterative correction procedure for the value of Eα. Here, one should be strongly advised against the frequently encountered practice of performing and reporting kinetic analyses based on the concurrent use of more than one form of Eq. (11). The concurrent use of two or more such equations only reveals the trivial difference in the Eα values computed by the methods of different accuracy. Since no kinetic information is produced from such comparison, the practice of the concurrent use of Eqs. (11)–(14) should be eliminated in favor of using only one more accurate equation. The ASTM E698 method uses the same predictive equation as ASTM E1641 (ASTM E1641-07 Standard Test Method for Decomposition Kinetics by Thermogravimetry, 2007). The value of E is determined either by the Kissinger method or by the Ozawa and Flynn and Wall methods (Eq. (12)). In the latter case, the method suggests replacing Tα in Eq. (12) with the peak temperature, Tp. The lifetime prediction by the E698 method is again based on the assumption of the first-order kinetics. The same assumption is used to determine the pre-exponential factor (Vyazovkin, et al., 2011):

$$A = \frac{\beta E}{RT_p^2} exp\left(\frac{E}{RT_p}\right) \tag{15}$$

The kinetic parameters can be obtained by substituting Arrhenius's equation:

$$k = A\ e^{-Ea/RT} \tag{16}$$

Due to the importance of understanding the thermal behavior of energetic material handled, this study aimed to determine the Arrhenius parameters (activation energy and pre-exponential factor) for the thermal decomposition reaction of a solid propellant containing nano and micro scale iron oxide. The comparison was made with a solid propellant grain formulated without burn rate catalyst additive (Silva, Iha, Cardoso, Mattos, & Dutra, 2010), (Lee, Hsu, & Chang, 2002). Among these catalysts, iron oxide is known to be structurally simple, highly stable, easy to synthesize and inert to side reactions. Above all, it imparts a high and reproducible burn rate for composite propellants (Satyawati, Prajakta, & Krishnamurthy, 2008). Catalysts are used in the majority of polyurethane formulations, and the selectivity of these systems is fundamental to obtaining materials with physical properties and processing characteristics desired. The action of the catalyst in the reaction between isocyanates and alcohols not only increases the reaction rate, but also influences the characteristics of the product that is formed after the reaction (Lima, 2007). Catalysts for the reaction between isocyanates and compounds containing active hydrogens (in practice, alcohols, amines, and water) fall into two general categories: (1) amines and (2) organotin compounds. Although there are numerous permutations, two common examples are DABCO (1,4-diazabicyclo-[2,2,2]-octane) and dibutyltin (IV) dilaurate. For reasons, these compounds are often used in combination. Other organometallic compounds based on lead, zinc, calcium, and magnesium can be used to catalyze the reactions between isocyanates and compounds containing active hydrogen compounds. These compounds have not, however, achieved the commercial significance of amines and organotin compounds. Organometallic compounds are generally more active than amine compounds and are particularly effective in catalyzing the reaction between isocyanates and alcohols. In most polyurethane systems, the reaction between the isocyanate and the alcohol group of the polyol is the most significant polymer forming reaction. As these reactions proceed, the molecular weight of the polymer increases as described by Flory (Lima, 2007), (Wegener, Brandt, Duda, & Weisbeck, 2001), (Flory, 1953), until a crosslinked system (in the common case of multifunctional monomers) is formed. This "gel point" is characterized by the rapid development of physical properties and mechanical strength (Lima, 2007), (Flory, 1953).

Experimental

A propellant formulation with good processing was chosen for this investigation and the general propellant composition that has been used was showed in Table 1. Propellant formulations were made using 0.6% of iron oxide and compared with the same propellant without a catalyst. Propellant compositions containing around 81.0% AP and 17.6% binder were prepared using slurry casting technique. Hydroxy terminated polybutadiene (HTPB), plasticizer dioctyl adipate (DOA), Tepanol 878, Ammonium perclorate (20 μm, 200 μm and 400 μm), aluminum and Iron oxide (nano and micro) were mixed in a mixer with continuous hot water circulation in the outer jacket. After the first mixing stage, the Isophorone diisocyanate (IPDI) was added in 1:1 stoichiometry rate (NCO:OH – 1: 1). The propellant was cured at 62 ± 2 °C for one week in a water jacketed oven.

The study was carried out with three samples of propellant, where the first sample (FA) was prepared with commercial iron oxide supplied by Lanxess Company. The second sample was prepared with nano iron oxide supplied by the Mach1 (FB), and the third sample was prepared without iron oxide (FC). In order to minimize variations introduced by sample preparation, three propellant batches were prepared in the same planetary mixer using the same raw materials for the binder, fuel and the oxidizing as well. Only the component iron oxide catalyst for each formulation was changed because it was the object of this study. The DSC analyses were obtained using DSC 7020 SII NanoTechnology Inc. equipment and

Table 1. Propellant formulations

Compound	Concentration (m/m %)		
	Formulation A	Formulation B	Formulation C
HTPB – binder	17.60	17.60	17.60
Al	1.00	1.00	1.00
AP	80.80	80.80	81.40
Burning rate catalyst	0.60%	0.60%	0,00%
Iron Oxide (Supplier)	Lanxess	Mach1	-

it was calibrated on the heating rates used, i.e., 10, 15, 20, 25 and 30 °C/min, with zinc and indium. The mass of the samples was approximately 1.0 to 2.0 mg. The analyses were performed in a temperature between 25 and 400°C. The tests were performed under flow of nitrogen gas (40 mL/min) and in atmospheric pressure, i.e., the lids of the DSC pans were perforated in order not to confine the sample. The thermogravimetric and differential thermal analysis (TG-DTA) were obtained using TG/DTA 6200 SII NanoTechnology Inc. Analysis of Scanning Electron Microscopy (SEM) with Energy dispersive analysis (SEM/EDS) techniques were used to evaluate the iron oxides morphology provided by LANXESS and Mach1. The mechanical properties were also studied to evaluate the effect of the catalyst in the binder reaction. Propellant mechanical properties were obtained in the equipment Kratos Model K-500 at 21°C and the test was performed with Jannaf type C specimen. The propellant viscosity behavior from the end of the mixing process until the end of casting was measured using Brookfield viscometer. The friction and impact propellant sensitivity were obtained with equipment BAM Friction Tester and BAM Drop Hammer model Reichel & Partner respectively. The burning rates were obtained in Strand burner equipment in order to know the constants "A" and "n" according to the Saint Robert and Vieille Law. Three bench firing tests were performed with small rocket motors to compare the propulsion parameters of each formulation.

RESULTS AND DISCUSSION

After cure of propellant composites FA, FB and FC, the thermal decomposition behavior of propellants i.e. heat release and exothermic nature was measured by differential scanning calorimetry (DSC), where 1 to 2 mg of sample were taken over a temperature range from 25 to 400 °C at different heating rates 10, 15, 20, 25 and 30 °C/min. An inert nitrogen flow of 50 ml/min was maintained throughout the DSC run. Differential Scanning Calorimetric Study of HTPB based Composite Propellants in presence of Iron Oxides FA - Lanxess (Micro), FB - Mach1 (Nano) and FC (without Iron Oxide) was presented in the figures 4, 5 and 6.

Figures 4, 5 and 6 show the DSC results for composites FA, FB and FC at different heating rates. Each DSC curve shows first an endothermic and then an exothermic stage. For all composites, FA, FB and FC, ammonium perchlorate shows an endothermic peak around 249 °C, which represents the transition from orthorhombic form to cubic form. Some studies for AP shows two exothermic peaks, one due to incomplete combustion at about 320 °C, that corresponds to the partial decomposition of AP and the formation of an intermediate and other at high temperature at about 475 °C. It was observed two high temperature exothermic peaks for both composites FA and FB, which indicates complete decomposition around 330°C. For composite FC exothermic peaks show different maximum temperatures (Tp) as 352.9 °C, 354.9 °C, 358.8 °C, 359.8 °C and 363.1 °C corresponding to 10 °C/min, 15 °C/min, 20 °C/min, 25 °C/min and 30 °C/min heating rates respectively. DSC of Propellant Composites according to FWO, Kissinger and Starink (Table 2).

Catalysts play a dual role in accelerating the decomposition of AP and promote the oxidation of the fuel by heterogeneous reactions (Prajakta, Krishnamurthy, & Satyawati, 2006), (Inami, Rajaphase, Shaw, & Wise, 1971). It is known in metal oxide catalyzed oxidation, that the metal ions in the lattice undergo redox cycle involving interactions of hydrocarbons with their high valence state followed by reoxidation by molecular oxygen. Hence significant alteration in the decomposition characteristics of AP was observed by addition of ultrafine iron oxide nanoparticles as burning rate modifier in composite

Figure 4. DSC of composite FA (Lanxess) at different scan rates (10 °C/min, 15 °C/min, 20 °C/min, 25 °C/min and 30 °C/min)

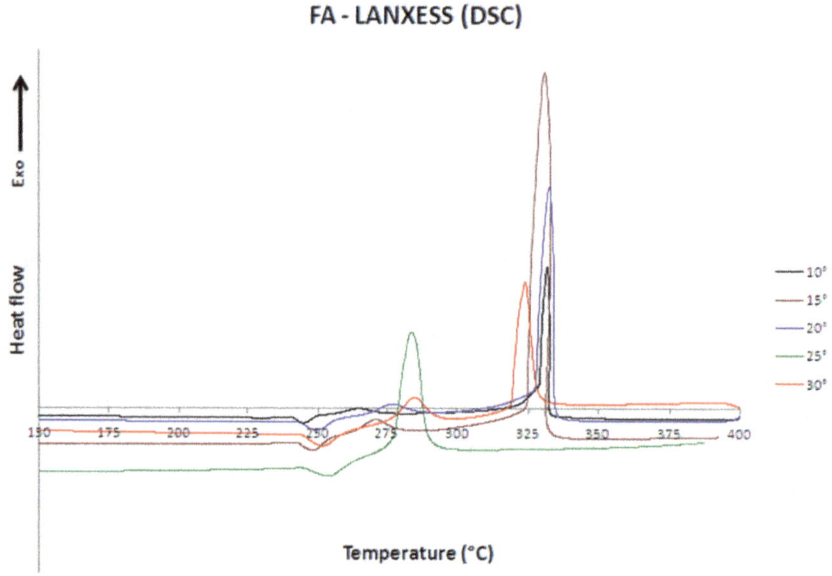

Figure 5. DSC of composite FB (Nano Iron Oxide) at different heating rates (10 °C/min, 15 °C/min, 20 °C/min, 25 °C/min and 30 °C/min)

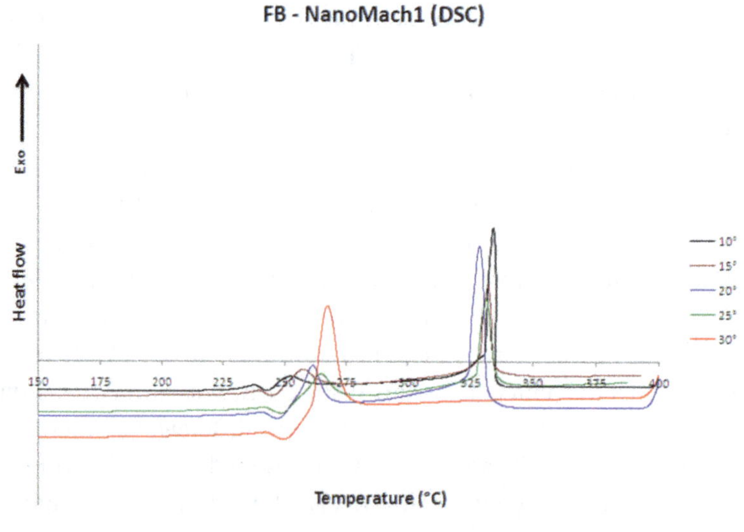

FB. The endothermic phase transition was unchanged by addition of nano Fe_2O_3 while large differences were observed in high temperature exothermic peak (Figures 4 and 5). The position of this exotherm strongly depends on the size of nanomaterials (Prajakta, Krishnamurthy, & Satyawati, 2006), (Liu, Li, Tan, Ming, & Yi, 2004). The addition of nano iron oxide lowered the high decomposition temperature of AP by 99.3°C, 98.1°C, 97.8°C, 95.8°C and 96.3 °C at 10 °C/min, 15 °C/min, 20 °C/min, 25 °C/min

Figure 6. DSC of composite FC (Nano Iron Oxide) at different heating rates (10 °C/min, 15 °C/min, 20 °C/min, 25 °C/min and 30 °C/min)

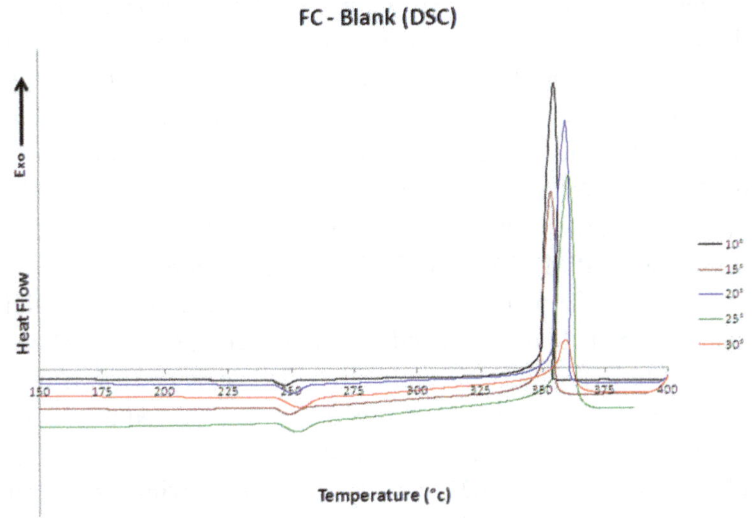

Table 2. DSC of FA composite

Propellant	ASTM-E698 -APDX 2 (FWO Method) Kinetic Parameters			ASTM-E698 -APDX 3 (Kissinger Method) Kinetic Parameters			Starink Method Kinetic Parameters		
	Ea_0 ($KJ.mol^{-1}$)	A (min^{-1}) at 20°C	k (s^{-1}) at 20°C	Ea_0 ($KJ.mol^{-1}$)	A (min^{-1}) at 20°C	k (s^{-1}) at 20°C	Ea_0 ($KJ.mol^{-1}$)	A (min^{-1}) at 20°C	k (s^{-1}) at 20°C
Formulation A (Fe_2O_3 - Micro)	114,5	7,39E+10	1,53E-02	132,7	4,64E+12	1,77E-02	132,4	4,33E+12	1,77E-02
Formulation B (Fe_2O_3 - Nano)	182	9,62E+17	2,56E-02	199,9	5,95E+19	2,81E-02	199,7	5,68E+19	2,81E-02
Formulation C (without catalyst)	337	1,45E+28	3,38E-02	358	8,39E+29	3,59E-02	357,9	8,23E+29	3,59E-02

and 30 °C/min heating rates respectively as compared to uncatalyzed AP (composite FC). The lowering in temperature is due to the catalytic effect of nano iron oxide. As the surface area of nano Fe_2O_3 is larger due to the high surface to volume ratio, there is less sputtering of particles during decomposition. This causes less mechanical loss and more efficient heat transfer within the composite. As a result, an interesting feature of a 32% higher heat release was observed in presence of nano Fe_2O_3 (FB) than in composite (FC). Another reason for the higher heat release may be attributed to higher concentration of Fe^{+3} moiety at the interphase between AP and organic hydrocarbon due to very strong ion-ion interactions. The values of Ea, A and k for high temperature decomposition are shown in Table 2. Many researchers have reported a higher activation energy (Ea) for catalysed AP than that of pure AP (Cavalheira, Gadiot, & Klerk, 1995), (Lima, 2007), (Wegener, Brandt, Duda, & Weisbeck, 2001), (Houghton & Mulvaney, 1996). Similarly, Dubey et al. (Flory, 1953) observed that the catalyst lowers Ea in the temperature range 220 – 260 °C and increases Ea in the temperature range 220 – 380 °C. The increase in activation energy and the corresponding increase in the A value is due to the kinetic compensation effect as reported earlier by Ninan et al. (Inami, Rajaphase, Shaw, & Wise, 1971). As a result, a direct correlation of Ea with the reaction rate becomes difficult since both Ea and A are altered in the reaction. Thus, the frequency factor plays an important role in increasing the reaction rate of the thermal decomposition of AP. Table 2 shows average values for the specific rate constant k. It was expected an increase in the rate constant by the addition of nano oxide, but this behavior did not occur. The concentration of this catalyst at the interphase and the catalytic decomposition of vapors of fuel and oxidizer further enhance the exothermic reaction and consequently increase the burning rate (Prajakta, Krishnamurthy, & Satyawati, 2006), (Nema, Jain, Sharma, Nema, & Verma, 2004). Scanning Electron Microscopy (SEM) with Energy dispersive spectroscopy analysis (SEM/EDS) were used to evaluate the iron oxides morphology provided by LANXESS and Mach1 with SEM TESCAN VEGA3 SB EasyProbe.

It is possible to observe that the particles of Fe_2O_3 Lanxess (6a and 6b) are slightly agglomerated. According to the supplier the particle size is around 3 to 5 μm. Figures 6c and 6d shows scanning electron microscopy of nano Fe_2O_3 Mach1. At higher magnification, well-separated nano iron oxide particles were observed with low agglomeration. However, in figure 6c it was observed the circle that can be a kind of necklace-like structure of Fe_2O_3 has been retained, which is a feature of α-Fe_2O_3 particles. Analysis with Transmission electron microscopy (TEM) can provide better viewing for the Iron Oxide nano-scale (Mach1). Scanning electron microscopy with Energy dispersive analysis (SEM / EDS) showed that the compositions of both Fe_2O_3 are very similar and the main chemical element is the α-Fe. The result was presented in the figure 8.

Both formulations exhibited AP particles in sizes from 20 μm to 400 μm embedded in HTPB binder to 100 x magnification. It is not possible to observe the Fe_2O_3 and aluminum particles embedded in the HTPB binder.

The TG/DTA was calibrated on the heating rate of 10 °C/min. The mass of the samples was approximately 2.0 mg. The analyses were performed in a temperature between 25 °C and 400°C. The tests were performed under flow of nitrogen gas (50 mL/min) with lids without pans in order not to confine the sample. Figure 10 shows typical thermogravimetric curves of composite FA, FB and FC obtained at heating rate 10 °C/min. A plasticizer agent like DOA (di-octyl adipate) is also and frequently used to improve the casting properties (Rocco, et al., 2005). Ammonium perchlorate (AP) is widely used as an oxidizer agent in composite solid propellants. In propellants with AP addition to the HTPB/DOA/IPDI

Figure 7. a) Fe_2O_3 Lanxess 1000 x Magnification, b) Fe_2O_3 Lanxess 1830 x Magnification, c) Fe_2O_3 Mach1 20000 x Magnification and d) Fe_2O_3 Mach1 29200 x Magnification

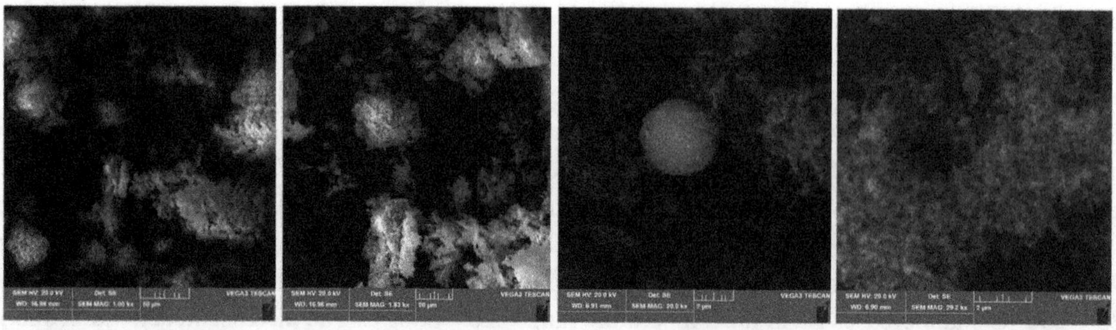

a) Iron Oxide Lanxess b) Iron Oxide Lanxess c) Iron Oxide Mach1 d) Iron Oxide Mach1

Figure 8. Fe_2O_3 Lanxess and Mach1 by SEM/EDS analysis

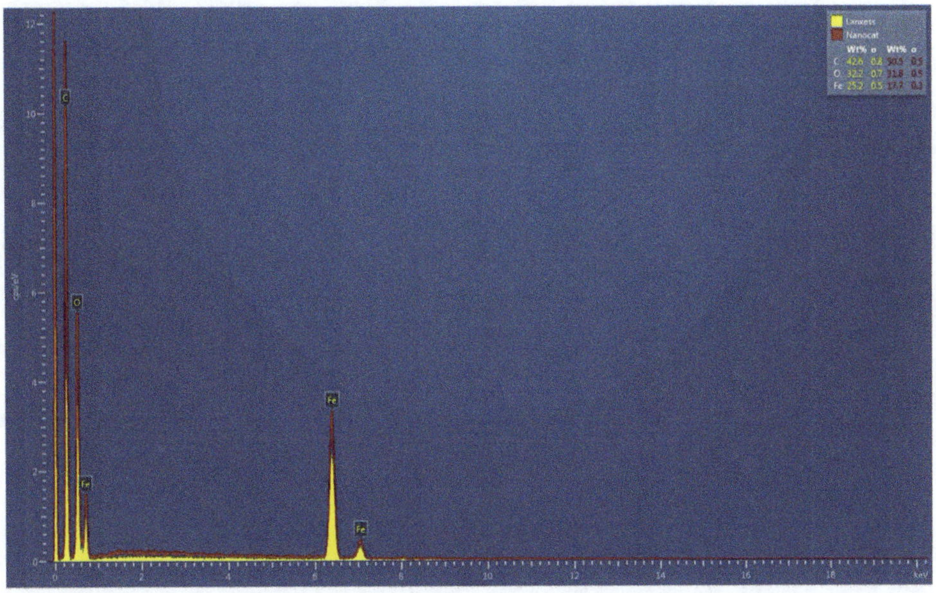

Figure 9. SEM of FA (Lanxess) and FB (Nano Mach1)

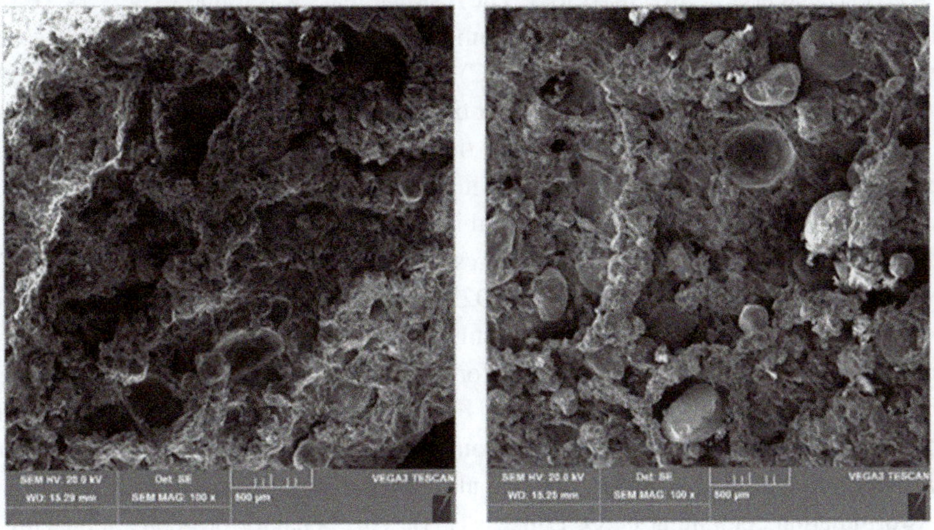

a) FA Propellant b) FB Propellant

binder the major loss steps may be primarily related to the ammonium perchlorate decomposition with some degradation of the polymer binder. Thin particles of aluminum powder (15 μm) are also added to the propellant formulation (1 – 20% mass) to increase the specific impulse (Isp) and it acts as a combustion instability suppressant in the solid rocket motor (Rocco, et al., 2005). The thermal degradation occurs in two main steps. The first step around 263.4 °C (FA), 253.6°C (FB) 243°C (FA) shows a mass loss of 6.8% and 7.5% respectively and it can be related to the plasticizer (DOA) raw material degradation. The

Figure 10. TG/DTA curves of formulation A in N2 atmosphere

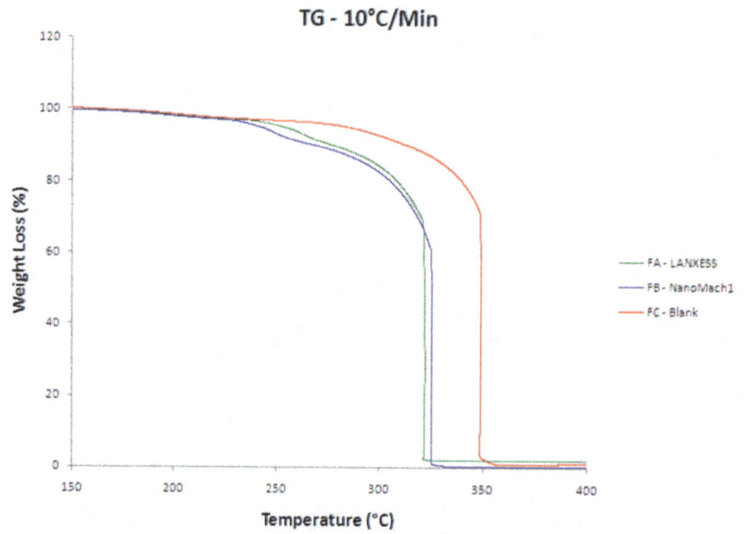

following step observed in the Figure 1 was found at 332.6 °C (FA) and 339.7 °C (FB) respectively. These steps are related to the partial HTPB/IPDI binder and total AP thermal decomposition and represent a total of 87.4% (FA) and 86.9% (FB) mass loss. The residual mass was found to be 3.7% (FA) and 4.2% (FB) and it probably may be attributed to aluminum oxidation during the combustion reaction leading to the formation of alumina. For the composite (FC) it was observed the first step around 251°C with a mass loss of 5.3% that can be related to plasticizer degradation as well. The second step was observed at 352 °C associated with 88.5% weight loss and this corresponds to the complete decomposition of AP. The residual mass was found around 3.4% and it was considered alumina. Tensile test data obtained for the formulations FA, FB and FC was presented in Table 3. The test was conducted with the specimen Jannaf type C and the tensile rate was 50 mm/s. Hardness tests were performed using Microstest Durometer according to ASTM D2240 (ASTM D2240 - 05(2010) Standard Test Method for Rubber Property - Durometer Hardness, 2007) and the results were presented in Table 3 as well.

It was observed that the tensile strength value for the sample containing Fe_2O_3 Lanxess was significantly increased and the elongation decreased (See Figure 11). This phenomenon might be related to an increase of binder reaction rate due to contamination by organometallic compounds as previously mentioned in this work. Regarding the viscosity, it was also observed a significant increase in the propellant viscosity of formulation containing Fe_2O_3 Lanxess (FA). This behavior promotes reduced pot life and lower casting time (See Figure 12).

Table 3. Results of mechanical properties FA, FB and FC

Sample	Stress at Break* (MPa)	Elongation* (%)	Hardness* (Shore A)
FA	11,4	47,2	78
FB	7,2	62,7	64
FC	7,6	59,7	62

* Test temperature: 21°C ± 2°C

Figure 11. Tensile strength and elongation of propellant formulations

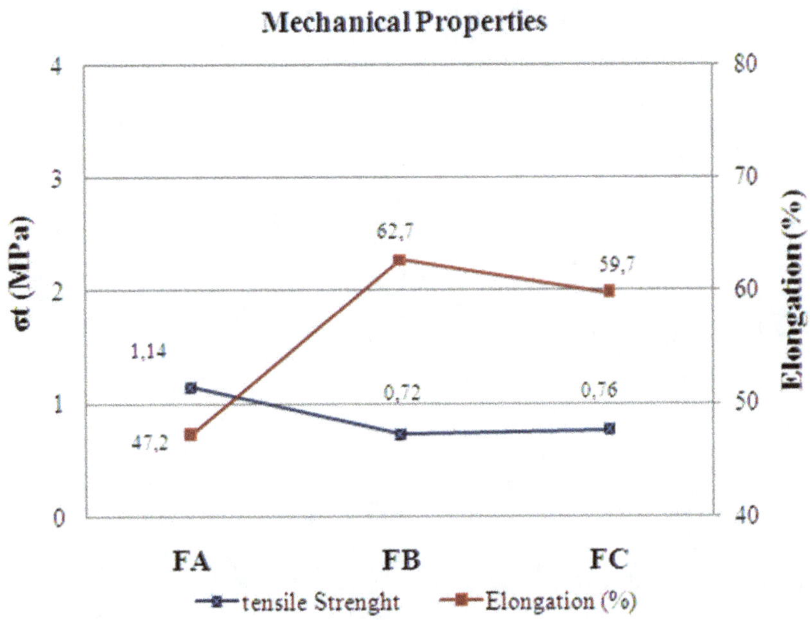

Figure 12. Tensile strength and elongation of propellant formulations

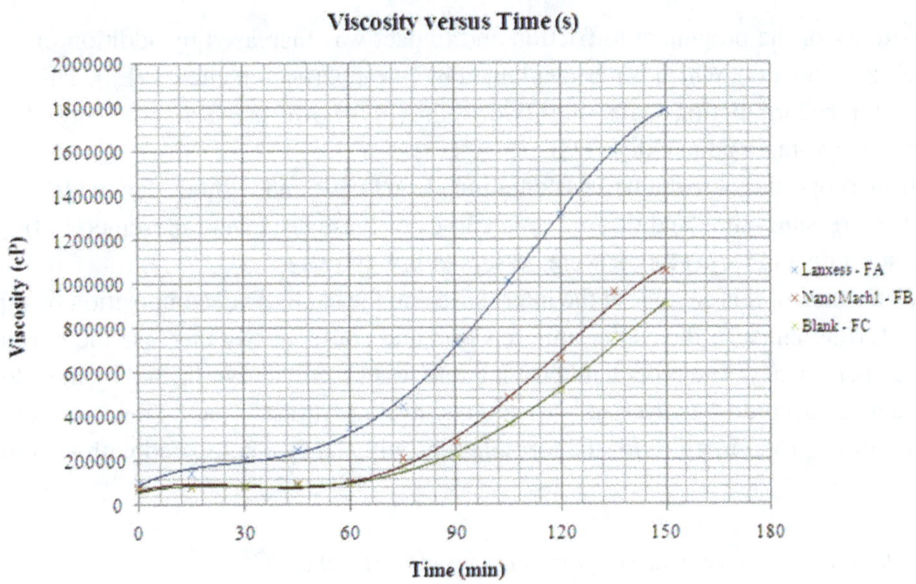

The burning rate results were presented in figure 13. Burning rate values for formulation FB (Nano Mach1) were significantly higher than the formulations FA (Lanxess) and FB (Blank). According to the burning rate laws (Saint Robert and Vieille), the pressure coefficient around 0.30 (FB) is desirable for rocket motors designers. Regarding pressure range from 40 to 120 kgf/cm^2, the burning rates of formulations FA and FB were 30.6% to 39.6% and 61.5% to 44.6% higher than the FC. See Figure.13.

Figure 13. Burning rates curves FA, FB and FC

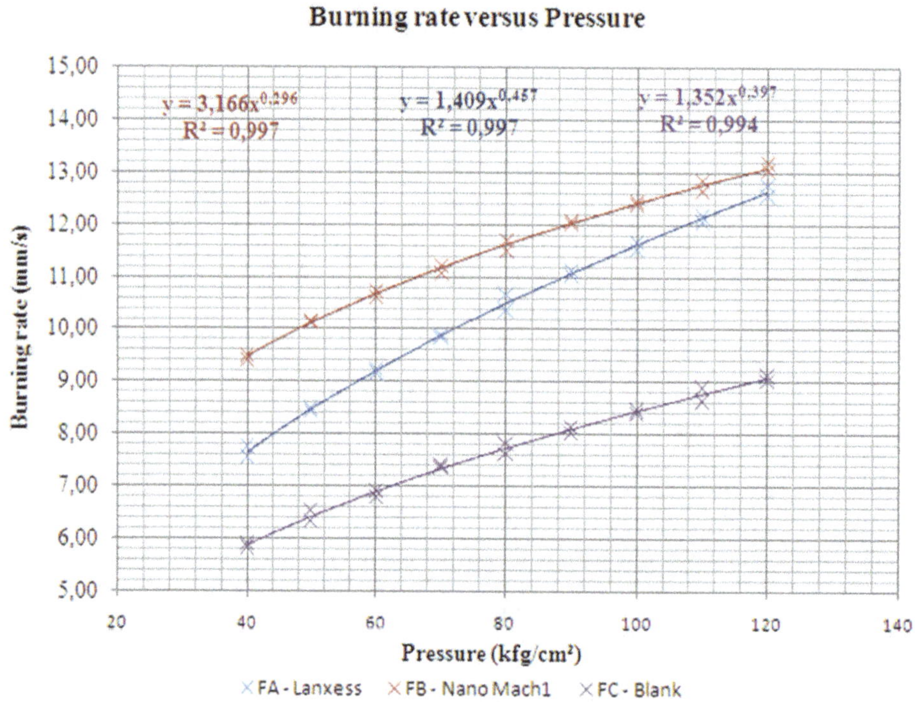

The sensitivity of the propellant to friction and impact was increased by addition of catalytic iron oxide micro size (Lanxess), which can be explained by lower activation energy (Ea). This behavior was also observed for the formulation containing Oxide Nano (FB), however in a discreet way when compared to the formulation without nano oxide (FC).

The ballistic properties of formulations containing Iron Oxide and without iron oxide were presented in table 5. The pressure –time and thrust - time curves of these tests were presented in figures 15 and 16. Firing test set up and rocket motor were presented in Figure 14.

The concentration of this catalyst at the interphase and the catalytic decomposition of vapors of fuel and oxidizer further enhance the exothermic reaction and consequently increase the burning rate and the combustion efficiency. The propellant burning rate was significantly affected by the addition of the iron oxide catalyst, according to average pressure and average thrust values. There was an expectation of an increase of the Total Impulse for the formulations with catalyst compared with formulation with-

Table 4. Friction and impact sensitivity formulations FA, FB, and FC

Propellant Sensitivity	FA - Lanxess	FB – Nano Mach1	FC - Blank
Friction Test (N) (BAM Friction Tester)	> 32	> 44	> 52
Impact Test (J) (BAM Drop Hammer)	> 35	> 44	> 57

Table 5. Rocket motor propulsion parameter

Propulsion Parameter		FA - Lanxess	FB – NanoMach1	FC - Blank
Average Pressure	(MPa)	7.38	6.01	2.54
Average Thrust	(kN)	12.67	10.22	4.00
Total Impulse	(kN.s)	25.74	25.02	17.82
Burning rate	(mm/s)	11.40	9.26	4.97
Burning time	(s)	2.16	1.75	4.02

Figure 14. Rocket motor (a) and Bench firing test set up (b)

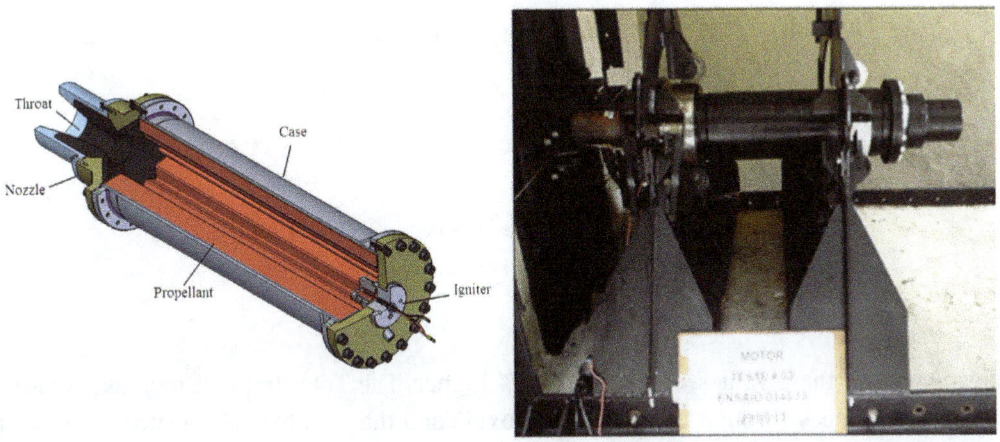

(a) (b)

Figure 15. Pressure-time curves for rockets motors with the formulations A, B, and C

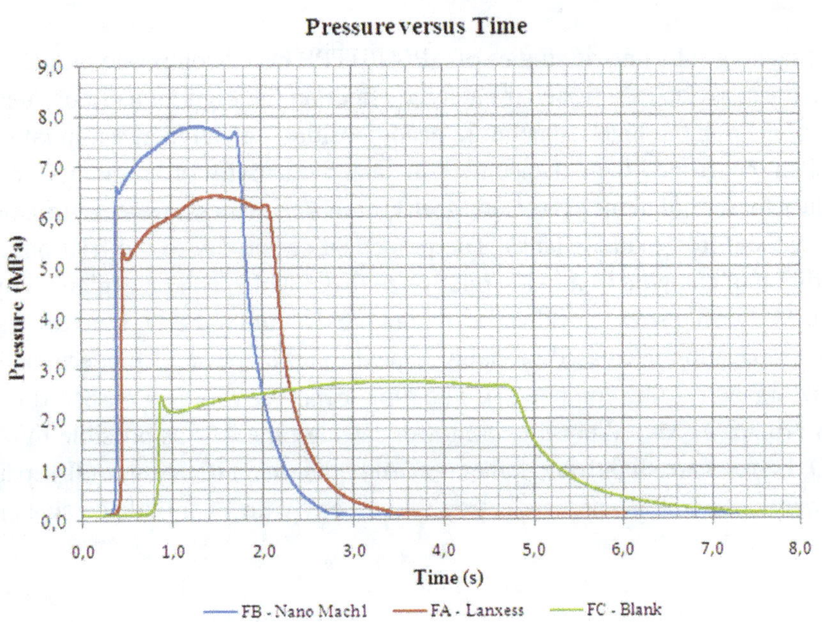

Figure 16. Thrust-time curves for rockets motors with the formulations FA, FB, and FC

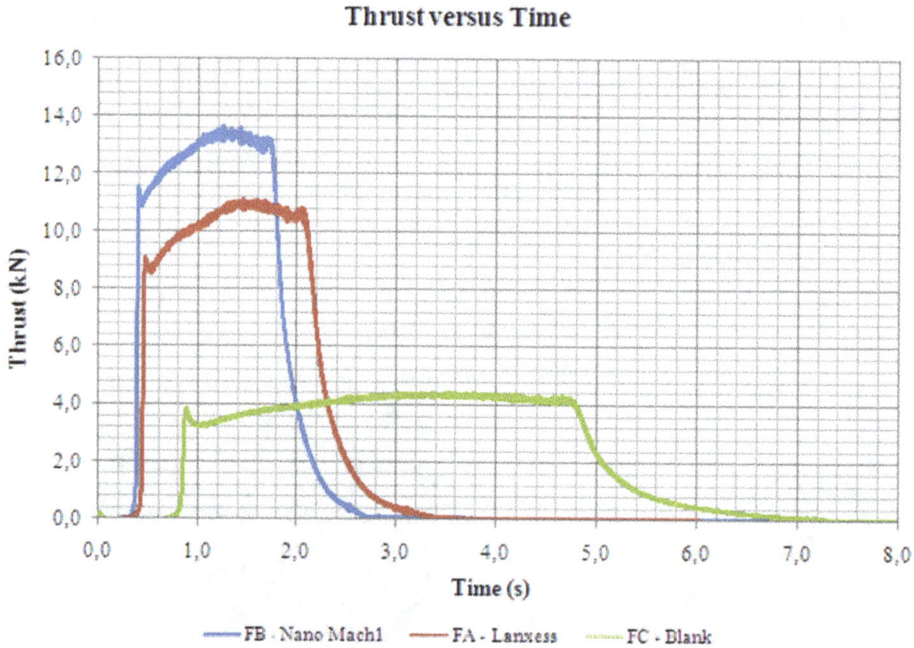

out catalyst, however these values were around 42% higher. The Total Impulse increase is partially due to combustion efficiency through the use of iron oxide and the suitable combustion chamber burning regime. The burning rates obtained in rocket motors were consistent with the burning rates obtained in strand burner.

CONCLUSION

The differential scanning calorimetric studies of AP/Al/HTPB based composite propellants in presence of nano and micro iron oxide exhibit better catalytic effect on the high temperature decomposition of AP. The lowering of Tp in the high temperature decomposition and an increase in rate constant in the catalyzed propellant confirmed the enhancement in catalytic activity of ammonium perchlorate. The Burning rate values for formulations FB (Nano Mach1) and FA (Laxness) were significantly higher than the formulation without iron oxide (Blank). The propellant sensitivity to the friction and impact was increased by addition of catalytic iron oxide micro size (FA), which can be explained by lower activation energy (Ea). The tensile strength value for the sample containing Fe_2O_3 Lanxess was significantly increased and the elongation decreased probably related to an increase of binder reaction rate due to contamination by organometallic compounds. The same behavior was also observed in the propellant viscosity, which promotes reduced pot life and lower casting time. It was possible to observe that the particles of Fe_2O_3 Lanxess are slightly agglomerated. At higher magnification, well-separated nano iron oxide particles were observed. Scanning electron microscopy with Energy dispersive analysis (SEM /

EDS) showed that the compositions of both Fe_2O_3 are very similar and the main chemical element is the α-Fe. The rockets motor ballistic results showed that the propellant burning rate was significantly affected by the addition of the iron oxide catalyst. There was an expectation of an increase of the Total Impulse for the formulations with catalyst compared with formulation without catalyst, however these values were surprisingly higher. The Total Impulse increase is partially due to combustion efficiency through the use of iron oxide and the suitable combustion chamber burning regime.

ACKNOWLEDGMENT

To Avibras Indústria Aeroespacial S.A., Chemistry Department of Technological Institute of Aeronautics (ITA).

REFERENCES

ASTM E1641-07. (n. d.). Standard Test Method for Decomposition Kinetics by Thermogravimetry. (2007). In *Annual Book of ASTM Standards* (Vol. 14.02). West Conshohocken, PA: ASTM International.

ASTM D2240 – 05. (2010). Standard Test Method for Rubber Property - Durometer Hardness. (2007). In *Annual Book of ASTM Standards.* West Conshohocken, PA: ASTM International.

Burnside, C. H. (1975). *Correlation of ferric oxide surface area and propellant burning rate*. Pasadena, CA, USA: AIAA. doi:10.2514/6.1975-234

Cavalheira, P., Gadiot, G. M., & Klerk, W. P. (1995). Thermal decomposition of phase-stabilised ammonium nitrate (PSAN), hydroxyl-terminated polybutadiene (HTPB) based propellants. The effect of iron (III) oxide burning-rate catalyst. *Thermochimica Acta, 269-270*, 273–293. doi:10.1016/0040-6031(95)02366-6

Engen, T. K., & Johannessen, T. C. (1990). The effects of two types of iron oxide on the burning rate of a composite propellant. In *Proceeding of the 21st International Annual Conference of Technology of Polymer Compounds and Energetic Materials*, Karlsruhe, Germany (pp. 81.1-81.12).

Flory, P. J. (1953). *Principles of Polymer Chemistry*. Ithaca: Cornell University Press.

Houghton, R. P., & Mulvaney, A. W. (1996). Mechanism of Tin (IV)-catalyzed urethane formation. *Journal of Organometallic Chemistry*, *518*(1-2), 21–27. doi:10.1016/0022-328X(96)06223-7

Inami, S. H., Rajaphase, Y., Shaw, R., & Wise, H. (1971). Solid Propellant Kinetics. I: The Ammonium Perchlorate-Copper Chromite-Fuel System. *Combustion and Flame*, *17*(2), 189–196. doi:10.1016/S0010-2180(71)80161-X

Kishore, K., Verneker, P., & Sunitha, M. R. (1980). Action of Transition Metal Oxides on Composite Solid Propellants. *AIAA Journal*, *18*(11), 1404–1405. doi:10.2514/3.7735

Kubota, N. (2007). *Propellants and explosives: thermochemical aspects of combustion.* Weinheim, Germany: Wiley-VCH Verlag GmbH & Co.

Lee, J. S., Hsu, C. K., & Chang, C. L. (2002). A Study on the Thermal Decomposition Behaviors of PETN, RDX, HNS and HMX. *Thermochimica Acta, 392-393*, 173–176. doi:10.1016/S0040-6031(02)00099-0

Li, W., & Cheng, H. (2007). Cu-Cr-O nanocomposites: Synthesis and characterization as catalysts for solid state propellants. *Solid State Sciences, 11*(2), 750–755. doi:10.1016/j.solidstatesciences.2007.05.011

Lima, V. (2007). *Estudo de Catalisadores Organometálicos na Síntese de Poliuretanos* [Master Degree Thesis]. Pontifícia Universidade Católica do Rio Grande do Sul.

Liu, L., Li, F., Tan, L., Ming, L., & Yi, Y. (2004). Effects of Nanometer Ni, Cu, Al, and NiCu Powders on the Thermal Decomposition of Ammonium Perchlorate. *Propellants, Explosives, Pyrotechnics, 29*(1), 34–38. doi:10.1002/prep.200400026

Ma, Z., & Li, F. (2006). Preparation and thermal decomposition behavior of TMOS/AP composite nanoparticles. *Nanoscience, 11*(2), 142–145.

Nema, A. K., Jain, S., Sharma, S. K., Nema, S. K., & Verma, S. K. (2004). Mechanistic Aspect of Thermal Decomposition and Burn Rate of Binder and Oxidizer of AP/HTPB Composite Propellants Comprising HYASIS-CAT. *Int. J. Plast. Technol., 8*, 344.

Pekel, F. E., Pinardag, E., & Türkan, A. (1990). An investigation of the Catalytic Effect of Iron (III) Oxide on the Burning rate of Aluminized HTPB/AP Composite Propellant. In *Proceedings of the 21st International Annual Conference of Technology of Polymer Compounds and Energetic Materials*, Karlsruhe, Germany (pp. 1-8).

Prajakta, R. P., Krishnamurthy, V. N., & Satyawati, S. J. (2006). Differential Scanning Calorimetric Study of HTPB based Composite Propellants in Presence of Nano Ferric. *Propellants, Explosives, Pyrotechnics, 31*(6), 442–446. doi:10.1002/prep.200600059

Rocco, J. A., Lima, J. E., Frutuoso, A. G., Iha, K., Ionashiro, M., Matos, J. R., & Suárez-Iha, M. E. (2005). Thermogravimetric Studies of a Composite Solid Rocket Propellant Based on HTPB-Binder. *Journal of Thermal Analysis and Calorimetry, 77*(3), 803–813. doi:10.1023/B:JTAN.0000041659.97749.fe

Satyawati, S. J., Prajakta, R. P., & Krishnamurthy, V. N. (2008). Thermal Decomposition of Ammonium Perchlorate in the Presence of Nanosized Ferric Oxide. *Defence Science Journal, 58*(6), 721–727. doi:10.14429/dsj.58.1699

Sciamareli, J., Takahashi, M. F., Teixeira, J. M., & Iha, K. (2002). Propelente sólido compósito polibutadiênico: I-influência do agente de ligação. *Quimica Nova, 25*(1), 107–110. doi:10.1590/S0100-40422002000100018

Silva, G., Iha, K., Cardoso, A. M., Mattos, E. C., & Dutra, R. C. (2010). Study of the thermal decomposition of 2,2,4,4,6,6- hexanitrostilbene. *Journal of Aerospace Technology and Management, 2*(1), 41–46. doi:10.5028/jatm.2010.02014146

Sutton, G. P., & Biblarz, O. (2017). *Rocket Propulsion Elements*. John Wiley & Sons.

Vyazovkin, S., Burnham, A. K., Criado, J. M., Pérez-Maqueda, L. A., Popescu, C., & Sbirrazzuoli, N. (2011). ICTAC Kinetics Committee recommendations for performing kinetic computations on thermal analysis data. *Thermochimica Acta*, *520*(1-2), 1–19. doi:10.1016/j.tca.2011.03.034

Wegener, G., Brandt, M., Duda, L., & Weisbeck, M. (2001). Trends in industrial catalysis in the polyurethane industry. *Applied Catalysis A, General*, *221*(1-2), 303–335. doi:10.1016/S0926-860X(01)00910-3

Zyl, G. J., & Keyser, R. (1996). Development of a composite propellant with a low Pressure Exponent suitable for Nozzleless Booster Motors. In *Proc. 27th Intl Ann. Conf. ICT.*, Karlsruhe, Germany (pp. 20/1-20/11).

Chapter 11
Study of Factors Influencing the Life Predictions of Solid Rocket Motor

Ricardo Viera Binda
Instituto de Aeronáutica e Espaço, Brazil

Roberta Jachura Rocha
Aeronautics Institute of Technology, Brazil

Luiz Eduardo Nunes Almeida
Avibrás Indústria Aeroespacial, Brazil

ABSTRACT

Storage of rocket motors loaded with composite solid propellant for long periods may change the propellant properties, thus causing failure and affecting the safety during launch. In this study, an accelerated aging assay was carried out, in order to predict the useful lifetime and to evaluate variations on the propellant properties with time by means of thermal analysis (TG/DSC). The aging temperatures used were 65°C, and samples were withdrawn after 3 months. Aging was also carried out at room temperature. There was significant variation in the activation energy of the solid propellant samples thermal decomposition in the two kinetic methods used – Ozawa or model-free isoconversional method and Kissinger method – during the aging period. There was significant decrease of enthalpy of aged propellant enthalpy causing changes in ballistics parameters of the solid propellant grain affecting the rocket's performance.

INTRODUCTION

Rocket engines are widely used in military missions, mainly for being the driving force of military artifacts of conventional use, including weapons and auxiliary devices. Currently, the peaceful intent and growing diplomatic relations between the UN countries make storage for long periods of time and readiness of use increasingly crucial for its proper functioning and safe operation. The high costs of acquisition and maintenance of solid propellant based weapons require the prediction and understanding of the useful life by the operators and manufacturers.

DOI: 10.4018/978-1-5225-2903-3.ch011

The solid composite propellant used in rocket motors is basically constituted by a polymeric matrix (MM 10% on average), rich in carbon and hydrogen, which serves as a binder and gas generator; an inorganic salt as the oxidant (70% average MM), rich in oxygen; and a ballistic auxiliary, generally a metallic additive, with the most commonly used is aluminum powder (20% on average MM). Once produced, the solid propellant grain installed in the engine shall attain appropriate ballistic characteristics and pre-defined mechanical behavior, which entails the use of raw materials and specific manufacturing processes for the propellant formulation design function motor. In addition, the proper storage, transport and handling is also of utmost importance for its proper functioning and safe operation (Sutton & Biblarz, 2010).

The chemistry composition of solid propellant changes with time, this is known as propellant aging. The parameters affected by aging are the chemical and the mechanical properties of the grain, compromising thus the ballistic properties of the motor. This phenomenon is critical to determine the window of use of use rocket engines. Among the parameters that affect aging, can be mentioned temperature, mechanical stress, environmental conditions (e.g., humidity) and even the contact with other organic materials (Kubota, 1984).

Studies conducted in order to obtain more accurate information about the propellant grain deteriorating conditions are an important issue in rocket engine applications, they allow the determination of the service life of each component. Due to the large surface area and its chemical potential energy, the composite propellants are subject to various tensions during handling, storage and operation, having consequences in ballistic results and in mechanical and chemical properties (Davenas, 2003).

The solid propellant composite is a complex and stable mixture of oxidizing and reducing ingredients. After ignition, the reagents react with each other, burning in a homogeneous, continuous and controlled manner, forming gas molecules at very high temperatures and pressures. The binder or polymeric matrix, most commonly used consists of a polyurethane elastomer obtained from hydroxylated polybutadiene

(Figure 1), commercially known as HTPB, which is classified as a liquid prepolymer with molar mass of the order of 2800 g.mol-1 containing terminal hydroxyl groups and reactive groups.

The HTPB is obtained by polymerization of butadiene (Figure 2), initiated by hydrogen peroxide using an alcohol as diluent (Vilar, 2005). A major advantage of the use of hydroxylated polybutadiene is the lack of secondary reactions and by-products during curing or forming of the propellant grain. The curing process consists of reacting the hydroxyl HTPB with compounds containing isocyanate groups, forming a polyurethane polymeric matrix.

The rocket engines loaded with solid propellants have a predetermined useful life, in which the security, the mechanical parameters and the ballistics can be guaranteed during launching. The aging of the propellant, despite being a slow and gradual phenomenon, can hamper adequate performance.

Recognizing the importance of the phenomenon, testing programs have been created to study the propellant grain aging. Due to the slow the aging process at room temperature, the accelerated aging is carried out. The aging is promoted above room temperatures, in in special chambers (Rezende, 2001), (Judge, 2003), (Sbriccoli, Saltarelli, & Martinucci, 1989).

In a study of accelerated aging of a propellant, one must identify the parameters that can affect and are affected by the process. In general, the mechanical properties of solid propellant grain, such as tensile strength, elongation and hardness, are the most affected. Parameters such as conductivity, density and thermal expansion may also be affected due to aging, but variations in their properties are rarely seen (Sbriccoli, Saltarelli, & Martinucci, 1989).

Figure 1. Structural form of hydroxylated polybutadiene (HTPB)

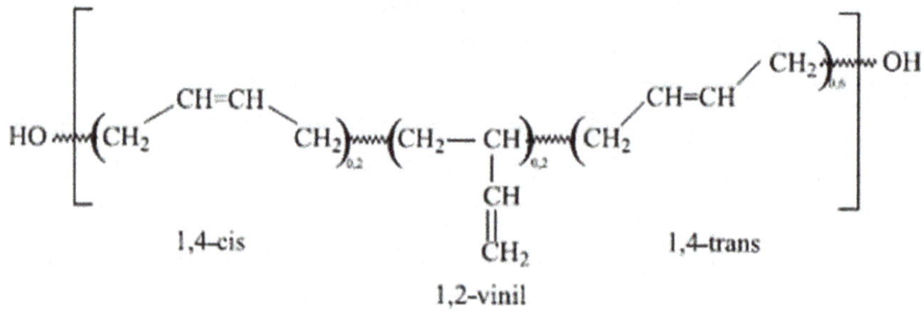

Figure 2. Obtaining reaction of hydroxylated polybutadiene (HTPB)

$$n\,H_2C{=}CH{-}CH{=}CH_2 + 2\,HO\cdot \rightarrow OH{-}[CH_2{-}\overset{H}{C}{=}\overset{H}{C}{-}CH_2]_n{-}OH$$

The temperature is regarded as the most important factor in the aging process, because storage at elevated temperatures tends to cause an acceleration in the velocity of chemical reactions according to the Arrhenius equation. Nevertheless, if the temperature is too high, there might be reactions that would be avoided in natural aging. For this reason, propellants aging studies based on HTPB are typically performed at temperatures not exceeding 70 °C (Judge, 2003).

An additional factor is the accumulation of mechanical stresses that may cause separation between the binder and solid fillers, inducing the formation of vacuoles and the rise of loose particles. These particles can act as a point of microcracks, magnifying the burning surface, thus heightening the internal pressure, which may result in detonation of the material.

The environmental conditions in the study of aging are also considered, because moisture and air can accelerate the material oxidation process. Besides, the contact with other organic materials can induce migration of any component of the propellant favoring, for instance, a weakening in the adhesion with the insulator or with the rubber (Davenas, 2003), (Sbriccoli, Saltarelli, & Martinucci, 1989), (Sciamareli, Takahashi, Teixeira, & Iha, 2002).

The changes deriving from propellant aging can be manifested in several ways such as increasing or decreasing grain hardness, swelling, discoloration, among others, in addition to changing kinetic properties due to increased crosslinking in the polyurethane chain (De la Fuente, 2009)· (Layton, 1974).

METHODOLOGY

Although physicochemical tests are available, composite propellant aging is most commonly studied using methods to measure and assess the variations of relevant mechanical properties such as Young's modulus and the tensile strength (De la Fuente, 2009).

Layton was one of the first researchers to study the aging propellant, he proposed an empirical equation that bears his name. Satisfying the rate of change of the mechanical properties, the Eq.1 can be written (Layton, 1974), (Hocaoglu, Özbelge, Pekel, & Özkar, 2001):

$$\frac{dP}{dt} = \frac{K}{t} \tag{1}$$

wherein P is the mechanical property; K is the aging constant and t is the aging time. With the integration of Eq. 1 we obtain:

$$\int_{P_0}^{P} dP = K \int_{t_0}^{t} \frac{1}{t} dt \tag{2}$$

which results in the Layton Equation:

$$P = K \ln \frac{t}{t_0} + P_0 \tag{3}$$

The K values can be obtained for various temperatures using the Arrhenius Equation:

$$K = Ae^{\left(-E_a/RT\right)} \tag{4}$$

where Ea is the activation energy; R is the universal gas constant; T is the absolute temperature, and A is the pre-exponential factor.

Note that the Layton equation should be used when there is only one single reaction or when there are competitive reactions. For this reason, the use of this equation is limited.

The Arrhenius method is also employed for this purpose, it assumes that the entire process consists entirely of first-order chemical reactions. Molecular decomposition reactions are basically molecular divisions; shock or contact with other species are. Therefore, a molecule decomposes to generate two or more (observed behavior in first-order reactions). For reactions or chemical processes of higher orders, it is necessary to consider the pre-exponential factor of each elementary reaction, directly related to the system temperature.

Kissinger's method estimates the kinetic parameters from thermogravimetric data. This method examines the processes and parameters involved in kinetics of decomposition, through experiments in dynamic conditions. To calculate the activation energy (Ea), the temperature at which the decomposition rate (or conversion) is maximum is used. Thus, the peak temperature (Tp) of the first derivative of the curves is required (Rezende, 2001).

The increased heating rate causes a displacement of the highest temperature peak, Kissinger method is based on this phenomenon. In this method, the activation energy value is obtained by linear regression, where ln (β / Tp^2) are plotted against 1000 / Tp for a series of experiments at different heating rates, where Tp is the temperature of the highest peak of the DTG curve.

Kissinger method is defined by Eq. 5:

$$\ln\left(\frac{\beta}{T_p^2}\right) = \ln\left(\frac{AR}{E_a}\right) - \frac{E_a}{RT_p} \tag{5}$$

where β is the heating rate (K s^{-1}), T_p is the peak temperature at each heating rate (K), A is the Arrhenius pre-exponential factor (s^{-1}), R is the universal gas constant (8.314 J mol^{-1} K^{-1}) and Ea is the reaction activation energy (J mol^{-1}).

The activation energy value is obtained from a graph of ln (β / Tp^2) to (1000 / Tp) for a series of experiments with different heating rates (β) as a function of the peak temperature (Tp) of the curve DTG (DTA or DSC). The activation energy (Ea) is obtained from the slope of the curve, which is equal to $-E_a/R$, and the pre-exponential factor is obtained by linear coefficient ln (AR / E_a).

In the method of Ozawa, there is a linear relationship between the heating rate (log β) and the inverse of the absolute temperature (Tp^{-1}) and the following linear equation can be established:

$$log\beta = a.T_p^{-1} + b \tag{6}$$

wherein a and b are the parameters of the linear equation, a equals to -0.4567Ea / R and b is a constant, R is the ideal gas constant.

Assuming that the rate constant follows the Arrhenius law and the exothermic reaction can be considered a process of one step, the conversion rate at the maximum ratio does not vary with heating when it is linear. Eq. 7 can then be applied to the maximum temperature of the exothermic peak, considering various heating rates. Therefore, performing analyzes at different heating rates, the log curve β vs. 1 / T can be plotted, allowing the direct calculation of the activation energy:

$$E_a = -2.19R\left[\frac{d\left(log\beta\right)}{d\left(T_p^{-1}\right)}\right] \tag{7}$$

The aging of the propellant grain, comprising post curing period, is generally observed when mechanical and ballistic characteristics differ from the design values, immediately after the engine production process.

In this study, the solid propellants aging behavior of composites based on HTPB / IPDI / AP / Al was investigated by means of changes in the thermal properties of the grain installed in the engine of a ballistic rocket, of 70 mm diameter. Following the curing period, one propellant grain was subjected to natural aging on the shelf and the other one, from the same lot, was subjected to an accelerated aging at 65 °C, for a period of 90 days, in a greenhouse.

The solid propellant composite formulation used in this study was HTPB (hydroxy terminated polybutadiene) as binder; curing agent IPDI (isophorone diisocyanate); oxidizer AP (ammonium perchlorate); metallic aluminum as an additive; and iron oxide as burn catalyst. The average composition of a composite solid propellant based on HTPB and AP and is shown in Table 1.

Table 1. Solid propellant composition AP/HTPB

Components	% Mass
HTPB	de 10.0 a 15.0
Bonding agent(RX)	de 1.0 a 1.5
Ballistic auxiliar (aluminum powder)	de 15.0 a 20.0
Oxidant (ammonium perchlorate)	de 69.0 a 75.0
additives	de 4.0 a 4.5

During thermal properties testing, data were collected in a range of three heating rates for the TG / DTG and four rates for the DSC.

RESULTS AND DISCUSSION

DSC Curves Analysis

The heat flow curves or DSC curve of the aged samples and unaged propellants showed endothermic peaks around 330 °C in all heating rates. This peak is attributed to the transition from orthorhombic to cubic phase of ammonium perchlorate (Kubota, 1984). After the occurrence of this particular endothermic peak, other exothermic peaks with several peaks occur, they are used in Ozawa Method to determine the thermal decomposition activation energy. During endothermic phase, the change in enthalpy (DH) was determined by drawing a reference line, as in Tables 2 to 5.

Applying the Ozawa method, it was possible to measure the activation energies (Ea) of propellant grain samples aged in an oven and in the shelf. The activation energy measured on unaged samples was equal to 158.4 kJ $mol^{-1.}$ The Ea measured on aged sample in the oven was 139.2 kJ mol^{-1}, a reduction of 12.12%. Samples of the two grains were measured with four heating rates: 2, 3, 4, and 5 °C min^{-1}.

For the same aging time, the figures show that there is an increasing trend in the variation of enthalpy with increasing heating rate. However, if the comparison is made between the aging time, there is a great dispersion in the results, and enthalpy values decrease with the passage of time.

Observing ΔH of naturally and artificially aged propellant (90 days in the greenhouse, at 65 °C), it can be noted that internal energies decrease during change phase, in different heating rates. With a heating rate of 2 °C min^{-1}, the ΔH measured on the endothermic phase of DSC curves in the naturally aged sample had a reduction of approximately 2.46% in comparison to the value obtained from the sample of propellant aged in an oven.

Table 2. Peak Temperatures and ΔH calculated from DSC for aged and not aged propellants in an oven at a heating rate of 2 °C min^{-1}

	2 °C.min^{-1}	
	Aged in Greenhouse	Not Aged in Greenhouse
Peak Temperature	315.70 °C	317.66 °C
Variation of enthalpy (ΔH)	-890.09 J.g^{-1}	-912.52 J.g^{-1}

Table 3. Peak Temperatures and ΔH calculated from DSC for aged and not aged propellants in an oven at a heating rate of 3 °C.min^{-1}

	3 °C.min^{-1}	
	Aged in Greenhouse	**Not Aged in Greenhouse**
Peak Temperature	325.28 °C	328.23 °C
Variation of enthalpy (ΔH)	-975.23 J.g^{-1}	-1062.13 J.g^{-1}

Table 4. Peak Temperatures and ΔH calculated from DSC for aged and not aged propellants in an oven at a heating rate of 4 °C.min^{-1}

	4 °C.min^{-1}	
	Aged in Greenhouse	**Not Aged in Greenhouse**
Peak Temperature	330.04 °C	332.47 °C
Variation of enthalpy (ΔH)	-1730.62 J.g^{-1}	-1872.66 J.g^{-1}

With the heating rate of 3 ° C min^{-1}, thc drop was 8.2%. With a ratio of 4 °C min^{-1}, the decrease was 7.6%.

The observed drop in the curve at a heating rate of 5 °C min^{-1} was 2.4%. However, in all graphs, a decrease of ΔH is observed. Whereas the rocket engine is a heat engine, this implies an enthalpy variation expressed in terms of work, with consequences in the engine thrust.

The percentages found in ΔH drops are significant, especially taking into account the accelerated aging time (just 90 days in an oven at 65 °C) which correlates to a natural aging time of 10 years in the shelf (Judge, 2003), (Shekhar, 2011). The confined accelerated aging performed with the propellant inside the motor tube showed smaller variations in comparison to the analyses of isolated aging samples (Rezende, 2001), (Judge, 2003).

Similarities With Literature Results

According to Judge (2003), oxidative cross-linking of the propellant binder is found to be the predominant degradation mechanism of unstressed, semi-confined composite propellant during accelerated ageing. The additional crosslinking of the binder resulting from oxidation manifests as significant changes in mechanical properties and soluble fraction over time. An appropriate accelerated ageing protocol to demonstrate a motor with a ten-year lifespan at ambient could therefore consist of 16 weeks ageing at

Table 5. Peak Temperatures and ΔH calculated from DSC for aged and not aged propellants in an oven at a heating rate of 5 °C.min^{-1}

	5 °C.min^{-1}	
	Aged in Greenhouse	**Not Aged in Greenhouse**
Peak Temperature	335.61 °C	333.03 °C
Variation of enthalpy (ΔH)	-1370.51 J.g^{-1}	-1404.63 J.g^{-1}

60 °C. In the present paper, ageing data was gathered from tests consisting of 13 weeks at 65 °C, arriving in similar propellant lifespan.

Shekhar, 2011, correlated the reduction in elongation to the propellant ageing, once the elongation has a measurable variation with time and is taken as the fastest degrading parameter for HTPB/AP/Al based composite solid rocket propellants. A reduction of the elongation to less than 50% was taken as the end-of-shelf life of the propellant, and it was found to be at 20 years. Using a factor of safety of 2, he estimated the useful life of 10 years.

In the literature, there are different methods to arrive at the ageing characteristics and at the shelf life, it was shown that the study can be focused on the oxidative reactions (Judge, 2003) or on the mechanical properties (Shekhar, 2011). The present work studies the decrease in ΔH and makes a correlation to the ageing time, arriving in the same conclusions acquired by other methods.

The innovation of this work is to study the variation of enthalpy, once it is the basis of work in heat engines. A reduction in enthalpy is related to the ageing of the propellant, and this reflects in the direct decrease of work and total power available in the engine.

Analysis of TG/DTA Curves

With the use of TG and DTA curves (Figures 3 to 5), the calculation of the activation energy can be accomplished in a similar manner to that previously mentioned for the DSC curve. The activation energy in this case was obtained in two distinct ways: firstly, using the value corresponding to the maximum peak temperature found in the DTA curve analysis; then, temperatures were used in accordance with the degree of conversion of the sample in terms of weight loss.

The calculation of activation energy based on the peak temperatures (TGA) was performed using the methods of Ozawa and Kissinger, for comparison purposes. The linearization of the two methods is shown in Figure 6. The fast analysis of the correlation coefficients shows greater linearity in the Ozawa method, therefore this method was selected for the analysis of activation energy per degree of conversion, which will subsequently be demonstrated.

Figure 3. Curves TG/DTA of aged propellant, with heating rates of 6 °C min^{-1}

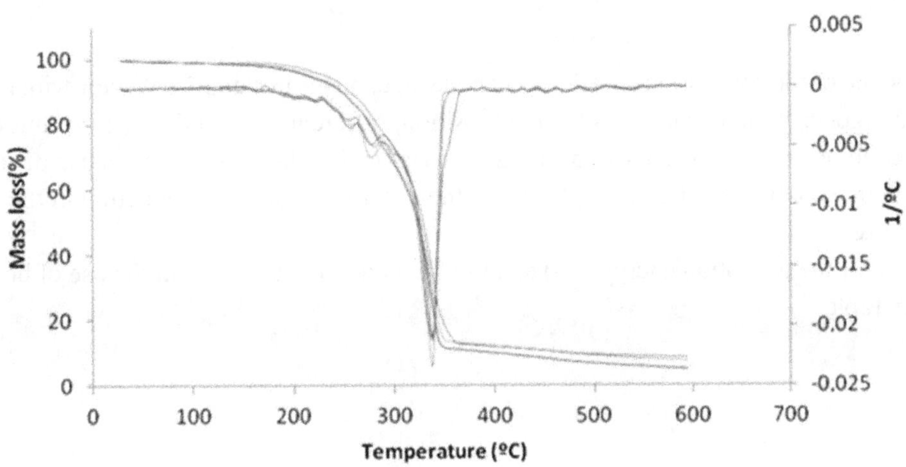

Figure 4. Curves TG/DTA of aged propellant, with heating rates of 8 °C min^{-1}

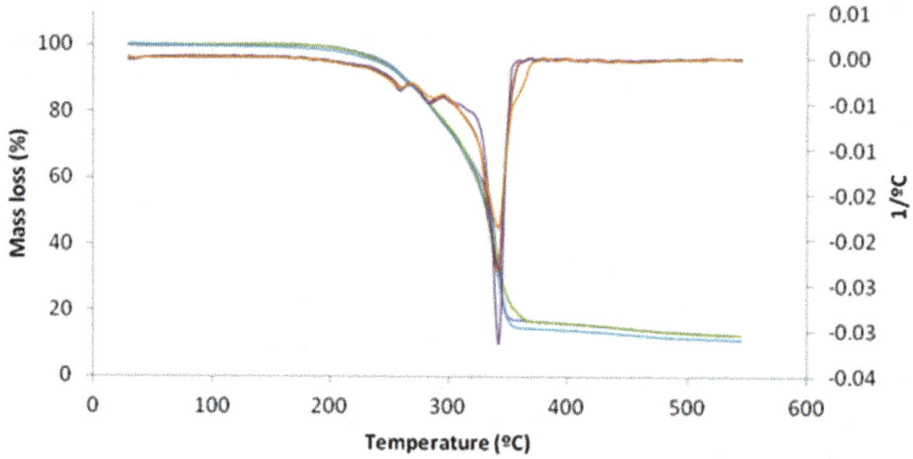

Figure 5. Curves TG/DTA of aged propellant, with heating rates of 10 °C min^{-1}

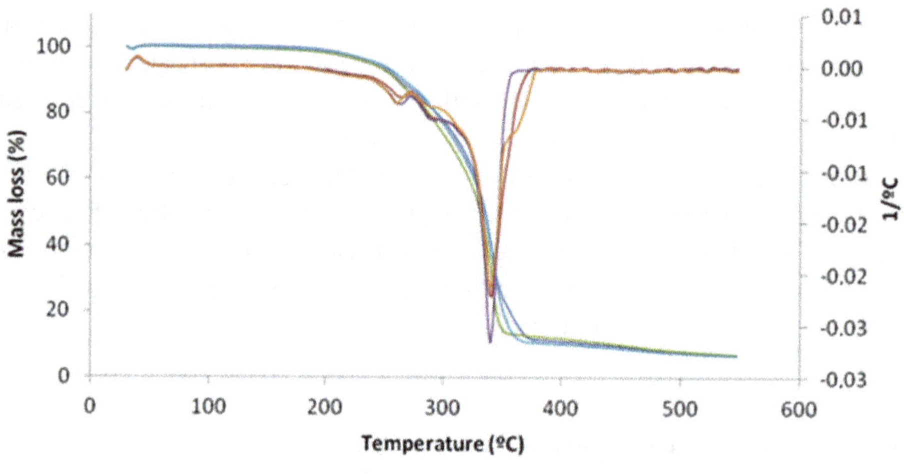

The low value of the correlation coefficient can mean: changes in sample behavior with heating rate, due to the emergence of new reactions of decomposition; different mass and shape of samples for each analysis; uncalibrated scale; or explosion of the material inside the crucible. Because of its energetic nature, it is more suitable to consider explosion in the crucible, producing reduction in the correlation between analyzes.

The values for the activation energy (Ea) and the frequency factor (A) with the use of both methods are shown in Table 6.

Table 6. Kinetic parameters from TG/DTA peak temperature curves

Method	Activation Energy (kJ mol^{-1})	Frequency Factor (.10^4)
Ozawa	80.58	6.87
Kissinger	81.99	7.00
Deviation (%)	1.72	1.77

Figure 6. Linearization of results using the following methods: a) Ozawa e b) Kissinger

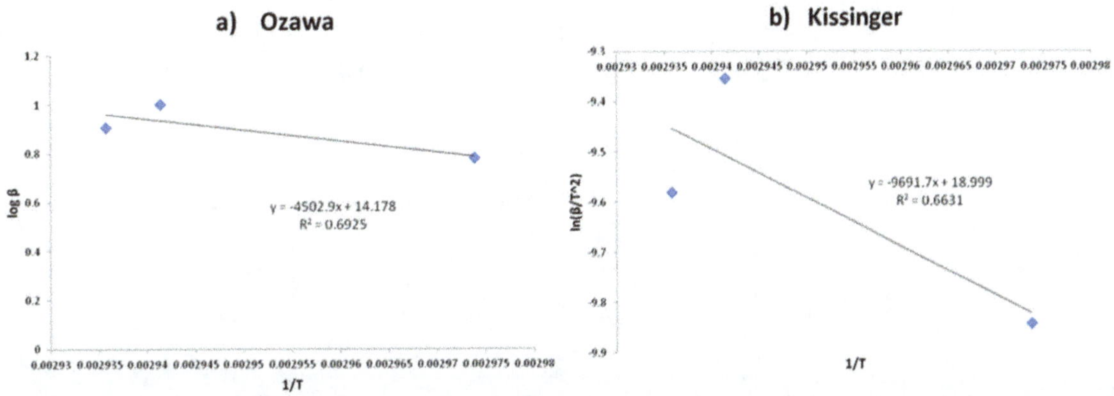

It is observed low disparity between the results of the two methodologies, but there is still a large discrepancy to the values obtained by DSC curves. When the activation energy is analyzed by varying degrees of conversion, better linearization is accomplished (Ozawa method). The curves are shown in Figure 7.

The kinetic parameters calculated for each degree of conversion are given in Table 7. The significant difference between the values are due to non-linearity of the TG curve, that is, a linear behavior is not observed during the single mass loss event of the sample. This lack of linearity demonstrates the existence of a complex mechanism of decomposition reactions of the material, due to its various components and structure. In a single event (the ideal case), governed only by a single elementary reaction, the weight loss would have a perfectly linear behavior, possibly with instant decay.

CONCLUSION

Samples of the solid composite propellant grain, subjected to accelerated aging program, showed changes in the thermal properties that match those found in the literature. These changes, mainly characterized by energetic properties, indicate that the composite has a predetermined life that takes into account the parameters of the corresponding engine design, the main one being the ballistic performance. The reduction of ΔH may alter flight characteristics when related to the work done by the device and can lead to an increase in the error of the ballistic trajectory.

Figure 7. Linearization of results using Ozawa method, with conversions of 5, 20, 35 e 50%

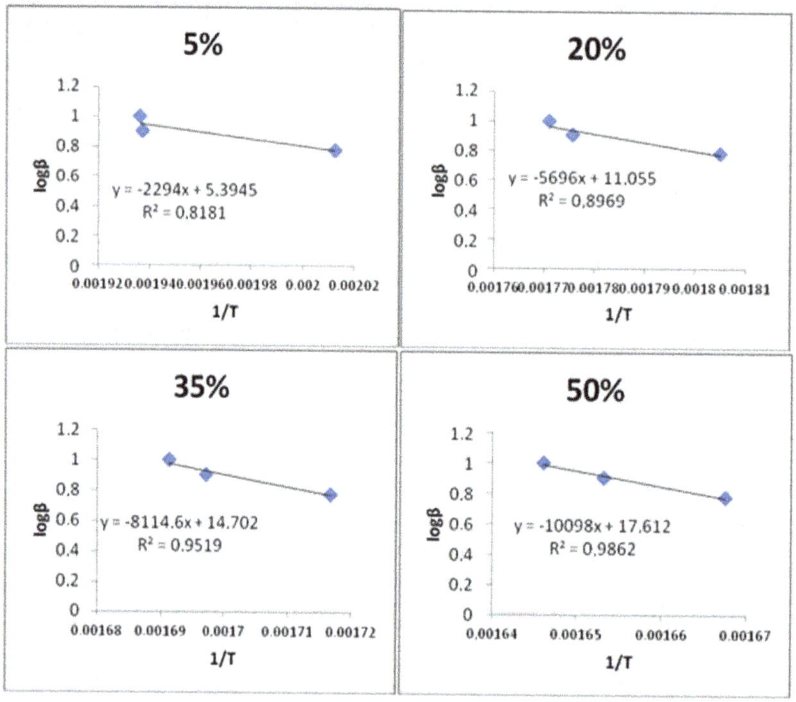

Table 7. Kinetic parameters from TG/DTA curves

Conversion Degree (%)	Activation Energy (kJ mol^{-1})	Frequency Factor ($.10^4$)
5	41.77	6.93
20	103.72	12.37
35	147.76	15.04
50	183.87	17.17

This grain aging process is directly related to the temperature and time of exposure. The aging mechanism is the growth of crosslinking reactions density by breaking the double bond of HTPB, which also decreases the activation energy, making the device unstable and compromising the risk in handling.

REFERENCES

Davenas, A. (2003). Development of modern solid propellants. *Journal of Propulsion and Power, 19*(6), 1108–1128. doi:10.2514/2.6947

De la Fuente, J. L. (2009). An Analysis of the thermal aging behavior in high-performance energetic composites through the glass transition temperature. *Polymer Degradation & Stability, 94*(4), 664–669. doi:10.1016/j.polymdegradstab.2008.12.021

Hocaoglu, Ö., Özbelge, T., Pekel, F., & Özkar, S. (2001). Aging of HTPB/AP-based composite solid propellants, depending on the NCO/OH and triol/diol ratios. *Journal of Applied Polymer Science*, *79*, 959–964. doi:10.1002/1097-4628(20010207)79:6<959::AID-APP10>3.0.CO;2-G

Judge, M. D. (2003). An Investigation of Composite Propellant Accelerated Ageing Mechanisms and Kinetics. *Propellants, Explosives, Pyrotechnics*, *28*(3), 114–119. doi:10.1002/prep.200390017

Kubota, N. (1984). Survey of Rocket Propellants and Their Combustion Characteristics. In K. K. Kuo (Ed.), Progress in Astronautics and aeronautics - Fundamentals of solid-propellant combustion (Vol. 90, p. 1). Summerfield: AIAA.

Layton, L. H. (1974). *Technical Report AFRPL-TR-74-16*. Brigham City, Utah: Thiokol Chemical Corporation.

Rezende, L. C. (2001). *Envelhecimento de Propelente Compósito à Base de Polibutadieno Hidroxilado* [PhD Thesis].

Sbriccoli, E., Saltarelli, R., & Martinucci, S. (1989). Comparison between Accelerated and Natural Aging in HTPB Composite Propellants. In *Proceedings of 20th International Annual Conference of ICT*, Karhsruhe, Germany.

Sciamareli, J., Takahashi, M. F., Teixeira, J. M., & Iha, K. (2002). Propelente sólido compósito polibutadiênico: I – Influência do agente de ligação. *Quimica Nova*, *25*(1), 107–110. doi:10.1590/S0100-40422002000100018

Shekhar, H. (2011). Prediction and Comparison of Shelf Life of Solid Rocket Propellants Using Arrhenius and Berthelot Equations. *Propellants, Explosives, Pyrotechnics*, *36*(4), 356–359. doi:10.1002/prep.200900104

Sutton, G. P., & Biblarz, O. (2010). Rocket Propulsion Elements (8th ed.). Wiley.

Vilar, W. D. (2005). *Química e Tecnologia dos Poliuretanos* (3rd ed.). Rio de Janeiro.

Chapter 12
Accelerated Aging Tests of HTPB-Based Propellants

Roberta Jachura Rocha
Aeronautics Institute of Technology, Brazil

ABSTRACT

In the late twentieth century, liquid and solid propulsion technologies have been integrated into hybrid engines currently apllied in propulsion launch vehicles and missiles. The reaction of polyol (HTPB) and diisocyanate (IPDI) provides the most versatile of the binders in the production of solid propellants due to its ability to withstand high loads combined with low cost and ease of processing. A propellant based on HTPB obtained in this study was submitted to natural and accelerated aging tests, seeking to evaluate the modifications of mechanical properties as tensile strength, elongation and hardness up to 360 days. The mechanism considered in the aging process is the increase of crosslink density by breaking the double bond contained in the HTPB molecule, which causes the instability of the propellant, increasing its handling risk. Samples of these propellants subjected to aging presented variations in their properties that match the values available in the literature.

INTRODUCTION

Propulsion Systems Operation

Propulsion by rocket engines followed two distinct paths throughout the twentieth century. One, according to Tsiolkovsky's view in Russia, Goddard's in the United States, Esnault-Pelterie's in France, and Oberth's in Germany, understood that rocket propulsion was the key to accessing outer space, and for the consolidation of space travel, we opted for rocket propellant (MFPL) propulsion because of its superior performance when compared to solid propellant rocket propulsion (MFPS) propulsion. The other route in which it followed the propulsion by rocket motors was towards the development of the systems of armaments, in which the propulsion was adopted by motor rocket to solid propellant. It was decided to use the MFPS based on the need to combine good performance with the other characteristics provided by this propulsion model, such as the possibility of immediate use, its compact dimensions, lower cost

DOI: 10.4018/978-1-5225-2903-3.ch012

and longer service life. At the end of the 20th century, liquid and solid propulsion technologies also found applications in an integrated form in launch vehicles and missiles, among others (Davenas, 2003).

Solid Propellant Rocket Engines

A rocket motor is the simplest form of chemical propulsion, capable of delivering high power in a relatively short period of time. Its main use is in the launching of satellites and probes used for space research.

Basically, the rocket motor consists of a metal cylinder, called a rocket motor tube, within which chemical reactions occur generating large amounts of gases at high temperature. In the rocket motor is the solid propellant, also called the propellant grain, which contains all the chemical elements necessary for a complete burning, since both the fuel and the oxidant are incorporated in a single solid. After ignition, the burning of the grain occurs without interruption on its surface at a predetermined speed, and the resulting gases from the burning are expelled back by a nozzle, which is an opening in the rear known as a nozzle (Vilar, 2005), as Illustrates Figure 1. Thus, the working principle of a rocket is based on Newton's Third Law, the Law of Action and Reaction: the mass of gases expelled at high speeds at the rear of the rocket generates a force in the opposite direction Known as a thrust force that, together with the expulsion of gases, acts by propelling the rocket (Sutton & Biblarz, 2010), (Klager, 1984), (De Luca, et al., 2013), (Paubel, 2002).

A rocket motor is the simplest form of chemical propulsion, capable of delivering high power in a relatively short period of time. Its main use is in the launching of satellites and probes used for space research.

The solid propellant is the only material of the rocket motor with the function of being a source of propulsive energy if presenting simultaneously as a structural component. In order to provide a better understanding of the structural problems related to the rocket motor with solid propellant, a detailed analysis is required of all steps involved in the manufacture of the propellant, such as its formulation, the preparation of the motor envelope that will contain it, its healing process and of finishing of the grain,

Figure 1. Solid propellant rocket engine
Paubel, 2002.

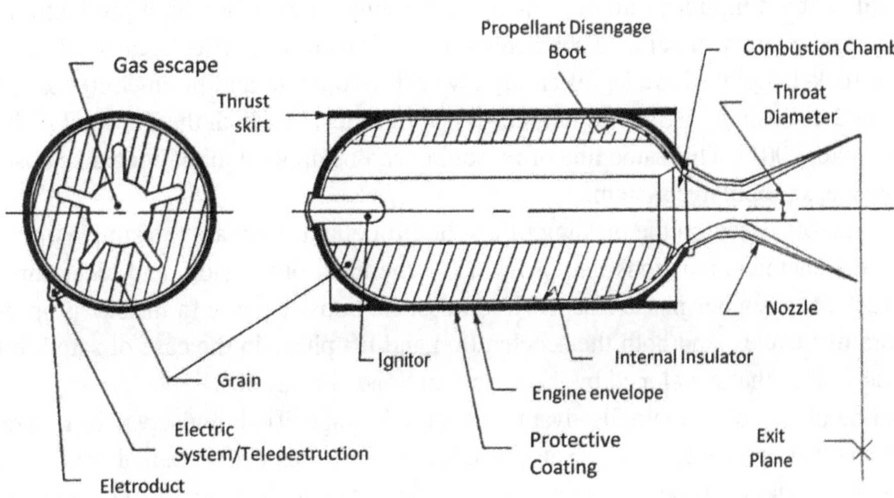

besides the requirements foreseen by the project for which it was developed. For example, at a time when a solid propellant rocket motor design is defined, its specifications must meet diverse expectations related to its performance, reliability and cost, which requires the involvement of large knowledge and techniques related to grain factors, Such as its nature, geometry, grain size, and the modes of operation of the rocket motor (Klager & Di Milo, Encyc1opedia of Polymer Science and Technology - Rocket Propellants, 1976).

Generally, the grain forms by loading the propellant into a motor envelope, pre-coated by thermal protections. Its configuration is defined by a mandrel, obtaining a central cavity that allows it to present diverse forms of geometry, for example, starry, circular or the specific format required by the design. The mass of the propellant in the engine can range from grams to tons and the operating times range from milliseconds to minutes.

Depending on its application, the motor envelope may be fabricated from metal or composite material based on fiber reinforced by various types of polymer matrices. Motors whose rocket motor tube (also called case) is based on composite materials, in addition to being lightweight, provides the same performance in relation to mechanical strength and can be manufactured directly on the coatings that make up the thermal protection, for example, Rubber, which allows better adhesion at the interface between the case and the solid propellant grain.

When compared to other types of aerospace engines, solid propellant rocket motors (MFPS) are more reliable when compared to other propulsion systems. Its advantages, especially in the case of tactical missiles and satellite launch vehicles, outweigh its disadvantages and for this reason are still widely used today. As a main advantage, we can mention the high density presented by the solid propellant grain when compared, for example, to the liquid propellant used in liquid propellant rocket engines (MFPL), because this factor allows the grain to have a higher energy content, which may be a limitation in the choice of propellant to be used depending on propulsive design characteristics. The propellant grain presents greater ease in its storage and, in addition, can be considered chemically stable, depending on the formulation used in its production, which allows to maintain its physicochemical characteristics little altered for a long period of time (useful life). At the current stage of development of MFPS-based propulsion systems integrated into tactical missiles, its useful life can be estimated at 15 years. Another important feature of MFPS is its relative simplicity compared to propellant systems by liquid propellant rocket motors (MFPL) or even gas turbines (Turbojet). Once again, considering warlike artifacts as tactical missiles, this simplicity can become a determining factor when high production scales and issues such as reproducibility of ballistic characteristics are imperative. Also, because of this simplicity, solid propellant rocket engines have significantly lower development and production costs; Except for the propulsion in the subsonic regime of medium-range missiles, in which the Turbojet is the option of lower cost (Fleeman, 2001). This same line of reasoning can be applied to maintenance costs when the MFPS is adopted as a propulsive system.

Thus, rocket motors based on the burning of a solid propellant grain are the simplest way to generate thrust from the chemical reactions that provide the chemical propulsion, and meet, among others, requirements such as: sustainer phase Rockets and missiles, booster phase in missile propulsion in the Ramjet and Scramjet cycles, and both the acceleration and lift phase in the case of satellite launch vehicles (VLS), as well as that developed by the Brazilian Space Program.

On the other hand, one of its main disadvantages is the low specific impulse value (Isp) reached. Isp is an important ballistic property of a rocket motor which indicates the propellant force available per second, given per unit mass of propellant consumed, and is related to the velocity of the ejected particles.

Another disadvantage is the difficulty of modulating the thrust in a large MFPS, since once the firing is started, it will be completed without the possibility of interruptions. However, great strides have been made in recent decades that tend to minimize MFPS limitations.

Hybrid Rocket Engine

The hybrid rocket engine is a chemical propulsion system in which the fuel is stored in two distinct phases, one liquid and the other solid. In the most common systems, the liquid acts as oxidant and the solid acts as a fuel (Figure 2). Several combinations of solid fuels and liquid oxidants, as well as solid liquid and oxidant fuels, have been tried for use in hybrid rocket engines (Vilar, 2005). Applications include rocket propulsion systems for space launching, satellite maneuvering systems and upper rocket stages.

For control of the dosage in the liquid oxidant system, a pressurizer or a turbo pump may be employed. The solid fuel is vaporized with heat from the combustion process and mixed with the vapor of the oxidant to produce further combustion. A mixing chamber is used to ensure that the fuel and oxidant are burning before leaving the nozzle outlet (Vilar, 2005).

Among the advantages of a hybrid rocket propulsion system, one can cite (Vilar, 2005), (Kuo, 1997), (Chiaverini, et al., 1992), (De Luca, Problemi energetici in propulsione aerospaziale, 1999):

1. Hybrid rockets have half of the systems for the supply and storage of bipropellant liquid propellants;
2. The control of the oxidant flow in the hybrid propellers allows the control of the burning of the solid fuel;
3. It is possible to start the system with the possibility of triggering its stop and also its starting again;
4. Provides greater safety during its manufacture and storage until the moment of its operation, because there is less risk of explosion or detonation by accident or operation in relation to the other systems;
5. Provides a higher specific thrust (Isp) when compared to solid fuel engines.

Among the disadvantages of a hybrid rocket propulsion system, we can mention:

1. It has a lower value of specific thrust density when compared to the solid propellant;
2. There is a loss of part of the fuel that may be trapped in the combustion chamber at the end of the burn due to the flame structure (in this case, classified as diffusive);
3. It is a system that has not yet been tested on a large scale.

The operation of a hybrid rocket motor basically takes place as follows: A propellant grain of a hybrid rocket engine is ignited by the heat supplied from a given source and then the solid fuel gasification at the end of the engine head is started. The oxidant flow provides the required flame by spreading the ignition throughout the engine, which ignition is typically accomplished by the injection of a hypergolic fluid into the combustion chamber of the engine.

Hybrid propulsion rocket motors are good candidates for various types of applications and may offer a less aggressive environmental proposal (Mead & Bornhorst, 1969). In addition, propulsion in hybrid rocket engines is considered especially suitable for private travel of humans to space, mainly due to the greater safety in its procedure and relatively low cost when compared to the engines that employ liquid propellants in relation to their development and control.

Figure 2. System of a rocket motor to hybrid propulsion

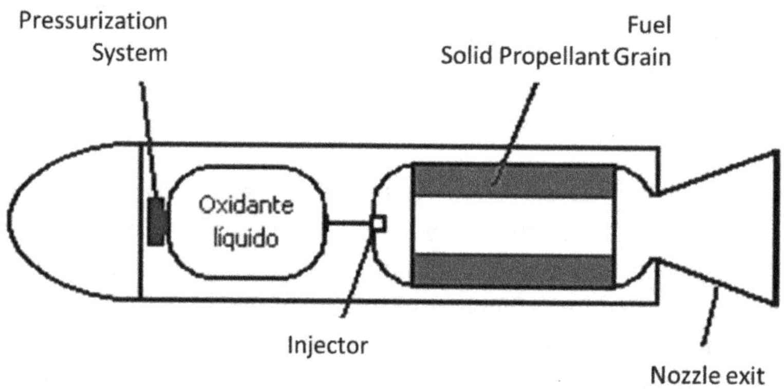

In hybrid propulsion engines, one has the option of selecting the best-known elements of solid and liquid rocket propulsion. The choice of liquid oxidant greatly affects the whole design of the propulsive system, but there is ample knowledge of the technology involved due to diverse research on the propulsion in liquid rocket engines made during the last decade. On the other hand, thc rocket propeller technology of the solid propellant still requires further studies and improvements (McFarlane, 1993).

Some propellants used in hybrid engines are composed of nitrogen tetroxide and nitrogen oxides as oxidant, and a hydrocarbon (e.g., polymethylacrylate) and a metal (e.g. magnesium) as fuel (Mead & Bornhorst, 1969). There have been developed some research linking high energy oxidants, such as mixtures of liquid oxygen and fluorine (FLOX), with chlorine and fluorine containing compounds such as chlorine trifluoride (CIF3) and CIFS; However, engines that employ these components have not yet been produced (Vilar, 2005).

A more practical engine that spends less energy, can be done with the use of hydrogen peroxide as fuel, set the *binder* based on PBLH. Hydrogen peroxide can be stored for as long as it needs to, and it has a relatively low cost. Another hybrid propellant propulsion system with application in rockets is what uses as oxidant liquid oxygen (LOX) and as fuel the PBLH. Liquid oxygen is widely used as an oxidizer in the space launch industry, being relatively safe and presenting high performance allied to low cost, besides being a propellant that does not produce toxic gases whose exhaustion is smoke free.

In the development of hybrid propellants, there has been much focus on engines thrust level of approximately 1,112,000 N, considered high compared to MFPS MPFL and whose large engines used as a *booster* develop thrusts from 100,000 in 1993 N., We carried out a test with this thrust level for the first time by the *American Rocket Company,* employing a propellant based on liquid oxygen (LOX) and PBLH (McFarlane, 1993). In 1999, a consortium of aerospace companies also tested several hybrid rocket prototypes also based on the use of LOX / PBLH (Laforce, 1967).

The manufacture, storage and operation of a rocket motor are involved in several phenomena related to chemistry, thermodynamics, geometry, combustion, gas dynamics and fluid mechanics, which makes it very difficult to fully analyze all aspects necessary for the definition of A propellant grain. Due to this fact, in this study the focus was given to aspects related to the mechanical properties of the propellant structural material.

PROPELLANTS

A propellant can be defined as a mixture of chemical compounds that, during its burning, generates gases and sufficient energy to propel artifacts such as rockets, missiles, projectiles, satellites, among others (Encyclopedia Of Chemical Technology, 1965), (Arendale, 1969). The energy to be released will depend on the thermodynamic properties of the products generated in the combustion reaction, the enthalpy of formation of the components of the formulation and the oxygen balance, which will determine the balance between the amount of fuel and the amount of oxidant needed to provide combustion Complete (Volk & Bathelt, 1997).

The grain solid propellant, generally, is shaped in the rocket motor tube *(bonded case)* adhered to a thin polymeric layer, called *liner,* which promotes chemical bonding of the propellant to the wall of the rocket engine tube or insulator, which acts as a barrier Thermal. The thermal barrier is installed in the rocket motor tube at critical locations exposed to high temperatures during the engine's operating time (burning), thereby exerting the function of thermally protecting the internal surface of the engine envelope of the combustion gases of the propellant.

The *case bonded* concept was adopted in the production of MFPS in the 50s and allowed the development of large engines with the molding of larger amounts of solid propellant in the rocket motor tube. The molding of the propellants in the rocket motor tube itself allowed the solid propellant grain to withstand the mechanical stresses during the firing process, as well as the effects of gravity in the MFPS storage period and the accelerations imposed by the flight that the molded grain outside the tube rocket engine *(case bonded* to the above process) could not endure the same. On the other hand, this molding introduced a series of internal stresses to the grain. In the production process, immediately after propellant mass is poured into the rocket engine tube, the MFPS undergoes a step called curing, which is obtained *binder* by polymerization effect of previously combined chemical species, at temperatures of the order of 70 ° C, for periods that may be days or weeks. It is at this stage that the composite acquires the mechanical properties necessary to withstand the stresses to which it is subject. However, this polymerization process is accompanied by a small contraction in the volume of the solid propellant grain, usually caused by "accommodations" of the polymer chains that form in the initial stages of the curing, which can subsequently impair its mechanical properties due to Internal stresses to the grain (Morais, 2000).

Currently, the most commonly used solid fuels and propellants are composed of three main components (Linnel & Miller, 2002):

1. The organic part rich in carbon and hydrogen is known as binder or polimer matrix (10 to 20% by mass), whose function is to act as a binder of several species present in the composite propellant, besides generating gas and contributing as fuel in the formula, ruling the reactivity of the propellant through its acting as carbon source in the burning process;
2. An inorganic oxidizer salt, rich in oxygen;
3. A ballistic additive, usually a finely divided metal with well-defined granulometric profile, which acts as fuel.

As a polymer network, the binder shows a typical viscoelastic behavior of their elastomeric nature, which allows the binder to better withstand the various working conditions to which is subjected. For their proper use, the binder should have the following characteristics in the formulations in which it is:

1. Display affinity to the constituent parts of the rocket engine, such as the thermal protection and metallic components (Klager, Polyurethanes, the Most Versatile Binder for Solid Composite Propellants, 1984), contributing in a direct manner in the assignment of the properties in the composite propellant;
2. Form a consistent elastic mass able to withstand the stresses caused by the mechanical and thermal stresses developed mainly during the flight phase of the rocket motor;
3. Reduce sensitivity to impact, friction and heat by forming a polymeric matrix that covers the explosive;
4. Check the characteristic geometrical stability of the propellants as well as the appropriate mechanical properties and strengths so that the resulting solid fuel / propellant grain does not suffer any mechanical failure during transportation, handling and combustion.

However, the binder tends to have its performance impaired by high mechanical loading rates that supports, i.e. the incorporation of high loads to which it is subjected during the manufacturing process of the solid propellant grain. For this reason, a number of flaws may arise in the propellant grain, mainly due to the presence of voids and the lack of adhesion between the polymer phase (binder) and loads, in particular oxidizing salt, which is found in high concentrations in the formulations (About 80% by mass). Furthermore, although the binder acting holding together the solid particles, which are usually ammonium perchlorate serving as oxidizer salt and finely divided aluminum, which acts as a ballistic additive, since stress concentrations during the firing process, the grain Solid propellant may not provide a uniform response over its full extent if deformations and breakdown of these solid particles are present. Thus, even if the grain is not submitted to an intense mechanical stress as occurs during the MFPS of flight, a series of accumulation of stresses can arise as a result of interactions between the solid particles and binder, where the grain is subjected to Intense temperature variations, among other requests. Depending on the way in which these stresses are absorbed, a microcracking of the solid propellant grain may occur, generating cracks and other types of failure, such as the emergence of voids. The presence of cracks in the solid propellant grain may favor its irregular burning during the combustion process, leading to an increase of the flame front. Consequently, there is also an increase in working pressure in the engine which eventually results in the explosion of the propellant.

The discontinuities represented, inter alia, by air bubbles incorporated during the manufacturing and loading process of the composite in the rocket motor tube may also cause a possible increase of the flame front during the burning of the fuel grain installed in the rocket motor tube. Thus, in order to avoid the formation of these discontinuities, a primary requirement in the manufacture of the solid composite propellant is that all ingredients used in the manufacture are free from moisture. The presence of water may cause the formation of bubbles in the material, affecting its ballistic properties (Bunyan, Cunliffe, Davis, & Kirby, 1993)[24]. The *pot life* of the *binder,* which represents the processing time of the mass in still liquid state, must also be adjusted so that continuity enables the removal of this mass by exposure to a reduced pressure environment. Also, antifoaming agents may be employed which decrease the surface tension of the bubble gas / solid interface, as well as decrease its concentration and size. These agents may be, among others, silicone-based chemical species or soy lecithin. Moreover, the introduction of plasticizers to the composite decreases the viscosity of the liquid mass improves its steep increases the *pot life* and contributes to the elimination of discontinuities during the manufacturing process (Linnel & Miller, 2002).

Also, it is necessary to determine the parameters like tension and deformation to which the grain will be subjected in order to predict if the MFPS will be able to reach the objectives of its mission in a specific environment, because, once this operational environment is previously defined, A previous analysis would make it possible to define the minimum safety factors required for the proper operation of the rocket motor by obtaining a detailed analysis of the structural integrity of the solid propellant grain.

The information necessary to conduct the analysis of the structural integrity of the solid propellant grain are: definitions of the forces to which the grain will be subjected in the field of operation and the conditions of its storage, the properties of the materials that compose the grain, the criterion of Failures and safety factors involved. In this way, a solid propellant grain used in rocket motors must present the ballistic characteristics and mechanical behavior pre-defined according to the design of the engine, which imposes the use of both raw materials and the specific manufacturing processes for the formulation of the fuel.

The efforts to which a propellant grain is subject can be classified by (Rocco, 2004):

1. **Thermal Origin:** Induce tensions and deformations to the grain by temperature gradients and differences between the coefficients of thermal expansion of materials of the walls of rocket motor, liner, thermal barriers and the propellant grain itself. These efforts arise due to the temperature cycles in the environment in which the MFPS is exposed, being all the more intense the higher the temperature variation reached in the medium. In the technical specifications of the rocket motor, the number of cycles and the temperature limits to which the MFPS is able to support without compromising its performance.
2. **Acceleration:** Also called slump, the efforts resulting from the effect of acceleration occur during storage, launch and maneuvers in flight. In storage, the slump causes deformation of the solid propellant grain due to the force of gravity acting through the vertical positioning of the MFPS. In the horizontal position also deformations can occur, resulting in the strangulation of certain internal regions the combustion chamber as a function of the geometric profile used by the grain. In flight, immediately after ignition, when the pressure in the combustion chamber reaches its operational value (design pressure), there is a voltage peak in the region between the solid propellant grain and the liner which represents the greatest effort due to acceleration at launch . However, with the advancement in the process of burning and consumption of the solid propellant grain, this effort tends to reduce.
3. **Transport and Handling:** Encompasses vibration stresses due to the positioning of the MFPS under the wings of vectors (aircraft) in the case of tactical missiles and during transportation, in addition to small shocks during transportation and handling. In general, it is efforts that induce low internal stresses to the solid propellant grain when compared to the other efforts, but should be considered in failure analyzes.
4. **Shrinkage During Curing:** Tensions and deformations are imposed by the contraction of volume sustained by the solid propellant grain due to the tendency of accommodation of stresses during the curing step in the manufacture of the solid propellant. These tensions tend to be larger the greater the concentration of the binder in the formulation.
5. **Pressurizing:** Efforts due to pressurizing of the MFPS combustion chamber occur during ignition and remain throughout the burning time of the propellant grain. In this condition, the solid propellant grain tends to accommodate the internal pressure with the relaxation at its terminal positions.

6. **Aging:** Another important factor to be considered in the applications in rocket motors and their safety parameters is the aging of the solid propellant, fundamental for the determination of its time of use. The rocket motors charged with solid propellant have a pre-determined lifetime, in which it is possible to guarantee the safety and other mechanical and ballistic parameters defined in the design during the launch operation. However, although it is a slow and gradual phenomenon, the aging of the propellant can make its performance inadequate to the design parameters, compromising its mechanical and ballistic properties, increasing the possibility of material failure by the appearance of microcracks, besides disintegration phenomena Of solid loads. In order to obtain more precise information about the conditions of deterioration of the propellant grain, it is essential to carry out studies that allow to determine its time of service in applications of rocket motors.

The mechanisms by which the aging process may occur are still poorly understood and are a limiting factor in predicting the life of rockets and missiles based on MFPS propulsion. The density of crosslinks and, consequently, the mechanical properties of the composite propellants, its affected by its aging. This property can be adjusted by changing the amount of the prepolymer, the curing agent and chain extender and the polyol used in formulating the binder. Equivalent proportions of the reactive species, such as the NCO / OH ratio, wherein the isocyanate content is determined in relation to the alcohol content in the formulation, and the triol / diol ratio which determines the level of crosslinking interfering with the properties of the polymer Are useful tools for adjusting the crosslinking density and for obtaining satisfactory mechanical and aging properties.

Temperature is considered the most important factor in the aging of a propellant, since its storage at elevated temperatures tends to cause acceleration in the chemical reactions of the material, according to the Arrhenius Equation. Another important factor is the accumulation of mechanical stresses that can cause the separation between the binder and some solid load, causing the formation of vacuoles and the presence of loose particles. These particles can act as a point of microcracks, causing an increase in the firing surface and, thus, greatly increasing the internal pressure of the motor, which may result in detonation of the material, as discussed above. Environmental conditions should also be considered in aging studies, since moisture and air may accelerate the oxidation process of the material. Also, contact with other organic materials can cause a migration of any component of the propellant causing, for example, a weakening in the interface with the insulator or liner (Sciamareli, Takahashi, Teixeira, & Iha, 2002), (Davenas, 2003), (Sbriccoli, Saltarelli, & Martinucci, 1989).

The observed changes in the propellant due to aging may manifest in several ways, such as increase or decrease in grain hardness, swelling, discoloration, among others, as well as altering its kinetic properties due to increased cross-links in the polyurethane chain (De la Fuente, 2009).

From these characteristics results the need for the elastomeric nature of formulations of fuels and solid propellants (Akhavan, 1998).

DEVELOPMENT OF PROPELLANTS

The first propellant to be used has been developed by the Chinese in the year 200 BC It was of black powder formed by physical mixture of sulfur and charcoal, which provides carbon, and potassium nitrate (KNO_3 or saltpetre), which supplies oxygen. The black powder is a substance that burns quickly and currently its use is restricted to rifles of large caliber and small cannon, and load initiation artillery ammunition due to their poor energy properties, the high production of smoke and formation of corrosive waste in their burning, as well as its mechanical fragility that generates load break when exposed to temperature cycles in which it is subjected (Akhavan, Introduction to explosives, 2004). After the development of black powder, all propellants began to show at least one polymer in its formulation, making the development of highly dependent propellants study of new materials (Bittencourt & Guanaes, 2008).

Propellants Based on Natural Polymers

The first-generation polymer had developed in its composition natural fibers such as silk, wool and cotton (Kumar & Gupta, 1998). Regarding the propellants, the same trend was followed and, in 1864, Schultze made use of nitrocellulose (NC) as the first polymer to be used in the manufacture of propellants. In 1838, it was developed by an advance in Pelouze use of this natural polymer, by nitration of paper and cotton, and 1846, used Schönbein pure cellulose (Akhavan, Introduction to explosives, 2004). Thus, nitrocellulose, whose structure is illustrated in Figure 3, was also known as "cotton powder" due to its natural origin.

The propellants which have the only energy as nitrocellulose base are known as "single base" (BS). In this case, this type of propellant is constituted by a homogeneous mixture of nitrocellulose and some additives such as plasticizers, stabilizers, flame arresters, etc., in the presence of suitable solvents which promote gelatinization of nitrocellulose and allow the mass conformation. Single base propellants to produce little smoke in his burning when compared to black powder and, therefore, are also called smokeless powder. In the years following its discovery, the use of propellants BS replaced the use of black powder in the military around the world.

After the development of nitroglycerin (NG) by Nobel Prize in 1863, it was found that it was obtained a gel on mixing with nitrocellulose, however, the use of this mixture as a propellant was detected only in the year of 1888, giving rise to a more energetic propellant to the BS, which is known as double-base (DB) by having two energy bases: nitrocellulose and nitroglycerine (Bittencourt & Guanaes, 2008).

The dual basis most commonly used polymers can generally be found in two forms:

Figure 3. Chemical structure of nitrocellulose

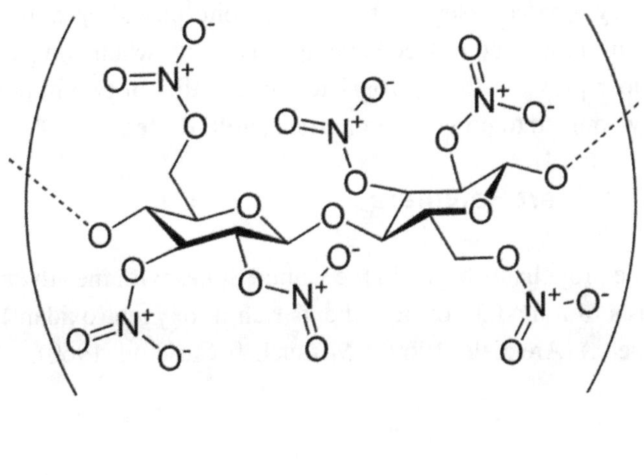

Figure 4. Chemical structure of nitroglycerin

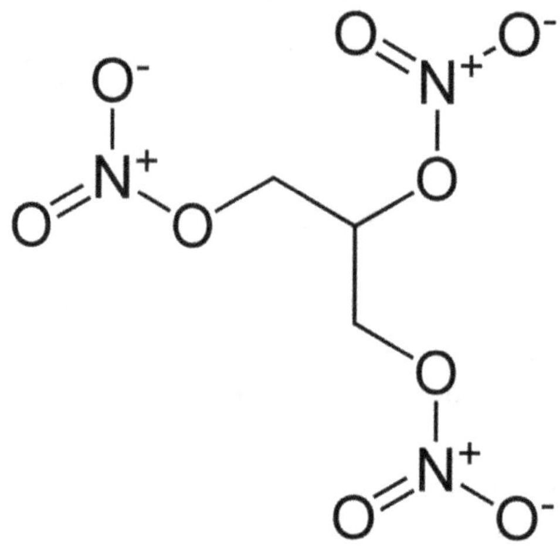

1. **Extruded Double Base (BDE):** After impregnation of nitrocellulose with nitroglycerin, usually in aqueous medium, a paste is obtained. Thereafter, the paste is added and, after removal of the water, extruded into the desired profile. From this process, the grain may still undergo some kind of treatment. In modern EDB production plants, the extrusion process occurs continuously;
2. **Molded Double-Base (BDM):** This type of propellant presents raw materials similar to the extruded base-double. It is obtained by molding a mixture of nitroglycerin and triacetin into a mold containing nitrocellulose in the solid state (powder).

The propellants Simple Base and Double Base are very used to this day. However, due to oxygen deficiency nitrocellulose and high temperature flame of nitroglycerin, the amounts of these components are restricted in current formulations (Singh, 1995). In order to solve this problem in the BD propellants have been studied alternative formulations such as triple base (which include nitrocellulose, nitroglycerine and nitroguanidine), those which are loaded with nitroamines, and employing energy polymers (Volk & Bathelt, 1997), (Singh, 1995), (Nakashita & Kubota, 1991). However, nitrocellulose is still considered in many energy - based research, because despite providing obtaining propellants which have lower sensitivity to initiation and reduced chemical stability when compared to the more modern formulations, nitrocellulose provides the appropriate combination of two important characteristics for propellant: power supply coupled to polymer properties (Volk & Bathelt, 1997).

Propellants Based on Inert Polymers

The research of composite propellants began in the United States with the advent of World War II. This type of propellant consists of a mixture of ingredients rich in oxygen (oxidant), binders and additives (plasticizers, stabilizers, etc.) (Arendale, 1969), (Mishuck & Carleton, 1960).

The use of organic materials as binders in composite solid propellants like, as they are currently known, was initiated by Theodore von Karman, ending the long reign of the black powder as a sole energy source in the propulsion rocket motor (Sutton & Biblarz, 2010). This substitution allowed in some decades, the energy accumulated in these solid propellants was tripled when compared to the energy obtained with the use of black powder.

Starting from this line of research, John Parsons developed the first composite propellant in 1942 with the use of high molecular weight asphalt as a binder and potassium perchlorate as the oxidizer (Hunley, 1999). However, high temperature uses or inclusion of hydrocarbons so that its viscosity was necessary to be decreased and, moreover, this propellant did not have adequate final properties nor a stable geometry. Therefore, it became necessary to search for alternative materials that can replace the asphalt in this type of formulation, the formulation being driven studies with the use of two classes of polymers: elastomers and thermoplastics (Bailey & Murray, 1989).

Currently, the most widely used propellants composites are based on polymeric binders (or *binders*) energy, typically a polybutadiene or polyurethane, and ammonium perchlorate (oxidizer salt), which may or may not contain finely divided aluminum in its formulation acting as fuel. These propellants are obtained by a maceration process of chemical species in the solid state (oxidants, fuels and additives) and liquid (binder), followed by molding the rocket motor tube and ultimately cure. In the case of formulations called *smokeless*, i.e. it occurs low smoke emission, the aluminum concentration is reduced or even eliminated, to avoid the appearance of smoke classified as primary arising due to the presence of aluminum oxide formed in the combustion of the composite. Moreover, the removal of aluminum in the formulation may cause burning phenomena of instability and lead to a decrease in the value of specific impulse (Isp).

Among the composite propellants currently used in more MFPS, there are two examples of double base:

1. Double base molded modified composite: is the dual basis molded which is incorporated cyclotrimethylenetrinitramine (RDX, *Research Department X)* or ciclotetramethylene-tetranitroamine (HMX, *High MElting point and eXplosive*), whose chemical structures are presented in Fig. 5, besides nitroglycerin in solid state (powder).
2. Elastomer is rich modified double base molded (EMBDM): is produced by the same process as the double-base molded and composite modified double base molded using the same basic formulation: does the addition of a prepolymer having hydroxyl terminations, policaprolactana type or polyester and an isocyanate type crosslinking agent, the liquid solvent used in forming the propellant grain.

Figure 5. Highly explosive Nitroamines: (a) RDX and (b) HMX

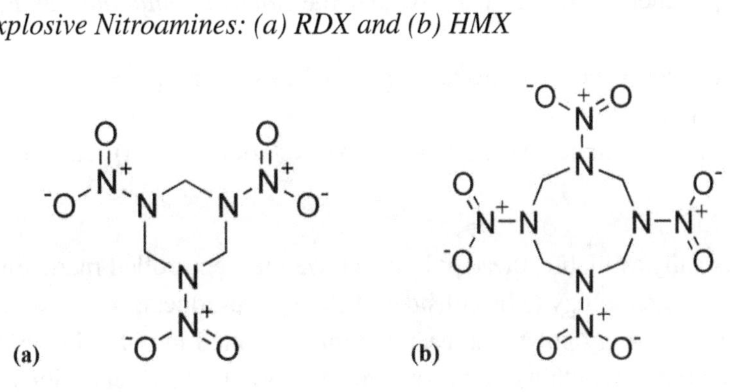

Propellants Based on Energy Polymers

The polymers used were hitherto considered inert due to the low total energy in the formulation, which hurt the performance of the propellant.

In order to achieve high performance and low vulnerability, which are critical requirements of the propellant, the following alternatives were considered:

1. Increase in propellant loading density: the amount of solid oxidants and hence energy is limited by the processing mode of the propellant and the characteristics of the binder employed (Oberth & Bruenner, 1969), (Leach & Debenham, 1998).
2. Use of more energetic or dense oxidants compared to RDX (Volk & Bathelt, 1997).
3. Use of energetic plasticizers (Leach, Plasticisers in energetic materials formulations a UK over view, 1998).
4. Use of energy polymers.

Energy polymers have been developed from this scenario. These polymers were obtained by inserting energy groups along the polymer chain, for example: Azido (N_2), nitro (NO_3), difluoramines (NF) and dinitrofluorates ($CF(NO2)_3$) (Arendale, 1969), (Liu, Hsiue, & Chiu, 1995), (Provatas, 2000), (Desai, 1996), (Desai, Synthesis of narrow molecular weight ?,?-hydroxy telechelic poly(glycidyl nitrate) and estimation of theoretical heat of explosion, 1996), (Agrawal, 1998). Such polymers are based on *binders* plasticized by liquid chemical species, such as nitrated esters, mixtures of nitrated esters and nitramines (RDX, HMX). It is also possible to insert ammonium perchlorate and aluminum in their formulations. The binder is crosslinked called double base (RDB) small amounts of nitrocellulose is contained in its composition.

Propellants and solid fuels based binders elastomeric assets, in addition to acting as a carbon source, release large amounts of energy to thermally decompose, are gradually being employed in loading rocket engines as tactical missiles of last generation. Once installed on the rocket motor tube, the composite must have the ability to work mechanically expand and contract without cracking or stand out of its inner walls. These mechanical stresses occur mainly during the process of burning the grain in the generation of thrust motor (Framcel, Grant, & Flanagan, 1989).

In recent years, polyols (polyether) with hydroxyl termination energy has become commercially available, the most cited in the literature (Bittencourt & Guanaes, 2008), (Thepenier, 2001):

- glycidyl azide polymer (GAP), 3M, USA, and the *Societe Nationale des Poudres et explosifs* (SNPE), France;
- nitratemethyl polymer 3-methyl-3-oxetane (polyNIMMO), and glycidyl nitrate polymer (polyG-LYN), ICI, England;
- 3-azidoethyl-3-methyloxetane (AMMO) and 3,3-bis (azido methyl) oxetane (BAMO), Thiokol, USA.

Although commercially available, these polymers have their controlled marketing and restricted by countries possessing the technology to be considered strategic products.

Synthetic methods for energy polymers resemble those applied in obtaining composite. Generally, these formulations can further contain other chemical species at low concentrations, such as stabilizers, burning instability suppression, post-burn suppressants, antioxidants, modifiers, additives burn rate and according to the combination of the chemical species in the formulation may have low smoke emission characteristics, because the gain of ballistic performance obtained by replacing binders inert by active allows the removal of ammonium oxidizing perchlorate formulation which minimizes the emission of chlorinated compounds from the products combustion.

Generation of Recyclable Polymers

In recent years, the development of new materials to be used in the manufacture of propellants sought to obtain characteristics such as greater availability of energy, thermal and chemical stability, and ease of handling combined with a lower cost, but also presented as the main approach to need to make them less harmful to the environment.

Studies that address the development of propellants based on derived from natural and renewable polyols such as vegetable oils, which have great ability to replace petrochemical polyols origin, have been currently highlighting (Desroches, Escouvois, Auvergne, Caillol, & Boutevin, 2012), (Guo, Demydov, Zhang, & Petrovic, 2002), (U.S. Patente Nº Patent 6107433 A., 2000). Formed by fatty acid triglyceride, vegetable oils such as linseed oil, castor bean, soybean, sunflower, among others, are excellent raw materials and promising sources for the synthesis of polyols and polymeric materials due to their high degree of introducing (Sharma & Kundu, 2006) and the presence of hydroxyl groups in its structure, being employed in the synthesis of interwoven polymers.

The *binders* based on vegetable oils, in addition to presenting advantageous in terms of renewable polyol, exhibit low toxicity, less residue formation during processing step of obtaining, and low cost. Moreover, the products obtained are generally biodegradable. These characteristics are a strong attraction for its use as a raw material in the chemical industry and, consequently, research in the academic and industrial field of vegetable oils based polyols have been recognized as attractive and viable (Desroches, Escouvois, Auvergne, Caillol, & Boutevin, 2012).

Studies on the production of polymer matrices based on plant sources have been made by research group in the Department of Energy Materials Chemistry ITA, using the process of urethane prepolymers from epoxidized vegetable oils (polyols). The results obtained with castor oil employment (Rocha, Lima, Gomes, Iha, & Rocco, 2013), and soybean oil (Clemente, Rocha, Iha, & Rocco, 2014) showed that the *binders* formulated based on these natural sources have similar performance to the *binder* based on the most commonly used polyol for this purpose (HTPB) in terms of processability and thermal decomposition characteristics. For example, the glass transition temperature (T_g) of the derivative of castor oil and soybean oil polyols, determined using the technique of differential scanning calorimetry (DSC) showed values equal to -37 ° C and -39 ° C respectively. These values approximate T_g provided by the polymeric matrix obtained from the HTPB from about -50 ° C, also lie close to the range T_g suited for *binders* and composite applications with solid fuels and propellants (-40 to -54 ° C) [6, 53]. Thus, the applications of natural resources synthesized as a solid fuel suggest promising results when used in polymeric matrices propellants and fuels. From about -50 ° C and also lie near range T_g suited for *binders* and composite applications with solid fuels and propellants (-40 to -54 ° C) [6, 53]. Thus, the applications of natural resources synthesized as a solid fuel suggest promising results when used in polymeric matrices propellants and fuels. From about -50 ° C and also lie near range T_g suited for *binders* and composite applications with solid fuels and propellants (-40 to -54 ° C) (Vilar, 2005), (Cohen, Fleming, & Derr,

1974). Thus, the applications of natural resources synthesized as a solid fuel suggest promising results when used in polymeric matrices propellants and fuels.

Vegetable oils consist predominantly of triglycerides, which are the condensation product of glycerol and unsaturated fatty acids. The soybean oil triglycerides contain saturated and unsaturated fatty acids, wherein the composition of unsaturated fatty acids in soybean oil is higher than 80% (Clemente, Rocha, Iha, & Rocco, 2014). In castor oil, approximately 90% of the mass of the fatty acid that constitutes its molecule is ricinoleic acid (R- (CH2)7-CH=CH-CH2-CHOH-(CH2)5-CH3), which it has a secondary hydroxyl group which can be used directly as an alternative raw material. The remaining 10% are not hydroxylated fatty acids such as oleic acid (3.1 to 5.9%) and linolenic (2.9 to 6.5%). This hydroxyl group in ricinoleína gives the castor oil exclusive property of solubility in alcohol, enabling the use of alcohol as fuel without causing major damage to the engine explosion. The functionality of castor oil is 2.7, which provides cross-reactivity and increased stiffness of the polymer obtained and its average hydroxyl value is 163 mg KOH / g with high purity, suitable for use in polyurethanes (Rocha, Lima, Gomes, Iha, & Rocco, 2013).

The requirement that the materials are "environmentally friendly" has motivated and directed research in this area, however, because the energy polymers and recyclable energy polymers present higher cost technology, its use is not feasible for applications where the appeal green is not considered critical, yet making the nitrocellulose and the HTPB polymers most used nowadays (Bittencourt & Guanaes, 2008).

METHODOLOGY

The aging tests have been employed as a support tool in an attempt to predict the useful life of an under the MFPS aspect of ballistic and mechanical behavior of solid propellant grain (Layton & Christiansen, 1979). As discussed above, the temperature is considered the most important factor in the aging of the propellant, because storage at elevated temperatures tends to cause acceleration of chemical reactions of the material by increasing the rate constant for the reaction (k) according to Equation Arrhenius:

$$k = Ae^{\frac{-Ea}{RT}} \tag{4.6}$$

where: A = frequency factor for the reaction, Ea = activation energy and R = universal gas constant and T = temperature.

If the temperature is too high, there may be reactions that would be prevented by the natural aging. For this reason, the propellant aging studies based on HTPB are typically performed at temperatures around 70 °C maximum [94] and the storage conditions in terms of temperature, temperature cycling and humidity become fundamental in control of an aging test.

This work was carried out the study of aging of the propellant formulated with the *binder* only HTPB-based polyol, a pattern searching for further studies with the use of propellant based on HTPB partially replaced by the selected polyols. The propellant-based *binder* default HTPB / IPDI / DOA employed ammonium perchlorate and aluminum as the metallic oxidant and fuel, respectively.

Natural weathering test was carried propellant at 20 °C. However, due to the slowness of experimental tests grain aging at room temperature compared to trials in which the temperature is raised in special chambers was also carried out accelerated aging, which consists of aging the material at temperatures above room temperature with the use of the greenhouse. In this case, accelerated aging tests were performed on samples of the solid composite propellant by continuous exposure to temperatures of 35 °C, 50 °C, 60 °C and 75 °C in an oven for period of 4 to 360 days. Initially, tests were conducted under conditions in which the grain was not subjected to any mechanical stress. however, grain solid propellant may be subject to a number of requests that must be taken into account in the assessment of their useful life, especially mechanical character, as when installed on the rocket motor tube.

RESULTS AND DISCUSSION

In this study, we also assessed the effects of natural aging and accelerated on the mechanical properties of tensile strength (σ), elongation at break (εr) and hardness (Hd) propellant based on the standard formulation the *binder* HTPB / IPDI / MOCA without replacement of HTPB polyol, filled with ammonium perchlorate (AP) (NH4ClO4) and metallic aluminum, acting respectively as fuel and oxidizing agent. Natural aging tests were conducted at a temperature of 20 ° C for period of 4 to 360 days. The results are shown in Table 1.

One can set the solid propellant grain aging based on HTPB / IPDI as increasing the crosslink density of the polyurethane polymer chain. The process of formation of the polymer chain, known as curing the solid propellant, can also influence the aging process. Another factor that can significantly affect the mechanical behavior of grain samples during the aging process there is a marked stoichiometric propor-

Table 1. Aging test of the binder formulated with HTPB without replacement at 20° C

Days	T = 20 °C		
	σ (kgf/cm^2)	ε (%)	Hd (Shore A)
4	125,4	47,9	71
8	126,4	45,8	71
12	125,4	42,5	72
16	116,2	39,8	71
24	123,4	39,7	71
32	129,5	37,8	71
40	129,5	43,8	71
50	125,4	43,7	71
60	128,5	42,4	72
80	124,4	42,0	72
100	128,5	42,0	72
120	127,5	39,9	73
140	124,4	39,7	73
170	126,4	39,5	74
200	130,5	39,4	75
240	128,5	39,5	75
360	128,5	39,6	76

tion between the reactants. This study was maintained at NCO / OH ratio equal to 1.0, as reported in the study of (Hocaoglu, Ozbelge, Pekel, & Ozkar, 2001), proving to be the best ratio to be used in the formulation of the samples.

Based on the data in Table 1, it is possible to note that the mechanical properties of tensile strength hardly undergo changes with aging at 20 °C, where one does not notice an increasing trend or decreasing this property throughout the study period, remaining values between 116.2 kgf / cm^2 (16 days) 130.5 kgf / cm^2 (200 days), presenting, at the end of the duration, an increase of 2.5% in tensile strength.

On the other hand, it is observed that the elongation of the material decreases over time, and its maximum value of 47.9% in 4 days, reaching the minimum value of 39.4% in 200 days, reaching 39, 6% after 360 days of aging.

Regarding the hardness values obtained show a tendency of the propellant become slightly more rigid with the advance in aging time. After 360 days aging, curing the propellant showed a 7% relative to the beginning of the aging period 4 days.

Based on these results, it is concluded that after a period of 360 days, the effect of natural aging (un-accelerated) does not change markedly the mechanical properties evaluated propellant based on HTPB / IPDI / MOCA studied, which demonstrates their stability on storage in these conditions. The data in Table 1 can be displayed as graphs (Figures 6 to 8), compared to other results of accelerated aging.

For studies of accelerated aging, the material was subjected to temperatures of 35 °C, 50 °C, 60 °C and 75 °C for the same period of 4 to 360 days. The values of the properties tensile strength (σ), elongation at break (εr) and hardness (Hd) are shown in Table 2.

As expected, the mechanical properties shown in Table 2 are modified with the same trends observed in natural aging test at 20 ° C.

The tensile strength shows an increase of the propellant that focuses more sharply the higher the temperature at which the material is subjected to the aging process. For example, at a temperature of 75 °C is the greatest increase in tensile strength after a period of 360 days, reaching a value of 149.90 kgf / cm^2, which means an increase of 17% compared to aging 4 days at the same temperature. After the same period (360 days) of aging at temperatures of 50 °C and 60 °C, resistance to draw propellant reaches values respectively equal to 136.6 kgf / cm^2 and 143.8 kgf / cm^2. At temperatures of 20 °C and 35 °C, does not notice a trend in the behavior of the variation of this property.

Conversely, the elongation of the propellant tends to decrease over time, reaching its lowest value, equal to 33.1%, when subjected to the high temperature cycle (75 °C). The trend of elongation reduction with advancing aging at all temperatures studied was similar, remaining between 9 and 10% elongation of the material as compared to lower and higher aging period studied. This decrease in elongation is also similar to the reduction obtained with elongation behavior in the natural aging which was 8% in the same period of 4 to 360 days.

It is also observed that the propellant has a tendency to become more rigid the greater the time and temperature of aging. Thus, the hardness reaches its highest value of 83 scale Shore A at 75 °C to 360 days of aging, which represents an increase of 14% compared to its hardness at the same temperature for 4 days of aging. This change is equivalent to twice the value corresponding to the same period of natural aging of 4 to 360 days at 20 °C, wherein the hardness increased only 7%.

According to the research related to aging of composite solid propellants, generalized hardening of the solid propellant grain occurred during an aging process by the effect of exposure to a temperature increase, considering the structural formula of liquid polybutadiene hydroxylated, it is a result of disruption of the double bond chain containing HTPB (Tuzun & Uysal, 2005). Thus, the unsaturated C = C

Table 2. Mechanical properties of the propellant formulated with the binder HTPB / IPDI / MOCA after accelerated aging at temperatures of 35 ° C, 50 ° C, 60 ° C and 75 ° C in periods of 4 to 360 days

Days	35 °C			50 °C			60 °C			75 °C		
	σ (kgf/cm²)	ε_r (%)	Hd (Sh A)	σ (kgf/cm²)	ε_r (%)	Hd (Sh A)	σ (kgf/cm²)	ε_r (%)	Hd (Sh A)	σ (kgf/cm²)	ε_r (%)	Hd (Sh A)
4	126,4	47,5	71	114,2	44,0	71	122,4	44,5	72	128,5	42,8	73
8	123,4	45,8	71	116,2	43,8	72	124,4	44,2	72	131,5	42,1	73
12	125,4	42,1	72	116,2	42,0	73	125,4	44,0	74	132,6	39,7	75
16	116,2	39,4	73	117,3	39,2	73	127,5	39,8	75	135,6	39,3	76
24	121,3	38,6	73	118,3	41,6	74	130,5	39,6	76	139,7	38,6	76
32	123,4	37,8	72	118,3	40,1	73	130,5	39,0	77	138,7	39,4	77
40	128,5	43,3	73	120,3	39,7	74	131,5	38,6	76	141,7	38,2	76
50	126,4	43,1	71	120,3	39,3	73	133,6	38,4	78	141,7	37,9	77
60	129,5	42,1	73	120,3	38,5	74	134,6	37,4	77	142,8	38,1	77
80	125,4	42,4	73	124,4	38,1	76	136,6	37,1	76	145,8	37,7	78
100	129,5	41,6	74	126,4	37,7	75	138,7	36,8	74	143,8	36,8	78
120	128,5	40,9	75	128,5	36,0	76	139,7	36,5	75	146,8	36,0	79
140	131,5	39,1	76	130,5	36,9	76	140,7	36,1	76	149,9	35,5	79
170	128,5	39,1	76	131,5	36,3	77	141,7	35,8	76	148,9	35,1	80
200	131,5	39,4	77	134,6	35,8	77	141,7	35,2	77	147,9	35,1	81
240	129,5	39,0	78	135,6	35,4	79	143,8	34,4	78	149,9	34,3	82
360	129,5	38,0	77	136,6	35,0	80	143,8	34,2	77	149,9	33,1	83

bond is responsible for the elimination of the hydrogen atom adjacent producing an allyl radical (H2C = CH-CH2). The allyl group can undergo oxidation or otherwise connecting adjacent polymeric chain to increase the crosslink density in the polymer matrix. In relation to this mechanism, it was demonstrated that perchloric acid ($HClO_4$) generated from the decomposition of ammonia traces of perchlorate (AP) catalyzes the process, especially in the presence of moisture (Gruntman, 2004).

Thus, these formulations, the aging process occurs with increasing crosslink density of the polymer matrix agglomerate the grain other than those specified in the design. Also, migration occurs as the plasticizer dioctyl adipate (DOA), among others, which causes the hardening of the solid propellant grain, influencing also the adhesion between the grain and the liner (Gruntman, 2004). The migration of ammonium perchlorate can lead the grain to a state of microcracking and different mechanical response when subjected to the designed firing process.

The obtained curves related to the variation of the mechanical tensile strength properties (σ), elongation at break (ε_r) and hardness (Hd) in relation to the aging time (in days) at 20 °C and accelerated aging at 35 °C, 50 °C, 60 °C and 75 °C are shown in Figures 6 to 8.

Figure 6. Effect of aging at different temperatures on the mechanical properties of tensile strength binder based on HTPB

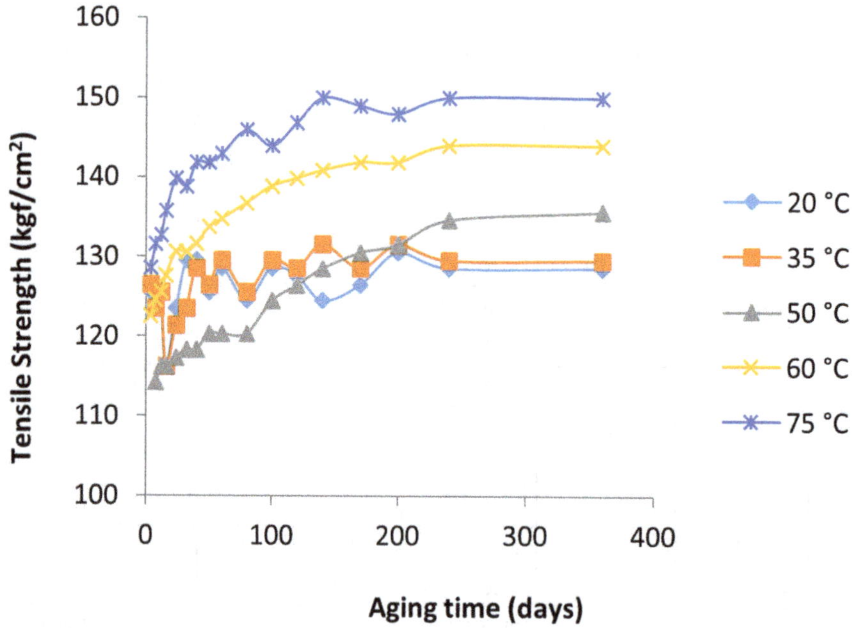

Figure 7. Effects of aging at different temperatures on the mechanical properties of elongation of the binder based on HTPB

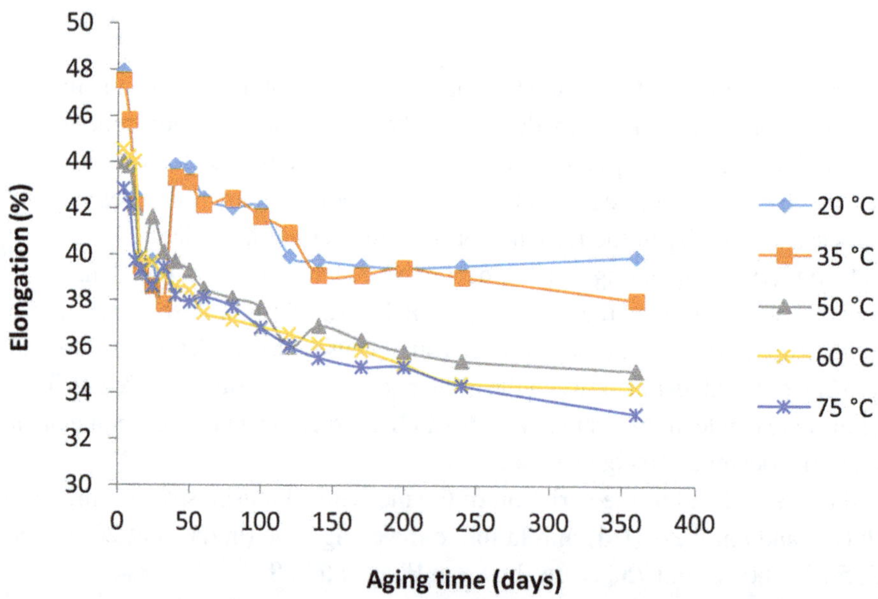

Figure 8. Effect of aging at different temperatures on the mechanical property of hardness of the binder based on HTPB

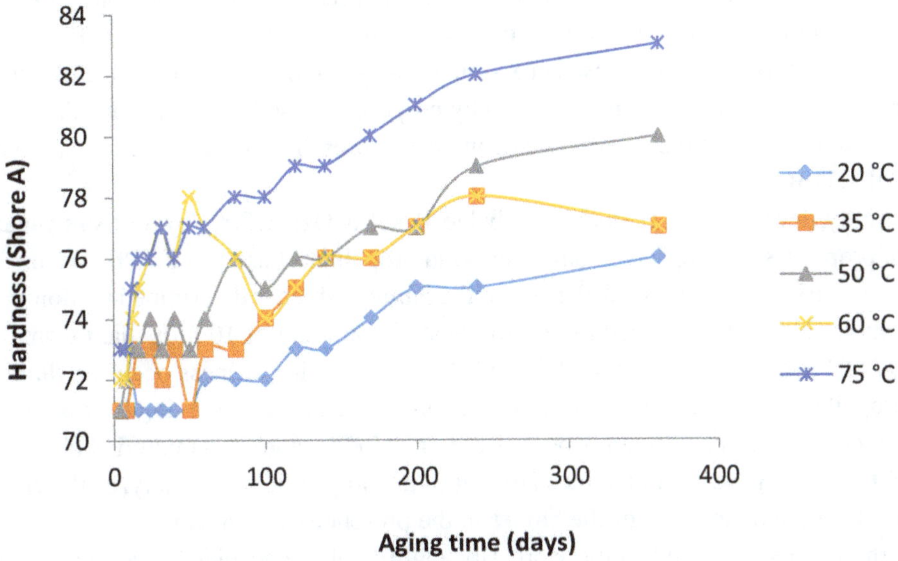

The aging behavior in composite solid propellant formulations determines a number of parameters life of solid rocket motors. As the aging defined as a process in which the mechanical and ballistic characteristics of the grain are modified from its original state, depend on many factors ranging from the process of getting the grain to the conditions of transport, handling and storage of solid engines. Originally, the crosslink density presented by the polymer matrix, or binder, is governed by the functionality HTPB, about 2.2, and optionally crosslinking the material added to the matrix. When employing chain extenders such as IPDI, bifunctional, or short chain polyols in urethane prepolymers, the trend is to obtain a linear polymer with high mechanical properties of elongation and low tensile strength. When the chain extender is trifunctional such as TEA, the opposite occurs. So, to equate these parameters depending on the engine design that is intended the formulation of solid propellant. The physicochemical characteristics of the polymer matrix lumped loads also exert their influence, for example, the form factor, physical profile and packaging. The physicochemical characteristics of the polymer matrix lumped loads also exert their influence, for example, the form factor, physical profile and packaging. The physicochemical characteristics of the polymer matrix lumped loads also exert their influence, for example, the form factor, physical profile and packaging.

Aging can be conducted studies solid propellant composite samples in several ways. In this aging process, the grain is directly related to exposure temperature to which the propellant is subjected during the corresponding period. The time parameter is also correlated. The mechanism considered in this aging is the increase in crosslink density by breaking the double bond of PBLH, making the device unstable and increases the risk in handling. Also, it can be considered an estimated lifetime of a composite solid propellant by employing thermal analysis calculations and kinetic decomposition. In the research group Energy Materials Department of ITA Chemistry, they were developed other studies related to the lifetime of solid propellants samples composites based on HTPB / AP and HTPB / PBX (Kirchoff, et al., 2016). However, using thermal analysis (DSC) to determine the kinetic parameters that provide aging projection of propellants, focusing on the energetic properties of the material. The results confirm the stability provided by propellant-based matrices formulated with HTPB, concluding the same way that the occurrence of increased crosslink density occurs due to breakage of double bond of HTPB. The results confirm the stability provided by propellant-based matrices formulated with HTPB, concluding the same way that the occurrence of increased crosslink density occurs due to breakage of double bond of HTPB. The results confirm the stability provided by propellant-based matrices formulated with HTPB, concluding the same way that the occurrence of increased crosslink density occurs due to breakage of double bond of HTPB.

In another study, developed by (Hocaoglu, Ozbelge, Pekel, & Ozkar, 2001) were also evaluated changes in mechanical properties of samples of composite solid propellants based on HTPB / AP during the cure and its relation to variations in crosslink density of the *binder*, which is determined predominantly by the relationships molar NCO / OH and triol / diol to the study of aging of 16 propellant formulations with various NCO / OH and triol / diol cured at 65 ° C. The increase in hardness occurred during the aging process the propellant was attributed to increased cross-link density by breaking the double bonds present in the main chain. This same conclusion was obtained in the study Sciamareli et al. it is shown that perchloric acid (HClO), generated in a trace amount of decomposition AP catalyzes the crosslinking of the double bonds of the main chain of the binder in the presence of moisture.

Based on the results generated in the work Hocaoğlu et al. on accelerated aging samples of solid propellant formulations under similar conditions, the changes in tensile strength propellant can also be explained by the relative triol (MOCA) / diol (HTPB) used in formulation of solid propellant grain studied. When high relationships are adopted diol / triol in formulating the *binder*, the material has markedly more aging than at low ratios (triol / diol). N study Hocaoğlu et al., although the curing period (7 days at 65 °C) samples of the composite was not the same as performed in this study, it can be considered that the formulation of binder which is based composite ratio adopted a triol / diol considered high, since MFPs design constraints so required. High triol concentrations in the starting formulation of the binder and hence the composite propellant, lead to a significant increase in crosslink density in the polymer network obtained by the curing process. However, the results show that the binder based on HTPB formulated in the described conditions can be considered stable, including the accelerated aging process in all analyzes. High triol concentrations in the starting formulation of the binder and hence the composite propellant, lead to a significant increase in crosslink density in the polymer network obtained by the curing process. However, the results show that the binder based on HTPB formulated in the described conditions can be considered stable, including the accelerated aging process in all analyzes. High triol concentrations in the starting formulation of the binder and hence the composite propellant, lead to a significant increase in crosslink density in the polymer network obtained by the curing process. However, the results show that the binder based on HTPB formulated in the described conditions can be considered stable, including the accelerated aging process in all analyzes. The results show that the binder based on HTPB formulated in the described conditions can be considered stable, including the accelerated aging process in all analyzes. The results show that the binder based on HTPB formulated in the described conditions can be considered stable, including the accelerated aging process in all analyzes.

These results are parity in the studies developed in this work. Thus, by analyzing the curves presented in Figures 6 to 8, it is noted that changes in mechanical properties of tensile and elongation caused by the accelerated aging process coincide with those reported in the literature.

Other studies related to aging propellants, performed by (Bunyan, Cunliffe, Davis, & Kirby, 1993) and (Cunliffe, Davis, & Tod, 1996) also conclude that the aging of the composite propellant occurs via oxidation of crosslinking the *binder* elastomer with the migration of additives and recrystallization of the oxidant, while the formation of the practically complete polyurethane during the period of curing and then does not give no significant contribution to the aging of the propellant. (Gottlieb, 1994) and (Probster & Schmucker, 1986) shown in their studies that the migration of the plasticizer in the propellant aging process generates an increased hardness in the propellant layer, influencing its adhesion. This process generates changes in mechanical properties of tensile strength (increase) and elongation (decrease) obtained soon after its manufacture (Celina, Minier, & Assink, 2002). Thus, the appearance of flaws and cracks in the grains is possible that, when asked during firing, have characteristics that differ from the original design in order to compromise the performance of MFPS.

CONCLUSION

Aging tests were performed based on the propellant grain pattern formulated with the HTPB binder, plasticizer, DOA and the IPDI loaded with additives ballistic ammonium perchlorate (oxidizing agent) and metallic aluminum (fuel). For this test, the propellant was exposed to temperatures of 20 °C, 35 °C, 50 °C, 60 °C and 75 °C, for periods ranging from 4 days to 360 days. It is concluded that the aging process is directly related to temperature exposure at which the propellant undergoes the corresponding period, and the most substantial changes in mechanical properties occur with aging at the higher temperature of 75 °C in this case study. In this way, it became possible to assess changes in the propellant grain samples based on HTPB as regards their mechanical properties of tensile strength, elongation and hardness. The observed increase in hardness during the aging process the propellant was attributed to increased cross-link density by breaking the double bonds present in the main chain. It can be concluded that the propellant grain-based binder formulated with HTPB / IPDI remained stable regarding its mechanical properties even when subjected to accelerated aging process during the period considered in this study. Aging tests performed demonstrate that the modifications in the mechanical properties of the composite aging advances, such as increased tensile strength and decreased elongation, coincide with the results available in the literature.

REFERENCES

Agrawal, J. P. (1998). Recent trends in high energy material. *Progress in Energy and Combustion Science*, *24*(1), 1–30. doi:10.1016/S0360-1285(97)00015-4

Akhavan, J. (1998). Investigation into the network struture of plasticized rocket propellant. *Polymer*, *39*(1), 215–221. doi:10.1016/S0032-3861(97)00259-0

Akhavan, J. (2004). *Introduction to explosives*. London: Royal Society of Chemistry. doi:10.1039/9781847552020-00001

Arendale, W. F. (1969). Chemistry of propellants based on chemically crosslinked binders. *Advances in Chemistry Series*, *88*, 67–83. doi:10.1021/ba-1969-0088.ch004

Bailey, A., & Murray, S. G. (1989). *Propellants, Explosives and Pyrotechnics*. Londres: Brassey's.

Bittencourt, E., & Guanaes, D. (2008). Propelentes sólidos: uma história ligada à evolução dos polímeros. *Revista Militar de Ciência e Tecnologia, 6*, 71-80.

Bunyan, P., Cunliffe, A. V., Davis, A., & Kirby, F. A. (1993). The degradation and stabilization of solid rocket propellants. In *Polym. Degrad. Stab.* (pp. 239-350).

Celina, M., Minier, L., & Assink, R. (2002). Development and application tool characterize the oxidative degradation of AP/HTPB/Al propellants in a propellant reability study. *Thermochimica Acta, 384*(1-2), 343–349. doi:10.1016/S0040-6031(01)00793-6

Chiaverini, M. J., Harting, G. C., Lu, Y., Kuo, K. K., Peretz, A., Stephen Jones, H., & Arves, J. P. et al. (1992). Regression rate behavior of HTPB-based solid fuels in a hybrid rocket 137 motor. *Journal of Propulsion and Power, 16*(1), 125–132. doi:10.2514/2.5541

Clemente, M., Rocha, R. J., Iha, K., & Rocco, J. A. (2014). Development of prepolymer technology in the synthesis of a polyurethane binder used in solid rocket fuels. *Quimica Nova*, 37(6), 982–988.

Cohen, N. S., Fleming, R. W., & Derr, R. L. (1974). Role of binders in solid propellant combustion. *AIAA Journal*, 1974, 212–218.

Cunliffe, A., Davis, A., & Tod, D. (1996). Aging and life prediction of composite propellant motors. In *Proceedings of the AGARD Conference* (p. 586).

Davenas, A. (2003). Development of modern solid propellants. *Journal of Propulsion and Power, 19*(6), 1108–1128. doi:10.2514/2.6947

De la Fuente, J. L. (2009). An analysis of the thermal aging behaviour in high-performance energetic composites through the glass transition temperature. *Polymer Degradation & Stability, 94*(4), 664–669. doi:10.1016/j.polymdegradstab.2008.12.021

De Luca, L. T. (1999). *Problemi energetici in propulsione aerospaziale.* Milan: Milano: SPLab, Dipartimento di Ingegneria Aerospazialem Politecnico do Milano, Campus Bovisa.

De Luca, L. T., Galfetti, L., Maggi, F., Colombo, G., Merotto, L., Boiocchi, M., & Fanton, L. et al. (2013). Characterization of HTPB-based solid fuel formulations: Performance, mechanical properties, and pollution. *Acta Astronautica, 92*(2), 150–162. doi:10.1016/j.actaastro.2012.05.002

Desai, H. J., Cunliffe, A. V., Hamid, J., Honey, P. J., Stewart, M. J., & Amass, A. J. (1996). Synthesis and characterization of?? -hydroxy and nitrato telechelic oligomers of 3,3-(nitratomethyl) methyl oxetane (NIMMO) and glycidyl nitrate (GLYN). *Polymer, 37*(15), 3461–3469. doi:10.1016/0032-3861(96)88498-9

Desai, H. J., Cunliffe, A. V., Lewis, T., Millar, R. W., Paul, N. C., Stewart, M. J., & Amass, A. J. (1996). Synthesis of narrow molecular weight?? -hydroxy telechelic poly(glycidyl nitrate) and estimation of theoretical heat of explosion. *Polymer, 37*(15), 3471–3476. doi:10.1016/0032-3861(96)88499-0

Desroches, M., Escouvois, M., Auvergne, R., Caillol, S., & Boutevin, B. (2012). From vegetable oils to polyurethanes: Synthetic routes to polyols and main industrial products. *Polymer Reviews, 52*(1), 38-79.

Encyclopedia of Chemical Technology (Vol. 8). (1965). Kirk-Othmer.

Fleeman, E. L. (2001). *Tacticcal Missile Design.* AIAA.

Framcel, M. B., Grant, L. R., & Flanagan, T. E. (1989). Historical development of GAP. In *Proceedings of the 25th Joint Propulsion Conference*, Monterey. AIAA. doi:10.2514/6.1989-2307

Gottlieb, L. (1994). Analysis of DOA migration in HTPB/AP composite propellants. In *Proceedings of the ICT Annual Conference,* Karlsruhe.

Gruntman, M. (2004). *Blazing the Trail: The Early History of Spacecraft and Rocketry*. Reston: American Institute of Aeronautics and Astronautics. doi:10.2514/4.868733

Guo, A., Demydov, D., Zhang, W., & Petrovic, Z. S. (2002). Polyols and polyurethanes from hydroformylation of soybean oil. *Journal of Polymers and the Environment*, *10*(1), 49-52.

Hocaoglu, O., Ozbelge, T., Pekel, F., & Ozkar, S. (2001). Aging of HTPB/AP-based composite solid propellants, depending on the NCO/OH and triol/diol ratios. *Journal of Applied Polymer Science*, 79(6), 959–964. doi:10.1002/1097-4628(20010207)79:6<959::AID-APP10>3.0.CO;2-G

Hunley, J. D. (1999). The history of solid-propellant rocketr y: what we do and do not know. In *Proceedings of the ASEE Joint Propulsion Conference*, Los Angeles (pp. 1-10).

Kirchoff, E., Rocha, R. J., Nakamura, N. M., Lapa, C. M., Pinheiro, G. F., Iha, K., & Gonçalves, R. F. et al. (2016). Estimativa de vida útil do pbx (plastic-bonded explosive) com envelhecimento acelerado. *Quimica Nova*, 39(6), 661–668.

Klager, K. (1984). Polyurethanes, the Most Versatile Binder for Solid Composite Propellants. In *Proceedings of 20th AIAA/SAE/ASME Joint Propulsion Conference*, Cincinnati, Ohio. AIAA. doi:10.2514/6.1984-1239

Klager, K., & Di Milo, A. J. (1976). *Encyc1opedia of Polymer Science and Technology - Rocket Propellants*. New York: Wiley and Sons.

Kumar, A., & Gupta, R. K. (1998). *Fundamentals of Polymer*. New York: McGraw-Hill.

Kuo, K. K. (1997). Importance and challenges of hybrid rocket propulsion beyond year 2000. In *Proceedings of the Israel Annual Conference on Aerospace Sciences* (p. 139). Israel Institute of Technology.

Laforce, P. D. (1967). *Technological Development of a Throttling Hybrid Propulsion System.* UTC 2215-FR.

Layton, L. H., & Christiansen, A. G. (1979). Effect of aging-strain on propellant mechanical properties. In *Proceedings of the 15th AIAA, SAE and ASM Joint Propulsion Conference*, Las Vegas. AIAA. doi:10.2514/6.1979-1245

Leach, C. (1998). Plasticisers in energetic materials formulations a UK over view. In *Proceedings of the International ICT Conference.*

Leach, C., Debenham, D., Kelly, J., & Gillespie, K. (1998). Advances in polyNIMMO composite gun propellants. *Propellants, Explosives, Pyrotechnics*, *23*(6), 313–316. doi:10.1002/(SICI)1521-4087(199812)23:6<313::AID-PREP313>3.0.CO;2-4

Linnel, J., & Miller, T. (2002). A preliminary design of a magnesium fueled marcian ramjet engine. *Proceedings of the 38th AIAA/ASME/SAE/ASEE Joint Propulsion Conference,* Indianápolis. AIAA. doi:10.2514/6.2002-3788

Liu, T. L., Hsiue, G. G., & Chiu, Y. S. (1995). Thermal characteristics of energetic polymers based on tetrahydrofuran and oxetane derivatives. *Journal of Applied Polymer Science*, *58*(3), 579–586. doi:10.1002/app.1995.070580313

McFarlane, J. S. (1993). Design and Testing of AMROC's 250,000 lbf Thrust Hybrid Motor. (*AIAA Paper* 93–2551).

Mead, F. B., & Bornhorst, B. R. (1969). *Certification Tests of a Hybrid Propulsion System for the Sandpiper Target Missile* (AFRPL-TR-69-73).

Mishuck, E., & Carleton, L. T. (1960). Chemical principles of solid propellants. *Industrial & Engineering Chemistry*, *52*(9), 754–760. doi:10.1021/ie50609a023

Morais, A. M. (2000). *Caracterização do comportamento mecânico do propelente sólido compósito como material estrutural* [PhD Thesis]. Campinas, Brazil: Universidade Estadual de Campinas.

Nakashita, G., & Kubota, N. (1991). Energetic of nitro/azide propellants. *Propellants, Explosives, Pyrotechnics*, *16*(4), 177–181. doi:10.1002/prep.19910160405

Oberth, A. E., & Bruenner, R. S. (1969). Polyurethane based propellants. *Advances in Chemistry Series*, *88*, 84–121. doi:10.1021/ba-1969-0088.ch005

Paubel, E. F. (2002). *Propulsão e Controle de Veículos Aeroespaciais - Uma Introdução*. Florianópolis: Editora UFSC.

Petrovic, Z., Guo, A., & Javni, I. (2000). U.S. Patente Nº Patent 6107433 A.

Probster, M., & Schmucker, R. H. (1986). Ballistic anomalies in solid rocket motors due to migration effects. *Acta Astronautica*, *13*(10), 599–605. doi:10.1016/0094-5765(86)90050-0

Provatas, A. (2000). Energetic Polymers for explosives formulation: a review of recent advances. Melbourne: DSTO Aeronautical and Maritime Researche Laboratory - DSTO-TR-0966.

Rocco, J. A. (2004). *Estudos sobre o envelhecimento de formulações de propelente sólido compósito baseadas em binders poliuretânico.* [PhD Thesis]. São José dos Campos, Brazil: Instituto Tecnológico de Aeronáutica.

Rocha, R. J., Lima, J. E., Gomes, S. R., Iha, K., & Rocco, J. A. (2013). Síntese de poliuretanos modificados por óleo de mamona empregados em materiais energéticos. *Quimica Nova*, *36*(6), 793–799. doi:10.1590/S0100-40422013000600009

Sbriccoli, E., Saltarelli, R., & Martinucci, S. (1989). Comparison between Accelerated and Natural Aging. In *Proceedings of the 20th International Annual Conference of ICT and 18th Annual Technical Meeting of GUS*, Karhsruhe, Germany.

Sciamareli, J., Takahashi, M. F., Teixeira, J. M., & Iha, K. (2002). Propelente sólido compósito polibutadiênico: I – Influência do agente de ligação. *Quimica Nova*, *25*(1), 107–110. doi:10.1590/S0100-40422002000100018

Sharma, V., & Kundu, P. P. (2006). Addition polymers from natural oils-A review. *Progress in Polymer Science*, *31*(11), 983–1008. doi:10.1016/j.progpolymsci.2006.09.003

Singh, H. (1995). High energy material research and development in India. *Journal of Propulsion and Power*, *11*(4), 848–855. doi:10.2514/3.23910

Sutton, G. P., & Biblarz, O. (2010). Rocket Propulsion Elements (8th ed.). Wiley.

Thepenier, J. (2001). Advanced technologies available for future solid propellant grains. *Acta Austronautica*, *48*(5), 245-255.

Tuzun, F. N., & Uysal, B. Z. (2005). The Effect of ammonium nitrate, coarse/fine ammonium nitrate ratio, plasticizer, bonding agent, and Fe2O3 content on ballistic and mechanical properties of hydroxyl terminated polybutadiene based composite propellants containing 20% AP. *Journal of ASTM International*, 2(6), 233–245.

Vilar, W. D. (2005). *Química e Tecnologia dos Poliuretanos* (3rd ed.). Rio de Janeiro.

Volk, F., & Bathelt, H. (1997). Influence of energetic materials on the energy-output of gun propellants. *Propellants, Explosives, Pyrotechnics*, *22*(3), 120–124. doi:10.1002/prep.19970220305

Chapter 13
Pre-Test and Analysis of a Reinforced Concrete Slab Subjected to Blast From a Non-Confined Explosive

Fausto B. Mendonça
Aeronautics Institute of Technology, Brazil

Girum S. Urgessa
George Mason University, USA

ABSTRACT

A large scale experimental program consisting of testing 10 RC slabs with different variations of concrete compressive strength, reinforcement ratio and retrofit was conducted in Brazil. As part of that test program, a small-scale blast pre-test setup and associated dynamic analysis were conducted in order to confirm the proper functioning of the blast test sensors (pressure gages, displacement meter and accelerometers). The results of the pre-test were compared with theoretical blast wave parameter predictions using established equations and maximum displacement predictions using simplified dynamic analysis. The pre-test experiment provided useful insights and was shown to be critical for the success of the subsequent large scale blast tests.

INTRODUCTION

In recent years, threats of terrorism, civil wars and accidental explosions have been heard routinely around the world. These actions can impact the performance of structures such as those made from reinforced concrete (RC). RC is one of the common materials used to construct most buildings and bridges around the world. Most of these structures are not designed to resist the effect of explosions. These explosions generate dynamic loads against essential supporting structural elements, such as slabs, columns or beams. In addition, protective structures made from RC can works as barrier, when the structure is designed to resist close-in detonation, or as shelter that is designed to prevent its content from blast effects (De-

DOI: 10.4018/978-1-5225-2903-3.ch013

partment of Defense, 2008). Understanding the behavior of these elements and the whole structure for the effects of blast is important because these structures hold people and private or government assets. Knowledge of blast wave parameters and how the energy coming from the blast generates damage to structures, will enable structural engineers to design RC structures efficiently incorporating short-duration dynamic effects. Many agencies have concerns about loses in both peaceful time and during conflicts generated from explosions close or in buildings (Di Stasio, 2016). Instructions giving guidance for better constructions to resist blast wave have been published due to those concerns (Department of Defense, 2008) in order to increase survival rates of structures and people after a blast event.

Full-scale tests having RC as a target have been tested with or without retrofits in order to increase the structures capacity to support blast loads (Jayasooriya et al., 2011; Maji et al., 2008; Mendonça et al., 2016; Zhao and Chen, 2013). These blast tests were conducted with proper care and were successful in collecting useful data that can increase our knowledge about the response of structures subjected to blast loads. In this book chapter, we present a small-scale blast pre-test setup and associated analysis in order to confirm the blast effect measuring sensors work properly (pressure gages, displacement meter and accelerometers) as part of a large-scale blast test program conducted in Brazil.

REPRESENTATIVE BLAST TEST EXPERIMENTS USING RC SLABS

The study of RC structures subjected to blast has been growing since the 20th century. Lessons from many accidents and two World Wars boosted research efforts in this area. In 1960, the first edition of the manual "Structures to Resist the Effects of Accidental Explosions" (Department of Defense, 2008) was published. Nevertheless, construction materials continue to evolve in composition and continue to provide more resistance to blast loads. Design codes and practice still lag behind in adopting or mandating requirements for blast analysis and design. Nonetheless, researchers have been conducting blast tests to improve our understanding of structural response for blast effects (Choi et al., 2008; Urgessa, 2009; Urgessa and Maji, 2010). Zhao and Chen (2013) conducted blast tests to verify the response of a thin RC slab with 42 MPa (6.09 ksi) having three different equivalent TNT mass and stand-off distance. Their tests verified that the higher the TNT mass and the lower the stand-off distance, the higher the damage recorded. In addition, their results pointed out that the response of concrete and reinforcement when subjected to blast have an increase in resistance, the dynamic increase factor (ASCE, 2010; Ngo et al., 2007; Urgessa, 2010).

Castedo et al. (2015) presented full-scale slab tests with variations in concrete mix composition. They showed that adding steel and polypropylene fibers to the slab resulted in a higher capacity to support blast waves. The steel fibers improved the tensile strength of the concrete and the polypropylene improved the capacity to absorb energy in opening cracks.

Li et al. (2016) conducted a full-scale field test with two RC slabs of 40 MPa (6.0 ksi) and five ultra-high performance concrete (UHPC) of 145 MPa (21 ksi). The slab with UHPC had micro steel fibers measuring 15 mm in length and 0.12 mm in diameter. The explosive in the test was positioned in contact to the upper face of the slabs and measured 0.1 kg and 1.0 kg of equivalent TNT. Slabs with UHPC had less damage than the slabs with normal RC. Diameter of perforation on the bottom surface of UHPC slabs were around 50% smaller than the slabs with normal concrete.

A large scale experimental program consisting of testing 10 RC slabs with different variations of concrete compressive strength, reinforcement ratio and retrofit was conducted in Brazil. As part of that test program, a small-scale blast pre-test setup and associated analysis were performed in order to confirm the proper functioning of the blast test sensors (pressure gages, displacement meter and accelerometers). The results of the pre-test were compared with theoretical blast wave parameter predictions using established equations and maximum displacement predictions using simplified dynamic analysis.

BLAST WAVE PARAMETERS

Explosions generate high quantities of energy in a very high velocity of propagation in the air. They can cause injury to people, and damages to equipments and buildings. In blast response analysis, it is necessary to consider factors such as vulnerability against blast wave, the explosive charge, the capacity of a strucutre to absorb the blast energy and its respective distance from the epicenter (stand-off distance).

Figure 1 shows a typical pressure-time history for a blast wave in free-field, where t_A is the time of arrival of the blast wave. When the blast wave hits the target, the ambient air pressure (P_0) is raised to the value of P_{SO}, the peak overpressure.

Goel et al. (2012) published a review of equations used to describe blast effects and recommend that it is important to understand the explosion event in order to choose a suitable mathematic model for predicting blast parameters for each type of test. The work summarized many equation including Bajic (2007); Brode (1955); Held (1983); Henrych (1979); Kingery and Bulmash (1984); Kinney and Graham (1985) and Sadovskiy (2004).

The P_{SO} value decays rapidly reaching negative values (below ambient air pressure). The pressure oscillates between positive and negative values up to stabilization. The duration of these oscillations is measured in milliseconds (ASCE, 2010). The exponential function that often describes the pressure-time history is the Friedlander equation (Goel et al., 2012; Jayasooriya et al., 2011; Kinney and Graham, 1985; Smith and Hetherington, 1994) as shown in Equation (1).

Figure 1. Typical time history curve for a blast wave

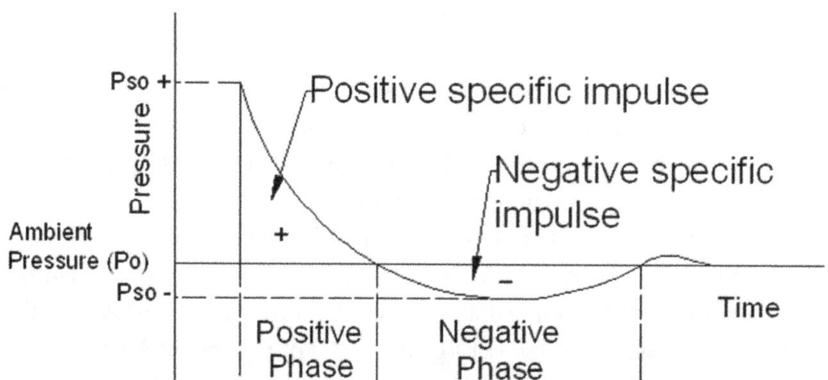

$$p(t) = P_0 + P_{SO}\left(1 - \frac{t}{t_0}\right)e^{-\alpha\frac{t}{t_0}} \tag{1}$$

The parameter α is the waveform parameter and depends on the peak overpressure (P_{SO}) and t_0 is the time of duration of the positive phase.

The positive specific impulse is another important parameter associated with blast effects. It is the area under the curve of Figure 1 before the curve crosses the ambient air pressure. It is given by Equation (2) where t_{0i} is t_A and t_{0f} is t_A+t_0. Higher impulses will result in more damages.

$$I = \int_{t_{oi}}^{t_{of}} P dt \tag{2}$$

Before tests, it is possible to predict the value of P_{SO} with a reasonable accuracy based on a scaled distance parameter (Z). This parameter depends on the stand-off distance (R) and the equivalent TNT mass of the explosive (W), as can be seen in Equation (3) (Brode, 1955; Goel et al., 2012). The US DoD recommends an increase of 20 percent of equivalent TNT mass for predicting loads during the design phase of a building (Department of Defense, 2008). This is due to some particularities of the constructions, methods and quality of materials, which may allow reflections of the blast wave or may reduce the structure's capacity to support the shock wave.

$$Z = \frac{R}{W^{1/3}} \tag{3}$$

As representative examples, the Kinney and Graham equation as shown in Equation (4) can be implemented to predict P_{SO} in free air, where P_0 is the ambient atmospheric pressure before explosion (Kinney and Graham, 1985).

$$\frac{P_{SO}}{P_O} = \frac{808\left[1 + \left(\frac{Z}{4.5}\right)^2\right]}{\sqrt{\left[1 + \left(\frac{Z}{0.048}\right)^2\right]} x \sqrt{\left[1 + \left(\frac{Z}{0.32}\right)^2\right]} x \sqrt{\left[1 + \left(\frac{Z}{1.35}\right)^2\right]}} \tag{4}$$

The time of duration of the positive phase (t_0) in milliseconds is given by Equation (5). (Goel et al., 2012; Kinney and Graham, 1985).

$$\frac{t_o}{W^{\frac{1}{3}}} = \frac{980\left[1+\left(\frac{Z}{0.54}\right)^{10}\right]}{\left[1+\left(\frac{Z}{0.02}\right)^{3}\right] x \left[1+\left(\frac{Z}{0.74}\right)^{6}\right] x \sqrt{\left[1+\left(\frac{Z}{6.9}\right)^{2}\right]}} \tag{5}$$

Another representative example is the use of equations and graphs developed by Kingery and Bulmash to predict values of P_{SO} (Kingery and Bulmash, 1984). The logarithmic curves of Kingery and Bulmash are given by Equation (6) and are valid for spherical TNT charge in a free air explosion. Y is the logarithm of the value of P_{SO}. This equation is applicable for explosions having scaled distances between 0.05 and 40 meters (Kingery and Bulmash, 1984).

$$Y = 2.611 - 1.690\,U + 0.008\,U^2 + 0.336\,U^3 - 0.005\,U^4 - 0.080\,U^5 - 0.004\,U^6 + 0.007\,U^7 + 0.0007\,U^8 \tag{6}$$

To get the value of Y, it is necessary to determine the value of U beforehand using T, which is the log of Z, as shown in Equation (7).

$$U = -0.214 + 1.350\,T \tag{7}$$

SIMPLIFIED DYNAMIC RESPONSE ANALYSIS

Most buildings are not designed to resist blast loads. However, they are designed for code specified lateral loads such as wind and earthquake loads. The most important difference between these lateral loads and blast loads is that the time duration of the blast loads is extremely small. The dynamic behavior of structures under blast event has been reported in literature (Department of Defense, 2008; Swisdak Jr, 1994).

As a simplification to design or analysis, a triangular pulse can be used as shown in Figure 2 (ASCE, 2010; Clough and Penzien, 2003). Integration of the area under the curve of the positive phase defines the specific positive impulse acting on the structure. Typically, the positive impulse is the main parameter that causes damage in blast events (UNODA, 2011). The higher the pressure and its time duration, the higher the total impulse and it will cause more damage to the target.

Each structure with an assumed single-degree-of-freedom has a natural period of vibration (T_n) as shown in Equation (8) (Ngo et al., 2007).

$$T_n = \frac{2\pi}{\omega_n} \tag{8}$$

The structure has a natural frequency that depends on its stiffness (k) and mass (m) as Equation (9).

Figure 2. Simplification of the time-history pressure

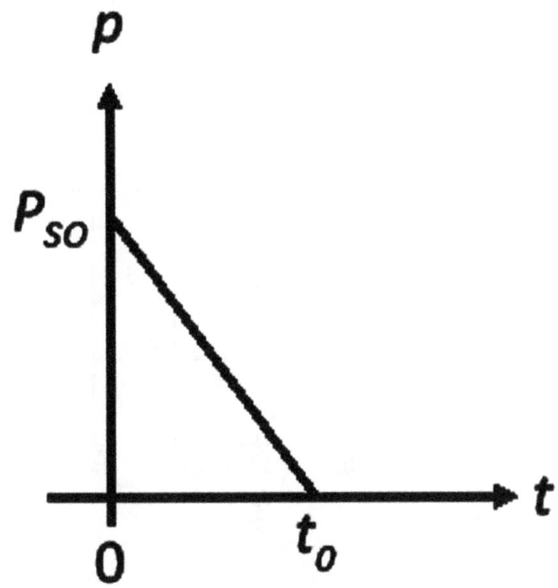

$$\omega_n = \sqrt{\frac{k}{m}} \tag{9}$$

The equation of motion as shown in Equation (10) defines the response of the structure where the displacement is represented by *u*, the velocity by $\dot{u}$, and the acceleration by $\ddot{u}$.

$$m\ddot{u} + C\dot{u} + ku = P(t) \tag{10}$$

The first component of the equation of motion depends on the mass of the structure (m), the second depends on the damping (C) and the third component depends on the stiffness (k). In blast analysis, the damping mechanism typically cannot be mobilized to absorb much energy, due to the very short time duration. Its value is assumed to be zero (Chopra, 2007; Clough and Penzien, 2003; Jayasooriya et al., 2011; Maji et al., 2008).

The maximum static displacement $(u_{st})_0$ can be used to determine the maximum dynamic displacement value (u_0). It is given by Equation (11) as a function of the blast load P_0 value and stiffness.

$$\left(u_{st}\right)_0 = \frac{P_0}{k} \tag{11}$$

The deformation response factor (R_d), also known as, the shock spectra for triangular pulse, is shown in Figure 3, where t_d is the time of duration of the positive phase. This is used to amplify the maximum static displacement as shown in Equation (12).

Figure 3. Shock spectra for triangular pulse

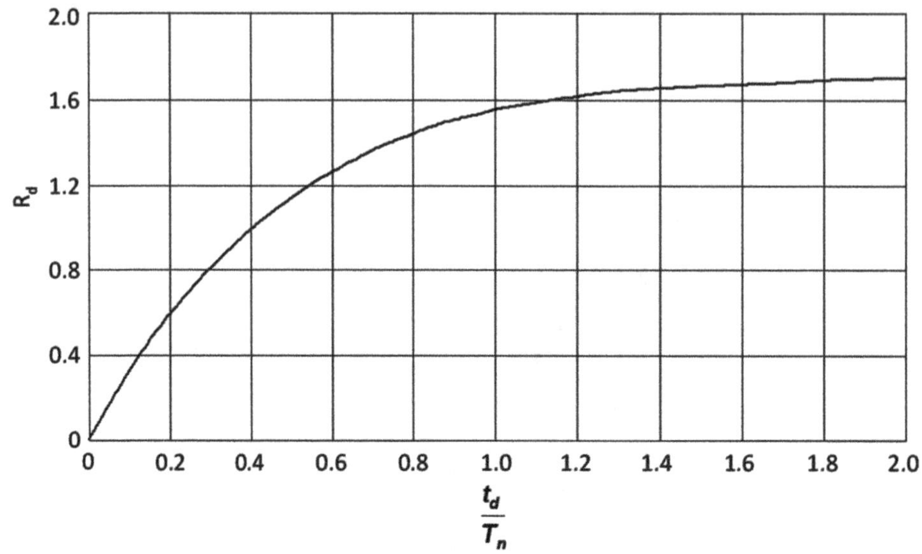

$$u_0 = R_d \left(u_{st}\right)_0 \tag{12}$$

A SMALL-SCALE BLAST PRE-TEST SETUP

It is worth to mention that the security of personnel during blast tests is of paramount importance. Applicable documents related to safe handling and operation of explosives should be consulted. If the chosen explosive is non-confined, it will not generate fragments and will be useful in verifying blast wave parameters. However, if the aim is to verify and study the effect of fragmentation on structures, the explosive is usually confined having an outer coating material. A small-scale blast test setup using subject to non-confined explosive is presented.

The specifications for the material of the blast target is important to guarantee the repeatability of the test. For the explosive, having the weight in TNT (W) is enough to determine P_{SO} and t_0 for a planned stand-off distance.

The specifications for the reinforced concrete (RC) slabs must be well articulated and must include good quality control. The steel and the concrete in the slab must obey patterns of usual construction practices. These are usually dictated by governing building codes of every country.

The transportation and handling of the RC slab is another important issue that needs to be addressed if the slab is cast off-site. If cracks are generated during transportation or handling, it will alter the resistance behavior of the slabs and will have unintended effect. Any hooks needed for carrying and handling must be prepared before casting the concrete to guarantee that the RC element integrity is not compromised.

As part of a full-scale blast experiment not included in this book chapter, a small-scale blast pre-test with one RC slab as target was conducted and results were compared to predicted values of P_{SO} and u_0. The slab measured 1.0x1.0 meter in plan and 0.08 m in thickness, 40 MPa (6.0 ksi) concrete compressive strength and reinforcement ratios (ρ) of 0.175% in two-way. The slab was simply supported on two sides and the explosive was suspended above the slab as shown in Figure 4. The stand-off distance, h, was 0.65 meters. The explosive weight was 0.188 kg of TNT, this corresponds to a scaled distance of 1.13 $m/kg^{1/3}$, and it was trigged by an electric fuze that was commanded by wires from a bunker where the research team was housed during the detonation.

Concrete is a suitable material to support blast effects because of its weight (ASCE, 2010). The slab was made with cast-in-place concrete and Qualitec® Company performed the tests of the static compressive strength, according to Brazilian Norms, NBR 12655 (ABNT, 1996), using 10x20 cm concrete test cylinders. The concrete was cured according to Brazilian Norms, NBR 6118 (ABNT, 2003) and NBR 14931(ABNT, 2004), keeping the slab with the wet surface within three days of casting. Figure 5 shows the slab fifteen days after casting.

The reinforcement was placed in the bottom of the slabs to carry positive moment as shown in Figure 6.

The reinforcement was held in place in the formworks before the concrete was cast with a 2.5 cm cover on the lower face of the slab (positive reinforcement). Figure 7 shows the formwork and the reinforcement before concrete was cast.

Pressure sensors were positioned 0.65 meters far from the explosive. Under the slab, there was a metallic box to protect the displacement meters, as shown in Figure 8.

Displacement meters were used to measure the displacement of the slab due to the blast waves. They were attached to the bottom surface of the slab by a wire emanating from a potentiometer that records the upward or downward movement of the wire during the explosion. The potentiometer was placed in a metallic box to protect it from the blast wave and surrounding debris generated during the explosion event. A hook was pasted on the lower surface of the slab using a two-part epoxy resin. The hook was needed to hold the wire in place. Two hooks with two potentiometers were attached to increase the likelihood of data collection in case there is a failure during the experiment.

Piezoelectric pencil pressure sensors were mounted to measure the blast pressure waves. All pressure sensors were placed around the same distance from the location of the explosive. There were four pressure sensors in contact with the slab, sensors A, B, C and D, and four additional pressure sensors at lower distances, sensors E, F, G and H, as shown in Figure 9.

Figure 4. Schematic of the test setup

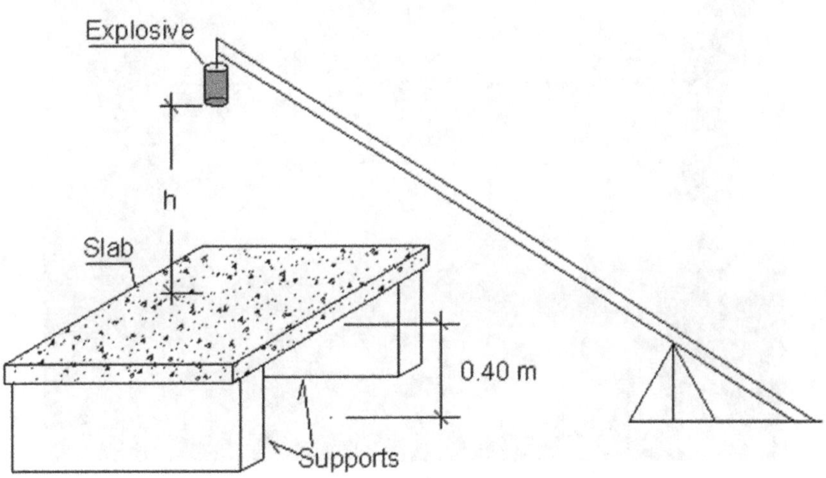

Figure 5. Slab fifteen days after casting

Figure 6. Cross section of the slab

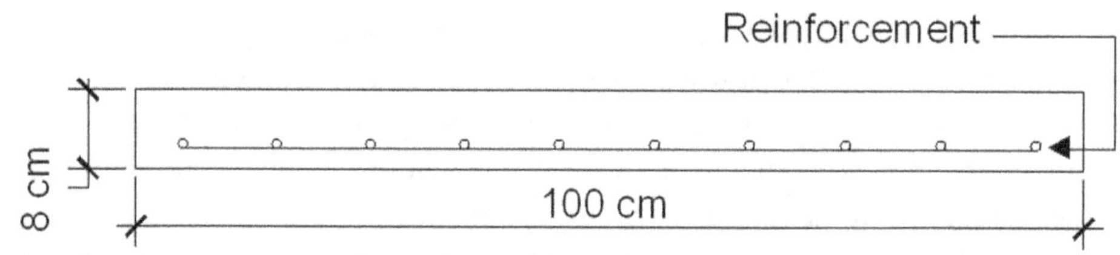

Figure 7. Formwork before casting. 5.0 mm diameter rebar both ways spaced at 15 cm

Figure 8. Set up ready for detonation during pre-test

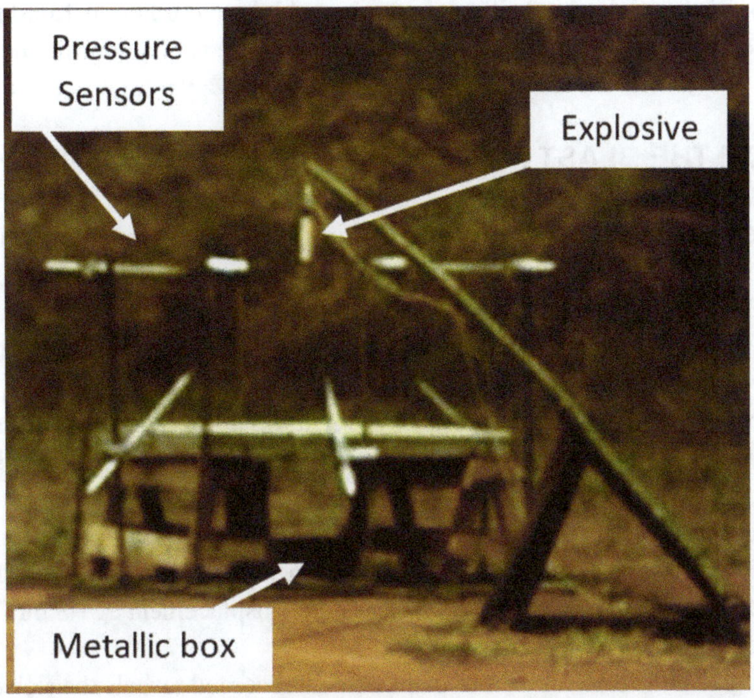

Figure 9. Pressure sensors close to the slab (A, B, C, and D) and in the same level as the explosive (E, F, G and H)

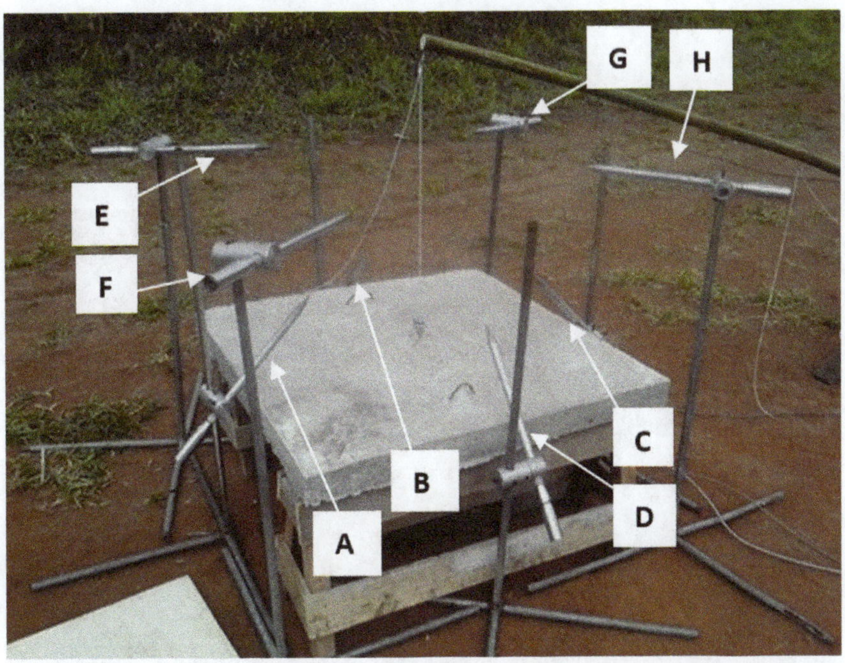

Olympus® high-speed digital camera was used to capture images from the explosion event in conjunction with a Toshiba® rugged notebook. The camera was able to record up to 1,000 frames per second (fps) and was placed 100 m far from the test. Figure 10 shows the fireball during the detonation.

RESULTS FROM THE BLAST PRE-TEST AND COMPARISON WITH ANALYSIS

Pressure sensors recorded overpressure during the detonation. Table 1 shows the recorded data and the distance of each sensor from the explosive.

The predicted P_{SO} value of 780 kPa using well-known equations for blast wave parameter determination was 12% lower than the recorded value (Goel et al., 2012; Kinney and Graham, 1985; Smith and Hetherington, 1994). Comparing to Kingery and Bulmash equation predicted a P_{SO} value of 760 kPa (Kingery and Bulmash, 1984; Swisdak Jr, 1994). However, it was just 15% lower than the average recorded value for sensors in the same distance as the explosive. As a representative example, Figure 11 shows the pressure time-history of sensor A. Peaks after the first points indicate re-reflections of the shock wave.

The displacement meters registered a maximum total displacement of 8.9 mm on the center of the slab as shown in Figure 12, once the box protecting the meter did not move. The slab returned to its original position after the explosion event is over. A residual displacement of 1.8 mm was observed due to support displacement in the surrounding clay soil.

It is worth to note that the total amount of displacement recorded reveals the slab displacement plus the support displacement in the surrounding clay soil. Therefore, the displacement of the slab was the total displacement measured minus the support displacement, which was 7.1 mm. This research reveals that the predicted maximum displacement value using the simplified dynamic response analysis with the recorded maximum displacement value are in good agreement as shown in Table 2. For this analysis, the time of duration was determined based on Equation (5).

Figure 10. Detonation recorded at 2.0 ms after trigger

Table 1. Pressure sensor positions and values of P_{so}

Sensor	Distance From the Explosive (cm)	P_{so} (kPa)
A	66	917
B	66	866
C	62	746
D	65	1052
Average		**895**
E	56	1365
F	57	961
G	59	637
H	56	879
Average		**960**

Figure 11. Recorded pressure time-history curve for sensor A

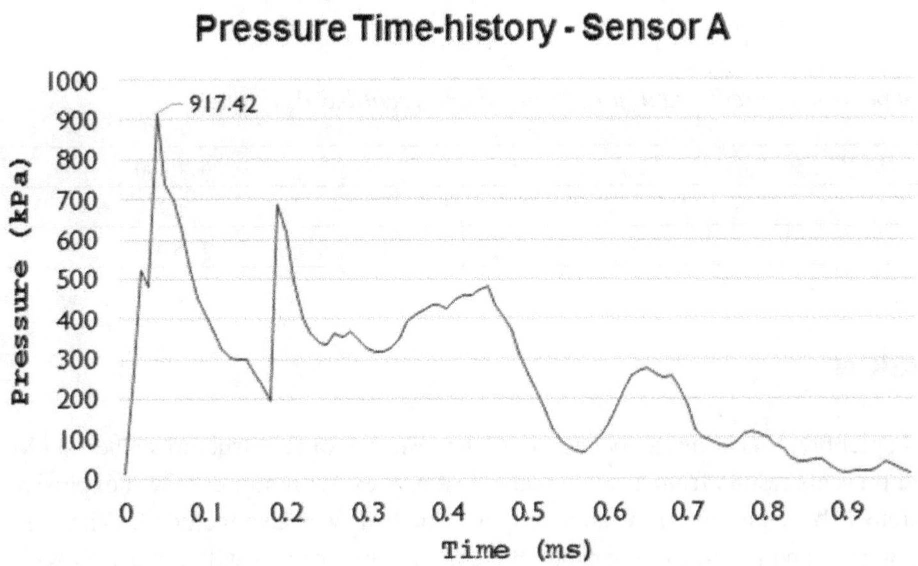

The predicted maximum displacement using the theoretical P_{SO} is almost the same as the maximum recorded displacement by the displacement meter during the blast pre-test. The maximum displacement using the recorded P_{SO} was 11% higher than the maximum recorded displacement.

The accelerometers failed during the pre-test indicating that the stick used to attach the accelerometers on the bottom surface of the slab did not work properly. Observing this phenomenon allowed us to specify and use a stronger stick ensuring that the sensors will be in the right position and capable of measuring test data.

Figure 12. Displacement meter records

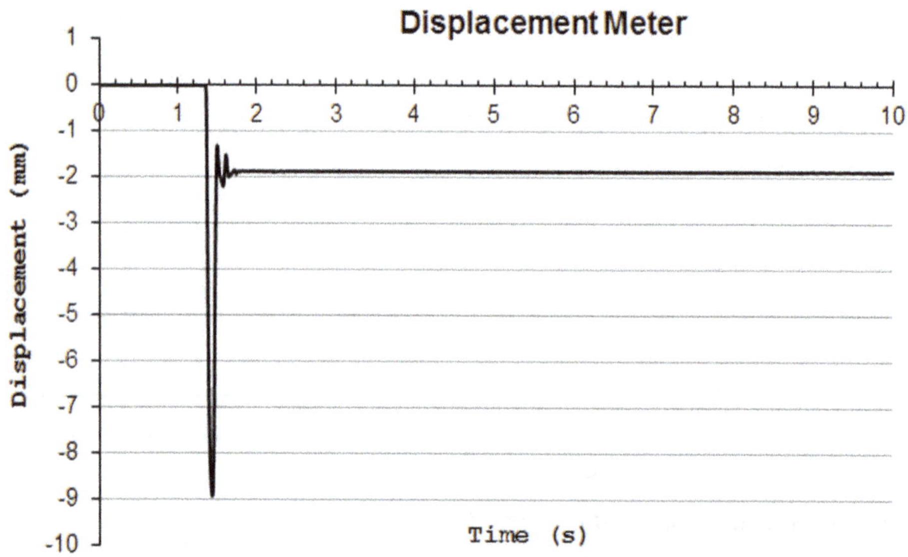

Table 2. Comparison of predicted u_0 for predicted and recorded P_{SO} values

P_{SO}	u_0 (mm)
Theoretical	6.7
Recorded	8.0

CONCLUSION

This book chapter presents issues associated with the response of RC structures when subjected to blast pre-tests and presents results from a small-scale blast pre-test with non-confined explosive and a reinforced concrete slab. A simplified dynamic response analysis was also included. When there is quality control in materials and accuracy of sensors, the experimental results will agree well with predictions from legacy equations. Having a good support during pre-test will ensure that the recorded displacement is only due the RC element deformation. Non-confined explosives can generate pressure without fragments, therefore these kinds of explosive are strongly recommended for tests to verify blast pressure waves and deformations of targets subject to blast.

The slab in our pre-test did not show cracks and kept its integrity due to the small-scale nature of the verification test presented. However, having a test with similar scaled distance do not denote that the slab would support, because the main effect to bring damage is the impulsive load. The simplified equations shown for predicting maximum displacement agree well with the pre-test experiments. For the amount of explosive used in the pre-test and the corresponding scaled distance, the slab was shown to works as good barrier following the DoD definition of function of a protective structure (Department of Defense, 2008).

Shielding sensitive equipment during the pre-test (and eventually the follow up actual test) is important, unless the sensor is made to measure pressures coming from explosion directly. However, the shielding should not be large because re-reflection from large shielding surfaces may mask reliable results.

REFERENCES

ABNT. (1996). *NBR 12655/96. Concreto - Preparo*. Rio de Janeiro: Controle E Recebimento.

ABNT. (2003). *NBR 6118/03. Projeto de Estruturas de Concreto - Procedimento*. Rio de Janeiro, Brasil: NBR 6118/03. Projeto de Estruturas de Concreto - Procedimento. NBR 6118/03.

ABNT. (2004). *NBR 14931/04. Execução de Estruturas de Concreto - Procedimento*. Rio de Janeiro, Brasil: NBR 14931 - Execução de Estruturas de Concreto - Procedimento.

ASCE. (2010). Design of Blast-Resistant Buildings in Petrochemical Facilities. 2nd ed. ed. W. L. Bounds. Reston: ASCE.

Bajic, Z. (2007). *Determination of TNT Equivalent for Various Explosives*. University of Belgrade.

Brode, H. L. (1955). Numerical Solutions of Spherical Blast Waves. *Journal of Applied Physics*, *26*(6), 766–775. doi:10.1063/1.1722085

Castedo, R., Segarra, P., Alañon, A., Lopez, L. M., Santos, A. P., & Sanchidrian, J. A. (2015). Air Blast Resistance of Full-Scale Slabs with Different Compositions: Numerical Modeling and Field Validation. *International Journal of Impact Engineering*, *86*, 145–156. doi:10.1016/j.ijimpeng.2015.08.004

Choi, K. K., Urgessa, G., Reda Taha, M. M., & Maji, A. (2008). Quasi-Balanced Failure Approach for Evaluating Moment Capacity of Concrete Beams Reinforced with FRP. *Journal of Composites for Construction*, *12*(3), 236–245. doi:10.1061/(ASCE)1090-0268(2008)12:3(236)

Chopra, A. K. (2007). Earthquake Spectra. In K. Scherwatzky (Ed.), Dynamics of Structures—Theory and Applications to Earthquake Engineering (3rd ed.). New Jersey: Prentice-Hall.

Clough, R. W., & Penzien, J. (2003). *Computer Methods in Applied Mechanics and Engineering Dynamics of Structures* (3rd ed.). Berkeley: Computers & Structures. Inc.

Department of Defense. (2008). UFC 3-340-02. Structures to Resist the Effects of Accidental Explosions. USA.

Di Stasio, A. (2016). JIMTP Strategy and Path Forward. In *Proceedings of the 42nd International Pyrotechnics Society Seminar* (pp. 102–105).

Goel, M. D., Matsagar, V. A., Gupta, A. K., & Marburg, S. (2012). An Abridged Review of Blast Wave Parameters. *Defence Science Journal*, *62*(5), 300–306. doi:10.14429/dsj.62.1149

Held, M. (1983). Blast Wave in Free Air. *Propellants, Explosives, Pyrotechnics*, *8*(1), 1–8. doi:10.1002/prep.19830080102

Henrych, J. (1979). *The Dynamic of Explosion and Its Use*. Amsterdan.

Jayasooriya, R., Thambiratnam, D. P., Pereira, N. J., & Kosse, V. (2011). Blast and Residual Capacity Analysis of Reinforced Concrete Framed Buildings. *Engineering Structures*, *33*(12), 3483–3495. doi:10.1016/j.engstruct.2011.07.011

Kingery, C. N., & Bulmash, G. (1984). *Airblast Parameters From TNT Spherical Air Bursts and Hemispherical Surface Bursts*. Maryland.

Kinney, G. F., & Graham, K. J. (1985). *Explosive Shocks in Air. 2 nd*. New York: Springer Science. doi:10.1007/978-3-642-86682-1

Li, J., Wu, C., Hao, H., Wang, Z., & Su, Y. (2016). Experimental Investigation of Ultra-High Performance Concrete Slabs under Contact Explosions. *International Journal of Impact Engineering*, *93*, 62–75. Retrieved from http://linkinghub.elsevier.com/retrieve/pii/S0734743X16300422 doi:10.1016/j.ijimpeng.2016.02.007

Maji, A., Brown, J. P., & Urgessa, G. (2008). Full-Scale Testing and Analysis for Blast-Resistant Design. *Journal of Aerospace Engineering*, *21*(4), 217–225. doi:10.1061/(ASCE)0893-1321(2008)21:4(217)

Mendonça, F. B., Rocco, J. A. F. F., Iha, K., & Urgessa, G. (2016). Blast Response of Reinforced Concrete Slab with Varying Standoff and Steel Reinforcement Ratios. In *Proceedings of the 42nd International Pyrotechnics Society Seminar* (pp. 291–292).

Ngo, T., Mendis, P., Gupta, A., & Ramsay, J. (2007). Blast Loading and Blast Effects on Structures - An Overview. *Electronic Journal of Structural Engineering*, *7*, 76–91.

Sadovskiy, M. A. (2004). Mechanical Effects of Air Shockwaves from Explosions according to Experiments. In M. A. Sadovskiy (Ed.), *Selected Works: Geophysics and Physics of Explosion*. Moscow: Nauka Press.

Smith, P. D., & Hetherington, J. G. (1994). *Blast and Ballistic Loading of Structures* (1st ed.). Burlington: Butterworth-Heinemann.

Swisdak, M.M., Jr. (1994). Simplified Kingery Airblast Calculations. Arlington: Naval Surface Warfere Center.

UNODA. (2011). *International Ammunition Technical Guideline* (2nd ed.). New York: United Nations SaferGuard.

Urgessa, G. (2009). Finite Element Analysis of Composite Hardened Walls Subjected to Blast Loads. *American Journal of Engineering and Applied Science*, *2*(4), 804–811. doi:10.3844/ajeassp.2009.804.811

Urgessa, G. (2010). "Vibration Properties of Beams Using Frequency-Domain System Identification Methods." *Journal of Vibration and Control* 17(9): 1287–1294. Retrieved October 21, 2014 from http://jvc.sagepub.com/cgi/doi/10.1177/1077546310378431

Urgessa, G., and Maji, A. (2010). Dynamic Response of Retrofitted Masonry Walls for Blast Loading. *Journal of Engineering Mechanics-Asce, 136*(7), 858–864.

Zhao, C. F., & Chen, J. Y. (2013). Damage Mechanism and Mode of Square Reinforced Concrete Slab Subjected to Blast Loading. *Theoretical and Applied Fracture Mechanics*, *63–64*, 54–62. doi:10.1016/j.tafmec.2013.03.006

Chapter 14
Characterization and Quantification of Polymer Content in Plastic Explosives Using FT-IR Techniques

Elizabeth da Costa Mattos
Instituto de Aeronáutica e Espaço, Brazil & Instituto de Tecnologia Aeronáutica, Brazil

Rita de Cássia Lazzarini Dutra
Instituto de Tecnologia Aeronáutica, Brazil

ABSTRACT

High explosives are tools used in different areas of research and are critical components in conventional and nuclear weapons. PBXs (plastic explosives) were developed to reduce the sensitivity of explosives, and are widely used in both applications, civil and military, where high performance is required. As the weapons may be used in aggressive thermal and mechanical environments, it is important to characterize the PBXs, in order to know its behavior and properties. Therefore, contributing for research of PBXs, is discussed in this chapter a methodology to characterize and to quantify the polymer content in PBX by Fourier Transform Infrared Spectroscopy (FT-IR), using the High-Performance Liquid Chromatography (HPLC) and Thermogravimetry (TG) as reference techniques to quantitative method. The proposed methodology for quantification of polymers in PBX uses the technique FT-IR of reflection ATR and UATR, being faster than the usual methodologies and eliminating the disadvantage of chemical residues generation.

INTRODUCTION

The development of space programs, drilling oil wells, etc., has designed a need to develop thermally stable explosives. The applications such explosives or explosive formulations must be reliable and safe at high temperatures. In this area, certain types of nitro compounds were given special attention be-

DOI: 10.4018/978-1-5225-2903-3.ch014

cause of their ability to withstand the high temperatures and low pressures found in space environments (Agrawal, 2005).

The ideal military explosive to be powerful, safe and easy to handle, can be stored for long periods of time in different climates and be difficult to detonate, unless under precisely circumstances specified (Mathieu & Stucki, 2004).

To start a detonation, a shock wave is required. This is accomplished with a small quantity of a primary explosive. Modern formulations, extremely insensitive have been developed in most countries to eliminate accidents with ammunition. Insensitive energetic compounds, new inert or energetic polymeric systems were introduced to improve the vulnerability of such weapons. The search for new energy materials with high performance characteristics is an ongoing task (Mathieu & Stucki, 2004).

The high explosives are tools used in many areas of research and are critical components in conventional and nuclear weapons. However, to be useful in most of these applications, the explosives must be manufactured to precise settings and safe.

One the most powerful explosives used in World War II was the cyclotrimethylene trinitramine (RDX). This was first prepared in 1899 by Henning (Bachmann & Sheehan, 1949), and was only used as explosive in 1920 (United States, 1979). The first discovery of cyclotetramethylene tetranitramine (HMX) date 1941 (Lukasavage, Nicolich, & Alster, 1993). Was obtained as impurity in the nitration of hexamethylenetetramine to produce RDX in small amount, approximately 10% (United States, 1979).

HMX is a high explosive, that by its high power is utilized by developed countries to carry out metal forming and energy transfer, and thus employed in plastic explosive, compressed and molded to the base of HMX (Calzia, 1969, Valença, 1998).

Pure explosive crystals are little used by the physicists and engineers because in almost all applications, the goal is to pack the greatest possible amount of explosives in a given space (high density) (Hayden, 2005). For obvious reasons, "loose explosive crystals" do not confer a high density. However, when crystals are coated with a polymeric binder, the raw material becomes very useful for the production of plastic explosives reducing sensitivity and dangerous stimuli such as shock, friction, impact, etc., the fitting them into a matrix crystals polymer (Hayden, 2005, Chen, Huang, Dai, & Ding, 2005, Singh, Felix, & Soni, 2003).

To get more energy in lower volume, it began the development of plastic explosives (PBX).

PBX is an acronym for "Plastic Bonded Explosive" (plastic explosive). A term applied to a variety of explosive mixtures which are characterized by high mechanical strength, good explosive properties (usually greater detonation velocity of 7800 m/s) excellent chemical stability, low sensitivity to shock and handling and low sensitivity to thermal initiation (Federoff & Sheffield, 1966). Many PBXs have been developed in the past half century, all usually with the goal of providing an engineering material for specific or wide range of applications [Hayden, 2005; Kim, & Park, 1999). These explosive mixtures contains a large percentage of secondary explosives such as RDX, HMX, hexanitrostilbene (HNS) or pentaerythritol-tetranitrate (PETN) in mixtures with a polymeric matrix such as polyester, polyurethane, nylon, polystyrene, various types of rubbers, nitrocellulose fluorinated polymers. In some cases, a plasticizer such as dioctylphthalate (DOP), dioctyladipate (DOA) or butyldinitrophenylamine (BDNPA) is added (Federoff & Sheffield, 1966, Thompson, Olinger, & Deluca, 2005).

PBXs were developed to reduce the sensitivity of the synthesized explosives, are widely used in both applications, civil and military, where high performance is required (Chen, Huang, Dai, & Ding, 2005).

The materials used in obtaining the PBX are generally a mixture bi-modal or tri-modal explosive crystals expressing two or three sizes of particles with a nominal specific distribution of the means. The compatibility and miscibility with other explosives, polymers, and additives should be evaluated before pressing, melting, cast cure, extruding, etc., to ensure the safe handling and storage of products [Mathieu & Stucki, 2004, Hayden, 2005).

Because weapons systems may be exposed to thermal and mechanical aggressive environments, it is important to characterize the PBX properties in order to satisfy their physical and chemical behavior (Thompson, Olinger, & Deluca, 2005).

The traditional method slurry for production of explosives coated has been used for over 45 years (Kasprzyk, Bell, Flesner, & Larson, 1999). In the "slurry" the explosive crystals are coated with a polymer. In this case the explosive is dispersed in water in a stirred tank reactor. A lacquer consisting of a polymer soluble in an organic solvent is added to the slurry. The combination of heat, agitation, and distillation remove the solvent. The explosive coated in the form of granules is filtered and can be pressed to obtain an explosive charge (Kasprzyk, Bell, Flesner, & Larson, 1999).

The pressing of explosive charges by means of hydraulic presses under high pressures represents the most important process for shaping high-explosive explosive charges (Wanninger, Wild, Kleinschmidt, & Spath, 1996). As the explosive is obtained in the form of crystals, it is necessary to involve them with polymers, performing a coating process.

The explosives are combined with polymers in order to minimize its sensitivity to shock, and which are less hazardous during the pressing process so as not to occur the friction between the crystals which could cause an explosion (Wanninger, Wild, Kleinschmidt, & Spath, 1996).

The basic objective of the polymer in the PBX is to reduce the sensitivity and stimuli such as friction, shock and impact, thus maintaining the integrity of the explosive charge.

The characterization of material in the PBX is essential not only from a safety point but also the development of new methodologies. Since there are a small number of information found in the literature about characterization by Infrared Spectroscopy Fourier Transform (FT-IR) on the coating energetic material is presented in this chapter the characterization and quantification of the polymer used in explosives coverage by FT-IR techniques, using the reference techniques such as high performance liquid chromatography (HPLC) and thermogravimetry (TG).

The PBX Characteristics

PBX compositions are based on polymeric systems typically loaded of explosive crystals, which can range from 85 to 95wt% (Thompson, Olinger, & Deluca, 2005). After World War II, old compositions TNT-based (Trinitrotoluene) begin to be replaced by the generation of the HMX explosive with polymeric binders (Skidmore, Phillips, & Crane, 1997). The first PBX composition was developed at the Los Alamos National Laboratory in 1952 in USA. The composition consisted of RDX crystals in a polystyrene polymer matrix. Since 1952, the Lawrence Livermore Laboratories, the US Navy and many other organizations have developed a series of PBX formulations (Akhavan, 2004).

Polymers are combined with explosives in order to minimize their sensitivity to shock then be less dangerous to handle. Regarding thermal properties Hoffman & Caley (1981) have tested the thermal cycle LX-10. They monitored the dynamic mechanical and molecular weight polymers in the PBX to check for degradation in order to determine the lifetime of the PBX. They found that although the LX-10 degrade at a very slow rate, the life of LX-10 may exceed 60 years, based on kinetic calculations from the changes in the molecular weight of the polymer (Burgess, Woodyard, Rainwater, Lightfoot, & Richardson, 1998).

PBXs contain explosive, polymer, and sometimes plasticizers and stabilizers, the analysis of this composition can be complicated. Most methods utilize HPLC and DSC to determine contamination or degradation of the polymers. The mechanical and explosive properties may also be observed before and after aging. Hoffman & Caley (1981) used HPLC (specifically SEC), DSC and mechanical properties (storage modulus, loss modulus, Tg) for determining degradation of fluorinated polymers in PBX- 9502 and LX-10-1. It was found small degradation by using these methods and have continued the work observing the changes in crystallinity and thermal aging, the aging occurred between 23 and 74 °C (Hoffman, Matthews, & Pruneda, 1989).

EXPLOSIVES CHARACTERISTICS

Various secondary explosives can be used for obtaining the PBX. This item will be presented the characteristics of explosive (RDX and HMX) that were used in this research.

RDX

RDX is an acronym for "Royal Demolition Explosive" is the chemical name is cyclotrimethylene trinitramine ($C_3H_6N_6O_6$), also known as cyclonite or hexogen. The density of cyclonite of the crystals is 1.820 g/cm^3, a white substance is practically insoluble in water and crystallizing in the orthorhombic system. The melting point is around 204 °C. It is an explosive widely used in military and industrial applications. It was of high importance during the Second World War as a constituent of many explosive mixtures from which a high power was required. Is usually used in mixtures with other explosives such as TNT, plasticizers and aluminum for use in bombs and torpedoes. It can be used as main charge in detonators. It is stable in storage at room temperature and is considered one of the most powerful military explosive (Urbanski, 1967).

HMX

HMX, also known as 1,3,5,7 - tetranitro - 1,3,5,7 - tetraazaciclooctano, cyclotetramethylene-tetranitramine ($C_4H_8N_8O_8$) or octogen, is the most important of modern energetic components for solid propellants due to its desirable properties, such as the absence of smoke, high specific impulse and thermal stability (Tang, Lee, Kudva, & Litzinger, 1999). It is used in certain propellants and explosives (Kohno, Maekawa, Tsuchioka, Hashizume, & Imamura, 1994). Propellants based on HMX can be found in weapons applications and solid rocket propulsion system (Tang, Lee, Kudva, & Litzinger, 1999).

The detonation of this material reaches a burn rate of 9100 m/s. It is a white crystal with molar mass of 296g/mol. It is high explosive of the high energy per unit volume, which has four different polymorphic forms (α HMX, β HMX, δ HMX, γ HMX) according to crystallization conditions, differing from one the other in terms of density and sensitivity to impact.

The HMX has been used in metal conformation, energy transference and in the composition of melting and pressing plastic bonded explosives (Calzia, 1969; Urbanski, 1984).

Characteristics of Polymers

Before 1960, mainly paraffin's and waxes used were as classic binder system for pressed explosive formulations. In modern formulations of PBXs polymers are used due to their properties are different [Mathieu, & Stucki, 2004). Since the early 50s a large variety of polymers was used in obtaining the PBX, among the most used has been fluoroelastomers and polyurethanes (Burgess, Woodyard, Rainwater, Lightfoot, & Richardson, 1998).

Were used in this research, Viton, Estane and EVA to perform the quantification methodology by FT-IR of these polymers in PBX compositions.

Viton

Fluoropolymers or fluoroelastomers, are used in the coating of energetic materials. They are known generically as FKM polymers, nomenclature adopted in ASTM D1418. Viton® is a trade name adopted for a series of fluoroelastomers manufactured by DuPont and is available in various formulations (co-polymer, terpolymer) and forms (pellets, slab, stick) (Hohmann & Tipton, 2000, Mattos, et al. 2008).

Hexafluoropropylene-vinylidene fluoride and tetrafluoroethylene, Viton B, is a terpolymer while Viton "A" is a copolymer of vinylidene fluoride and hexafluoropropylene (Wanninger, Wild, Kleinschmidt, & Spath, 1996). The building block of Viton consists in repeating subunits called "mers." Mers can be likened to links in a chain. The more links the chain has the longer the chain is. In Viton the majority of monomers contain carbon and fluorine [Hohman & Tipton, 2000, Mattos, et al. 2008). During the manufacture of polymers, several factors determine the number of links each molecular string will have in its chain. Mixtures of monomers are used to modify the chemical and physical properties of the polymers. Fluoroelastomers are rubbers, not a simple monomer (Hohmann & Tipton, 2000). They can be classified by the fluorine amount, 66, 68, and 70%. Figure 1 shows the structural unit of a fluorinated terpolymer.

Fluoroelastomers, such as Viton shows excellent chemical stability and become very used polymers for the coating energetic materials in propellants. Their presence is very important in relation to the oxidation time, as cited in the literature (Hohmann & Tipton, 2000).

Estane

Estane is another kind of polymer used in obtaining plastic explosive (PBX) for manufacturing of pressed charges can be based on polyurethanes. It is a thermoplastic polyurethane polymer (TPU), based on aromatic polyester prepared by long chains of a copolymer of 1,4-butandiol, 4,4'-phenylmethylene

Figure 1. Terpolymer Fluoroelastomer - Viton B (FKM)
MATTOS, et al.

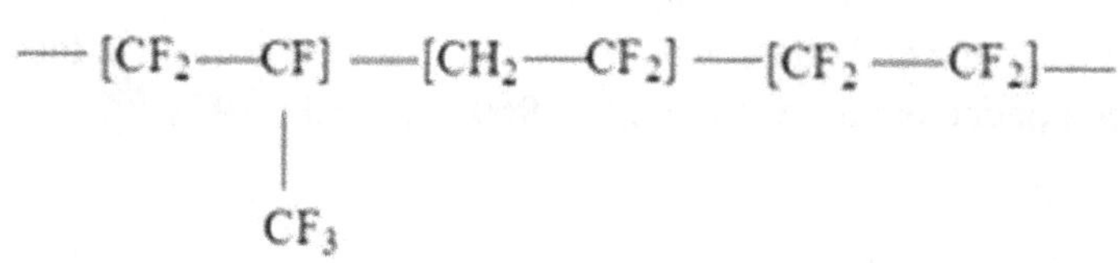

diisocyanate, and a macrodiol made from adipic acid and1,4-butandiol (Burgess, Woodyard, Rainwater, Lightfoot, & Richardson, 1998).

These polymers contain two functional groups which are particularly vulnerable to chemical attack: the ester linkage in the polyester block and a urethane bonds formed by the reaction of MDI with the diol components. Ester bonds polyester urethanes have notably poor hydrolytic stability, and the urethane structure has been show to be susceptible to oxidative attack (Wewerka, Loughran, & Williams, 1976). The breaking of one or both of these functional groups types could be the responsible for degradation of the Estane when stored in the PBX (Wewerka, Loughran, & Williams, 1976).

Campbell, Garcia, & Idar, (2000) investigated the glass transition of the polymers, including Estane 5703, and found that there is a small change in mixing the polymer with explosives. The results also showed that the pressing does not change the soft segment glass transition temperature of the polymer, but appears to have an effect on the hard segment and crystallinity of the polymer (Campbell, Garcia, & Idar, 2000).

EVA

Poly (ethylene-co-vinyl acetate) (EVA) is formed by connecting random sequences of polyethylene and poly (vinyl acetate) (PVA), Figure 2. Their properties are typically intermediate when compared to the pure components, which is due to its complex morphology, which consists of a crystalline phase containing methylene units, an interfacial region with methylene segments and segments of vinyl acetate (VAc) and an amorphous phase with methylene segments and VAc units (Zattera, Bianchi, Zeni, & Ferreira, 2005). If compared with a low-density polyethylene (LDPE) with the same molecular weight, the EVA has a higher elongation value at rupture and impact strength and lower modulus of elasticity (Zattera, Bianchi, Zeni, & Ferreira, 2005).

It has a large variety of industrial applications. These include its use in films, hoses, coatings and adhesives (Allen, Edge, Rodriguez, Liauw, & Fontan, 2001). EVA has dependent upon optical and mechanical properties of the VAc content (Dutra, 1997).

Dagley and colleagues used EVA with different explosives (RDX, PETN and TATB) to simulate the thermal effects in plastic explosives pressed caused by the variation of the components (Dagley, Parker,

Figure 2. Poly(ethylene-co-vinyl acetate)
Allen, at al. 2001.

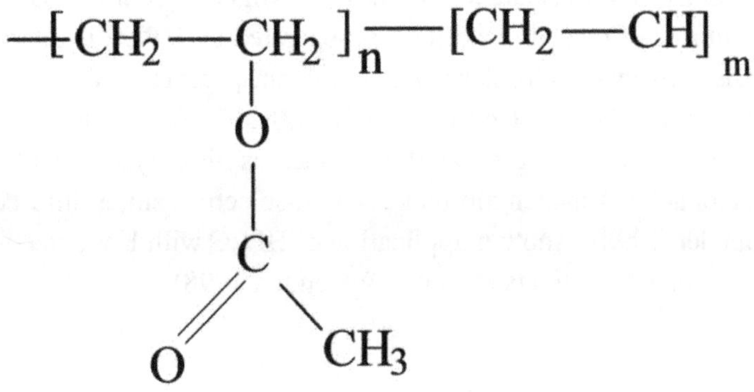

Jones, & Montelli, 1996). Authors concluded that the simulation are not able to predict the violence thermal response, but reproduce exactly radial heat flow in the test assembly and satisfactorily predict the thermal response time and surface temperature in response for the mixed explosive compositions.

EXPLOSIVE COVERED OBTAINING

The coating process of the crystalline explosives, such as RDX and HMX, uses the slurry process. The explosive is slurred in water (aqueous phase) and the polymer Viton, Estane or EVA is dissolved in an organic solvent with low boiling point, not miscible with water, that is added during the process. This process is carried out under strong stirring (500 – 700 rpm) (Campbell, Garcia, & Idar, 2000, Benziger, 1973). The solvent is removed by distillation under continuous stirring, and the polymer precipitates on the surface of the crystals of the explosive at the end of the solvent elimination (Campbell, Garcia, & Idar, 2000).

After the coating process the granular product formed basically by RDX/HMX can be used in pressed explosive charges with high-energetic content (Wanninger, Wild, Kleinschmidt, & Spath, 1996).

MEASUREMENT METHODS

Chromatography (HPLC)

Chromatography permits the separation, detection and quantification of components, explosive or not of energetic material, allowing then to be characterized. The importance of the use of this method in the investigation of explosives, that allows to separate small amounts of these materials without reacting, in a wide variety of matrix samples (Mccord, & Bender, 1998).

The most suitable chromatographic technique for analyzing explosives, is the HPLC for being an non-destructive technique, requiring minimal of the sample.

The most commonly used HPLC separation system explosives analysis is reversed phase, in which the explosives are separated in mixtures of methanol or acetonitrile and water. The separation can be performed by isocratic mode (working with a solvent and needs only one pump) or by using a gradient elution in which the eluent composition is changed by increasing the concentration of the organic solvent as the run progresses. Reversed phase HPLC has many advantages for explosives analysis, including system stability, low toxicity and transparency to ultraviolet light (Mccord, & Bender, 1998).

The most common means of detecting explosives separated by HPLC is ultraviolet radiation (UV) absorption. Reverse phase solvents (high polarity) are more transparent to UV radiation than normal phase solvents (type of column, low-polarity solvent) (Ciola, R. 2003, Mccord, & Bender, 1998). This becomes particularly important when considering nitrated ester such as nitroglycerin and PETN. However, the operation of a UV detector at low wavelengths makes it non-selective, since almost every compound will respond (Mccord & Bender, 1998). Another application of HPLC with UV detection is the quantitative analysis of explosives in aqueous extracts (Mccord & Bender, 1998).

HPLC has been used by several researchers (Kaiser, 1998, Bunte, Pontius, Kaiser, 1999, Groom, Halasz, Paquet, D'Cruz, & Hawari, 2003, Mattos, et al., 2004a, Silva, 2007) for identification of different explosives (HMX, RDX, HNS, TNT, CL20) on the content of explosives, the presence of different explosives and its polymorphs (Kaiser, 1998, Bunte, Pontius, Kaiser, 1999), traces of explosives in soil (Marple & Lacourse, 2005, Pan, Tian, Jones, & Cobb, 2006).

HPLC analysis was used to quantify the polymer in the PBX as a reference method for the quantification methodology development FT-IR in systems with HMX/Viton and HMX/EVA.

Thermogravimetry (TG)

Thermogravimetry is a technique basically quantitative by its nature, in which weight loss can be determined accurately. However, temperature ranges in which such weight variations occur are qualitative and depend on the characteristics of the sample and apparatus (Monthé, & Azevedo, 2002).

Lee and Hsu (2002) analyzed PBX with different amounts of polymer and types of explosives by means of different techniques thermal analysis (TG, DSC and DTA) to check the military artefact lifetime with plastic explosive. Research has shown that shelf life of HMX-based formulations is longer than 60 years at ambient temperature. Also analyzed the compatibility of explosives with silicone rubber and the kinetic parameters by means of TG and DSC techniques and found that PBX silicone rubber is compatible with PETN, RDX, HMX and HNS (Lee, Hsu, Chang, 2002).

The thermal stability, specific heat and thermal diffusivity of the plastic explosive PBXW-11 was evaluated by Tompa and Boswell (2000) by TG and DSC in order to calculate the explosion time and it was concluded that this decomposition is first order.

Singh et al. (2003, 2005) and coworkers (Felix, Singh, Sikder, & Aggrawal, 2005) have used the TG technique (isoconversional method) to perform kinetic studies in plastic explosives with RDX and HMX and concluded that the thermal stability of the explosive is not affected when polymer (Estane or Viton) are present in PBXs corresponding.

Burnham and Weese (2005) determined the kinetic parameters of the thermal decomposition of polymers of specific interest, used in plastic explosives compositions such as Viton, Estane and Kel-F 800 (copolymer of vinylidene clorotrifluoretano) by TG.

In this chapter, TG technique was used to calculate the polymer content in the explosive composition, and reference technique in the development of FT-IR quantitative methodology for HMX/Viton system.

Espectroscopia FT-IR

Infrared spectroscopy (IR) technique is an identification and characterization of organic, inorganic and polymeric materials, and therefore may be useful for a variety of studies.

The IR spectrum of a pure explosive is often a reliable basis for their identification. The functional groups in organic molecules are identified in the IR spectrum, often because of the value of "fingerprint" that can identify the whole molecule (Zitrin, S., 1998).

Infrared radiation corresponding to the part of the spectrum situated between the regions of the visible and microwave, being the best known spectral range for characterization of organic molecules is among 4000-600 cm^{-1}, usually called mid-infrared (MIR) . The region

14000-4000 cm^{-1}, known as near infrared (NIR) is a little explored field, and thus arouses interest for material analysis applications, especially explosives (Graf, Koenig & Ishida, 1987)

There are various sample preparation techniques to FT-IR analysis. Transmission is the oldest technique for obtaining spectra (Graf, Koenig & Ishida, 1987). In this technique, an infrared beam passes through surface and in the interior of species (Ishida, 1987).

Surface analysis techniques may also be used to aid in transmission techniques for a more complete evaluation of material, for example, the diffuse reflectance (DRIFT), where diffusely scattered radiation is collected by mirrors, and focused to the detector (Pandey & Kulshreshtha, 1993).

The combination of infrared spectroscopy with the reflection theories, produced advancements in surface analysis. The FT-IR/ATR technique combines the power of IR spectroscopy with the optics of attenuated total reflection (Nogueira, Rosa & Santana, 2000, Kwan, 1998).

In Spectroscopy FT-IR/ATR can be used crystals of germanium, KRS-5, diamond and ZnSe, which differ mainly due to refractive index ranging from 2.38 (KRS-5) 4.01 (germanium) (Kwan, 1998).

Universal ATR (UATR) is an accessory of internal reflection FT-IR, used to simplify the analysis of solids, powders, liquids and gels folders. It is a non-destructive technique that consists of placing the sample on the crystal composed of diamond and ZnSe with a high refractive index. As beam does not penetrate the sample much, this technique can be useful for analyzing solid, solid surfaces and coatings for solid samples (Abidi & Hequet, 2005).

The photoacoustic spectroscopy (PAS), a less conventional technique, according to data from the literature (Graf, Koenig & Ishida, 1987, Pandey, & Kulshreshtha, 1993) and expertise in developing methodologies, provides several advantages over transmission techniques: the first is the ability to analyze opaque materials, explored in composites, carbon fibers, etc. PAS spectroscopy can be applied to samples where other IR techniques, not may be appropriate to provide quality spectrum seeking more complete evaluation of the material (Mattos, Viganó, Dutra, Diniz & Iha, 2002, Ishida, 1987, Pandey & Kulshreshtha, 1993, Mattos, 2001).

FT-IR spectroscopy by combine the merit of identifying and characterizing structures associated with the speed of analysis is a powerful tool to examine the various polymorphic structures of HMX and mixtures thereof with other explosives. Among several studies, can highlight the following research:

Litch (1970) obtained the four polymorphic forms of HMX by methods involving recrystallization of HMX with high boiling solvents. Used, among other methods, FT-IR spectroscopy for characterization of HMX forms of according to transmission technique with KBr pellet.

Achuthan & Jose (1990) used FT-IR spectroscopy (KBr pellet technique) in the MIR region, Raman spectroscopy, and DSC to distinguish the different polymorphic forms of energetic material HMX and understand the transition mechanisms between them.

McAuley, Joubert & Steyn (1985) used spectroscopy dispersive IR for evaluation of polymorphic forms of HMX, according to the FT-IR transmission technique, have concluded that the lowest α content and γ HMX which can be detected by the direct method, i.e. without use of standard internal.

Wenhui, Zhanning, & Jianwu, (1994) used the FT-IR reflection technique (DRIFT) to characterize the surface of HMX samples treated with coupling agents, searching the proper selection of these agents for use in propellants. The goal of the study was to investigate the interaction mechanism in the interface based on level of molecular structure and assist in the selection of suitable couplings agents for propellants containing HMX as load.

Gjersøe (2000) used the DRIFT technique for detecting contents between 0.1 to 1.0% of alpha in beta HMX in the MIR region using the band at 712 cm^{-1}. The method was quick and with good linear relationship with the concentration of alpha HMX.

Mattos (2001) in his research used FT-IR techniques (transmission, reflection and photoacoustic), qualitative and quantitative, in large spectral range in the characterization and quantification of explosive mixtures of HMX and RDX.

According the literature, several studies conducted using FT-IR with different accessories for characterization of explosives, but are not observed the use of FT-IR technique for quantification of PBX. Therefore the IR spectroscopy can be used to identify explosive such as RDX and HMX, and quantifying polymers such as Viton, EVA and Estane in the PBX samples with these components.

DISCUSSION

PBX samples were prepared with different polymers (Estane®, Viton® and EVA) and the HMX and RDX explosives, constituting the systems: HMX/Viton, HMX/EVA, RDX/Viton and RDX/Estane. For developing the FT-IR quantitative method proposed to measure the polymer concentration in PBX, explosives covered were prepared using EVA and Viton the following contents: 2, 5, 8, 10, 13 and 15%.

Quantification TG Analysis

Thermogravimetric analyses were carried out in the TG/ SDTA Mettler (Model TGA/SDTA851e), which was calibrated in a heat ratio of 10 ºC/min, with aluminum and indium, under nitrogen gas flow (40 mL/mim). About 2.0 mg of the sample was used and the heating range was between 50 and 700ºC.

TG Analysis of HMX and Viton

The analysis of HMX by thermogravimetry (TG) shows that the maximum weight loss occurs 285ºC as can be observed in Figure 3.

Viton analyzed under the same conditions shows maximum weight loss at about 487.8 °C, according to Figure 4 that illustrates the thermal behavior of this polymer

PBX Characterization (HMX/Viton) by TG Analyses

PBX samples were prepared according to the concentrations indicated in Table 1, and the results obtained by TG are the same table. Figure 5 shows the thermal behavior of these mixtures, as well as the pure product

TG Analysis of EVA

The EVA thermal curve (Figure 6) shows the loss of mass in two levels, the first loss occurs at 356 °C (peak) related to the degradation of vinyl acetate (acetic acid formation), and the second loss at a temperature of 476ºC (peak)which refers to the degradation of the olefinic copolymer (C-H and C-C bonds).

Figure 3. TG and DTG curves of HMX sample obtained in the heating rate 10 °C/min and under a dynamic atmosphere with N_2 flow 40 ml / mim

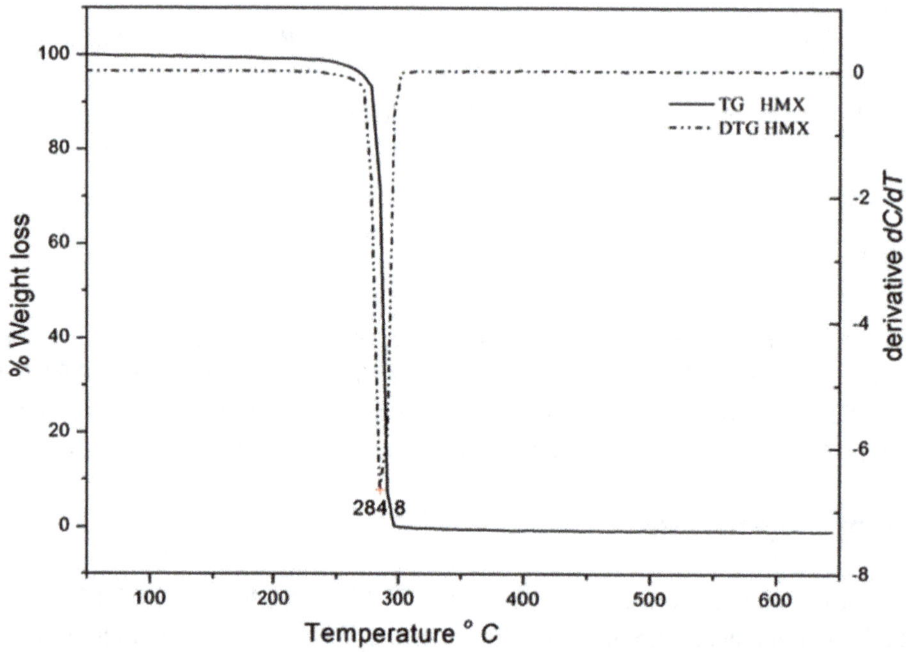

Figure 4. Viton sample TG and DTG curves obtained in the heating rate 10 °C/min and under a dynamic atmosphere with N_2 flow 40 ml/mim

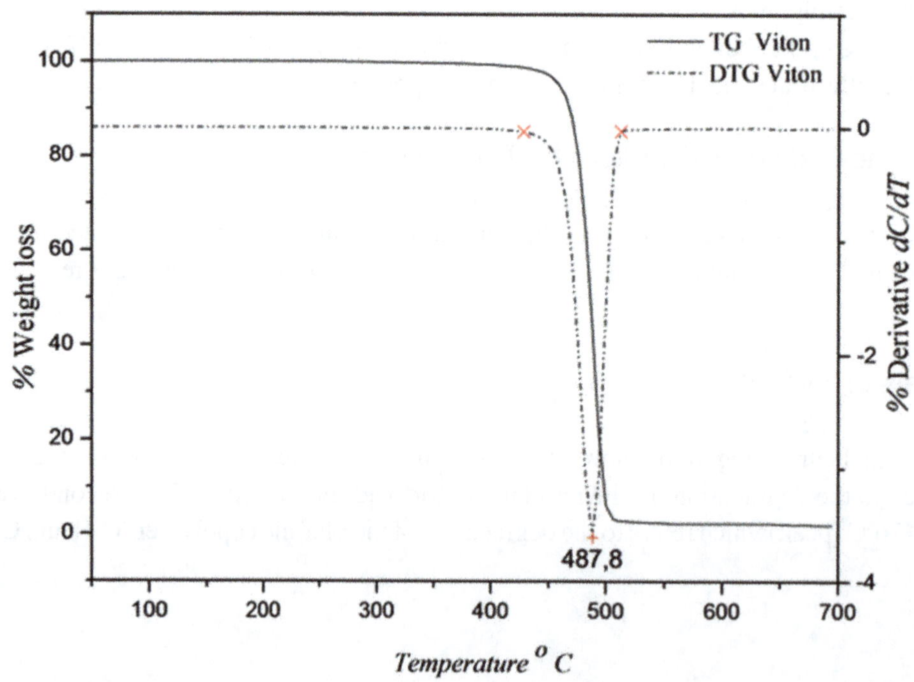

Table 1. HMX contents in the HMX/Viton samples obtained by TG

Code	Theoretical HMX/ Viton Content (%)	HMX Contents by TG (%)	Viton Contents by TG (%)	Standard Deviation (%)
Cob - 01/06	98/2	97.87	2.13	0.68
Cob - 01/95	95/5	95.24	4.76	0.25
Cob - 05/06	92/8	93.67	6.33	0.93
Cob - 04/06	90/10	91.10	8.90	1.06
Cob - 06/06	87/13	89.88	10.12	0.79
Cob - 02/06	85/15	89.98	10.02	1.06

Figure 5. TG curves of the system HMX / Viton in heating rate 10 °C/min and under a dynamic atmosphere with N_2 flow 40 ml / min

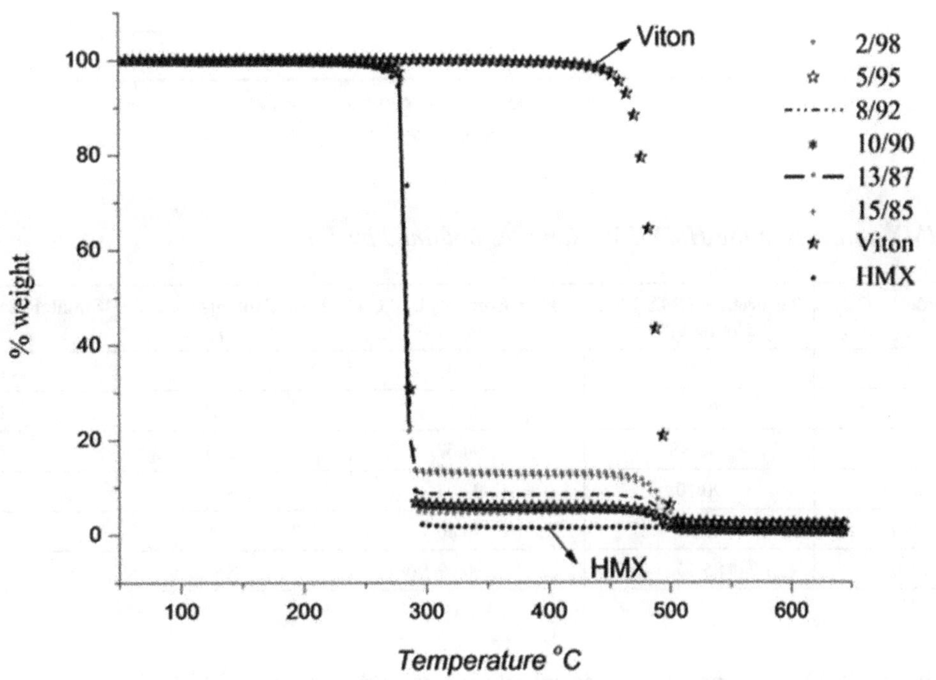

PBX Characterization (HMX/EVA) by TG Analyses

PBX samples were prepared according to the concentrations indicated in Table 2, and the results obtained by TG are the same table. Figure 7 shows the thermal behavior of these mixtures, as well as the pure product

It is possible to observe, by means of Figure 7 (expanded TG curve), the PBX samples show different contents of EVA, but it is not possible to observe the first weight loss step of EVA that probably occurs due to the large supply of heat the HMX releases in its thermal decomposition, which occurs around 280°C. This fact should accelerate the first weight loss step of EVA which can be associated with the

Figure 6. EVA sample TG and DTG curves obtained in the heating rate 10 °C/min and under a dynamic atmosphere with N_2 flow 40 ml / min

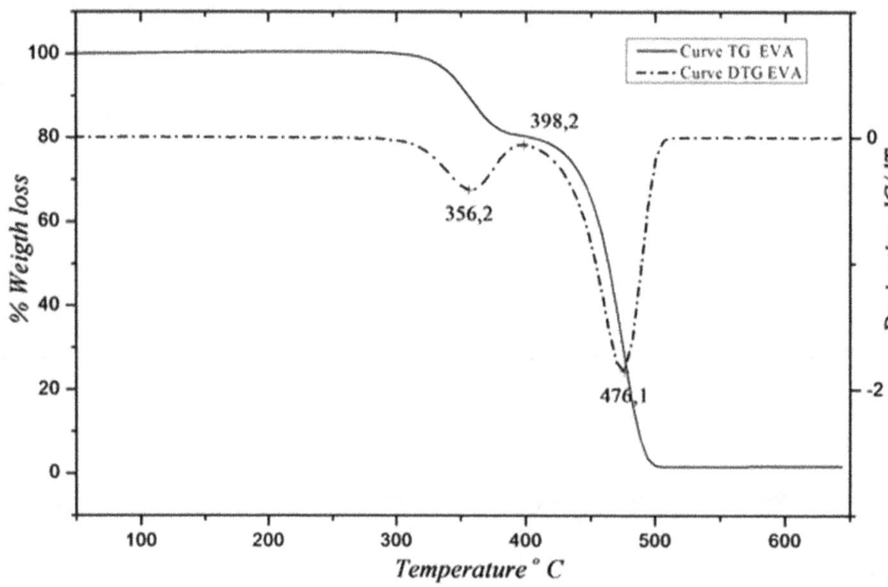

Table 2. HMX contents in the HMX/EVA samples obtained by TG

Code	Theoretical HMX/EVA Content (%)	HMX Contents by TG (%)	EVA Contents by TG (%)	Standard Deviation (%)
Cob - 09/06	98/2	97.37	1.07	0.15
Cob - 03/05	95/5	95.52	1.85	0.36
Cob - 01/07	92/8	94.67	2.71	0.17
Cob - 02/07	90/10	93.05	3.77	0.39
Cob - 05/07	87/13	90.80	6.09	1.27
Cob - 04/07	85/15	90.69	6.52	0.36

decomposition of HMX, compromising the calculation of EVA content by TG analysis and that way it was not possible to use the values obtained by this technique as a reference for quantitative analysis by FT-IR.

HPLC METHODOLOGY FOR PBX QUANTIFICATION

HPLC analyzes were conducted in the Liquid chromatograph equipped with a Waters 1515 binary pump, a manual injector (Breeze 7725i, Rheodyne with 10-μL loop), a Waters 2487 UV-VIS wavelength detector at the λmax of 230 nm. It was used a RP C18 column (μ Bondapack), the mobile phase used was methanol:water (40:60) with isocratic flow (1.2 mL/min). The Results were acquired and processed by internal software Breeze version 3.30.

Figure 7. Amplified TG curve of the system HMX / EVA obtained at heating rate of 10 °C/min and under a dynamic atmosphere with N_2 flow 40 ml / min

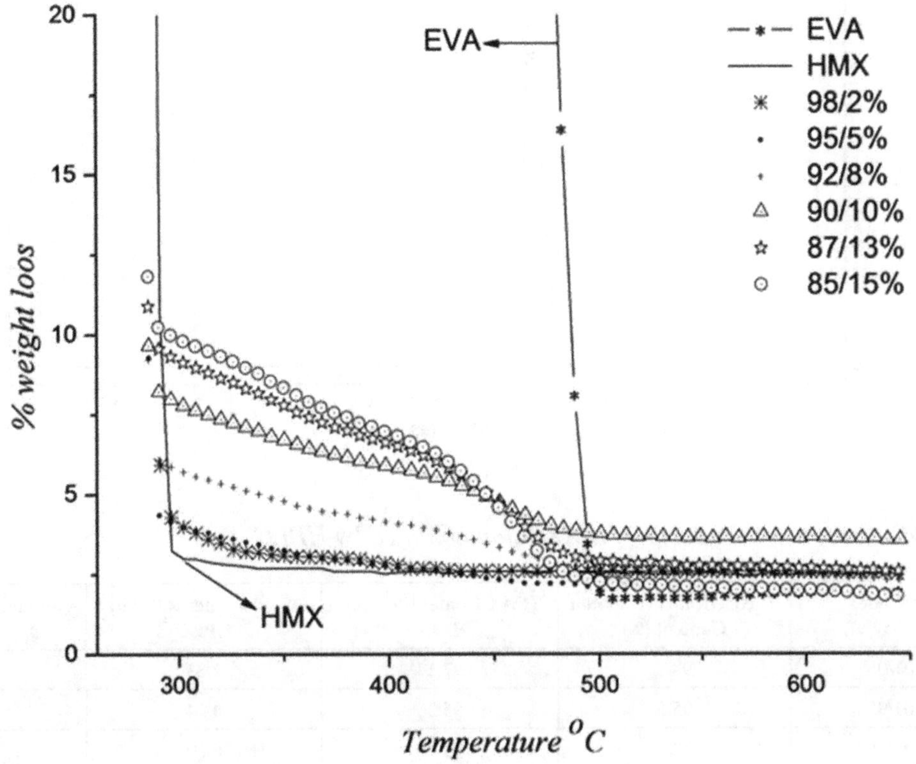

The polymer content in the PBX when analyzed by HPLC is calculated by the difference between the content of HMX used as a reference and HMX content found in the sample. The explosive is dissolved in a proper solvent, and as the polymer is not soluble in the medium it is separated during the filtration process solution. The solution containing the HMX is injected and then calculates the difference between the real sample and the reference, the result is the polymer content (Figure 8).

It is possible to observe the HMX chromatogram shows the peak retention in approximately 3.7 minutes and the concentration is calculated on the peak area of the sample. As higher the polymer contents in the sample, the lower the peak area of HMX. This behavior is described in all samples of systems HMX/Viton and HMX/EVA, since the HPLC technique quantifies and identifies only the explosive present in the composition. Samples of HMX/Viton and HMX/EVA systems were prepared and analyzed according to described conditions and results obtained are shown in Tables 3 and 4. This methodology showed excellent results.

Figure 8. Chromatogram of the sample running of HMX/Viton (85/15%)

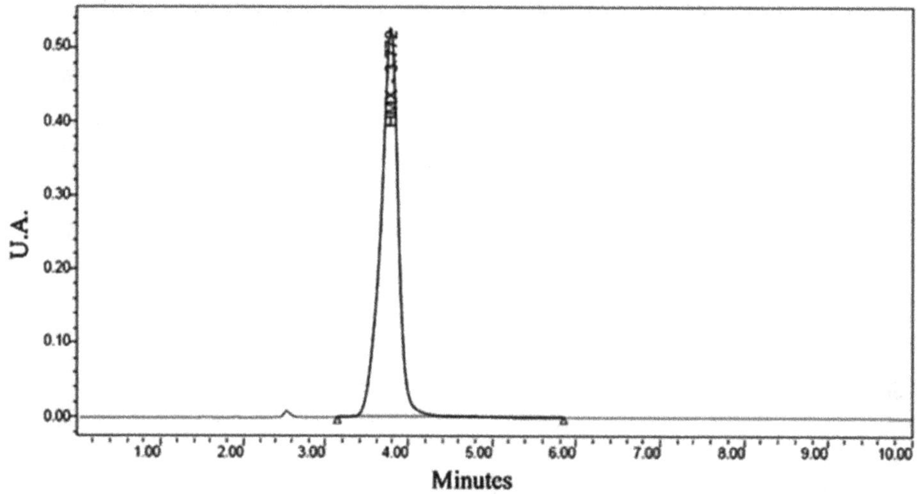

Table 3. HMX contents in the samples HMX/Viton obtained by HPLC

Sample Code	Theoretical HMX/Viton Content (%)	HMX Contents (%) by HPLC	Viton Contents (%) by HPLC	Standard Deviation (%)
Cob - 01/06	98/2	97.17	2.83	2.17
Cob - 01/95	95/5	95.26	4.74	1.79
Cob - 05/06	92/8	93.20	6.80	3.56
Cob - 04/06	90/10	91.96	8.04	1.06
Cob - 06/06	87/13	89.20	10.80	2.25
Cob - 02/06	85/15	84.53	15.47	1.01

Table 4. HMX contents in the samples HMX/ EVA obtained by HPLC

Sample Code	Theoretical HMX/ EVA Content (%)	HMX Contents (%) by HPLC	EVA Contents (%) by HPLC	Standard Deviation (%)
Cob - 09/06	98/2	97.86	2.14	1.91
Cob - 03/05	95/5	93.53	6.47	3.23
Cob - 01/07	92/8	92.63	7.37	0.53
Cob - 02/07	90/10	90.54	9.46	1.79
Cob - 05/07	87/13	87.44	12.56	2.94
Cob - 04/07	85/15	84.75	15.25	0.94

ANALYSIS BY FT-IR

Qualitative Analysis in RDX and HMX by Transmission FT-IR

The average and intense absorptions observed in FT-IR spectra of RDX and HMX were chosen to characterize these explosives, in agreement with literature (Smith, 1979, Achuthan & Jose, 1990, Chasan & Norwitz, 1971, Mattos, et al., 2004b). The most probable assignments of these bands are shown in Table 5.

FT-IR Qualitative Analysis of RDX/Viton by Transmission

It can be seen by Figure 9, the spectrum A (technique KBr) relating to the polymer coating, shows basically the absorption bands for RDX, according to the reference spectrum of the energetic material (Mattos, et al., 2008) (spectrum B). It is not possible to identify the presence of the polymer. This result is in accordance with the transmission technique characteristics which show the higher content, in this case, the RDX. A treatment with solvent to separate the polymer of the energetic material, and analyze, is required.

After treatment with methanol, a residue was obtained which was analyzed in the form of cast film (spectrum C, Figure 9.) and then possible to compare it to spectrum D, Figure 9 (Viton reference spectrum) by identifying presence of a polymer that coats the explosive

The presence of C-F groups is detected in the IR medium based on the intensive absorptions at 1397 - 1074 cm $^{-1}$ (C-F ν) (Mattos, et al. 2008, Urbański, Andrzej, & Cameron, 1977). The characteristic bands of C-F_2 bonds are found at 1273, 1191, 1134 and 1111 cm^{-1} (C- $F_2\nu$) and 635 and 610 cm^{-1} (C-F_3 ν) (Mattos, et al.,2008, Silverstein, Bassler & Morril, 1981), of medium intensity.

Table 5. Probable assignments of functional groups of the RDX and HMX in the corresponding vibrational mode IR

Wavenumber (cm^{-1}) β HMX	Wavenumber (cm^{-1}) RDX	Vibrational Mode/Functional Group
3035	3073	ν CH_2
1564	1592	ν_a NO_2
1462 1432	1459 1433	δ_s CH_2
1396	1390	δ_s CH_2
1347	1351	δ_s CH_2
1279 1202	1270	ν_s NO_2 + ν N-N
1145 1087		ν N-N + ν ring
964 946	1039 945	Ring stretch
830 761		δ and γ (NO_2)
625 600	604	τ + γ (NO_2)

Chasan & Norwitz, 1971, Mattos, et al., 2004b.

Figure 9. FT-IR spectra of transmission: a) Coverage RDX / Viton 03/91 (KBr disc); B) RDX (KBr disc); C) residue obtained after treatment with hot methanol of Coverage RDX / Viton; D) Viton B

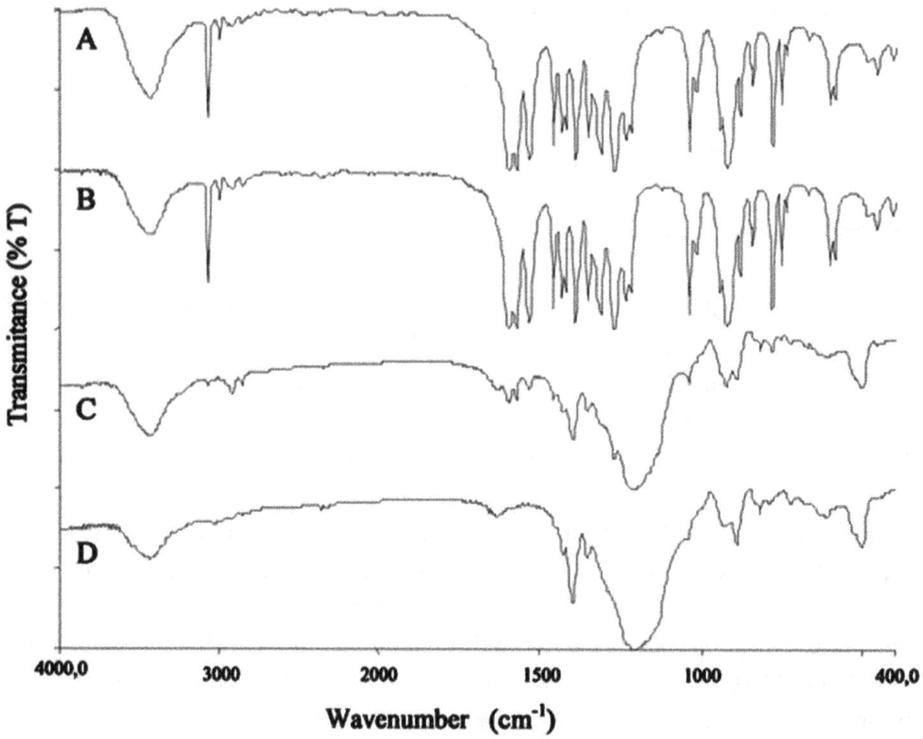

FT-IR Qualitative Analysis of RDX/Estane by Transmission

The analysis of the samples by transmission of RDX/Estane demonstrates that this technique can be applied to this system. It is possible to observe, Figure 10, in the spectrum A relating to the polymer coating, shows basically the absorption bands for RDX, however, it can be observed in this spectrum a weak absorption at 1730 cm^{-1}, characteristic of $\nu C = O$, suggesting the presence other material probably the polymer covering the energetic material (D spectrum). This result is in accordance with the characteristics of transmission technique, which shows that at higher concentration, in this case, the RDX (Mattos, Dutra, Diniz, & Iha, 2004).

After appropriate treatment with carbon tetrachloride, the residue obtained was analyzed in the form of cast film (C spectrum), and was then possible to compare this C spectrum (residue) to the D spectrum (Estane reference spectrum) by identifying presence of a polymer that coats the explosive. The bands that show the presence of the polymer were marked C and D spectra with asterisks (Mattos, Dutra, Diniz, & Iha, 2004).

The presence of urethane structure is detected in the IR medium (Mattos, Dutra, Diniz, & Iha, 2004, Urbański, Andrzej, & Cameron, 1977) based on the absorptions at 3345 cm^{-1} (ν NH), 1732cm^{-1} (ν CO -amida I) and 1534 cm^{-1} (δ NH group - amide II). The polyester bands in polyurethanes can be detected by the presence of absorption at 1223 cm^{-1} (ν group –COOC), and 1143 cm^{-1}(ν COC) to the polyether-urethane.

Figure 10. Spectra FT-IR transmission: A) Covering 4/91; B) RDX C) residue obtained after treatment with carbon tetrachloride of Covering 4/91; D) Estane 5702-F

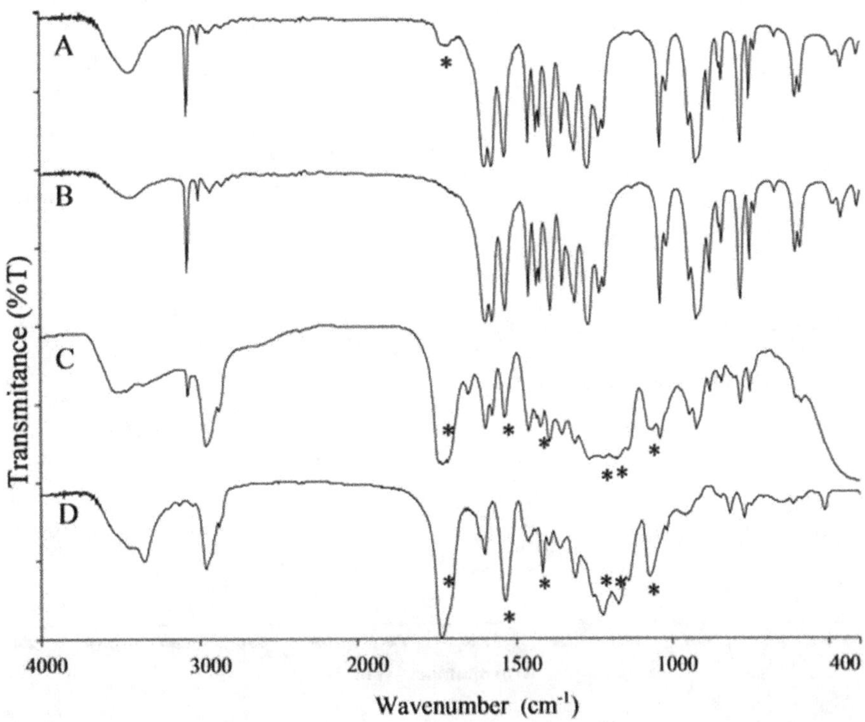

FT-IR Analysis Qualitative by PAS of Systems RDX/Viton and RDX/Estane

The PAS signal is generated by converting the absorbed radiation into heat, which leads to temperature changes in the sample surface and resulting in pressure changes in the surrounding gas, generating a sound wave that can be detected by a sensitive microphone. The sampling depth is related to the modulation frequency of the radiation and the signal amplitude is directly related to the intensity of radiation absorption. PAS allows varying the frequency modulation different speeds and thereby obtaining spectra different layers of the sample. Slower speeds, in this case, reach an underlayer. Since the use of higher speeds, allows analysis of a surface layer (Zhang, Lowe & Smith, 2009).

This technique was applied to systems of RDX/Viton and RDX/Estane, where may be possible to check the presence of the polymer that is in a more superficial layer is more easily identified when using slow scan speed (0.05 cm/s).

Development of Quantitative Methodology by FT-IR/ATR of the System HMX/Viton

Several tests performed with PBX samples in different accessories showed that the technique ATR using germanium crystal with sample on both sides of the crystal at 45° presented quality spectra and more detailed so that is possible to view changes in the characteristics bands in the polymer (ν C-F, ν $C\text{-}F_2$ and ν $C\text{-}F_3$) according to the increase of the same in the explosive mixture and a more defined baseline

Figure 11. FT-IR/ATR spectra using germanium crystal: Viton B and Cob. 02/06-85/15%

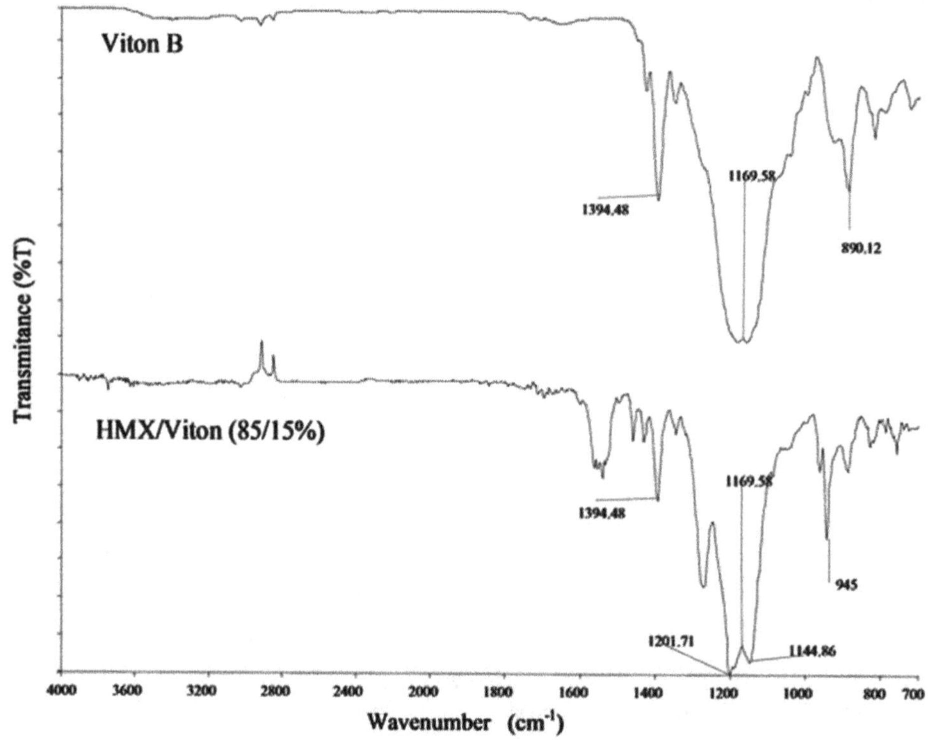

As observed in the spectra of Figure 11, the spectrum of the HMX/Viton (85/15%) sample presents characteristic bands of Viton in 1169 cm^{-1} and 890 cm^{-1}. In this way, the best analytical band for the Viton is chosen (Mattos, Nakamura, Diniz, & Dutra, 2009).

All the covering samples (HMX/Viton) were analyzed by ATR with Germanium crystal at 45°. Therefore, with this technique better results were obtained for indicating the analytical band and quantitative determination, as represented in Figure 12 (Mattos, Nakamura, Diniz, & Dutra, 2009).

According to the set of absorption spectra FT-IR/ATR in Figure 12, characteristic bands for the HMX and Viton can be observed. The band at approximately 1200 cm^{-1} is the characteristic groupings C-F existing in the Viton. It can be observed that while the Viton concentration increases in the mixture, there is a widening in this region, indicating the presence of polymer in the energetic material composition of HMX/Viton. The band at 945 cm^{-1} characteristic for the HMX is also evidenced in this set of spectra. And while the Viton content increases, it is observed the reduction in the intensity of HMX band due the presence of the polymer (Mattos, Nakamura, Diniz, & Dutra, 2009).

Other characteristic Viton bands were chosen at 890 cm^{-1} and 1170 cm^{-1}. For the HMX, the bands in 945 cm^{-1} and 1145 cm^{-1} can be observed. Since there are Viton bands that increase the intensity with the polymer concentration in the system, it was possible to initiate the quantification of these samples for FT-IR. TG was used as reference for IR methodology in this system, (Table 1). Next, analytical IR bands for determining the Viton percentage in the HMX/Viton system were analyzed. The base line chosen for the bands at 945 and 890 cm^{-1} was: 986 at 853 cm^{-1}, and for the bands of 1145 and 1170 cm^{-1}: 1476 at 984 cm^{-1} (Mattos, Nakamura, Diniz, & Dutra, 2009).

Figure 12. FT-IR/ATR spectra for HMX/Viton using germanium crystal with sample on both sides of the crystal at 45°

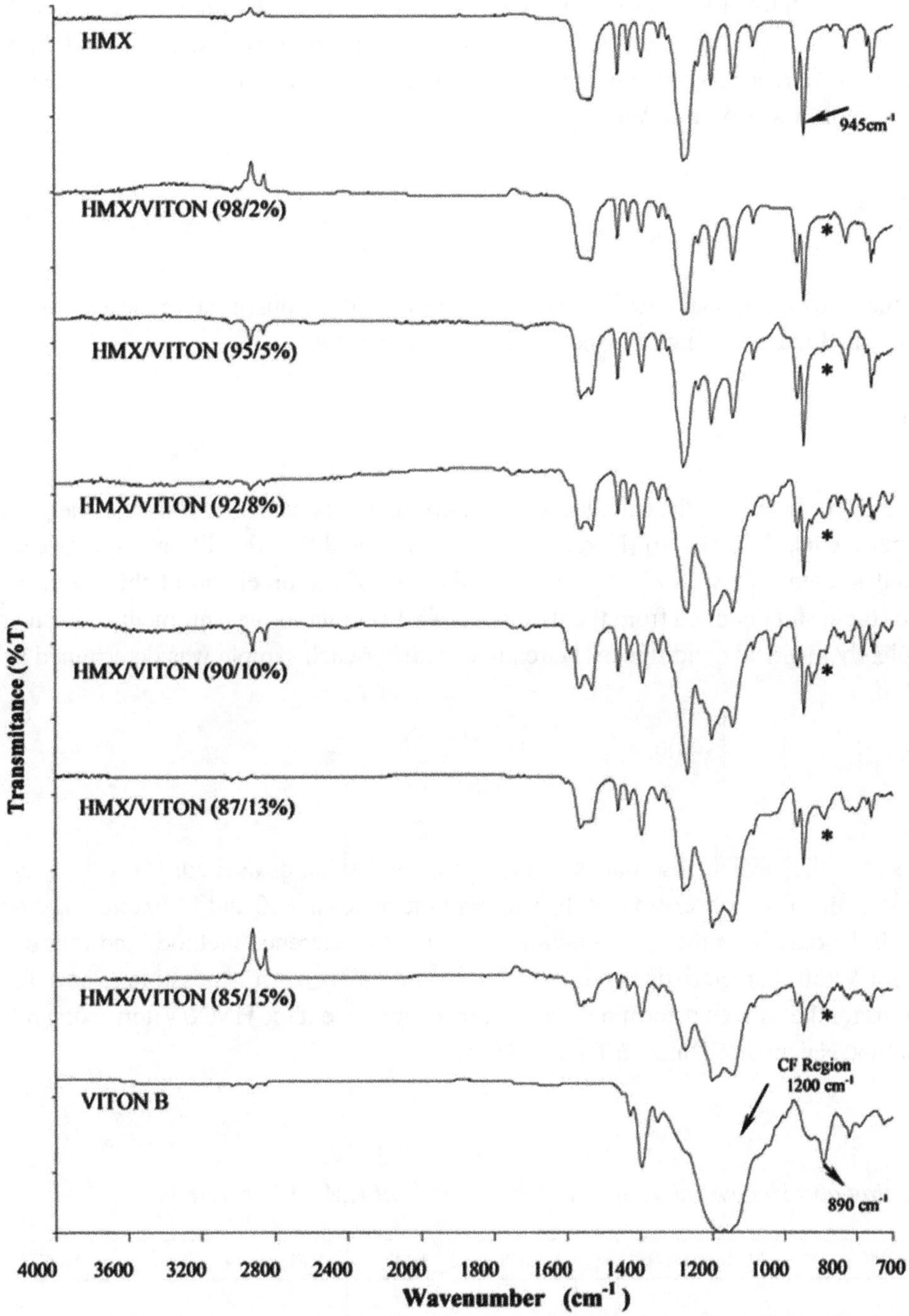

The absorbance values represent the median (m) of five analysis. According to Hórak & Vitek (1978), it is recommended to work with median numbers when working with a low number of experimental values. It may happen that the values of parameters $\hat{\mu}$ and $\hat{\sigma}$ thus determined are subject to larger errors. These errors are difficult to determine due to a non-uniform distribution of random errors in the set. Therefore, a difference assessment is made. The standard deviation, $\hat{\sigma}$ μ of the median absorbance is calculated as follows (Hórak & Vitek, 1978) equation:

$$\hat{\sigma}_{\mu} = \hat{\sigma} \Big/ \sqrt{n} \quad (1)$$

where $\hat{\sigma}$ is the assessed standard deviation of the basic set and is a quantitative measure of the precision for each individual measurement; n is the number of experiments.

$$\hat{\sigma} = K_R \cdot R \quad (2)$$

where K_R is the coefficient for the calculation of the average standard deviation from the variation range (for five experiments, $K_R = 0.430$) (Hórak & Vitek, 1978); and R is the difference between the largest and the smallest element ($X_n - X_1$). $\hat{\sigma}$ μ is an evaluation of the precision of this median, that is, an evaluation of the result obtained from the treatment of a finite set of measurements, which are repeated under completely identical conditions. The relative error for each sample was determined as follows:

$$relative\ error(\%) = \left(\hat{\sigma}_{\mu} \Big/ \mu \right) \times 100 \quad (3)$$

Table 6 shows the median absorbance values for analytical bands availed.

Analyzing Table 6, it is observed that the values of the bands in 890 and 1170 cm^{-1} must obey a linear relation. With the data from the termogravimetric analyses (reference method) and relative to the FT-IR analysis for Viton characteristic bands in 890 cm^{-1} and 1170 cm^{-1} it is possible to have the following linear relations for the content determination of Viton in agreement the HMX/Viton, using relative bands method (Mattos, Nakamura, Diniz, & Dutra, 2009).

Table 6. Median absorbance values for analytical bands of HMX/Viton system

HMX/Viton (%)	Median A_{945} HMX	Median A_{1145} HMX	Median A_{890} Viton	Median A_{1170} Viton
98/2	0.060	0.032	0.002	0.008
95/5	0.041	0.032	0.005	0.017
92/8	0.027	0.042	0.006	0.035
90/10	0.034	0.040	0.007	0.032
87/13	0.030	0.049	0.010	0.044
85/15	0.031	0.064	0.014	0.059

Viton Content by TG/FT-IR Using Relative Band

Viton content by TG/FT-IR using relative band it is known that the relative bands method (relative absorbance) can be used for thickness correction of the specimens (Gedeon, & Ngyuen, 1985). Thus, for quantitative FT-IR analysis in HMX/Viton samples, the absorbance values at 945 cm^{-1} (HMX) were related to those corresponding to the absorption at 890 cm^{-1} (Viton), in an attempt to evaluate whether there would be an improvement in the precision of the methodology developed. With regards to the points of the base line at 986 cm^{-1}, the 853 cm^{-1} was established for the calculations of absorbance values for the vibration absorption of the ring 945 cm^{-1} and C-F band at 890 cm^{-1}. To apply the Lambert-Beer law in this analysis, the relative absorbance data of A_{945}/A_{890} was placed as a function of explosive/plastic composition (percentage in weight), as presented in Table 7(Mattos, Nakamura, Diniz, & Dutra, 2009).

Figure 13 shows the relative graph of the medium values of A_{945}/A_{890} as a function of the relation of HMX/Viton concentration (%w/w); a good linear relation is obtained (R=0.993). From the analytical curve obtained by FT-IR/TG analysis (Table 7), the following relation is proposed;

$$y = -3.95 + 0.7x$$

where y is the median value of the relative absorbance (A_{945}/A_{890}) as a function of the relative concentration of HMX/Viton.

Another possibility is to use bands relative to A_{1145} / A_{1170}, Figure 14 shows the relative graph or the medium values of A_{1145} / A_{1170} as a function of the relation of HMX/Viton concentration (%w/w); a good linear relation is obtained (R=0.991). From the analytical curve obtained by FT-IR/TG analysis (Table 8), the following relation is proposed;

$$y = 0.344 + 0.0789x$$

where y is the median value of the relative absorbance (A_{1145}/A_{1170}) as a function of the relative concentration of HMX/Viton (Mattos, Nakamura, Diniz, & Dutra, 2009).

With the analytical curves obtained (IR/TG) for the different IR bands used (relative band) the concentrations of Viton were calculated. The values of the linear correlation coefficients (R) were evaluated and are presented in Table 9 (Mattos, Nakamura, Diniz, & Dutra, 2009).

Table 7. FT-IR data (relative band A_{945}/A_{890}) PBX samples of (HMX/Viton)-reference: data TG

HMX Content/Viton Content by TG	Relative Bad A_{945}/A_{890} (Median)	Standard Deviation (σ_μ)	Relative Error (%)
45.95	28.50	6.50	22.81
20.01	9.60	1.11	11.56
14.72	4.28	0.15	3.5
10.24	4.62	0.16	3.46
8.88	3.00	0.12	4.00
8.98	2.21	0.07	3.17

Figure 13. Medians values of the relative absorbance A_{945}/A_{890} as a function of HMX/Viton content obtained by TG

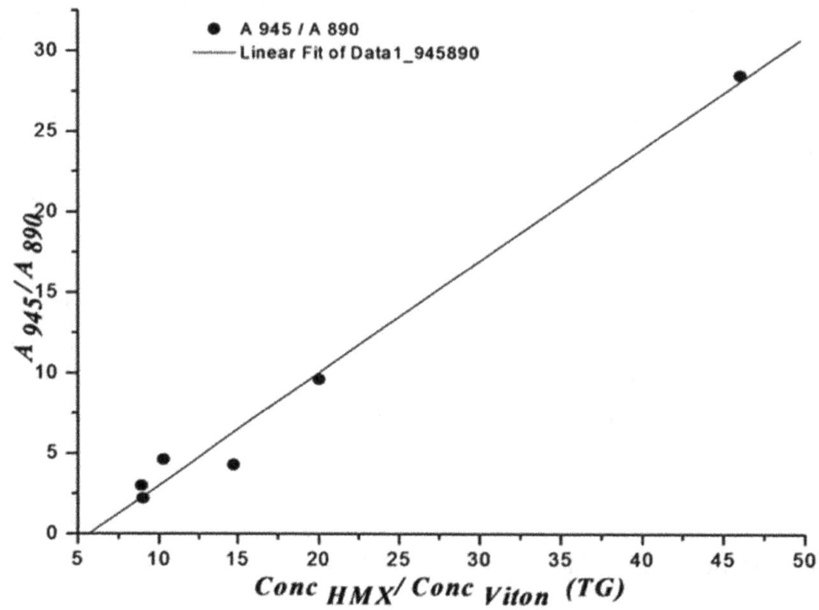

Figure 14. Medians values of the relative absorbance A_{1145} / A_{1170} as a function of HMX/Viton content obtained by TG

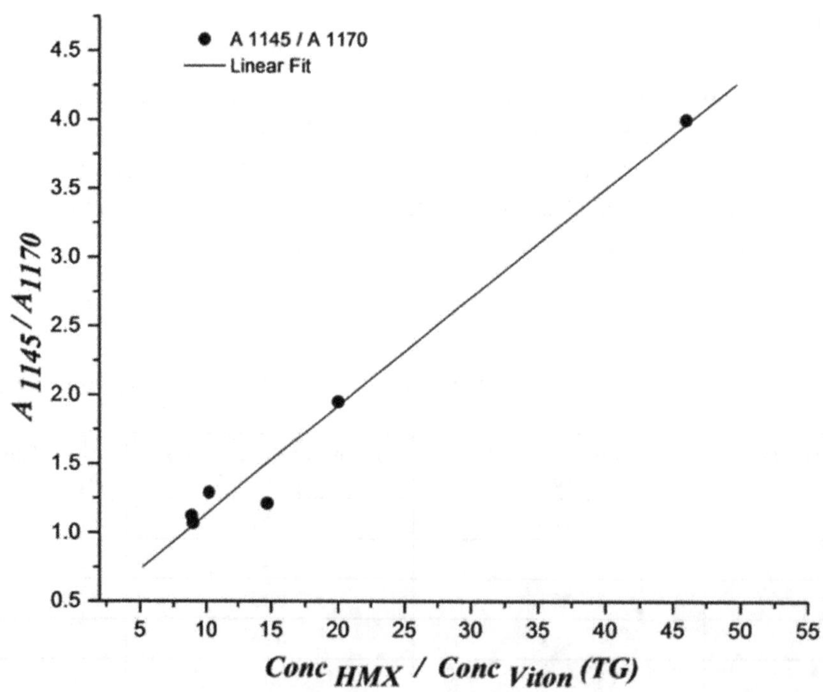

Table 8. FT-IR data (relative band A_{1145}/A_{1170}) of the PBX samples (HMX/Viton)-reference: data TG

HMX Content/Viton Content by TG	Relative Band A_{1145}/A_{1170} (Median)	Standard Deviation (σ_{μ})	Relative Error (%)
45.95	4.000	0.500	12.5
20.01	1.947	0.036	1.85
14.72	1.206	0.005	0.41
10.24	1.289	0.016	1.24
8.88	1.120	0.007	0.62
8.98	1.071	0.005	0.47

Table 9. Viton contents by TG and FT-IR technique (relative bands at 1145/ 1170 cm^{-1} and 945/890 cm^{-1})

Viton Content by TG Analysis (%)	Viton Content IR Relative Band 1145/1170 cm^{-1} TG Reference (%)	Viton Content IR Relative Band 945/890 cm^{-1} TG Reference (%)
2.13	2.11	2.10
4.76	4.68	4.90
6.33	8.36	7.82
8.90	7.71	7.53
10.12	9.24	9.13
10.02	9.82	10.18
Linear Correlation	R = 0.991	R = 0.993

The data in Table 9 showed that the use of the relative band was the best adjusted for Viton content determination by FT-IR for PBX (HMX/Viton), presenting excellent results, that is, better coefficients of linear correlation in relation to the obtained values of Table 9. The best relative band was of 945/890 cm^{-1} using the TG technique as a reference technique, with a linear correlation of 0.993, and furthermore it's possible to better visualize the bands in the spectra in Figure 10 (Mattos, Nakamura, Diniz, & Dutra, 2009).

ANALYSIS OF HMX/EVA BY UATR TECHNIQUE

Samples of HMX/EVA were analyzed by UATR (ZnSe crystal covered with diamond) and showed excellent baseline, operation practicality and the result for calculating the EVA content of this technique is the most suitable.

The UATR spectrum (Figure 15) reveals the details in relation to the polymer, the coverage of HMX with a high content EVA (15%), but are even previewed the bands of HMX, since there are parts of the explosive that are not completely covered with polymer.

Can be observed that in the set of spectra of Figure 16 that with the increase in the EVA content the system, the analytical band at 1737 cm^{-1} (C = O), characteristic of EVA sample also becomes more intense evidencing the quantitative merit of this technique.

Figure 15. Spectra UATR sample of HMX / EVA - 85/15% and pure EVA

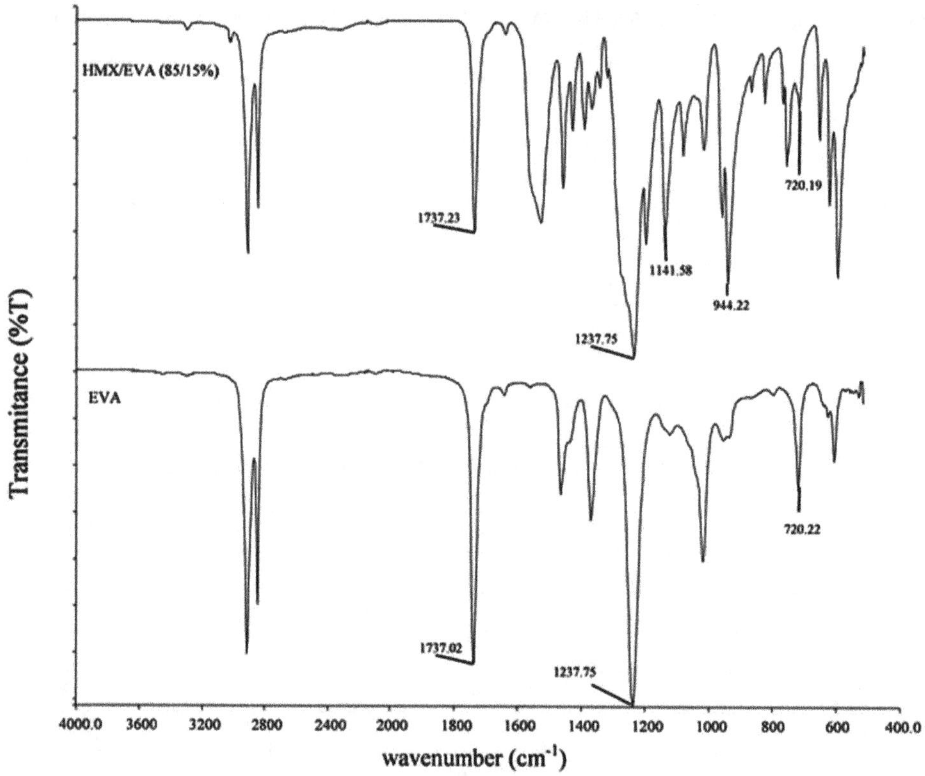

The application of torque in UATR technique may be considered a differential because can be controlled the same intensity (100 N) in all samples. The analysis conditions were: Region 4000-515 cm^{-1}, 10% energy, 40 scans, and wherein the "background" was obtained in the ZnSe crystal and samples were placed directly on the accessory Crystal.

In this system it was possible to observe the following characteristics bands for the EVA:

(ν C = O) 1737 cm^{-1} (ν CO) in 1238 cm^{-1} and (ρ CH_2) at 720 cm^{-1} and the HMX (ν N-N) in 1141 cm^{-1} and (ring vibration) at 943 cm^{-1}.

The analytical band used for determining the percentage of EVA was in 1737 cm^{-1} and the baseline chosen at1780 to 1630 cm^{-1}.

Using the data obtained from the chromatographic analyzes (reference method) and those associated to FT-IR analysis to EVA characteristic band at 1737 cm^{-1} it was possible to have the following linear relationship to determine the EVA content of HMX/EVA system as shown in Table 10.

Figure 17 shows the relative graph of the median values of A_{1737} as a function of EVA concentration (%w); a good linear relation is obtained (R=0.989). From the analytical curve obtained by FT-IR/HPLC analysis (Table 10), the following relation is proposed

$$y = 0.0391 + 0.0136x$$

Figure 16. UATR Spectra of the samples HMX/EVA system

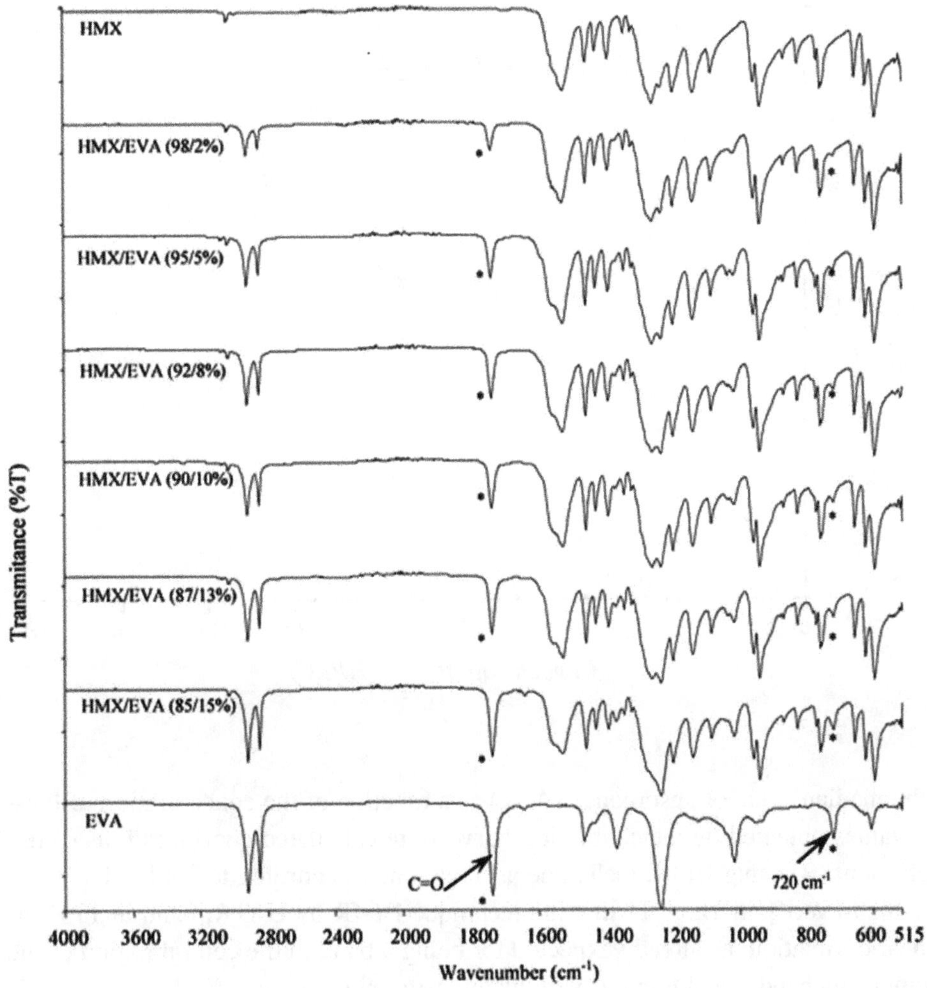

Table 10. FT-IR data (band at 1737 cm^{-1}) of the samples PBX (HMX / EVA) - Reference: HPLC data

EVA Content by HPLC (%)	A_{1737} (Median)	Standard Deviation (σ_{μ})	Relative Error (%)
2.14	0.061	0.004	6.58
6.47	0.134	0.010	7.41
7.37	0.153	0.010	6.75
9.46	0.159	0.014	8.58
12.56	0.203	0.013	6.49
15.25	0.250	0.016	2.89

Figure 17. Medians of values the absorbance A_{1737} as a function EVA content obtained by HPLC

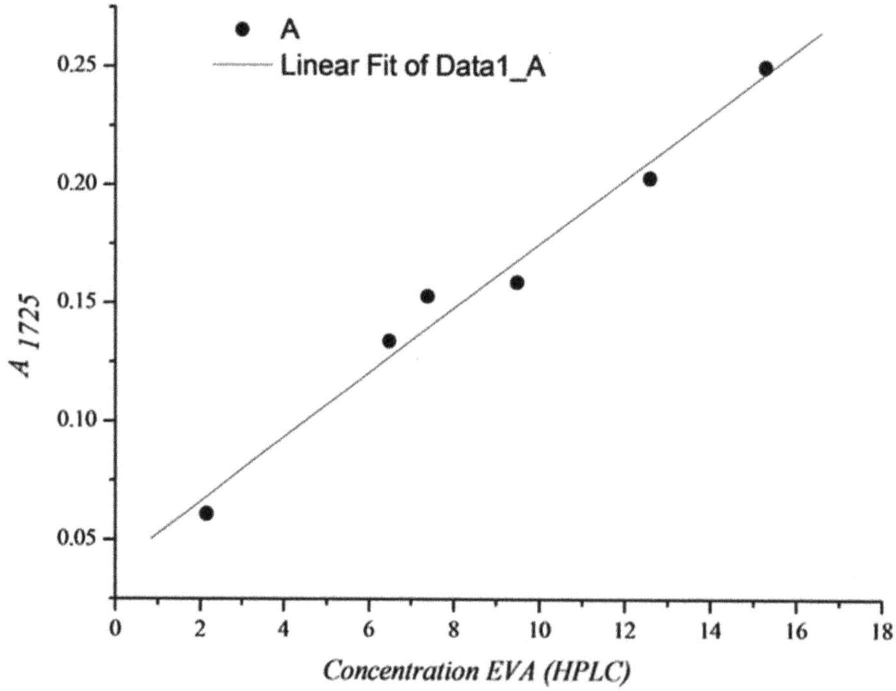

where y is the median value of absorbance (A_{1737}) as a function of the concentration of EVA

With the values obtained by the calibration curve were calculated concentrations of EVA and correlation coefficient (R) using HPLC technique as a reference, according to Table 11.

It is possible to verify in Table 11 that the technique FT-IR by UATR, band in 1737 cm^{-1}, showed excellent linear correlation, as already expected for being a band in the central region of the spectrum, without neighboring bands, enabling measure its intensity more precisely.

The same investigation was performed with the ATR technique using KRS-5 crystal to 45°, but the values obtained with UATR technique showed excellent linear correlation showing to be most suitable for quantitative determination for HMX/EVA system using HPLC as a reference technique.

Table 11. EVA contents obtained by means of HPLC techniques and FT-IR (bands at 1737 cm^{-1})

EVA Content By HPLC Analysis (%)	EVA Content Band 1737 cm^{-1} Reference HPLC (%)
2.14	1.59
6.47	6.97
7.37	8.37
9.46	8.82
12.56	12.02
15.25	15.46
Linear Correlation	**R = 0.989**

CONCLUSION

The FT-IR transmission analysis enables the characterization of explosives employed in producing the PBX samples, indicating its main bands. Using these results, it is possible follow up the research. In employing FT-IR techniques by transmission for characterization of Viton in the coverage of energetic material (RDX), it was found that it is not suitable to identify the presence of the polymer, unless performing treating the sample with a suitable solvent, for obtaining a polymeric film which can be characterized by transmission technique, a casting film. In those circumstances it is possible to verify the presence of Viton in the PBX.

The same procedure applied to the characterization of RDX/Estane system shows basically the presence of RDX and only suggests the presence of polymeric compound containing C = O, for the presence of band at 1730cm^{-1}. For visualization of polymer bands coating, it is also necessary to treat the material with suitable solvent, to obtain a polymeric film, and analyze by transmission technique, a casting film.

PAS is an alternative technique, since the FT-IR transmission technique does not identify the polymer in PBX without treatment. By the technical PAS characteristics, it is possible to identify the polymer used in the coating process and suggest that this polymer is a superficial layer of few microns (RDX/Estane).

For development quantitative methodology FT-IR, the use of auxiliary techniques such as HPLC and TG is required. In this chapter, these techniques were employed for systems of HMX/Viton and HMX/EVA to be used as a reference for quantitative analysis FT-IR, TG technique for determining the Viton content and HPLC to determine the content EVA.

For HMX/Viton system, the ATR technique (Germanium crystal) with sample on both sides of the crystal at 45° is the most suitable and present quality spectra. It is possible to view the changes in the characteristic bands of the polymer according to the increase of this content in the explosive mixture, and also to indicate the analytical band for quantitative determination, which allows for the quantification PBX (HMX/Viton) by FT-IR. The use of relative band for this system is more suitable for determination of Viton content by FT-IR in PBX (HMX/Viton), and the best on the relative band is 945/890 cm^{-1} using TG technique as the technique of reference.

The FT-IR analysis by UATR technique of the PBX samples (HMX/EVA system), exhibit good practicality operation and excellent results for preparation of the quantitative methodology of EVA content by spectroscopy. In this system, the band 1737 cm^{-1} (C = O), was chosen as the analytical band, which is most suitable for HMX/EVA system for quantitative determination by FT-IR.

The torque application in UATR technique can be considered an advantage, since it can be controlled with the same intensity in all aliquots, and thus avoid absorption intensity variations.

Facing the results attained with samples of HMX/Viton and HMX/EVA systems, it is concluded that the proposed methodology of quantification of polymers using FT-IR/ATR and

FT-IR/UATR for quantification of polymers PBX displays excellent results, being the faster technique and does not generate chemical waste.

REFERENCES

Abidi, N., & Hequet, E. (2005). Fourier Transform Infrared Analysis of Trehalulose and Sticky Cotton Yarn Defects Using ZnSe–Diamond Universal Attenuated Total Reflectance. *Textile Research Journal*, *9*(75), 645–652. doi:10.1177/0040517505057527

Achuthan, C. P., & Jose, C. L. (1990). Studies on Octahydro-1,3,5,7-Tetranitro-1,3,5,7-Tetrazocine (HMX) polymorphism. *Propellants, Explosives, Pyrotechnics*, *5*(6), 271–275. doi:10.1002/prep.19900150609

Agrawal, J. P. (2005). Some new high energy materials and their formulations for specialized applications. *Propellants, Explosives, Pyrotechnics*, *30*(5), 316–328. doi:10.1002/prep.200500021

Akhavan, J. (2004). *The Chemistry of Explosives. 2*. Cambridge, United Kingdom: The Royal Society of Chemistry.

Allen, N. S., Edge, M., Rodriguez, M., Liauw, C. M., & Fontan, E. (2001). Aspects of the thermal oxidation, yellowing and stabilization of ethylene vinyl acetate copolymer. *Polymer Degradation & Stability*, *71*, 1–14. doi:10.1016/S0141-3910(00)00111-7

Bachmann, W., & Sheehan, J. C. (1949). A new method of preparing the hing explosive RDX. *Journal of the American Chemical Society*, *71*, 1842–1845.

Benziger, T. M. (1973). *High-Energy Plastic-Bonded Explosive*. U.S. Patent US3,778,319.

Bunte, G., Pontius, H., & Kaiser, M. (1999). Analytical characterization of impurities or byproducts in new energetic materials. *Propellants, Explosives, Pyrotechnics*, *24*(3), 149–155. doi:10.1002/(SICI)1521-4087(199906)24:03<149::AID-PREP149>3.0.CO;2-4

Burgess, C. E., Woodyard, J. D., Rainwater, K. A., Lightfoot, J. M., & Richardson, B. R. (1998). Literature review of the lifetime of DOE Materials: aging of plastic bonded explosive and the explosive and polymers contained therein (Technical Report ANRCP-1998-12). Austin: Amarilo National Resource Center for Plutonium, The University of Texas.

Burnham, A. K., & Weese, R. K. (2005). Kinetics of thermal degradation of explosive binders Viton A, Estane, and Kel-F. *Thermochimica Acta*, *426*(1-2), 85–92. doi:10.1016/j.tca.2004.07.009

Calzia, J. (1969). *Les substances explosives et leurs nuisances*. Paris: Dunod.

Campbell, M. S., Garcia, D., & Idar, D. (2000). Effects of temperature and pressure on the glass transitions of plastic bonded explosives. *Thermochimica Acta*, *357-358*, 89–95. doi:10.1016/S0040-6031(00)00372-5

Chasan, D. E., & Norwitz, G. (1971). Qualitative analysis of primes, tracers, igniters, incendiaries, boosters, and delay compositions on a micro scale by use of Infrared Spectroscopy (Test report T71-6-1). Philadelphia: Department of the Army.

Chen, P., Huang, F., Dai, K., & Ding, Y. (2005). Detection and characterization of long-pulse low-velocity impact damage in plastic bonded explosives. *International Journal of Impact Engineering*, *1*(5), 497–508. doi:10.1016/j.ijimpeng.2004.01.008

Ciola, R. (2003). *Fundamentos da cromatografia a líquido de alto desempenho HPLC*. São Paulo: Edgard Blücher.

DUTRA. R. C. L. (1997). *Modificação de fibras de polipropileno com EVA funcionalizado*. Unpublished doctoral dissertation, Universidade Federal do Rio de Janeiro, Rio de Janeiro.

Federoff, B. F., & Sheffield, O. E. (1966). *Encyclopedia of explosives and related itens* (Vol. 8). Dover: Picantinny Arsenal. doi:10.21236/AD0653029

Felix, S. P., Singh, G., Sikder, A. K., & Aggrawal, J. P. (2005). Studies on energetic compounds part 33: Thermolysis of keto-RDX and its plastic bonded explosives containing thermally stable polymers. *Thermochimica Acta*, *426*(1-2), 53–60. doi:10.1016/j.tca.2004.06.020

Gedeon, B. J., & Ngyuen, R. H. (1985). Computerization of ASTM D 3677 – Rubber identification by infrared Spectrophotometry.

Gjersøe, R. (2000). Detection of α-HMX in β-HMX with FTIR diffuse reflection technique. Energetic materials: analysis, diagnostics and testing. In *Proceedings of the International Annual Conference of ICT* (pp. 96-1 – 96-6).

Graf, R. T., Koenig, J. L., & Ishida, H. (1987). Introduction to optics and infrared spectroscopic techniques. *Polymer of Science Technology*, *36*, 1–31.

Groom, C. A., Halasz, A., Paquet, L., DCruz, P., & Hawari, J. (2003). Cyclodextrin-assisted capillary electrophoresis for determination of the cyclic nitramine explosives RDX, HMX and CL-20 Comparison with high-performance liquid chromatography. *Journal of Chromatography*, *999*(1-2), 17–22. doi:10.1016/S0021-9673(03)00389-3 PMID:12885047

Hayden, D. J. (2005). *An analytic tool to investigate the effect of binder on the sensitivity of HMX-based plastic bonded explosives in the skid test* [Master's thesis of Science Department of Mechanical Engineering]. Institute of Mining and Technology, Socorro, New Mexico.

Hoffman, D. M., & Caley, L. E. (1981). Dynamic mechanical and molecular weight measurements on polymer bonded explosives from thermally accelerated aging tests I. Fluoropolymer binders. *Journal Organic Coatings and Plastics Chemistry*, *44*, 680–685.

Hoffman, D. M., Matthews, F. M., & Pruneda, C. O. (1989). Dynamic mechanical and thermal analysis of crystallinity development in Kel-F 800 and TATB/Kel-F 800 plastic bonded explosives: Part I. Kel-F 800. *Thermochimica Acta*, *156*(2), 365–372. doi:10.1016/0040-6031(89)87203-X

Hohmann, C., & Tipton, B., Jr. (2000). Viton's impact on NASA standard initiador propellants properties (Tech. Report, NASA/TP 210187). Houston: NASA.

Hórak, M., & Vitek, A. (1978). *Interpretation and processing of vibrational spectra*. New York: John Wiley & Sons.

Humphrey, J. R. (1974). LX-10-1: a high-energy plastic-bonded explosive (Technical Report UCRL-51629). Livermore: Lawrence Livermore Lab, California University.

Imamura, Y. Y. (2002). *Avaliação do defeito da granulometria sobre a transição cristalina de β-HMX por calorimetria exploratória diferencial e microscopia eletrônica de varredura* [Master's thesis, Physics and Chemistry of Aerospace Materials]. Instituto Tecnológico de Aeronáutica, São José dos Campos, SP.

Ishida, H. (1987). Quantitative surface FTIR spectroscopy analysis of polymers. *Rubber Chemistry and Technology*, *60*(3), 497–554. doi:10.5254/1.3536139

Kaiser, M. (1998). HPLC Investigation of explosives and nitroaromatic compounds with a cyano-propyl phase. *Propellants, Explosives, Pyrotechnics*, *23*(6), 309–312. doi:10.1002/(SICI)1521-4087(199812)23:6<309::AID-PREP309>3.0.CO;2-L

Kasprzyk, D. J., Bell, D. A., Flesner, R. L., & Larson, S. A. (1999). Characterization of a slurry process used to make a plastic-bonded explosive. *Propellants, Explosives, Pyrotechnics*, *24*(6), 333–338. doi:10.1002/(SICI)1521-4087(199912)24:6<333::AID-PREP333>3.0.CO;2-T

Kim, H.-S., & Park, B.-S. (1999). Characteristics of the insensitive pressed plastic bonded explosive DXD-59. *Propellants, Explosives, Pyrotechnics*, *24*(4), 217–220. doi:10.1002/(SICI)1521-4087(199908)24:4<217::AID-PREP217>3.0.CO;2-A

Kohno, Y., Maekawa, K., Tsuchioka, T., Hashizume, T., & Imamura, A. (1994). A relationship between the impact sensitivity and the electronic structures for the unique N-N bond in the HMX polymorphs. *Combustion and Flame*, *96*(4), 343–350. doi:10.1016/0010-2180(94)90103-1

Kotova, I. P., Ivanova, M. P., Berends, L. K., & Chechetkina, L. N. (1986). Identification of textile materials in rubbers and finished products. *International Polymer Science and Technology*, *8*(13), 66–70.

Kwan, K. S. (1998). *The role of penetrant structure in the transport and mechanical properties of a thermoset adhesive*. Unpublished doctoral dissertation, Faculty of the Virginia Polytechnic Institute, Blacksburg.

Lee, J.-H., Hsu, C.-K., & Chang, C.-L. (2002). A study on the thermal decomposition behaviors of PETN, RDX, HNS and HMX. *Thermochimica Acta*, *392-393*, 173–176. doi:10.1016/S0040-6031(02)00099-0

Lee, J.-S., & Hsu, C.-K. (2002). Thermal properties and shelf life of HMX–HTPB based plastic-bonded explosives. *Thermochimica Acta*, *392-393*, 152–156. doi:10.1016/S0040-6031(02)00095-3

Litch, H. H. (1970). HMX (octogen) and its polymorphic forms. In *Proceedings of the Symposium on Chemistry Problems With The Stability of Explosives* (*Vol. 2*, pp.168 - 179).

Lukasavage, W. J., Nicolich, S., & Alster, J. (1993). *Process of making impact Alfa-HMX*. U.S. Patents, US5,268,469.

Marple, R. L., & Lacourse, W. R. (2005). A platform for on-site environmental analysis of explosives using high performance liquid chromatography with UV absorbance and photo-assisted electrochemical detection. *Talanta*, *66*(3), 581–590. doi:10.1016/j.talanta.2004.11.034 PMID:18970024

Mathieu, J., & Stucki, H. (2004). Military high explosives. *Chimia, 58*(6), 383–389. doi:10.2533/000942904777677669

Mattos, E. C. (2001). *Síntese de HMX e avaliação da aplicabilidade de técnicas FT-IR na sua caracterização e quantificação* [Master's thesis, Plasma Physics]. Instituto Tecnológico de Aeronáutica, São José dos Campos, SP.

Mattos, E. C., Dutra, R.C.L., Diniz, M.F., & Iha, K. (2004). Avaliação do Uso de Técnicas FT-IR para Caracterização de Cobertura Polimérica de Material Energético. *Polímeros: Ciência E Tecnologia,* 14(2), 63-68.

Mattos, E. C., Dutra, R. C. L., Diniz, M. F., Silva, G., Iha, K., & Teipel, U. (2008). Characterization of Polymer Coated RDX and HMX Particles. *Propellants, Explosives, Pyrotechnics, 33*(1), 44–50. doi:10.1002/prep.200800207

Mattos, E. C., Moreira, E. D., Dutra, R. C. L., Diniz, M. F., Ribeiro, A. P., & Iha, K. (2004a). Determination of the HMX and RDX content in synthesized energetic material by HPLC, FT-MIR, and FT-NIR spectroscopies. *Quimica Nova, 27*(4), 540–544. doi:10.1590/S0100-40422004000400005

Mattos, E. C., Nakamura, N. M., Diniz, M. F., & Dutra, R. C. L. (2009). Determination of polymer content in energetic materials by FT-IR. *Journal of Aerospace Technology and Management, 1*(2), 167–175. doi:10.5028/jatm.2009.0102167175

Mattos, E. C., Viganó, I., David, L. H., Diniz, M. F., Dutra, R. C. L., & Iha, K. (2004b). Application of FT-IR techniques for identification of different polymers used in the coating process of energetic materials. In *Proceedings of the International Annual Conference of ICT* (pp. 174-1 - 147-13). Karlsruhe.

Mattos, E. C., Viganó, I., Dutra, R. C. L., Diniz, M. F., & Iha, K. (2002). Aplicação de metodologias FTIR de transmissão e fotoacústica à caracterização de materiais altamente energéticos - parte II. *Quimica Nova, 25*(5), 722–728. doi:10.1590/S0100-40422002000500003

McAuley, D. C., Joubert, A. L., & Steyn, L. T. (1985). Methods for the determination of (a) RDX in HMX and (b) polymorphs of HMX in β HMX. In *Proceedings of the Symposium on Chemistry Problems With The Stability Off Explosives* (pp. 181-196).

McClelland, J. F., Jones, R. W., Luo, S., & Seaverson, L. M. (1992). *A Practical Guide to FT-IR Photoacoustic Spectroscopy. Center for Advanced Technology Development and Ames Laboratory. Iowa State University: MTEC Photoacoustics,* Catalog.

Mccord, B., & Bender, E. C. (1998). Chromatography of explosives. *Forensic Investigation of Explosives, 8*, 231–265.

Monthé, C. G., & Azevedo, A. D. (2002). *Análise térmica de materiais*. São Paulo: Ieditora.

Nogueira, D. A. R., Rosa, P. T. V., & Santana, C. C. (2000). Adsorção de proteínas na superfície de polímeros: quantificação com FTIR/ATR. In *I brasilian congresso phase equilibrium na fluid propeties for chemical process desin (1). Águas de São Pedro*. Campinas: UNICAMP.

Pan, X., Tian, K., Jones, L. E., & Cobb, G. P. (2006). Method optimization for quantitative analysis of octahydro-1,3,5,7-tetranitro-1,3,5,7-tetrazocine (HMX) by liquid chromatography-electrospray ionization mass spectrometry. *Talanta, 70*(2), 455–459. doi:10.1016/j.talanta.2006.03.005 PMID:18970792

Pandey, G. C., & Kulshreshtha, A. K. (1993). Fourier transform infrared spectroscopy as a quality control tool. *Process Control and Quality, 4*, 109–123.

Silva, G. (2007). *Síntese, caracterização e quantificação de 2,2',4,4',6,6'-hexanitroestilbeno.* Unpublished doctoral dissertation, Instituto Tecnológico de Aeronáutica, São José dos Campos, SP.

Silverstein, R. M., Bassler, G. C., & Morril, T. C. (1981). *Spectrometric identification of organic compounds. 5.* New York: John Wiley & Sons.

Singh, G., Felix, S. P., & Soni, P. (2003). Studies on energetic compounds part 28: Thermolysis of HMX and its plastic bonded explosives containing Estane. *Thermochimica Acta, 399*(1-2), 153–165. doi:10.1016/S0040-6031(02)00460-4

Singh, G., Felix, S. P., & Soni, P. (2005). Studies on energetic compounds part 31: Thermolysis and kinetics of RDX and some of its plastic bonded explosives. *Thermochimica Acta, 426*(1-2), 131–139. doi:10.1016/j.tca.2004.07.013

Skidmore, C. B., Phillips, D. S., & Crane, N. B. (1997). *Microscopical examination of plastic-bonded explosives* (Technical Report LA-UR-97-2807). Los Alamos: Los Alamos National Laboratory.

Smith, A. L. (1979). *Applied infrared spectroscopy.* New York: John Wiley & Sons.

Tang, C.-J., Lee, Y.-J., Kudva, G., & Litzinger, T. A. (1999). A study the gas-phase chemical strutcture during CO_2 laser assisted combustion of HMX. *Combustion and Flame, 117*(1-2), 170–188. doi:10.1016/S0010-2180(98)00094-7

Thompson, D. G., Olinger, B., & Deluca, R. (2005). The effect of pressing parameters on the mechanical properties of plastic bonded explosives. *Propellants, Explosives, Pyrotechnics, 30*(6), 391–396. doi:10.1002/prep.200500030

Tompa, A. S., & Boswell, R. F. (2000). Thermal stability of a plastic bonded explosive. *Thermochimica Acta, 357-358*, 169–175. doi:10.1016/S0040-6031(00)00386-5

United States. (1979). Department of the Army. *Military Explosives.* Washington, DC. (TM 9-1300-214 C2).

Urbański, J., Andrzej Skup, A., & Cameron, G. G. (1977). *Handbook of analysis of synthetic polymers and plastics.* New York: John Wiley & Sons.

Urbanski, T. (1967). *Chemistry and technology of explosives* (Vol. 3). Warszawa: PWN Polish Scientific Publishers.

Urbanski, T. (1984). *Chemistry and Technology of Explosives* (Vol. 4). Great Britain: Pergamon Press.

Valença, U. S. V. (1998). *Curso de emprego de explosivos industriais na engenharia. Unpublished apostille,* Instituto Militar de Engenharia, Rio de Janeiro.

Wanninger, P., Wild, R., Kleinschmidt, E., & Spath, H. (1996). *Pressable explosives granular product and pressed explosive charge.* U.S. Patents, US5,547,526.

Wenhui, W., Zhanning, J., & Jianwu, D. 1994. Surface characterization of the coupling interaction on HMX particles by FTIR DR spectroscopy. Energetic materials: analysis, characterization and techniques. In *Proceedings of the International Annual Conference of ICT* (pp. 85-1 – 85-12) Karlsruhe.

Wewerka, E. M., Loughran, D. E., & Williams, M. J. (1976). The effects of long term storage at elevated temperatures on small cylinders of PBX 9501 (Report, Tech. LA-6302-MS). Los Alamos: Los Alamos Scientific Laboratory.

Zattera, A. J., Bianchi, O., Zeni, M., & Ferreira, C. A. (2005). Caracterização de resíduos de copolímeros de etileno-acetato de Vinila – EVA. *Polímeros: Ciência e Tecnologia, 15*(1), 73–78.

Zhang, W. R., Lowe, C., & Smith, R. (2009). Depth profiling of coil coating using step-scan photoacoustic FTIR. *Progress in Organic Coatings, 65*(4), 469–476. doi:10.1016/j.porgcoat.2009.04.005

Zitrin, S. (1998). Forensic Investigation of Explosions. In *Analysis of Explosives by Infrared Spectrometry and Mass Spectrometry* (pp. 267–314). Taylor & Francis Ltd.

Chapter 15
An Analysis of the Initiation Process of Electroexplosive Devices

Paulo C. C. Faria
Aeronautics Institute of Technology, Brazil

ABSTRACT

Electroexplosive devices, EEDs or squibs (an electric resistance encapsulated by a primary explosive), fundamentally convert electrical energy into heat, solely to start off an explosive chemical reaction. Obviously, the EED activation shall not happen by accident or, even worse, by intentional exogenous influence. From an ordinary differential equation (ODE), which describes this device thermal behaviour for both continuous and pulsed electrical excitation, a remarkable, but certainly not intuitive, dependence of the temperature response on the heat transfer process time-constant is verified: the EED temperature profile dramatically changes as the time-constant spans a wide range of values, from much lesser than the pulse width to much greater than the pulse period. On the basis of this dependence, important recommendations, concerning the EED safety (and efficient) operation, are presented.

INTRODUCTION

In order to study the EED ignition process one needs a consistent model, one capable of reproducing coherent temperatures in response to the electrical power applied to a bridgewire. In fact, the EED only converts energy from one type to another, which means that the EED thermal model is a direct consequence of the energy conservation principle.

All the theory behind such a model already exists and is founded on Rosenthal's (Rosenthal, 1961) and Leew's (Prince & Leeuw, 1988) work. But, as the focus here is on the application of the model, not in the theory sustaining it, only the necessary level of abstraction will be retained from the original effort of those contributors (for a deeper understanding of thermal models, Hoberman's (Hoberman, 1965) and Potter (Potter & Scott, 2004) are the indicated references).

DOI: 10.4018/978-1-5225-2903-3.ch015

From this model solution, an outstanding advantage, in terms of temperature gain, will be noticed in favor of the electrical pulsed excitation. Thus the choice of the power source most appropriate to drive an specific EED is primarily affected by its thermal time-constant.

In addition, the knowledge of those many energy conversion significant aspects manifests itself as a reliable and a risk free operation of EEDs.

Before analyzing the electroexplosive as an energy converter device, one starts, for instructive purposes, with the investigation of a topic much more familiar to everyone: the rain-barrel problem (Moran & Shapiro, 2008).

A CONVENIENT ANALOGY

Consider a barrel (a cylinder with transversal area πr^2, height H_b, and top surface removed), initially empty, having a small hole in the center of its base. Assume that it is raining (uniform precipitation). Naturally, a question arises: how is the level of the water inside the barrel going to evolve as a function of time?

To facilitate the analysis of this problem, let the hole at the barrel base be modeled by a linear hydraulic resistance R (actually, $R = R_0 H^{(1-\gamma)}$, with γ adjusted to each specific case (Cochin, 1980)), and let the water flux through the hole, Q_o, be proportional to H:

$$Q_o = \frac{H}{R} = \frac{H}{R_0 H^{(1-\gamma)}} = \left.\frac{H^{\gamma}}{R_0}\right|_{\gamma=1} = \frac{H}{R_0} \tag{1}$$

On the other hand, the water volumetric flow that enters the barrel (it is raining) is constant and equal to Q_i. From the cylindrical geometry, $\frac{dV}{dt} = \pi r^2 \cdot \frac{dH}{dt}$, and from the mass-rate balance principle, $\rho.\frac{dV}{dt} = \rho.(Q_i - Q_o)$. Combining these two expressions, it results the ordinary differential equation (ODE):

$$\frac{dH}{dt} = \frac{1}{\pi r^2}\cdot\left(Q_i - Q_o\right) = \frac{1}{\pi r^2}\cdot\left(Q_i - \frac{H}{R_0}\right) \Rightarrow \frac{dH}{dt} + \frac{H}{\tau_b} = \frac{Q_i}{\pi r^2}\left(\tau_b = \pi r^2 \cdot R_0\right) \tag{2}$$

This ODE has $H = Q_i \cdot R_0 \cdot \left(1 - e^{-\frac{t}{\tau_b}}\right)$ for solution $\left(t > 0\right)$. Obviously, the necessary condition for the water to overflow is $Q_i > \frac{H_b}{R_0}$.

THE ANALYSIS OF THE EED THERMAL MODEL

By coating an electric resistance with a thick layer of high sensitivity explosive and applying a electric current to this circuit (bridgewire), it will be possible to initiate the tiny explosive mass, but what is the best way to do that, by continuous or by pulsed excitation? To answer this question, one is going to present a simplified representation of the thermal phenomena involved in the EED ignition.

Applying the energy rate balance principle to the EED, one has that (injected energy)' = (stored energy)' + (dissipated energy)' or

$$P_i = C_T \cdot \frac{d\theta}{dt} + \frac{\theta}{R_T} \Rightarrow \frac{d\theta}{dt} + \frac{\theta}{\tau_{EE}} = \frac{P_i}{C_T} \tag{3}$$

This differential equation is the mathematical equivalent of the EED thermal model given in Figure 1 (for ingenious experimental procedures to determine $\tau_{EE} = R_T \cdot C_T$, see (Rosenthal, 1961) and (Prince & Leeuw, 1988)).

Comparing (2) with (3), the conclusion is evident: the EED temperature behaves exactly as the level of the fluid accumulated inside the barrel, since the differential equations that model both phenomena, hydraulic and thermal, are analogous (they have the same structure). As a matter of fact, this similarity will be very useful in grasping the subtleties of the heat transfer process involved in the EED ignition.

One already knows, from the introduction, that

Figure 1. EED thermal model

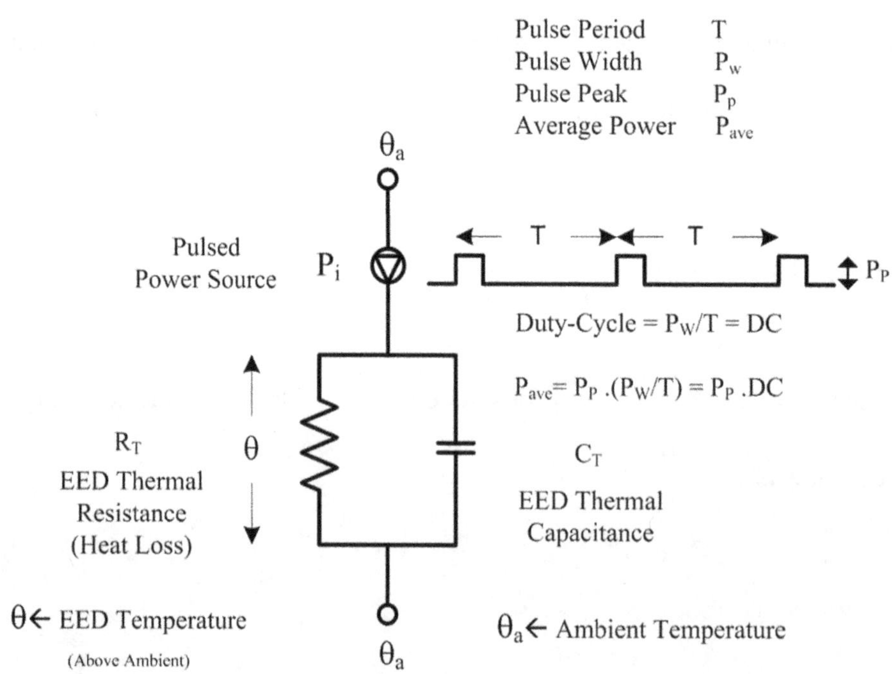

$$\theta(t) = P_i . R_T . \left(1 - e^{-\frac{t}{\tau_{EE}}}\right) \tag{4}$$

is the solution of (3), when P_i is constant. From (4), with $t \to \infty$, the maximum temperature (continuous input) is given by

$$\theta_{MAXCONT} = P_i \cdot R_T = P_{ave} \cdot R_T \tag{5}$$

Therefore, for fixed P_i, the EED temperature is driven by the injected power. See that the EED temperature, due to the heat dissipation to the environment, via R_T, does not increase indefinitely—it is bounded (Figure 2). Nevertheless, it suffices to make $\theta_{MAXCONT} > \theta_{THRESHOLD}$, by adjusting P_i, for ignition to occur.

What happens then when $P_i(t)$ is a train of power pulses? Despite the incontestable simplicity of the model adapted to our purposes, in this case, a certain mathematical expertise is required to solve it (for the completed solution of the EED ODE, with pulsed forcing function, using the Laplace transform method, see (Wylie, 1975)). Here, one is going to derive an approximated solution.

From the theory of differential equations (Wylie, 1975), the total temperature can be factored into transient and periodic components. However, after a large number of pulse cycles ($n >> 1$, n is the cycle counting), certainly the transient portion will have faded into insignificance (Figure 3). The problem was thus reduced to solve the following system of two ODEs with their associated boundary conditions (considering only the periodic factor):

$$\begin{cases} \dfrac{d\theta}{dt} + \dfrac{\theta}{\tau_{EE}} = \dfrac{P_P}{C_T}, with \left[0 \le t \le P_W\right] \wedge \left[\theta(0) = \theta_1, \theta(P_W) = \theta_2\right] \\ \\ \dfrac{d\theta}{dt} + \dfrac{\theta}{\tau_{EE}} = 0, with \left[P_W \le t \le T\right] \wedge \left[\theta(P_W) = \theta_2, \theta(T) = \theta_1\right] \end{cases} \tag{6}$$

Figure 2. Signal flow graph (sfg) corresponding to eq. (3)

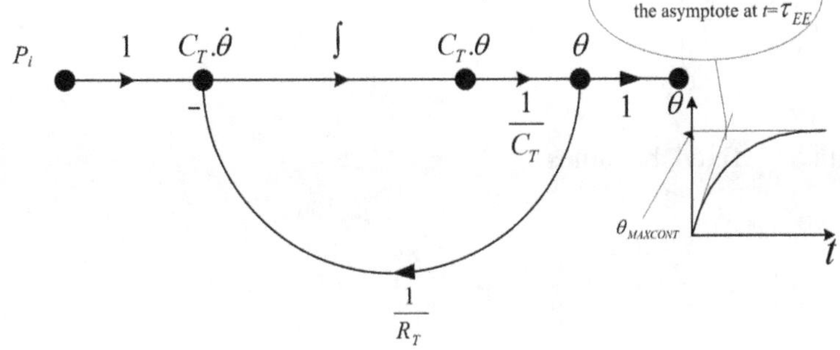

Figure 3. EED temperature after the settle-down of the transient phase (Pulsed Excitation)

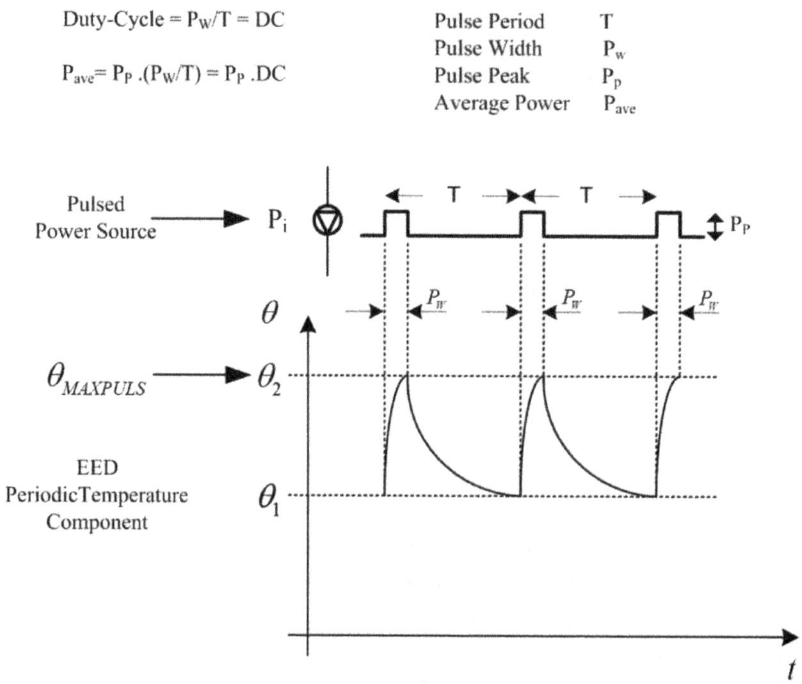

From (6), it is not difficult to show, that θ_1 and θ_2 must satisfy (7) below:

$$\begin{cases} \theta_2 - \left(e^{-\frac{P_w}{\tau_{EE}}}\right) \cdot \theta_1 = P_P \cdot R_T \cdot \left(1 - e^{-\frac{P_w}{\tau_{EE}}}\right) \\ \left(e^{-\frac{(T-P_w)}{\tau_{EE}}}\right) \cdot \theta_2 - \theta_1 = 0 \end{cases} \tag{7}$$

Solving (7) for $\theta_2 = \theta_{MAXPULS}$ (maximum temperature):

$$\theta_{MAXPULS} = P_P \cdot R_T \cdot \left(\frac{1 - e^{-\frac{P_W}{\tau_{EE}}}}{1 - e^{-\frac{T}{\tau_{EE}}}} \right) \tag{8}$$

Equation (8), with P_{ave} fixed, becomes

$$\theta_{MAXPULS} = \theta_{MAXCONT} \cdot \frac{1}{DC} \cdot \left(\frac{1 - e^{-\frac{T}{\tau_{EE}}.DC}}{1 - e^{-\frac{T}{\tau_{EE}}}} \right) \tag{9}$$

because (see Figure 1 and Figure 3)

$$P_P = P_P \cdot \frac{\frac{P_W}{T}}{\frac{P_W}{T}} = \frac{P_P.\frac{P_W}{T}}{\frac{P_W}{T}} = \frac{P_{ave}}{\frac{P_W}{T}} = \frac{P_{ave}}{DC},$$

$$P_P.R_T = \frac{P_{ave}}{DC}.R_T = P_{ave}.R_T.\frac{1}{DC} = \theta_{MAXCONT}.\frac{1}{DC},$$

$\theta_{MAXCONT} = P_{ave}.R_T$, and

$$\frac{P_W}{\tau_{EE}} = \frac{1}{\tau_{EE}} \cdot T \cdot \frac{P_W}{T} = \frac{T}{\tau_{EE}}.DC.$$

Now, defining the operational efficiency η (a ratio of two temperatures, a kind of temperature gain) as

$$\eta = \frac{\theta_{MAXPULS}}{\theta_{MAXCONT}} = \frac{1}{DC} \cdot \left(\frac{1 - e^{-\frac{T}{\tau_{EE}}.DC}}{1 - e^{-\frac{T}{\tau_{EE}}}} \right) \tag{10}$$

It is evident that, for certain combinations of DC and $\frac{T}{\tau_{EE}}$, η can be made >> 1 (Figure 4). Apparently, the pulsed excitation of the EED can be much more efficient, in terms of the maximum attainable temperature (this is only true if T is an independent variable, therefore under project control). In principle, for high efficiency, one just needs to operate the EED in the high $\frac{T}{\tau_{EE}}$, and low $DC = \frac{P_W}{T}$ domain. These conditions are equivalent to $T >> \tau_{EE} >> P_W$, since

$$\begin{cases} \left(\frac{T}{\tau_{EE}} >> 1\right) \Rightarrow \left(T >> \tau_{EE}\right) \\ \left(\frac{P_W}{T} << 1\right) \Rightarrow \left(\frac{P_W}{\tau_{EE}}.\underbrace{\frac{\tau_{EE}}{T}}_{<<1} << 1\right) \Rightarrow \left(\frac{P_W}{\tau_{EE}} << 1\right) \Rightarrow \left(\tau_{EE} >> P_W\right) \end{cases}$$

But be careful with misleading math manipulations. Considering that both $P_{ave} = P_P.DC = P_P.\frac{P_W}{T}$ and T are fixed, short pulses ($P_W \downarrow$) also mean high peak power ($P_P \uparrow$). So, this specific arrangement of $\frac{T}{\tau_{EE}}$ and DC makes the train of pulses function as a train of impulses. The EED is then quasi-instantaneously responding to the energy of each incoming pulse. Now, from (8) and from $T >> \tau_{EE} >> P_W$, it follows that

$$\theta_{MAXPULS} = P_P.R_T.\left(\frac{1 - e^{\overbrace{-\frac{P_W}{\tau_{EE}}}^{\cong 1 - \frac{P_W}{\tau_{EE}}}}}{1 - \underbrace{e^{-\frac{T}{\tau_{EE}}}}_{\cong 0}} \right) \cong P_P.R_T.\left[\frac{1 - \left(1 - \frac{P_W}{\tau_{EE}}\right)}{1} \right] = P_P.R_T.\frac{P_W}{\underbrace{\tau_{EE}}_{R_T.C_T}} \tag{11}$$

$$= \frac{1}{C_T}.\overbrace{P_P.P_W}^{Pulse\ \ Energy} \Rightarrow \theta_{MAXPULS} \propto Pulse\ \ Energy$$

There is another way (11) can be verified: from (6), provided $T >> \tau_{EE} >> P_W$, it happens that

$$\frac{d\theta}{dt} + \frac{\theta}{\tau_{EE}} = \frac{P_P}{C_T} \Rightarrow d\theta + \frac{\theta}{\tau_{EE}}.dt = \frac{P_P}{C_T}.dt \Rightarrow$$

$$d\theta + \underbrace{\theta.\overbrace{\frac{dt}{\tau_{EE}}}^{<<1}}_{\cong 0} = \frac{P_P}{C_T}.dt \Rightarrow d\theta \cong \frac{P_P}{C_T}.dt \Rightarrow$$

$$\theta_{MAXPULS} = \int_0^{P_W} d\theta \cong \int_0^{P_W} \frac{P_P}{C_T}.dt = \frac{P_P}{C_T}.\int_0^{P_W} dt \Rightarrow$$

$$\theta_{MAXPULS} = \frac{1}{C_T}.P_P.P_W \propto (P_P.P_W) \Rightarrow \theta_{MAXPULS} \propto Pulse\ \ Energy \tag{12}$$

which is the same as (11).

Note that, to calculate $\int_{\theta_1}^{\theta_2} d\theta = \theta_2 - \theta_1 = \theta_{MAXPULS}$, the EED temperature rise θ_1 (above ambient temperature), at the beginning of the power pulse, was considered null (the EED was not able to stockpile thermal energy during the previous cycles), a valid supposition, as $T >> \tau_{EE}$.

Similar reasoning applies to electrical capacitive discharges (electrostatic discharges–ESDs–as well) actuating as $P_i(t)$, whenever C_T is small, and R_T is large: a narrow pulse of high energy, an impulse for all practical purposes, can provoke a rapid squib temperature increase, since the EED will not be able to dissipate quasi-instantaneously, through $\frac{1}{R_T}$, the energy delivered to it. The operation of electrical capacitive discharge fuses, a particular case of the pulsed excitation (a pulsed excitation with $T \to \infty$, $P_P \to \infty$, and $P_W \to 0$), is now clear: almost all the thermal energy transferred to the EED will be converted, in just a single pulse, into a temperature increase, that is,

$$\begin{aligned}&\theta_{MAXPULS} \cong \theta_{IMPULSE} \cong \frac{1}{C_T} \cdot \int_0^{P_W} P_P \cdot dt \\ &= \frac{1}{C_T} \cdot P_P \cdot P_W \propto (P_P.P_W) \Rightarrow \theta_{MAXPULS} \cong \theta_{IMPULSE} \propto Pulse\ \ Energy\end{aligned} \tag{13}$$

Once again, if $\theta_{MAXPULSE} > \theta_{THRESHOLD}$, ignition is assured (it suffices only one short high energy pulse—one shot trigger impulse—to initiate the explosive charge).

Equations (11)–(13) show that the EED is rapidly responding to the energy of each pulse (or impulse) in all those situations, since the feedback loop in Figure 2 can be ignored (by rapidly discharging a huge volume of water into a reasonable barrel, the liquid reaches the top almost instantaneously, because the leak through the hole at its base is, in this case, certainly irrelevant).

But this is not all: if $\tau_{EE} >> T > P_W$, then, from (6),

$$\frac{d\theta}{dt} + \frac{\theta}{\tau_{EE}} = \frac{P_P}{C_T} \Rightarrow d\theta + \underbrace{\frac{\theta}{\tau_{EE}}}_{negligible}.dt = \frac{P_P}{C_T}.dt \Rightarrow d\theta \cong \frac{P_P}{C_T}.dt \xrightarrow{\int} \Delta\theta \cong \frac{P_P}{C_T}.\Delta t$$

and the EED temperature increases progressively by $\Delta\theta = \frac{P_P}{C_T}.P_W$, whenever an electric pulse of peak power P_P and duration P_W is applied to it (between pulses, $\frac{d\theta}{dt} \cong 0$, and the EED temperature, θ, remains approximately constant). After many such pulses, therefore relatively slowly, the temperature θ can certainly cross the ignition threshold, causing, without notice, the squib initiation. This adiabatic effect, temperature stacking (the EED is storing thermal energy), is nothing else than a gradual thermal energy accumulation, cycle after cycle. The EED has become, via C_T, a good long term temperature integrator (a large, in diameter, barrel with a very small role in its base can gradually hold water inside it, since $\tau_b = \pi r^2.R_0$).

Finally, if τ_{EE} is very small ($\tau_{EE} << P_W$), the EED temperature practically follows the source of electrical power pulses $P_i(t)$, once

$$\underbrace{\frac{d\theta}{dt}}_{negligible} + \underbrace{\frac{\theta}{\tau_{EE}}}_{dominant\ factor} = \frac{P_i(t)}{C_T} \Rightarrow \frac{\theta}{\tau_{EE}} \cong \frac{P_i(t)}{C_T} \Rightarrow \theta \cong R_T.P_i(t)$$

Figure 4. Operational efficiency (η) related to duty-cycle (DC) and $\frac{T}{\tau_{EE}}$ ratios

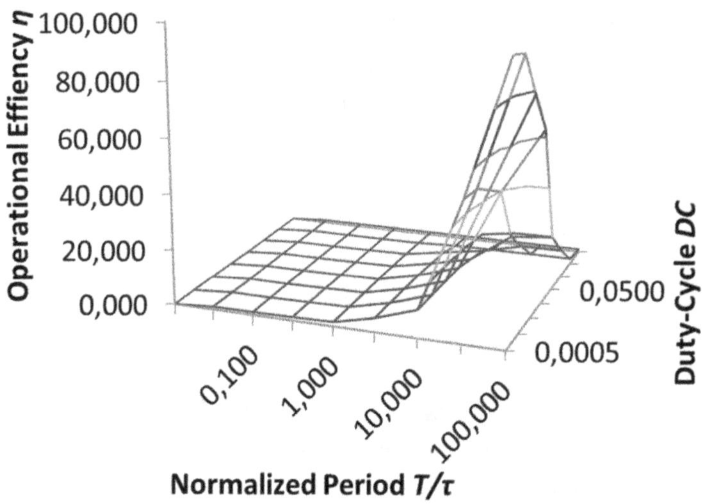

Now, forcing $R_T.P_P > \theta_{THRESHOLD}$, which will require high peak power P_P, ignition is assured (this corresponds to a barrel of small cross-section with a large hole in its base—the liquid level follows the input pattern).

CONCLUSION AND FURTHER RESEARCH

Under strict conditions, the pulsed excitation of EEDs can be advantageous. Namely: If $T >> \tau_{EE} >> P_W$ (with T being an independent variable, therefore under project control), the temperature gain can be even ten to hundred times higher. If $\tau_{EE} >> T$, the EED temperature can softly reach the trigger threshold (either by friend or foe action), after many T cycles, due to the gradual integration of the energy conveyed by each pulse, a kind of discrete electrothermal energy pumping - the temperature stacking effect. Finally, if $\tau_{EE} << P_W$, the EED temperature simply follows the source of electrical power pulses $P_i(t)$.

It is also possible to initiate the EED after a single high energy impulse (one shot fuse operation), particularly when C_T is small, and R_T is large.

For optimized operation, the electrical driving source shall match, in terms of either energy or power, the EED characteristics.

For the sake of safe operations, EEDs have to be handled with extra care: coded ignition commands shall always be the rule. This will shield EEDs from the environment (each EED must have its own local driving circuit module; only the coded commands shall come from the outside).

A matter that, for sure, deserves further investigation is the EED (one without any kind of protection) sensitivity to electromagnetic radiation (radar-like pulsed modulated RF). Although very interesting, this task is out of the scope of this article (Thompson, 1989).

REFERENCES

Cochin, I. (1980). *Analysis and Design of Dynamic Systems*. New York: Harper & Row.

Hoberman, C. M. (1965). *Engineering Systems Analysis*. Ohio: C.E. Merrill Books.

Moran, M. J., & Shapiro, H. N. (2008). *Fundamentals of Engineering Thermodynamics*. New York: John Wiley & Sons.

Potter, M. C., & Scott, E. P. (2004). *Thermal Sciences: An Introduction to Thermodynamics, Fluid Mechanics, and Heat Transfer*. Kentucky: Brooks & Cole.

Prince, W. C., & Leeuw, M. W. (1988). Analysis of the Functioning of the Bridgewire Igniters Based on the Fitted Wire Model. *Propellants, Explosives, Pyrotechnics*, *13*(4), 120–125. doi:10.1002/prep.19880130406

Rosenthal, L. A. (1961). Thermal Response of Bridgewires Used in Electroexplosive Devices. *The Review of Scientific Instruments*, *32*(9), 1033–1036. doi:10.1063/1.1717607

Thompson, R. (1989). RF Sensitivity of Electroexplosives to Pulsed Sources and the Effect of Thermal Time Constants, Explosives and Pyrotechnics. *The Newsletter of Explosives, Pyrotechnics and Their Devices*.

Wylie, C. R. (1975). *Advanced Engineering Mathematics*. Tokyo: McGraw-Hill.

APPENDIX: LIST OF SYMBOLS

$a = \dfrac{1}{\tau_{EE}}$: Inverse of the EED thermal model time-constant

C_T : EED thermal capacitance

DC : Duty Cycle = $\dfrac{P_W}{T}$

H : Height of the water column inside the barrel

H_b : Height of the barrel

n : Cycle counting (pulsed excitation)

P_{ave} : Average power

P_i : EED electrical input power

P_P : Peak power

P_W : Pulse width

Q_i : Water volumetric flow input rate

Q_o : Water volumetric flow output rate

r : Radius of the barrel

R : Hydraulic resistance

R_T : Thermal resistance

V : Volume of water inside the barrel

t : Time

T : Pulse period

$\wedge$: And

Greek Symbols

γ : Exponent in Eq. (1) (rain-barrel analogy)

η : Operational efficiency (temperature gain)

θ : EED temperature increment (above ambient)

π : 3.1415…

ρ: Water density

τ_b: Barrel time-constant

τ_{EE} : EED time-constant

Subscripts

ave : Average
b : Barrel
EE : Electroexplosive
i : Input
$MAXCONT$: EED Maximum temperature above ambient (Continuous Excitation)
$MAXPUL$: EED Maximum temperature above ambient (Pulsed Excitation)
O : Output
T : Thermal
$THRESHOLD$: Ignition temperature threshold

Compilation of References

Abdel-Aal, H. K., Sadik, M., Bassyouni, M., & Shalabi, M. (2005). A new approach to utilize hydrogen as a safe fuel. *International Journal of Hydrogen Energy*, *30*(13-14), 1511–1514. doi:10.1016/j.ijhydene.2005.07.007

Abidi, N., & Hequet, E. (2005). Fourier Transform Infrared Analysis of Trehalulose and Sticky Cotton Yarn Defects Using ZnSe–Diamond Universal Attenuated Total Reflectance. *Textile Research Journal*, *9*(75), 645–652. doi:10.1177/0040517505057527

ABNT. (1996). *NBR 12655/96. Concreto - Preparo.* Rio de Janeiro: Controle E Recebimento.

ABNT. (2003). *NBR 6118/03. Projeto de Estruturas de Concreto - Procedimento.* Rio de Janeiro, Brasil: NBR 6118/03. Projeto de Estruturas de Concreto - Procedimento. NBR 6118/03.

ABNT. (2004). *NBR 14931/04. Execução de Estruturas de Concreto - Procedimento.* Rio de Janeiro, Brasil: NBR 14931 - Execução de Estruturas de Concreto - Procedimento.

Abramovich, G. N. (1944). The Teory of Swirl Atomizers. In *Industrial Aerodynamics* (pp. 114–121). BNT ZAGI.

Achuthan, C. P., & Jose, C. L. (1990). Studies on Octahydro-1,3,5,7-Tetranitro-1,3,5,7-Tetrazocine (HMX) polymorphism. *Propellants, Explosives, Pyrotechnics*, *5*(6), 271–275. doi:10.1002/prep.19900150609

AEB. (2017, April 03). *Veículos Lançadores, Brasília.* Retrieved from http://www.aeb.gov.br/programa-espacial/veiculos-lancadores/

Agrawal, J. P. (1998). Recent trends in high energy material. *Progress in Energy and Combustion Science*, *24*(1), 1–30. doi:10.1016/S0360-1285(97)00015-4

Agrawal, J. P. (2005). Some new high energy materials and their formulations for specialized applications. *Propellants, Explosives, Pyrotechnics*, *30*(5), 316–328. doi:10.1002/prep.200500021

Akansu, S. O., Kahraman, N., & Ceper, B. (2007). Experimental study on a spark ignition engine fueled by methane-hydrogen mixtures. *International Journal of Hydrogen Energy*, *32*(17), 4279–4284. doi:10.1016/j.ijhydene.2007.05.034

Akhavan, J. (1998). Investigation into the network struture of plasticized rocket propellant. *Polymer*, *39*(1), 215–221. doi:10.1016/S0032-3861(97)00259-0

Akhavan, J. (2004). *Introduction to explosives.* London: Royal Society of Chemistry. doi:10.1039/9781847552020-00001

Akhavan, J. (2004). *The Chemistry of Explosives. 2.* Cambridge, United Kingdom: The Royal Society of Chemistry.

Alajbegovic, A., Meister, G., Greif, D., & Basara, B. (2002). Three phase cavitationg flows in high-pressure swirl injectors. *Experimental Thermal and Fluid Science*, *26*(6-7), 677–681. doi:10.1016/S0894-1777(02)00179-6

Al-Baghdadi, M. A. R. S. (2004). Effect of compression ratio, equivalence ratio and engine speed on the performance and emission characteristics of a spark ignition engine using hydrogen as a fuel. *Renewable Energy*, *29*(15), 2245–2260. doi:10.1016/j.renene.2004.04.002

Alemasov, V. E. (1980). *Teoria dos motores-foguete*. Machinostroenie.

All3DP. (2016, November 30). *STL File Format for 3D Printing - Explained in Simple Terms.* Retrieved from https://all3dp.com/what-is-stl-file-format-extension-3d-printing/

Allen, N. S., Edge, M., Rodriguez, M., Liauw, C. M., & Fontan, E. (2001). Aspects of the thermal oxidation, yellowing and stabilization of ethylene vinyl acetate copolymer. *Polymer Degradation & Stability*, *71*, 1–14. doi:10.1016/S0141-3910(00)00111-7

Altman, D. (2001). Rocket Motors, Hybrid. In D. Altman (Ed.), *The Encyclopedia of Physical Science and Technology*.

Altman, D., & Humble, R. (1995). Hybrid rocket propulsion systems. In R. Humble, G. Henry, & W. Larson (Eds.), *Space Propulsion Analysis and Design* (pp. 365–441). New York: McGraw-Hill.

Aluker, E. D., Krechetov, A. G., Loboiko, B. G., Nurmukhametov, D. R., Filin, V. P., & Kazakova, E. A. (2008). Effect of temperature on the laser initiation of pentaerythritol tetranitrate (PETN). *Russian Journal of Chemical Physics*, *27*(5), 53–55.

Amrouche, F., Erickson, P., Park, J., & Varnhagen, S. (2014). An experimental investigation of hydrogen-enriched gasoline in a Wankel rotary engine. *International Journal of Hydrogen Energy*, *39*(16), 8525–8534. doi:10.1016/j.ijhydene.2014.03.172

Anflo, K., & Bergman, G. (2014). Dual mode chemical rocket engine, and dual mode propulsion system comprising the rocket engine. *World Intellectual Property Organization*, *WO2014193300*, A1.

Anflo, K., & Wingborg, N. (2002). Ammonium dinitramide based liquid monopropellants exhibiting improved combustion stability and storage life. *World Intellectual Property Organization*, *WO02096832*, A1.

Araos, M. (2013). *Composition explosive amelioree*. CA: Office de la Propriété Intellectuelle du Canada.

Arendale, W. F. (1969). Chemistry of propellants based on chemically crosslinked binders. *Advances in Chemistry Series*, *88*, 67–83. doi:10.1021/ba-1969-0088.ch004

Arkhipov, V. A., Korotkikh, A. G., Kuznetsov, V. T., Razdobreev, A. A., & Evseenko, I. A. (2011). Influence of the Dispersity of Aluminum Powder on the Ignition Characteristics of Composite Formulations by Laser Radiation. *Russian Journal of Physical Chemistry*, *5*(4), 616–624. doi:10.1134/S1990793111040026

ASCE. (2010). Design of Blast-Resistant Buildings in Petrochemical Facilities. 2nd ed. ed. W. L. Bounds. Reston: ASCE.

Assovskiy, I. G. (1992). Effect of Absorbing Microinclusions on the Interaction of Light with a Reacting Substance. *Dokladi Physical Chemistry (Proc. Rus. Acad. Scien.)*, *324*(1-3), 205-210.

Assovskiy, I. G. (1994). Interaction of Laser Radiation with Reagents. Critical Diameter of Laser Beam. *Doklady Physical. Chemistry, Proc. RAS*, *337*(6), 752-756.

Assovskiy, I. G. (2000). Heat Transfer in Laser Pulse Interaction with Reactive Substances. In A. E. Bergles, & I. Golobic (Eds.), Thermal Sciences 2000 (pp. 107-112). Ljubljana, Slovenia. doi:10.1615/ICHMT.2000.TherSieProcVol2TherSieProcVol1.130

Assovskiy, I. G., Zemskih, V. I., Lejpunskij, O. I., & Marshakov, V. N. (1981). *Experimental Method for Determination of Ignition Characteristics of Solid Propellants* (Report of ICP AS USSR No. 75063377).

Assovskiy, I. G. (2005). *Physics of Combustion and Interior Ballistics*. Moscow: Nauka.

Assovskiy, I. G., & Leipunskiy, O. I. (1980). Theory of Propellant Ignition by Pulse Radiation. *Combustion, Explosion, and Shock Waves*, *16*(1), 310.

Assovskiy, I. G., Melik-Gaikazov, G. V., & Kuznetsov, G. P. (2015). Direct laser initiation of open secondary explosives. *Journal of Physics: Conference Series*. doi:10.1088/1742-6596/653/1/012014

Astapov, N. S., Nikolaev, Y. A., & Ul'yanitskii, V. Y. (1984). Detonation Parameters of Hydrogen-Oxygen and Hydrogen-Air Mixtures at High Initial Density. *Fizika Gorenia i Vzryva*, *20*(1), 98–105.

ASTM D2240 – 05. (2010). Standard Test Method for Rubber Property - Durometer Hardness. (2007). In *Annual Book of ASTM Standards*. West Conshohocken, PA: ASTM International.

ASTM E1641-07. (n. d.). Standard Test Method for Decomposition Kinetics by Thermogravimetry. (2007). In *Annual Book of ASTM Standards* (Vol. 14.02). West Conshohocken, PA: ASTM International.

Attili, A. (2008). *Numerical simulation of internal ballistic of solid rocket motors* [PhD thesis]. Università degli Studi di Roma, Rome.

Atwood, A. I., Boggs, T. L., Curran, P. O., Parr, T. P., Hanson-Parr, D. M., Price, C. F., & Wiknich, J. (1999). Burning Rate of Solid Propellant Ingredients, Part 1: Pressure and Initial Temperature Effects. *Journal of Propulsion and Power*, *15*(6), 740–747. doi:10.2514/2.5522

Ayala, F., Gerty, M., & Heywood, J. B. (2006). SI engine performance. Influences of relative air–fuel ratio, load, compression ratio, and combustion phasing (SAE technical paper 2006-01-0229).

Bachmann, W., & Sheehan, J. C. (1949). A new method of preparing the hing explosive RDX. *Journal of the American Chemical Society*, *71*, 1842–1845.

Bailey, A., & Murray, S. G. (1989). *Propellants, Explosives and Pyrotechnics*. Londres: Brassey's.

Bajic, Z. (2007). *Determination of TNT Equivalent for Various Explosives*. University of Belgrade.

Bari, S., & Esmaeil, M. M. (2010). Effect of H2/O2 addition in increasing the thermal efficiency of a diesel engine. *Fuel*, *89*(2), 378–383. doi:10.1016/j.fuel.2009.08.030

Bauer, C. G., & Forest, T. W. (2001). Effect of hydrogen addition on the performance of methane-fueled vehicles. Part I: Effect of hydrogen addition on S.I. engine performance. *International Journal of Hydrogen Energy*, *26*, 55–70. doi:10.1016/S0360-3199(00)00067-7

Bauer, P., Dabora, E. K., & Manson, N. (1991). Chronology of the Early Research on Detonation Wave. In B. A. L. Kuhl et al. (Eds.), *Dynamics of Detonations and Explosions: Deflagrations, Progress in Astronautics and Aeronautics* (Vol. 133, pp. 3–18). Washington, DC: AIAA.

Bauer, P., Krishnan, S., & Brochet, C. (1979). *"Detonation Characteristics of Gaseous Ethylene, Oxygen, and Nitrogen at High Initial Pressures", Progress in Astronautics and Aeronautics* (Vol. 75, pp. 408–421). AIAA.

Bayvel, L., & Orzechowski, Z. (1993). *Liquid Atomization*. Washington: Taylor & Francis.

Bazarov, V., Yang, V., & Puri, P. (2004). Design and Dynamics of Jet and Swirl Injectors. In V. Yang, M. Habiballah, J. Hulka, & M. Popp (Eds.), *Liquid Rocket Thrust Chambers: Aspects of modeling, analysis and design* (pp. 19–103). Massachusetts: Progress in Astronautics and Aeronautics.

Beckstead, M. W. (2000). Overview of combustion mechanisms and flames structures for advanced solid propellant. In V. Yang, T. B. Brill, & W. Z. Ren (Eds.), Solid Propellant Chemistry, Combustion, and Motor Interior Ballistics (pp. 267-285). New York, USA: Progress in Astronautics and Aeronautics, AIAA. doi:10.2514/6.2000-3325

Beckstead, M. W. (2006, November). Recent Progress in Modeling Solid Propellant Combustion. *Combustion, Explosion, and Shock Waves*, *42*(6), 623–641. doi:10.1007/s10573-006-0096-5

Beckstead, M. W., Puduppakkam, K., Thakre, P., & Yang, V. (2007). Modeling of combustion and ignition of solid-propellant ingredients. *Progress in Energy and Combustion Science*, *33*(6), 497–551. doi:10.1016/j.pecs.2007.02.003

Beltrame, A., Porshnev, P., Merchan, W., Saveliev, A., Fridman, A., & Kennedy, L. A. (2001). Soot and NO formation in methane-oxygen enriched diffusion flames. *Combustion and Flame*, *124*(1-2), 295–310. doi:10.1016/S0010-2180(00)00185-1

Benziger, T. M. (1973). *High-Energy Plastic-Bonded Explosive*. U.S. Patent US3,778,319.

Beyer, R. B., & Fishman, N. (1960). Solid Propellant Ignition Studies with High Flux Radiant Energy as a Thermal Source. In *Progress in Astronautics and Rocketry: Solid Propellant Rocket Research* (Vol. *1*, pp. 673–692).

Bianchi, D., Nasuti, F., Onofri, M., & Martelli, E. (2009). Thermochemical erosion analysis of Carbon-Carbon nozzles in solid-propellant rocket motors. In *Proceedings of the 45th AIAA/ASME/SAE/ASEE Joint Propulsion Conference & Exhibit*, Denver, Colorado, USA. AIAA.

Bittencourt, E., & Guanaes, D. (2008). Propelentes sólidos: uma história ligada à evolução dos polímeros. *Revista Militar de Ciência e Tecnologia, 6*, 71-80.

Blazek, J. (2001). *Computational Fluid Dynamics*. Baden-Daettwil, Switzerland: Alstom Power Ltd.

Bohn, M. A. (2015). Review of some peculiarities of the stability and decomposition of HNF and ADN. In *Proceedings of the 18th seminar new trends in research of energetic material*, University of Pardubice, Czech Republic, April 15–17.

Braeunig, R. A. (2014, June 14). *Rocket and Space Technology, 2012*. Retrieved from http://braeunig.us/space/index.htm

Brill, T. B., Ren, W., Yang, V. (2000) Solid Propellant Chemistry, Combustion, and Motor Interior Ballistics. *Progress in Astronautics and Aeronautics, 185*.

Brill, T. B., & Budenz, B. T. (2000). Flash Pyrolysis of Ammonium Perchlorate-Hydroxyl-Terminated-Polybutadiene Mixtures Including Selected Additives. *Progress in Aeronautics and Astronautics, 185*, 3.

Brish, A. A., Galeev, I. A., & Zaitsev, B. N. (1969). The initiation mechanism in condensed explosives by optical quantum generator radiation. *Combustion, Explosion, and Shock Waves*, *5*(4), 475. doi:10.1007/BF00742068

Brode, H. L. (1955). Numerical Solutions of Spherical Blast Waves. *Journal of Applied Physics*, *26*(6), 766–775. doi:10.1063/1.1722085

Bunte, G., Pontius, H., & Kaiser, M. (1999). Analytical characterization of impurities or byproducts in new energetic materials. *Propellants, Explosives, Pyrotechnics*, *24*(3), 149–155. doi:10.1002/(SICI)1521-4087(199906)24:03<149::AID-PREP149>3.0.CO;2-4

Bunyan, P., Cunliffe, A. V., Davis, A., & Kirby, F. A. (1993). The degradation and stabilization of solid rocket propellants. In *Polym. Degrad. Stab.* (pp. 239-350).

Burgess, C. E., Woodyard, J. D., Rainwater, K. A., Lightfoot, J. M., & Richardson, B. R. (1998). Literature review of the lifetime of DOE Materials: aging of plastic bonded explosive and the explosive and polymers contained therein (Technical Report ANRCP-1998-12). Austin: Amarilo National Resource Center for Plutonium, The University of Texas.

Burnham, A. K., & Weese, R. K. (2005). Kinetics of thermal degradation of explosive binders Viton A, Estane, and Kel-F. *Thermochimica Acta*, *426*(1-2), 85–92. doi:10.1016/j.tca.2004.07.009

Burnside, C. H. (1975). *Correlation of ferric oxide surface area and propellant burning rate*. Pasadena, CA, USA: AIAA. doi:10.2514/6.1975-234

Cai, G., Sun, W., Fang, J., Mao, L., Han, L., & Wu, J. (2011). Nitrous oxide monopropellant thruster and design method thereof. *State Intellectual Property Office of P.R.*

Calabro, M. (2012). Evaluation of the Interest of New ADN Solid Propellants for the Vega Launch Vehicle. In Proceedings of 12-IWCP, Politecnico di Milan, Milan, Italy, June 9-10.

Calzia, J. (1969). *Les substances explosives et leurs nuisances*. Paris: Dunod.

Campbell, M. S., Garcia, D., & Idar, D. (2000). Effects of temperature and pressure on the glass transitions of plastic bonded explosives. *Thermochimica Acta*, *357-358*, 89–95. doi:10.1016/S0040-6031(00)00372-5

Carmicino, C., & Russo Sorge, A. (2003). Investigation of the Fuel Regression Rate dependence on Oxidizer Injection and Chamber Pressure in a Hybrid Rocket. In *Proceedings of the 39th AIAA/ASME/SAE/ASEE Joint Propulsion Conference and Exhibit*, Huntsville, Alabama. doi:10.2514/6.2003-4591

Castedo, R., Segarra, P., Alañon, A., Lopez, L. M., Santos, A. P., & Sanchidrian, J. A. (2015). Air Blast Resistance of Full-Scale Slabs with Different Compositions: Numerical Modeling and Field Validation. *International Journal of Impact Engineering*, *86*, 145–156. doi:10.1016/j.ijimpeng.2015.08.004

Catapano, F., Iorio, S. D., Magno, A., Sementa, P., & Vaglieco, B. M. (2015). A comprehensive analysis of the effect of ethanol, methane and methane-hydrogen blend on the combustion process in a PFI (portfuel injection) engine. *Energy*, *88*, 101–110. doi:10.1016/j.energy.2015.02.051

Cavalheira, P., Gadiot, G. M., & Klerk, W. P. (1995). Thermal decomposition of phase-stabilised ammonium nitrate (PSAN), hydroxyl-terminated polybutadiene (HTPB) based propellants. The effect of iron (III) oxide burning-rate catalyst. *Thermochimica Acta*, *269-270*, 273–293. doi:10.1016/0040-6031(95)02366-6

Cavallini, E., Favini, B., Di Giacinto, M., & Serraglia, F. (2009). SRM internal ballistic numerical simulation by SPINBALL model. In *Proceedings of the 45th AIAA/ASME/SAE/ASEE Joint Propulsion Conference & Exhibit*, Denver, Colorado, USA. AIAA. doi:10.2514/6.2009-5512

Cebeci, T., Rshao, J., Kafyeke, F., & Laurendeau, E. (2005). *Computational Fluid Dynamics for Engineers*. Long Beach, California: Horizons Publishing Inc.

Celina, M., Minier, L., & Assink, R. (2002). Development and application tool characterize the oxidative degradation of AP/HTPB/Al propellants in a propellant reability study. *Thermochimica Acta*, *384*(1-2), 343–349. doi:10.1016/S0040-6031(01)00793-6

Cerri, S. (2011). Characterization of the Ageing of Advanced Solid Rocket Propellants and First Step Design of Green Propellants [PhD. Thesis]. Politecnico di Milano, Milan, Italy.

Cerri, S., Bohn, M. A., Menke, K., & Galfetti, L. (2014). Characterization of ADN/GAP-Based and ADN/Desmophen®-Based Propellant Formulations and Comparison with AP Analogues. *Propellants, Explosives, Pyrotechnics*, *39*(2), 192–204. doi:10.1002/prep.201300065

Chan, M. L., Parr, T., Hanson-Parr, D., Bui, T., & Mason, M. (2004). Characterization of a boron containing propellant, In L.T. DeLuca, L. Galfetti, R.A. Pesce-Rodriguez (Eds.), Novel energetic materials and applications. Grafiche GSS, Bergamo.

Chankapoe, S., Winya, N., & Pittayaprasertkul, N. (2013). Performance Investigation of Solid-Rocket Motor with Nozzle Throat Erosion. *International Journal of Mechanical, Aerospace, Industrial and Mechatronics Engineering, 7*(9).

Chasan, D. E., & Norwitz, G. (1971). Qualitative analysis of primes, tracers, igniters, incendiaries, boosters, and delay compositions on a micro scale by use of Infrared Spectroscopy (Test report T71-6-1). Philadelphia: Department of the Army.

Chaturvedi, S., & Dave, P. N. (2015). Solid propellants: AP/HTPB composite propellants. *Arabian Journal of Chemistry*.

Chen, P., Huang, F., Dai, K., & Ding, Y. (2005). Detection and characterization of long-pulse low-velocity impact damage in plastic bonded explosives. *International Journal of Impact Engineering*, *1*(5), 497–508. doi:10.1016/j.ijimpeng.2004.01.008

Chiaverini, M. J., Harting, G. C., Lu, Y., Kuo, K. K., Peretz, A., Stephen Jones, H., & Arves, J. P. et al. (1992). Regression rate behavior of HTPB-based solid fuels in a hybrid rocket 137 motor. *Journal of Propulsion and Power*, *16*(1), 125–132. doi:10.2514/2.5541

Chintala, V., & Subramanian, K. A. (2015). Experimental investigations on effect of different compression ratios on enhancement of maximum hydrogen energy share in a compression ignition engine under dual-fuel mode. *Energy*, *87*, 448–462. doi:10.1016/j.energy.2015.05.014

Choi, K. K., Urgessa, G., Reda Taha, M. M., & Maji, A. (2008). Quasi-Balanced Failure Approach for Evaluating Moment Capacity of Concrete Beams Reinforced with FRP. *Journal of Composites for Construction*, *12*(3), 236–245. doi:10.1061/(ASCE)1090-0268(2008)12:3(236)

Chopra, A. K. (2007). Earthquake Spectra. In K. Scherwatzky (Ed.), Dynamics of Structures—Theory and Applications to Earthquake Engineering (3rd ed.). New Jersey: Prentice-Hall.

Chorin, A. (1985). Curvature and Solidification. *Journal of Computational Physics*, *57*(3), 472–490. doi:10.1016/0021-9991(85)90191-3

Chung, T. J. (2002). *Computational fluid dynamics*. Cambridge University Press. doi:10.1017/CBO9780511606205

Ciola, R. (2003). *Fundamentos da cromatografia a líquido de alto desempenho HPLC*. São Paulo: Edgard Blücher.

Clemente, M., Rocha, R. J., Iha, K., & Rocco, J. A. (2014). Development of prepolymer technology in the synthesis of a polyurethane binder used in solid rocket fuels. *Quimica Nova*, 37(6), 982–988.

Clough, R. W., & Penzien, J. (2003). *Computer Methods in Applied Mechanics and Engineering Dynamics of Structures* (3rd ed.). Berkeley: Computers & Structures. Inc.

Cochin, I. (1980). *Analysis and Design of Dynamic Systems*. New York: Harper & Row.

Cohen, N. S., Fleming, R. W., & Derr, R. L. (1974). Role of binders in solid propellant combustion. *AIAA Journal*, 1974, 212–218.

Consaga, J. P. (1990). Bonding Agents for Composite Propellants. U.S. Patent 4,944,815.

Cooney, C. P., Worm, Y. J., & Naber, J. D. (2009). Combustion characterization in an internal combustion engine with ethanol gasoline blended fuels varying compression ratios and ignition timing. *Energy & Fuels*, *23*(5), 2319–2324. doi:10.1021/ef800899r

Correa, S. M. (1993, January). A review of NOx formation under gas-turbine combustion conditions. *Combustion Science and Technology*, *87*(1-6), 329–362. doi:10.1080/00102209208947221

Couto, H. S., Carvalho, J. A., & Netto, D. B. (1997). Theoretical formulation for sauter mean diameter of pressure-swirl atomizers. *Journal of Propulsion and Power*, *13*(5), 691–696. doi:10.2514/2.5221

Cunliffe, A., Davis, A., & Tod, D. (1996). Aging and life prediction of composite propellant motors. In *Proceedings of the AGARD Conference* (p. 586).

Dauch, F., & Ribereau, D. (2002). A software for SRM grain design and internal ballistics evaluation, PIBAL. In *Proceedings of the 38th AIAA/ASME/SAE/ASEE Joint Propulsion Conference & Exhibit*, Indianapolis, Indiana, USA. AIAA. doi:10.2514/6.2002-4299

Davenas, A. (2003). The Development of Modern Solid Propellants. *Journal of Propulsion and Power*, *19*(6), 1108–1128. doi:10.2514/2.6947

de Flon, J., Andreasson, S., Liljedahl, M., Oscarson, C., Wanhatalo, M., & Wingborg, N. (2011). Solid Propellants Based on ADN and HTPB. *AIAA Paper*.

De la Fuente, J. L. (2009). An Analysis of the thermal aging behavior in high-performance energetic composites through the glass transition temperature. *Polymer Degradation & Stability*, *94*(4), 664–669. doi:10.1016/j.polymdegradstab.2008.12.021

De Luca, L. T. (1999). *Problemi energetici in propulsione aerospaziale*. Milan: Milano: SPLab, Dipartimento di Ingegneria Aerospazialem Politecnico do Milano, Campus Bovisa.

De Luca, L. T., Galfetti, L., Maggi, F., Colombo, G., Merotto, L., Boiocchi, M., & Fanton, L. et al. (2013). Characterization of HTPB-based solid fuel formulations: Performance, mechanical properties, and pollution. *Acta Astronautica*, *92*(2), 150–162. doi:10.1016/j.actaastro.2012.05.002

DeLuca L. T. (2016) Innovative Solid Formulations for Rocket Propulsion. *Eurasian Chemico-Technological Journal*, *18*(3), 181-196.

DeLuca L. T., Galfetti L, Maggi F, Colombo G, Reina A, Dossi S, Consonni D, Brambilla M (2012). Innovative metallized formulations for solid or hybrid rocket propulsion. *Hanneng Cailiao [Chin. J. Energetic Mater.]*, *20*(4), 465–474.

DeLuca, L. T., Maggi, F., Dossi, S., Weiser, V., Franzin, A., Gettwert, V., & Heintz, T. (2013). High-energy metal fuels for rocket propulsion: Characterization and performance. *Chin. J. Explos. Propellants*, *36*, 1–14.

DeLuca, L. T., Shimada, T., Sinditskii, V. P., & Calabro, M. (Eds.). (2016). Chemical rocket propulsion: A comprehensive survey of energetic materials. doi:. Cham, Switzerland: Springer International Publishing. doi:10.1007/978-3-319-27748-6

DeLuca, L.T., Rossettini, L., Kappenstein, C., Weiser, V. (2009). Ballistic Characterization of AlH_3-Based Propellants for Solid and Hybrid Rocket Propulsion (AIAA Paper 2009-4874).

DeLuca, L. T., Galfetti, L., Severini, F., Meda, L., Marra, G., Vorozhtsov, A. B., & Babuk, V. A. et al. (2005). Burning of Nano-Aluminized Composite Rocket Propellants. *Combustion, Explosion, and Shock Waves*, *41*(6), 680–692. doi:10.1007/s10573-005-0080-5

DeLuca, L. T., Galfetti, L., Severini, F., Rossettini, L., Meda, L., Marra, G., & Pavlovets, G. J. et al. (2007). Physical and ballistic characterization of AlH_3-based space propellants. *Aerospace Science and Technology*, *11*(1), 18–25. doi:10.1016/j.ast.2006.08.010

DeLuca, L. T., Palmucci, I., Dossi, S., Maggi, F., Franzin, A., Trache, D., & Weiser, V. et al. (2014). *Combustion Features of ADN-Based Solid Rocket Propellants for Space Applications, Zel'dovich Memorial* (Vol. 1, pp. 108–116). Moscow, Russia: Torus Press.

DeLuca, L. T., Palmucci, I., Franzin, A., Weiser, V., Gettwert, V., Wingborg, N., & Sjöblom, M. (2014, February). *New Energetic Ingredients for Solid Rocket Propulsion*. In Proceedings of *HEMCE '14* (pp. 13–15).

DeLuca, L., Ohlemiller, T. J., Caveny, L. H., & Summerfield, M. (1976). Radiative Ignition of Double Base Propellants: II. Pre-ignition Events and Source Effects. *AIAA Journal*, *14*(8), 1111–1117. doi:10.2514/3.7193

DeLuca, L., Ohlemiller, T. J., Caveny, L. H., & Summerfield, M. (1976). Radiative Ignition of Double-Base Propellants. *AIAA Journal*, *14*(7), 940–946. doi:10.2514/3.7167

Department of Defense Interface Standard. (2010). *Eletromagnetic Environmental Effects requirements for Systems, MIL-STD-464*. Philadelphia: Author.

Department of Defense. (2008). UFC 3-340-02. Structures to Resist the Effects of Accidental Explosions. USA.

Depcik, C., Mangus, M., & Ragone, C. (2014). Ozone-Assisted Combustion—Part I: Literature Review and Kinetic Study Using Detailed n-Heptane Kinetic Mechanism. *Journal of Engineering for Gas Turbines and Power*, *136*(9), 091507. doi:10.1115/1.4027068

Desai, H. J., Cunliffe, A. V., Hamid, J., Honey, P. J., Stewart, M. J., & Amass, A. J. (1996). Synthesis and characterization of?? -hydroxy and nitrato telechelic oligomers of 3,3-(nitratomethyl) methyl oxetane (NIMMO) and glycidyl nitrate (GLYN). *Polymer*, *37*(15), 3461–3469. doi:10.1016/0032-3861(96)88498-9

Desai, H. J., Cunliffe, A. V., Lewis, T., Millar, R. W., Paul, N. C., Stewart, M. J., & Amass, A. J. (1996). Synthesis of narrow molecular weight?? -hydroxy telechelic poly(glycidyl nitrate) and estimation of theoretical heat of explosion. *Polymer*, *37*(15), 3471–3476. doi:10.1016/0032-3861(96)88499-0

Desroches, M., Escouvois, M., Auvergne, R., Caillol, S., & Boutevin, B. (2012). From vegetable oils to polyurethanes: Synthetic routes to polyols and main industrial products. *Polymer Reviews*, *52*(1), 38-79.

Dhole, A. E., Yarasu, R. B., & Lata, D. B. (2016). Effect of hydrogen and producer gas as secondary fuels on combustion parameters of a dual fuel diesel engine. *Applied Thermal Engineering*, *108*, 764–773. doi:10.1016/j.applthermaleng.2016.07.157

Dhole, A. E., Yarasu, R. B., & Lata, D. B. (2016). Investigations on the combustion duration and ignition delay period of a dual fuel diesel engine with hydrogen and producer gas as secondary fuels. *Applied Thermal Engineering*, *107*, 524–532. doi:10.1016/j.applthermaleng.2016.06.151

Di Stasio, A. (2016). JIMTP Strategy and Path Forward. In *Proceedings of the 42nd International Pyrotechnics Society Seminar* (pp. 102–105).

Doble, S. M., & Halton, E. M. (1947). The Application of Cyclone Theory to Centrifugal Spray Nozzles. *Institute of Machenical Engineers*, *154*(1), 111–119. doi:10.1243/PIME_PROC_1947_157_015_02

Dombrowski, N., & Johns, W. R. (1963). The aerodynamic instability and disintegration of viscous liquid sheets. *Chemical Engineering Science*, *18*(3), 203–214. doi:10.1016/0009-2509(63)85005-8

Dossi, S., Reina, A., Maggi, F., & DeLuca, L. T. (2012). Innovative metal fuels for solid rocket propulsion. *Int. J. Energy Mater. Chem. Propuls.*, *11*(4), 299–322.

Dremin, A. N. (1999). *Toward Detonation Theory*. New York: Springer. doi:10.1007/978-1-4612-0563-0

Dumouchel, C., Ledoux, M., Bloor, M. I., Dombrowski, N., & Ingham, D. B. (1990). The Design of Pressure Swirl Atomizers. In *Proceedings of the Twenty-Third Symposium on Combustion* (pp. 1461-1467). The Combustion Institute.

DUTRA. R. C. L. (1997). *Modificação de fibras de polipropileno com EVA funcionalizado.* Unpublished doctoral dissertation, Universidade Federal do Rio de Janeiro, Rio de Janeiro.

Eisenklam, P. (1964). On ligament formation from spinning discs and cups. *Chemical Engineering Science*, *19*(9), 693–694. doi:10.1016/0009-2509(64)85056-9

Encyclopedia of Chemical Technology (Vol. 8). (1965). Kirk-Othmer.

Engen, T. K., & Johannessen, T. C. (1990). The effects of two types of iron oxide on the burning rate of a composite propellant. In *Proceeding of the 21st International Annual Conference of Technology of Polymer Compounds and Energetic Materials*, Karlsruhe, Germany (pp. 81.1-81.12).

Erikson, W. W., & Beckstead, M. W. (1997) Modeling unsteady monopropellant combustion with full chemical kinetics. In *Proceedings of the 36th AIAA Aerospace Sciences Meeting and Exhibit, Aerospace Sciences Meetings*, Reno, NV.

Federoff, B. F., & Sheffield, O. E. (1966). *Encyclopedia of explosives and related itens* (Vol. 8). Dover: Picantinny Arsenal. doi:10.21236/AD0653029

Felix, S. P., Singh, G., Sikder, A. K., & Aggrawal, J. P. (2005). Studies on energetic compounds part 33: Thermolysis of keto-RDX and its plastic bonded explosives containing thermally stable polymers. *Thermochimica Acta*, *426*(1-2), 53–60. doi:10.1016/j.tca.2004.06.020

Fickett, W., & Davis, W. C. (2000). *Detonation Theory and Experiment.* New York: Dover Publications.

Fleeman, E. L. (2001). *Tacticcal Missile Design.* AIAA.

Flory, P. J. (1953). *Principles of Polymer Chemistry.* Ithaca: Cornell University Press.

Fontana, A., Galloni, E., Jannelli, E., Minutillo, M. (2002). Performance and fuel consumption estimation of a hydrogen enriched gasoline engine at part-load operation (technical paper).

Framcel, M. B., Grant, L. R., & Flanagan, T. E. (1989). Historical development of GAP. In *Proceedings of the 25th Joint Propulsion Conference*, Monterey. AIAA. doi:10.2514/6.1989-2307

Frank-Kamenetskiy, D. A. (1969). *Diffusion and Heat Transfer in Chemical Kinetics.* New York: Plenum.

Franzin, A. (2013). *Burning and Agglomeration of Aluminized ADN/GAP Solid Propellants* [MSc. Thesis]. Politecnico di Milano, Milan, Italy.

Fraser, R. P., Dombrowski, N., & Routley, J. H. (1963). The atomization of a liquid sheet by an impinging air stream. *Chemical Engineering Science*, *18*(6), 339–353. doi:10.1016/0009-2509(63)80027-5

Fraser, R. P., Dombrowski, N., & Routley, J. H. (1963). The filming of liquids by spinning cups. *Chemical Engineering Science*, *18*(6), 323–337. doi:10.1016/0009-2509(63)80026-3

Fuchs, H., Ritz, J., Thomas, E., & Weiss, F. J. (1988). Verfahren zur Herstellung von Hydroxylammoniumsalzen. European Patent Office, EP0287952 A2.

Fujiwara, S., Shibamoto, H., Nakamura, T., Furukawa, K., Matsuo, T., Hori, K., . . . Katsumi, T. (2012). Propellant for thruster. Japan Patent Office, JP2012046390 A.

Gany, A. (2007). Similarity and Scaling Effects in Hybrid Rocket Motors. In M. J. Chiaverini, & K. K. Kuo (Eds.), Progress in Astronautics and Aeronautics - Fundamentals of Hybrid Rocket Combustion and Propulsion (Vol. 218). AIAA.

Gao, X., Zhang, Y., Adusumilli, S., Seitzman, J., Sun, W., Ombrello, T., & Carter, C. (2015). The effect of ozone addition on laminar flame speed. *Combustion and Flame*, *162*(10), 3914–3924. doi:10.1016/j.combustflame.2015.07.028

Gealer, R. L., & Churchill, S. W. (1960). Detonation Characteristics of Hydrogen-Oxygen Mixtures at High Initial Pressures, A.1. *Ch.E. Journal*, *6*(3), 501–505. doi:10.1002/aic.690060330

Gedeon, B. J., & Ngyuen, R. H. (1985). Computerization of ASTM D 3677 – Rubber identification by infrared Spectrophotometry.

Gettwert, V., Fischer, S., & Menke, K. (2013). Aluminized ADN/GAP Propellants – Formulation and Properties. In *Proceedings of the 44th Int'l Annual Conference of ICT*, Karlsruhe, Germany.

Gettwert, V., Tagliabue, C., Weiser, V., & Imiolek, A. (2015). Green Advanced High Energy Propellants for Launchers (GRAIL) - First results on the Burning Behavior of AN/ADN Propellants. In *Proceedings of the 6th European Conference for Aeronautics and Space Sciences* (EuCASS 2015), Krakow, Poland.

Gjersøe, R. (2000). Detection of α-HMX in β-HMX with FTIR diffuse reflection technique. Energetic materials: analysis, diagnostics and testing. In *Proceedings of the International Annual Conference of ICT* (pp. 96-1 – 96-6).

Glassman, I. (1996). *Combustion* (3rd ed.). California: Academic Press.

Goel, M. D., Matsagar, V. A., Gupta, A. K., & Marburg, S. (2012). An Abridged Review of Blast Wave Parameters. *Defence Science Journal*, *62*(5), 300–306. doi:10.14429/dsj.62.1149

Goldshleger, U. I., Barzykin, V. V., & Mezhanov, A. G. (1971). On mechanism and regularities of solid propellant ignition by disperse flow. *Combustion, Explosion, and Shock Waves*, (3), 319–332.

Golovina, N. I., Nechiporenko, G. N., Lempert, D. B., & Lempert, D.B. et al. (2006). Phase stabilization of ammonium nitrate by means of determined classes of nitrate heterocyclic compounds. In Condensed Energetic Systems.

Gomes, S. R., Rocco, L. J., Rocco, J. A., & Iha, K. (2010). Gaseous Oxygen Injection Effects in Hybrid Labscale Rocket Motor Operations. In *Proceedings of the 46th AIAA/ASME/SAE/ASEE Joint Propulsion Conference and Exhibit*, Nashville, Tennessee. doi:10.2514/6.2010-6545

Gonçalves, R. F. B., Iha, K., & Rocco, J. A. F. F. (2015) Enhancement of Ammonium Perchlorate/Hydroxyl-Terminated Polybutadiene Combustion Kinetics Using Ethanol-Doped Air. In *Proceedings of the 51st AIAA/SAE/ASEE Joint Propulsion Conference, Propulsion and Energy Forum*. doi:10.2514/6.2015-3869

Gonzáles, C. A. Q. (2010). Propagation of Combustion Waves in Real Gases [Master Dissertation] (In Portuguese). INPE, São José dos Campos, Brazil.

Gottlieb, L. (1994). Analysis of DOA migration in HTPB/AP composite propellants. In *Proceedings of the ICT Annual Conference,* Karlsruhe.

Graf, R. T., Koenig, J. L., & Ishida, H. (1987). Introduction to optics and infrared spectroscopic techniques. *Polymer of Science Technology*, *36*, 1–31.

Grassie, N., & Scott, G. (1985). *Polymer Degradation and Stabilization*. Cambridge: Cambridge University Press.

Greenwood, J. B., Erickson, P. A., Hwang, J., & Jordan, E. A. (2014). Experimental results of hydrogen enrichment of ethanol in an ultra-lean internal combustion engine. *International Journal of Hydrogen Energy*, *39*(24), 12980–12990. doi:10.1016/j.ijhydene.2014.06.030

Gromov, A. A., Korotkikh, A. G., Il'in, A., DeLuca, L. T., Arkhipov, V. A., Monogarov, K. A., & Teipel, U. (2016). Nanometals: Synthesis and Application in Energetic Systems. In V.E, Zarko & A.A. Gromov (Eds.), Energetic Nanomaterials: Synthesis, Characterization, and Application (Ch. 3). Elsevier. doi:10.1016/B978-0-12-802710-3.00003-9

Gronland, T. A., Westerberg, B., Bergman, G., Anflo, K., Brandt, J., & Lyckfeldt, O. ... Wingborg, N. (2006). *Reactor for decomposition of ammonium dinitramide-based liquid monopropellant and process for the decomposition.* U.S. Patents 7,137,244 B2.

Groom, C. A., Halasz, A., Paquet, L., DCruz, P., & Hawari, J. (2003). Cyclodextrin-assisted capillary electrophoresis for determination of the cyclic nitramine explosives RDX, HMX and CL-20 Comparison with high-performance liquid chromatography. *Journal of Chromatography*, *999*(1-2), 17–22. doi:10.1016/S0021-9673(03)00389-3 PMID:12885047

Gruntman, M. (2004). *Blazing the Trail: The Early History of Spacecraft and Rocketry*. Reston: American Institute of Aeronautics and Astronautics. doi:10.2514/4.868733

Gueyffier, D., Roux, F. X., Fabignon, Y., & Chaineray, G. (2014). High-order computation of burning propellant surface and simulation of fluid flow in solid rocket chamber. In *Proceedings of the 50th AIAA/ASME/SAE/ASEE Joint Propulsion Conference & Exhibit*, Cleveland, USA. AIAA. doi:10.2514/6.2014-3612

Guo, A., Demydov, D., Zhang, W., & Petrovic, Z. S. (2002). Polyols and polyurethanes from hydroformylation of soybean oil. *Journal of Polymers and the Environment*, *10*(1), 49-52.

Hansen, K. G., & Madsen, J. (2001). *A Computational and Experimental Study of the Internal Flow in a Scaled Pressure-Swirl Atomizer* [MSc Thesis]. Denmark: Aalborg University Esbjerg.

Hawkins, T. W., Schneider, S., Drake, G. W., Vaghjiani, G., & Chambreau, S. (2011). *Hypergolic fuels.* U.S. Patents 8,034,202 B1.

Hayden, D. J. (2005). *An analytic tool to investigate the effect of binder on the sensitivity of HMX-based plastic bonded explosives in the skid test* [Master's thesis of Science Department of Mechanical Engineering]. Institute of Mining and Technology, Socorro, New Mexico.

Held, M. (1983). Blast Wave in Free Air. *Propellants, Explosives, Pyrotechnics*, *8*(1), 1–8. doi:10.1002/prep.19830080102

Henrych, J. (1979). *The Dynamic of Explosion and Its Use*. Amsterdan.

Hertfield, R., Jenkins, P., Burkhalter, J., & Foster, W. (2004). Analytical methods for predicting grain regression in tactical solid-rocket motors. *Journal of Spacecraft and Rockets*, *41*(4), 689–693. doi:10.2514/1.3177

Heywood, J. (1988). *Internal Combustion Engine Fundamentals*. McGraw-Hill Education.

Hoberman, C. M. (1965). *Engineering Systems Analysis*. Ohio: C.E. Merrill Books.

Hocaoglu, Ö., Özbelge, T., Pekel, F., & Özkar, S. (2001). Aging of HTPB/AP-based composite solid propellants, depending on the NCO/OH and triol/diol ratios. *Journal of Applied Polymer Science*, *79*, 959–964. doi:10.1002/1097-4628(20010207)79:6<959::AID-APP10>3.0.CO;2-G

Hoffman, D. M., & Caley, L. E. (1981). Dynamic mechanical and molecular weight measurements on polymer bonded explosives from thermally accelerated aging tests I. Fluoropolymer binders. *Journal Organic Coatings and Plastics Chemistry*, *44*, 680–685.

Hoffman, D. M., Matthews, F. M., & Pruneda, C. O. (1989). Dynamic mechanical and thermal analysis of crystallinity development in Kel-F 800 and TATB/Kel-F 800 plastic bonded explosives: Part I. Kel-F 800. *Thermochimica Acta*, *156*(2), 365–372. doi:10.1016/0040-6031(89)87203-X

Hohmann, C., & Tipton, B., Jr. (2000). Viton's impact on NASA standard initiador propellants properties (Tech. Report, NASA/TP 210187). Houston: NASA.

Hórak, M., & Vitek, A. (1978). *Interpretation and processing of vibrational spectra.* New York: John Wiley & Sons.

Hora, T. S., & Agarwal, A. K. (2015). Experimental study of the composition of hydrogen enriched compressed natural gas on engine performance, combustion and emission characteristics. *Fuel*, *160*, 470–478. doi:10.1016/j.fuel.2015.07.078

Hora, T. S., & Agarwal, A. K. (2016). Effect of varying compression ratio on combustion, performance, and emissions of a hydrogen enriched compressed natural gas fueled engine. *Journal of Natural Gas Science and Engineering*, *31*, 819–828. doi:10.1016/j.jngse.2016.03.041

Houghton, R. P., & Mulvaney, A. W. (1996). Mechanism of Tin (IV)-catalyzed urethane formation. *Journal of Organometallic Chemistry*, *518*(1-2), 21–27. doi:10.1016/0022-328X(96)06223-7

Humphrey, J. R. (1974). LX-10-1: a high-energy plastic-bonded explosive (Technical Report UCRL-51629). Livermore: Lawrence Livermore Lab, California University.

Hunley, J. D. (1999). The history of solid-propellant rocketr y: what we do and do not know. In *Proceedings of the ASEE Joint Propulsion Conference*, Los Angeles (pp. 1-10).

Ilyushin, M. A., Tselinsky, I. V., & Shugalei, I. V. (2012). Environmentally Friendly Energetic Materials for Initiation Devices. *Central European Journal of Energetic Materials*, *9*(4), 292–327.

Imamura, Y. Y. (2002). *Avaliação do defeito da granulometria sobre a transição cristalina de β-HMX por calorimetria exploratória diferencial e microscopia eletrônica de varredura* [Master's thesis, Physics and Chemistry of Aerospace Materials]. Instituto Tecnológico de Aeronáutica, São José dos Campos, SP.

Inami, S. H., Rajaphase, Y., Shaw, R., & Wise, H. (1971). Solid Propellant Kinetics. I: The Ammonium Perchlorate-Copper Chromite-Fuel System. *Combustion and Flame*, *17*(2), 189–196. doi:10.1016/S0010-2180(71)80161-X

Ishida, H. (1987). Quantitative surface FTIR spectroscopy analysis of polymers. *Rubber Chemistry and Technology*, *60*(3), 497–554. doi:10.5254/1.3536139

Ivanic, Z., Ayala, F., & Goldwitz, J. (2005) Heywood JB. Effects of hydrogen enhancement on efficiency and NOx emissions of lean and EGR-diluted mixture in a SI engine (SAE paper 2005-01-0253).

Jayasooriya, R., Thambiratnam, D. P., Pereira, N. J., & Kosse, V. (2011). Blast and Residual Capacity Analysis of Reinforced Concrete Framed Buildings. *Engineering Structures*, *33*(12), 3483–3495. doi:10.1016/j.engstruct.2011.07.011

Jhang, S., Chen, K., Lin, S., Lin, Y., & Cheng, W. L. (2016). Reducing pollutant emissions from a heavy-duty diesel engine by using hydrogen additions. *Fuel*, *172*, 89–95. doi:10.1016/j.fuel.2016.01.032

Ji, C., Liu, X., Gao, B., Wang, S., & Yang, J. (2014). A laminar burning velocity correlation for combustion simulation of hydrogen-enriched ethanol engines. *Fuel*, *133*, 139–142. doi:10.1016/j.fuel.2014.05.013

Ji, C., Su, T., Wang, S., Zhang, B., Yu, M., & Cong, X. (2016). Effect of hydrogen addition on combustion and emissions performance of a gasoline rotary engine at part load and stoichiometric conditions. *Energy Conversion and Management*, *121*, 272–280. doi:10.1016/j.enconman.2016.05.040

Jones, C. C., Myre, D. D., & Cowart, J. S. (2009). Performance and Analysis of Vortex Oxidizer Injection in a Hybrid Rocket Motor. In *Proceedings of the 45th AIAA/ASME/SAE/ASEE Joint Propulsion Conference and Exhibit*, Denver, Colorado. doi:10.2514/6.2009-4938

Judge, M. D. (2003). An Investigation of Composite Propellant Accelerated Ageing Mechanisms and Kinetics. *Propellants, Explosives, Pyrotechnics*, *28*(3), 114–119. doi:10.1002/prep.200390017

Kahraman, N., Çeper, B., Akansu, S. O., & Aydin, K. (2009). Investigation of combustion characteristics and emissions in a spark-ignition engine fueled with natural gas-hydrogen blends. *International Journal of Hydrogen Energy*, *34*(2), 1026–1034. doi:10.1016/j.ijhydene.2008.10.075

Kaiser, M. (1998). HPLC Investigation of explosives and nitroaromatic compounds with a cyanopropyl phase. *Propellants, Explosives, Pyrotechnics*, *23*(6), 309–312. doi:10.1002/(SICI)1521-4087(199812)23:6<309::AID-PREP309>3.0.CO;2-L

Karabanov, Y. F., & Bobolev, V. K. (1981). Experimental Study of Laser Pulse Ignition of Explosives. In *Proceedings of the U.S.S.R. Academy of Sciences. Atmospheric and Oceanic Physics* (Vol. *256*, pp. 1152–1154).

Karagöz, Y., Güler, I., Sandalcı, T., Yüksek, L., & Dalkılıc, A. S. (2016). Effect of hydrogen enrichment on combustion characteristics, emissions and performance of a diesel engine. *International Journal of Hydrogen Energy*, *41*(1), 656–665. doi:10.1016/j.ijhydene.2015.09.064

Karagöz, Y., Güler, I., Sandalcı, T., Yüksek, L., Dalkılıc, A. S., & Wongwises, S. (2016). Wongwises, S. (2016) Effects of hydrogen and methane addition on combustion characteristics, emissions, and performance of a CI engine. *International Journal of Hydrogen Energy*, *41*(2), 1313–1325. doi:10.1016/j.ijhydene.2015.11.112

Kartte, K., Fuchs, H., Jockers, K., Kurt, K., & Meier, H. (1972). *Procédé de preparation de nitrate d'hydroxylammonium* (p. 2121005). FR: Institut Nacional de la Propriété Industrielle.

Kasprzyk, D. J., Bell, D. A., Flesner, R. L., & Larson, S. A. (1999). Characterization of a slurry process used to make a plastic-bonded explosive. *Propellants, Explosives, Pyrotechnics*, *24*(6), 333–338. doi:10.1002/(SICI)1521-4087(199912)24:6<333::AID-PREP333>3.0.CO;2-T

Kawaguchi, J., Tokudome, S., Habu, H., Yagishita, T., & Hotta, M. (2009). *Thruster using nitrous oxide*. U. S. Patents, US20090007541 A1.

Khalil, A. E. E., & Gupta, A. K. (2013). Hydrogen addition effects on high intensity distributed combustion. *Applied Energy*, *104*, 71–78. doi:10.1016/j.apenergy.2012.11.004

Khavkin, Y. I. (2004). *Theory and Practice of Swirl Atomizers*. New York: Taylor & Francis.

Kim, H.-S., & Park, B.-S. (1999). Characteristics of the insensitive pressed plastic bonded explosive DXD-59. *Propellants, Explosives, Pyrotechnics*, *24*(4), 217–220. doi:10.1002/(SICI)1521-4087(199908)24:4<217::AID-PREP217>3.0.CO;2-A

Kingery, C. N., & Bulmash, G. (1984). *Airblast Parameters From TNT Spherical Air Bursts and Hemispherical Surface Bursts*. Maryland.

Kinney, G. F., & Graham, K. J. (1985). *Explosive Shocks in Air. 2 nd*. New York: Springer Science. doi:10.1007/978-3-642-86682-1

Kirchoff, E., Rocha, R. J., Nakamura, N. M., Lapa, C. M., Pinheiro, G. F., Iha, K., & Gonçalves, R. F. et al. (2016). Estimativa de vida útil do pbx (plastic-bonded explosive) com envelhecimento acelerado. *Quimica Nova*, 39(6), 661–668.

Kishore, K., Verneker, P., & Sunitha, M. R. (1980). Action of Transition Metal Oxides on Composite Solid Propellants. *AIAA Journal*, *18*(11), 1404–1405. doi:10.2514/3.7735

Kızıltan, E. (1988) *Effect of alcohol addition to motor fuels on engine performance* [MS Thesis]. University of Karadeniz Technical University.

Klager, K. (1984). Polyurethanes, the Most Versatile Binder for Solid Composite Propellants. In *Proceedings of 20th AIAA/SAE/ASME Joint Propulsion Conference*, Cincinnati, Ohio. AIAA. doi:10.2514/6.1984-1239

Klager, K., & Di Milo, A. J. (1976). *Encyclopedia of Polymer Science and Technology - Rocket Propellants*. New York: Wiley and Sons.

Kliachko, L. A. (1962). Theory of Centrifugal Injectors. *Teploenergetica*, 34-38.

Knuth, W. H., Chiaverini, M., Gramer, D., & Sauer, J. (1998). Experimental investigation of a vortex-driven high-regression rate hybrid rocket engine. In *Proceedings of the 34th AIAA/ASME/SAE/ASEE Joint Propulsion Conference and Exhibit*, Cleveland, Ohio. doi:10.2514/6.1998-3348

Kohno, Y., Maekawa, K., Tsuchioka, T., Hashizume, T., & Imamura, A. (1994). A relationship between the impact sensitivity and the electronic structures for the unique N-N bond in the HMX polymorphs. *Combustion and Flame*, *96*(4), 343–350. doi:10.1016/0010-2180(94)90103-1

Kökkülünk, G., Parlak, A., Ayhan, V., Cesur, I., Gonca, G., & Boru, B. (2014). Theoretical and experimental investigation of steam injected diesel engine with EGR. *Energy*, *74*, 331–339. doi:10.1016/j.energy.2014.06.091

Kondrikov, B. N., Ohlemiller, T. J., & Summerfield, M. (1970). Ignition and Gasification of a Double-Base Propellant Induced by CO2 Laser Radiation. In *Proceedings of the Thirteenth International Symposium on Combustion*. Salt Lake City, Utah.

Koomey, J., Berard, S., Sanches, M., & Wong, H. (2011). Implications of Historical Trends in the Electrical Efficiency of Computing. *IEEE Annals of the History of Computing*, *33*(3), 46–54. doi:10.1109/MAHC.2010.28

Köse, H., & Ciniviz, M. (2013). An experimental investigation of effect on diesel engine performance and exhaust emissions of addition at dual fuel mode of hydrogen. *Fuel Processing Technology*, *114*, 26–34. doi:10.1016/j.fuproc.2013.03.023

Kotova, I. P., Ivanova, M. P., Berends, L. K., & Chechetkina, L. N. (1986). Identification of textile materials in rubbers and finished products. *International Polymer Science and Technology*, *8*(13), 66–70.

Koval'skiy, A. A., Khlevnoy, S. S., & Mikheev, V. F. (1967). On ignition of doublebased propellant. *Combustion, Explosion, and Shock Waves*, *3*(4), 527–541.

Kreitz, K., Petersen, E., Reid, D., & Seal, S. (2012). Scale-Up Effects of Nanoparticle Production on the Burning Rate of Composite Propellant. *Combustion Science and Technology*, *184*(6), 750–766. doi:10.1080/00102202.2012.663026

Krishnan, S., & Jeenu, R. (2012). Combustion characteristics of AP/HTPB propellants with burning rate modifiers. *Journal of Propulsion and Power*, *8*(4), 748–755. doi:10.2514/3.23545

Kubota, N. (1984). Survey of Rocket Propellants and Their Combustion Characteristics. In K. K. Kuo (Ed.), Progress in Astronautics and aeronautics - Fundamentals of solid-propellant combustion (Vol. 90, p. 1). Summerfield: AIAA.

Kubota, N. (2007). *Propellants and explosives: thermochemical aspects of combustion*. Weinheim, Germany: Wiley-VCH Verlag GmbH & Co.

Kumar, A., & Gupta, R. K. (1998). *Fundamentals of Polymer*. New York: McGraw-Hill.

Kuo, K. K., & Acharya, R. (2012). Fundamentals of Turbulent and Multiphase Combustion. In Solid Propellants and Their Combustion Characteristics (Ch. 9). John Wiley & Sons, Inc.

Kuo, K. K. (1997). Importance and challenges of hybrid rocket propulsion beyond year 2000. In *Proceedings of the Israel Annual Conference on Aerospace Sciences* (p. 139). Israel Institute of Technology.

Kuo, K. K. (2005). *Principles of combustion* (2nd ed.). New Jersey: John Wiley & Sons.

Kwan, K. S. (1998). *The role of penetrant structure in the transport and mechanical properties of a thermoset adhesive.* Unpublished doctoral dissertation, Faculty of the Virginia Polytechnic Institute, Blacksburg.

Kwon, S.J., Lee, J.S., Jang, D.W. (2013). *Environmental hydrogen peroxide monopropellant.* Korean Intellectual Property Office, KR20130064931.

Laforce, P. D. (1967). *Technological Development of a Throttling Hybrid Propulsion System.* UTC 2215-FR.

Langlet, A., Östmark, H., & Wingborg, N. (1999). *Method of preparing dinitramide acid and salts thereof.* U.S. Patents 5,976,483 A.

Layton, L. H. (1974). *Technical Report AFRPL-TR-74-16.* Brigham City, Utah: Thiokol Chemical Corporation.

Layton, L. H., & Christiansen, A. G. (1979). Effect of aging-strain on propellant mechanical properties. In *Proceedings of the 15th AIAA, SAE and ASM Joint Propulsion Conference*, Las Vegas. AIAA. doi:10.2514/6.1979-1245

Leach, C. (1998). Plasticisers in energetic materials formulations a UK over view. In *Proceedings of the International ICT Conference.*

Leach, C., Debenham, D., Kelly, J., & Gillespie, K. (1998). Advances in polyNIMMO composite gun propellants. *Propellants, Explosives, Pyrotechnics, 23*(6), 313–316. doi:10.1002/(SICI)1521-4087(199812)23:6<313::AID-PREP313>3.0.CO;2-4

Lee, J. H. S. (2008). The Detonation Phenomenon (illustrated ed.). Cambridge: University Press – UK.

Lee, J. S., Hsu, C. K., & Chang, C. L. (2002). A Study on the Thermal Decomposition Behaviors of PETN, RDX, HNS and HMX. *Thermochimica Acta, 392-393*, 173–176. doi:10.1016/S0040-6031(02)00099-0

Lee, J.-S., & Hsu, C.-K. (2002). Thermal properties and shelf life of HMX–HTPB based plastic-bonded explosives. *Thermochimica Acta, 392-393*, 152–156. doi:10.1016/S0040-6031(02)00095-3

Lee, S., Kim, C., Choi, Y., Lim, G., & Park, C. (2014). Emissions and fuel consumption characteristics of an HCNG-fueled heavy-duty engine at idle. *International Journal of Hydrogen Energy, 39*(15), 8078–8086. doi:10.1016/j.ijhydene.2014.03.079

Lefebvre, A. H. (1989). *Atomization and Sprays.* West Lafayette: Purdue University.

Levinthal, M. L., Willer, R. L., Park, D. J., & Bridges, R. (1993). *Process for making high purity hydroxylammonium nitrate.* U.S. Patents, US5,266,290 A.

Liang, X., Wang, Z., Weng, W., Zhou, Z., Huang, Z., Zhou, J., & Cen, K. (2013). Study of ozone-enhanced combustion in H2/CO/N2/air premixed flames by laminar burning velocity measurements and kinetic modeling. *International Journal of Hydrogen Energy, 38*(2), 1177–1188. doi:10.1016/j.ijhydene.2012.10.075

Liau, Y. C., & Yang, V. (1995). Analysis of RDX monopropellant combustion with two-phase subsurface reactions. *Journal of Propulsion and Power, 11*(4), 729–739. doi:10.2514/3.23898

Librovich, V. B. (1963). Ignition of Powders and Explosives. *Zhurnal Prikladnoi Mekhaniki i Technicheskoi Fiziki, 6*, 74.

Liew, C., Li, H., Nuszkowski, J., Liu, S., Gatts, T., Atkinson, R., & Clark, N. (2010). An experimental investigation of the combustion process of a heavy-duty diesel engine enriched with H2. *International Journal of Hydrogen Energy, 35*(20), 11357–11365. doi:10.1016/j.ijhydene.2010.06.023

Li, J., Wu, C., Hao, H., Wang, Z., & Su, Y. (2016). Experimental Investigation of Ultra-High Performance Concrete Slabs under Contact Explosions. *International Journal of Impact Engineering, 93*, 62–75. Retrieved from http://linkinghub.elsevier.com/retrieve/pii/S0734743X16300422 doi:10.1016/j.ijimpeng.2016.02.007

Lilik, G. K., Zhang, H. D., Herreros, J. M., Haworth, D. C., & Boehman, A. L. (2010). Hydrogen assisted diesel combustion. *International Journal of Hydrogen Energy, 35*(9), 4382–4398. doi:10.1016/j.ijhydene.2010.01.105

Lima, V. (2007). *Estudo de Catalisadores Organometálicos na Síntese de Poliuretanos* [Master Degree Thesis]. Pontifícia Universidade Católica do Rio Grande do Sul.

Linnel, J., & Miller, T. (2002). A preliminary design of a magnesium fueled marcian ramjet engine. *Proceedings of the 38th AIAA/ASME/SAE/ASEE Joint Propulsion Conference,* Indianápolis. AIAA. doi:10.2514/6.2002-3788

Litch, H. H. (1970). HMX (octogen) and its polymorphic forms. In *Proceedings of the Symposium on Chemistry Problems With The Stability of Explosives* (*Vol. 2*, pp.168 - 179).

Liu, L., Li, F., Tan, L., Ming, L., & Yi, Y. (2004). Effects of Nanometer Ni, Cu, Al, and NiCu Powders on the Thermal Decomposition of Ammonium Perchlorate. *Propellants, Explosives, Pyrotechnics, 29*(1), 34–38. doi:10.1002/prep.200400026

Liu, T. L., Hsiue, G. G., & Chiu, Y. S. (1995). Thermal characteristics of energetic polymers based on tetrahydrofuran and oxetane derivatives. *Journal of Applied Polymer Science, 58*(3), 579–586. doi:10.1002/app.1995.070580313

Li, W., & Cheng, H. (2007). Cu-Cr-O nanocomposites: Synthesis and characterization as catalysts for solid state propellants. *Solid State Sciences, 11*(2), 750–755. doi:10.1016/j.solidstatesciences.2007.05.011

Loehr, R. D. (2009). *Hydroxyl amine based staged combustion hybrid rocket motor.* U.S. Patents, US20090211227 A1.

Louwers, J., & Heijden, A.E.D.M. (2003). *Hydrazinum nitroformate.* U. S. Patents, US6,572,717 B1.

Louwers, J., Heijden, A. E. D. M., & Elands, P. J. M. (2005). *Hydrazinum nitroformate based high performance solid propellants.* U.S. Patents, US6,916,388 B1.

Louwers, J., Gadiot, G. M. H. J. L., Landman, A. J., Peeters, T. W. J., van der Meer, Th. H., & Roekaerts, D. (1999). Combustion and Flame Structure of HNF Sandwiches and Propellants (*AIAA Paper* 99–2359).

Low, G. M., & Haury, V. E. (1972). *Hydrazinum nitroformate propellant stabilized with nitroguanidine.* U.S. Patents, US3,658,608.

Lukasavage, W. J., Nicolich, S., & Alster, J. (1993). *Process of making impact Alfa-HMX.* U.S. Patents, US5,268,469.

MACH I, Inc. (2009). A New Class of Bonding Agents for High Energy Propellants (Technical Bulletin).

Ma, F., Liu, H., Wang, Y., Li, Y., Wang, J., & Zhao, S. (2008). Combustion and emission characteristics of a port-injection HCNG engine under various ignition timings. *International Journal of Hydrogen Energy, 33*(2), 816–822. doi:10.1016/j.ijhydene.2007.09.047

Ma, F., Wang, Y., Liu, H., Li, Y., Wang, J., & Zhao, S. (2007). Experimental study on thermal efficiency and emission characteristics of a lean burn hydrogen enriched natural gas engine. *International Journal of Hydrogen Energy, 32*(18), 5067–5075. doi:10.1016/j.ijhydene.2007.07.048

Maher, A. R., & Haroun, A. K. (2003). A prediction study of a spark ignition supercharged hydrogen engine. *Energy Conversion and Management, 44*(20), 3143–3150. doi:10.1016/S0196-8904(03)00127-4

Maji, A., Brown, J. P., & Urgessa, G. (2008). Full-Scale Testing and Analysis for Blast-Resistant Design. *Journal of Aerospace Engineering, 21*(4), 217–225. doi:10.1061/(ASCE)0893-1321(2008)21:4(217)

Manson, N., & Dabora, E. K. (1993). Chronology of Research on Detonation Waves: 1920-1950. In B. A. L. Kuhl et al. (Eds.), *Dynamic Aspects of Detonations, Progress in Astronautics and Aeronautics* (Vol. 153, pp. 3–42). Washington, DC: AIAA.

Marchionni, M., Aggarwal, S. K., Puri, I. K., & Lentini, D. (2007). The Influence of Real-gas Thermodynamics on Simulations of Freely Propagating Flames in Methane/Oxygen/ Inert Mixtures. *Combustion Science and Technology, 179*(9), 1777–1795. doi:10.1080/00102200701259999

Marmureanu, M. I. (2014). Solid rocket motor internal ballistics simulation using different burning rate models. *U.P.B. Sci. Bull., 76*, 49–56.

Marple, R. L., & Lacourse, W. R. (2005). A platform for on-site environmental analysis of explosives using high performance liquid chromatography with UV absorbance and photo-assisted electrochemical detection. *Talanta, 66*(3), 581–590. doi:10.1016/j.talanta.2004.11.034 PMID:18970024

Marxman, G. A., & Gilbert, G. M. (1963). Turbulent boundary layer combustion in the hybrid rocket. In *Proceedings of the Ninth Symposium (Intern.) on Combustion* (pp. 371-383). doi:10.1016/B978-1-4832-2759-7.50048-0

Mathieu, J., & Stucki, H. (2004). Military high explosives. *Chimia, 58*(6), 383–389. doi:10.2533/000942904777677669

Mattos, E. C. (2001). *Síntese de HMX e avaliação da aplicabilidade de técnicas FT-IR na sua caracterização e quantificação* [Master's thesis, Plasma Physics]. Instituto Tecnológico de Aeronáutica, São José dos Campos, SP.

Mattos, E. C., Dutra, R.C.L., Diniz, M.F., & Iha, K. (2004). Avaliação do Uso de Técnicas FT-IR para Caracterização de Cobertura Polimérica de Material Energético. *Polímeros: Ciência E Tecnologia,* 14(2), 63-68.

Mattos, E. C., Viganó, I., David, L. H., Diniz, M. F., Dutra, R. C. L., & Iha, K. (2004b). Application of FT-IR techniques for identification of different polymers used in the coating process of energetic materials. In *Proceedings of the International Annual Conference of ICT* (pp. 174-1 – 147-13). Karlsruhe.

Mattos, E. C., Dutra, R. C. L., Diniz, M. F., Silva, G., Iha, K., & Teipel, U. (2008). Characterization of Polymer Coated RDX and HMX Particles. *Propellants, Explosives, Pyrotechnics, 33*(1), 44–50. doi:10.1002/prep.200800207

Mattos, E. C., Moreira, E. D., Dutra, R. C. L., Diniz, M. F., Ribeiro, A. P., & Iha, K. (2004a). Determination of the HMX and RDX content in synthesized energetic material by HPLC, FT-MIR, and FT-NIR spectroscopies. *Quimica Nova, 27*(4), 540–544. doi:10.1590/S0100-40422004000400005

Mattos, E. C., Nakamura, N. M., Diniz, M. F., & Dutra, R. C. L. (2009). Determination of polymer content in energetic materials by FT-IR. *Journal of Aerospace Technology and Management, 1*(2), 167–175. doi:10.5028/jatm.2009.0102167175

Mattos, E. C., Viganó, I., Dutra, R. C. L., Diniz, M. F., & Iha, K. (2002). Aplicação de metodologias FTIR de transmissão e fotoacústica à caracterização de materiais altamente energéticos - parte II. *Quimica Nova, 25*(5), 722–728. doi:10.1590/S0100-40422002000500003

Mayer, A. E. H. J., & Nardini, F. T. (2014). Fuel composition for hypergolic bipropellant. *World Intellectual Property Organization, WO2014193235*, A1.

Ma, Z., & Li, F. (2006). Preparation and thermal decomposition behavior of TMOS/AP composite nanoparticles. *Nanoscience, 11*(2), 142–145.

McAuley, D. C., Joubert, A. L., & Steyn, L. T. (1985). Methods for the determination of (a) RDX in HMX and (b) polymorphs of HMX in β HMX. In *Proceedings of the Symposium on Chemistry Problems With The Stability Off Explosives* (pp. 181-196).

McClelland, J. F., Jones, R. W., Luo, S., & Seaverson, L. M. (1992). *A Practical Guide to FT-IR Photoacoustic Spectroscopy. Center for Advanced Technology Development and Ames Laboratory. Iowa State University: MTEC Photoacoustics,* Catalog.

Mccord, B., & Bender, E. C. (1998). Chromatography of explosives. *Forensic Investigation of Explosives*, *8*, 231–265.

McFarlane, J. S. (1993). Design and Testing of AMROC's 250,000 lbf Thrust Hybrid Motor. (*AIAA Paper* 93–2551).

McGrattan, B. J. (1994). Decomposition of Ethylene—Vinyl Acetate Copolymers Examined by Combined Thermogravimetry, Gas Chromatography, and Infrared Spectroscopy. In T. Provder, M. W. Urban, & H. G. Barth (Eds.), Hyphenated Techniques in Polymer Characterization (pp. 103–115). Washington, D.C.: Americal Chemical Society. doi:10.1021/bk-1994-0581.ch008

McLellan, B., Shoko, E., Dicks, A. L., & Diniz da Costa, J. C. (2005). Hydrogen production and utilization opportunities for Australia. *International Journal of Hydrogen Energy*, *30*(6), 669–679. doi:10.1016/j.ijhydene.2004.06.008

Mead, F. B., & Bornhorst, B. R. (1969). *Certification Tests of a Hybrid Propulsion System for the Sandpiper Target Missile* (AFRPL-TR-69-73).

Melik-Gaykazov, G. V., Kuznetsov, G. P., & Assovskiy, I. G. (2016). On mechanism of laser initiation of secondary explosives. In *Space Technologies, Materials, and Devices* (pp. 255-259).

Mendonça, F. B., Rocco, J. A. F. F., Iha, K., & Urgessa, G. (2016). Blast Response of Reinforced Concrete Slab with Varying Standoff and Steel Reinforcement Ratios. In *Proceedings of the 42nd International Pyrotechnics Society Seminar* (pp. 291–292).

Merzhanov, A. G., & Averson, A. E. (1971). The Present State of Thermal Ignition Theory: An Invited Review. *Combustion and Flame*, *16*(1), 89–124. doi:10.1016/S0010-2180(71)80015-9

Milos, F., & Rasky, D. (1994). Review of Numerical Procedures for Computational Surface Thermochemistry. *J. Thermophys. Heat Transfer*, *8*, 24–34.

Mishuck, E., & Carleton, L. T. (1960). Chemical principles of solid propellants. *Industrial & Engineering Chemistry*, *52*(9), 754–760. doi:10.1021/ie50609a023

Mohammadi, A., Kawanabe, H., Ishiyama, T., Shioji, M., & Komada, A. (2006) Study on Combustion Control in Natural-Gas PCCI Engines with Ozone Addition into Intake Gas (SAE Technical Paper No. 2006-01-0419).

Monthé, C. G., & Azevedo, A. D. (2002). *Análise térmica de materiais*. São Paulo: Ieditora.

Morais, A. M. (2000). *Caracterização do comportamento mecânico do propelente sólido compósito como material estrutural* [PhD Thesis]. Campinas, Brazil: Universidade Estadual de Campinas.

Moran, M. J., & Shapiro, H. N. (2008). *Fundamentals of Engineering Thermodynamics*. New York: John Wiley & Sons.

Moreno, F., Arroyo, J., Muñoz, M., & Monné, C. (2012). Combustion analysis of a spark ignition engine fueled with gaseous blends containing hydrogen. *International Journal of Hydrogen Energy*, *37*(18), 13564–13573. doi:10.1016/j.ijhydene.2012.06.060

Mungas, G. S., Fisher, D. J., & Mungas, C. (2012). *Nitrous oxide flame barrier.* U. S. Patents, US20120279197 A1.

Nagamachi, M. Y., Oliveira, J. I. S., Kawamoto, A. M., & Dutra, R. C. L. (2009). ADN – The new oxidizer around the corner for an environmentally friendly smokeless propellant. *Journal of AerospaceTechnology and Management*, *1*(2), 153–160. doi:10.5028/jatm.2009.0102153160

Nakashita, G., & Kubota, N. (1991). Energetic of nitro/azide propellants. *Propellants, Explosives, Pyrotechnics*, *16*(4), 177–181. doi:10.1002/prep.19910160405

NASA. (1971). *Solid rocket motor performances analysis and prediction (NASA SP-8039)*. Cleveland, Ohio, USA: NASA.

Nema, A. K., Jain, S., Sharma, S. K., Nema, S. K., & Verma, S. K. (2004). Mechanistic Aspect of Thermal Decomposition and Burn Rate of Binder and Oxidizer of AP/HTPB Composite Propellants Comprising HYASIS-CAT. *Int. J. Plast. Technol.*, *8*, 344.

Nettleton, M. A. (1987). *Gaseous Detonations: their Nature, Effects and Control*. London: Chapman and Hall Ltd. doi:10.1007/978-94-009-3149-7

Ngo, T., Mendis, P., Gupta, A., & Ramsay, J. (2007). Blast Loading and Blast Effects on Structures - An Overview. *Electronic Journal of Structural Engineering*, *7*, 76–91.

Nogueira, D. A. R., Rosa, P. T. V., & Santana, C. C. (2000). Adsorção de proteínas na superfície de polímeros: quantificação com FTIR/ATR. In *I brasilian congresso phase equilibrium na fluid propeties for chemical process desin (1). Águas de São Pedro*. Campinas: UNICAMP.

Norbye, J. P. (1971). *The Wankel Engine*. Philadelphia: Chilton Book Company.

Oates, G. C. (1984). *Aerothermodynamics of Gas Turbine and Rocket Propulsion*. New York: American Institute of Aeronautics and Astronautics.

Oberth, A. E., & Bruenner, R. S. (1969). Polyurethane based propellants. *Advances in Chemistry Series*, *88*, 84–121. doi:10.1021/ba-1969-0088.ch005

Ohlemiller, T. J., & Summerfield, M. A. (1968). A critical Analysis of arc ignition of solid propellants. *AIAA Journal*, *6*(5), 878–886. doi:10.2514/3.4613

Oliveira, J. I. S. (2014). *Síntese e caracterização do oxidante ADN (dinitramida de amônio), usado em propelentes para aplicação aeroespacial*. Unpublished doctoral dissertation, Instituto Tecnológico de Aeronáutica, São José dos Campos, Brazil.

Osher, S., & Sethian, J. A. (1988). Fronts propagating with curvature-dependent speed: Algorithms based on Hamilton-Jacobi formulations. *Journal of Computational Physics*, *79*(1), 12–49. doi:10.1016/0021-9991(88)90002-2

Pak, Z. P. (1993). Some Ways to Higher Environmental Safety of Solid Rocket Propellant Application. (*AIAA Paper* 93–755).

Palmucci, I. (2014, July 25). ADN-Based Double Oxidizer Solid Propellants Formulations [MSc. Thesis]. Politecnico di Milano, Milan, Italy.

Pandey, G. C., & Kulshreshtha, A. K. (1993). Fourier transform infrared spectroscopy as a quality control tool. *Process Control and Quality*, *4*, 109–123.

Pang, W. Q. (2012). Effects of ADN on the Properties of Nitrate Ester Plasticized Polyether (NEPE) Solid Rocket Propellants. In Proceedings of 12-IWCP, Politecnico di Milan, Milan, Italy, June 9-10.

Pang, W. Q., DeLuca, L. T., Fan, X. Z., & Liu, F. L. (2014). Effects of ADN on the Properties of NEPE Solid Propellant. In *Proceedings of the International Workshop on New Energetic Materials and Propulsion Techniques for Space Exploration*, Milan, Italy, June 9-10.

Pang, W. Q., Fan, X. Z., Zhang, W., Xu, H. X., Wu, S. X., Liu, F. L., & Yan, N. et al. (2013). Effect of ADN on the Characteristics of HTPB Based Composite Solid Propellant. *Journal of Chemical Science and Technology*, *2*(2), 53–60.

Pan, X., Tian, K., Jones, L. E., & Cobb, G. P. (2006). Method optimization for quantitative analysis of octahydro-1,3,5,7-tetranitro-1,3,5,7-tetrazocine (HMX) by liquid chromatography-electrospray ionization mass spectrometry. *Talanta*, *70*(2), 455–459. doi:10.1016/j.talanta.2006.03.005 PMID:18970792

Park, C., Choi, Y., & Lee, J. (2013). Operating strategy for exhaust gas reduction and performance improvement in a heavy-duty hydrogen-natural gas blend engine. *Energy*, *50*, 262–269. doi:10.1016/j.energy.2012.10.048

Park, C., Lee, S., Lim, G., Choi, Y., & Kim, C. (2014). Full load performance and emission characteristics of hydrogen-compressed natural gas engines with valve overlap changes. *Fuel*, *123*, 102–106. doi:10.1016/j.fuel.2014.01.041

Patil, K. R., Khanwalkar, P. M., Thipse, S. S., Kavathekar, K. P., Rairikar, S. D. (2010) Development of HCNG blended fuel engine with control of NO x emissions. *Journal of Computer Information Systems and Industrial Management Applications*, *2*, 87–95.

Paubel, E. F. (2002). *Propulsão e Controle de Veículos Aeroespaciais - Uma Introdução*. Florianópolis: Editora UFSC.

Pekel, F. E., Pinardag, E., & Türkan, A. (1990). An investigation of the Catalytic Effect of Iron (III) Oxide on the Burning rate of Aluminized HTPB/AP Composite Propellant. In *Proceedings of the 21st International Annual Conference of Technology of Polymer Compounds and Energetic Materials*, Karlsruhe, Germany (pp. 1-8).

Peterson, J. D., Vyazovkin, S., & Wight, C. A. (2001). Kinetics of the Thermal and Thermo-Oxidative Degradation of Polystyrene, Polyethylene and Poly(propylene). *Macromolecular Chemistry and Physics*, *202*(6), 775–784. doi:10.1002/1521-3935(20010301)202:6<775::AID-MACP775>3.0.CO;2-G

Petrovic, Z., Guo, A., & Javni, I. (2000). U.S. Patente Nº Patent 6107433 A.

Pinazzi, P. M., & Foucher, F. (in press). Influence of injection parameters, ozone seeding and residual NO on a Gasoline Compression Ignition (GCI) engine at low load. In *Proceedings of the Combustion Institute*.

Pisa, I., Lazaroiu, G., & Prisecaru, T. (2014). Influence of hydrogen enriched gas injection upon polluting emissions from pulverized coal combustion. *International Journal of Hydrogen Energy*, *39*(31), 17702–17709. doi:10.1016/j.ijhydene.2014.08.119

Pokhil, P. F. (1982). On combustion mechanism of double based propellants. *J. Physics of Combustion and Explosion*, 1982, 117-140.

Pokhil, P. F., Maltsev, V. M., & Zaitsev, V. M. (1969). *Methods of Experimental Study of Combustion and Detonation*. Moscow: Nauka.

Potter, M. C., & Scott, E. P. (2004). *Thermal Sciences: An Introduction to Thermodynamics, Fluid Mechanics, and Heat Transfer*. Kentucky: Brooks & Cole.

Prajakta, R. P., Krishnamurthy, V. N., & Satyawati, S. J. (2006). Differential Scanning Calorimetric Study of HTPB based Composite Propellants in Presence of Nano Ferric. *Propellants, Explosives, Pyrotechnics*, *31*(6), 442–446. doi:10.1002/prep.200600059

Prince, W. C., & Leeuw, M. W. (1988). Analysis of the Functioning of the Bridgewire Igniters Based on the Fitted Wire Model. *Propellants, Explosives, Pyrotechnics*, *13*(4), 120–125. doi:10.1002/prep.19880130406

Probster, M., & Schmucker, R. H. (1986). Ballistic anomalies in solid rocket motors due to migration effects. *Acta Astronautica*, *13*(10), 599–605. doi:10.1016/0094-5765(86)90050-0

Provatas, A. (2000). Energetic Polymers for explosives formulation: a review of recent advances. Melbourne: DSTO Aeronautical and Maritime Researche Laboratory - DSTO-TR-0966.

Qin, F., Guoqing, H., Peijin, L., & Jiang, L. (2006). Algorithm study on burning surface calculation of solid rocket motor with complicated grain based on level set methods. In *Proceedings of the 42th IAA/ASME/SAE/SEE Joint Propulsion Conference & Exhibit*, Sacramento, CA. doi:10.2514/6.2006-4774

Radwan, M. S. (1985) Performance and knock limits of ethanol-gasoline blends in spark-ignited engines (SAE Paper 850213).

Rahn, M. (2010). *Green Propellants.* (Unpublished doctoral dissertation). Royal Institute of Technology, Stockholm, Sweden.

Ramakrishna, P. A., Paull, P. J., & Mukunda, H. S. (2012). Revisiting the Modeling of Ammonium Perchlorate Combustion: Development of an Unsteady Model. *Journal of Propulsion and Power, 22*(3), 661–668. doi:10.2514/1.15502

Rezende, L. C. (2001). *Envelhecimento de Propelente Compósito à Base de Polibutadieno Hidroxilado* [PhD Thesis].

Riahi, Z., Bounaouara, H., Hraiech, I., Mergheni, M. A., Sautet, J., & Nasrallah, S. B. (in press). Combustion with mixed enrichment of oxygen and hydrogen in lean regime. *International Journal of Hydrogen Energy.*

Risha, G. A., Evans, B. J., Boyer, E., & Kuo, K. K. (2007). Metals, Energetic Additives, and Special Binders Used in Solid Fuels for Hybrid Rockets. In M. J. Chiaverini, & K. K. Kuo (Eds.), Progress in Astronautics and Aeronautics - Fundamentals of Hybrid Rocket Combustion and Propulsion (pp. 423-442). AIAA.

Ritchie, D. (1993). The Development of the C Language. In *Proceedings of the Second History of Programming Languages conference.* Cambridge, Mass.

Rocco, J. A. (2004). *Estudos sobre o envelhecimento de formulações de propelente sólido compósito baseadas em binders poliuretânico.* [PhD Thesis]. São José dos Campos, Brazil: Instituto Tecnológico de Aeronáutica.

Rocco, J. A., Lima, J. E., Frutuoso, A. G., Iha, K., Ionashiro, M., Matos, J. R., & Suárez-Iha, M. E. (2005). Thermogravimetric Studies of a Composite Solid Rocket Propellant Based on HTPB-Binder. *Journal of Thermal Analysis and Calorimetry, 77*(3), 803–813. doi:10.1023/B:JTAN.0000041659.97749.fe

Rocha, R. J., Lima, J. E., Gomes, S. R., Iha, K., & Rocco, J. A. (2013). Síntese de poliuretanos modificados por óleo de mamona empregados em materiais energéticos. *Quimica Nova, 36*(6), 793–799. doi:10.1590/S0100-40422013000600009

Rosenthal, L. A. (1961). Thermal Response of Bridgewires Used in Electroexplosive Devices. *The Review of Scientific Instruments, 32*(9), 1033–1036. doi:10.1063/1.1717607

Roßmann, Ch., Heintz, T., & Herrmann, M. (2014) Generation of Very Fine ADN Particles by Bead Milling Technology. In *Proceedings of the 45th International Annual Conference of ICT*, Karlsruhe, Germany.

Rypl, D., & Bittnar, Z. (2006). Generation of computational surface meshes of STL models. *Journal of Computational and Applied Mathematics, 192*(1), 148–151. doi:10.1016/j.cam.2005.04.054

Sadovskiy, M. A. (2004). Mechanical Effects of Air Shockwaves from Explosions according to Experiments. In M. A. Sadovskiy (Ed.), *Selected Works: Geophysics and Physics of Explosion.* Moscow: Nauka Press.

Sahin, Z., Durgun, O., Kurt, M. (2015). Experimental investigation of improving diesel combustion and engine performance by ethanol fumigation-heat release and flammability analysis. *Energy Conversion and Management, 89*, 175–187.

Sahin, Z., & Durgun, O. (2009). Numerical investigation of the effects of injection parameters on performance and exhaust emissions in a D I diesel engine. In *Proceedings of the 4th International Exergy, Energy and Environment Symposium*, Sharjah, UAE.

Sandalcı, T., & Karagöz, Y. (2014). Experimental investigation of the combustion characteristics, emissions and performance of hydrogen port fuel injection in a diesel engine. *International Journal of Hydrogen Energy, 39*(32), 18480–18489. doi:10.1016/j.ijhydene.2014.09.044

Saravanan, N., & Nagarajan, G. (2008). An experimental investigation of hydrogen-enriched air induction in a diesel engine system. *International Journal of Hydrogen Energy*, *33*(6), 1769–1775. doi:10.1016/j.ijhydene.2007.12.065

Saravanan, N., & Nagarajan, G. (2010). Performance and emission studies on port injection of hydrogen with varied flow rates with diesel as an ignition source. *Applied Energy*, *87*(7), 2218–2229. doi:10.1016/j.apenergy.2010.01.014

Saravanan, N., Nagarajan, G., Dhanasekaran, C., & Kalaiselvan, K. M. (2007). Experimental investigation of hydrogen port fuel injection in DI diesel engine. *International Journal of Hydrogen Energy*, *32*(16), 4071–4080. doi:10.1016/j.ijhydene.2007.03.036

Sarigul-Klijn, M. M., & Sarigul-Klijn, N. (2007). *High propellant mass fraction hybrid rocket propulsion*. U. S. Patents, US20070144140 A1.

Sastri, M. V. C. (1987). Hydrogen Energy research-and-development in India - an overview. *International Journal of Hydrogen Energy*, *12*(3), 137–145. doi:10.1016/0360-3199(87)90145-5

Satyawati, S. J., Prajakta, R. P., & Krishnamurthy, V. N. (2008). Thermal Decomposition of Ammonium Perchlorate in the Presence of Nanosized Ferric Oxide. *Defence Science Journal*, *58*(6), 721–727. doi:10.14429/dsj.58.1699

Sbriccoli, E., Saltarelli, R., & Martinucci, S. (1989). Comparison between Accelerated and Natural Aging in HTPB Composite Propellants. In *Proceedings of 20th International Annual Conference of ICT*, Karhsruhe, Germany.

Sbriccoli, E., Saltarelli, R., & Martinucci, S. (1989). Comparison between Accelerated and Natural Aging. In *Proceedings of the 20th International Annual Conference of ICT and 18th Annual Technical Meeting of GUS*, Karhsruhe, Germany.

Schefer, R. W., Wickall, D. M., & Agrawal, A. K. (2002). Combustion of hydrogen-enriched methane in lean premixed swirl stabilized burner. *Proceedings of the Combustion Institute*, *29*(1), 843–851. doi:10.1016/S1540-7489(02)80108-0

Schifter, L., Diaz, R., Rodriguez, J. P., & Gomez Gonzalez, U. (2011). Combustion and emissions behavior for ethanol gasoline blends in a single cylinder engine. *Fuel*, *90*(12), 3586–3592. doi:10.1016/j.fuel.2011.01.034

Schlichting, H. (1995). *Boundary Layer Theory*. New York: McGraw-Hill Book Company Inc.

Schmitt, R. J., Bottaro, J. C., Penwell, P. E., & Bomberger, D. (1994). *Process for forming ammonium dinitramide salt by reaction between ammonia and a nitronium-containing compound.* U.S. Patents US5,316,749 A.

Schmitt, R. G., & Butler, P. B. (1995a). Detonation Properties of Gases at Elevated Initial Pressures. *Combustion Science and Technology*, *106*(1-3), 167–193. doi:10.1080/00102209508907773

Schmitt, R. G., & Butler, P. B. (1995b). Detonation Wave Structure of Gases at Elevated Initial Pressures. *Combustion Science and Technology*, *107*(4-6), 355–386. doi:10.1080/00102209508907811

Schneider, S., Hawkins, T. W., Ahmed, Y., & Rosander, M. (2014). *Catalytic hypergolic bipropellants*. U. S. Patents, US8,758,531 B1.

Schönborn, A., Hellier, P., Aliev, A. E., & Ladommatos, N. (2010). Ignition control of homogeneous-charge compression ignition (HCCI) combustion through adaptation of the fuel molecular structure by reaction with ozone. *Fuel*, *89*(11), 3178–3184. doi:10.1016/j.fuel.2010.06.005

Sciamareli, J. (2012). *Síntese, caracterização e aplicação do polímero metil azoteto de glicidila (GAP) no desenvolvimento de novos propelentes para o programa aeroespacial brasileiro.* (Unpublished doctoral dissertation). Instituto Tecnológico de Aeronáutica, São José dos Campos, Brazil.

Sciamareli, J., Takahashi, M. F., Teixeira, J. M., & Iha, K. (2002). Propelente sólido compósito polibutadiênico: I-influência do agente de ligação. *Quimica Nova*, *25*(1), 107–110. doi:10.1590/S0100-40422002000100018

Sethian, J. A. (1985). Curvature and the Evolution of Fronts. *Communications in Mathematical Physics*, *101*(4), 487–499. doi:10.1007/BF01210742

Sethian, J. A. (1994). Level Set Techniques for Tracking Interfaces, Fast Algorithms, Multiple Regions, Grid Generation, and Shape/Character Recognition. In *Proceedings of the International Conference on Curvature Flows and Related Topics*, Trento, Italy (pp. 215-231).

Shang, Dong-qin, & Huang, Hong-yong (2010). Combustion Properties of PGN/ADN Propellants. Chinese Journal of Energetic Materials, 18(4), 372–376.

Shanwu, W., Yang, V., Hsiao, G., Hsieh, S., & Mongia, C. (2007). Large-eddy simulations of gas-turbine swirl injector flow dynamics. *Journal of Fluid Mechanics*, 583, 99–122.

Sharma, V., & Kundu, P. P. (2006). Addition polymers from natural oils-A review. *Progress in Polymer Science*, *31*(11), 983–1008. doi:10.1016/j.progpolymsci.2006.09.003

Shekhar, H. (2011). Prediction and Comparison of Shelf Life of Solid Rocket Propellants Using Arrhenius and Berthelot Equations. *Propellants, Explosives, Pyrotechnics*, *36*(4), 356–359. doi:10.1002/prep.200900104

Silva, G. (2007). *Síntese, caracterização e quantificação de 2,2',4,4',6,6'-hexanitroestilbeno.* Unpublished doctoral dissertation, Instituto Tecnológico de Aeronáutica, São José dos Campos, SP.

Silva, G., Iha, K., Cardoso, A. M., Mattos, E. C., & Dutra, R. C. (2010). Study of the thermal decomposition of 2,2,4,4,6,6-hexanitrostilbene. *Journal of Aerospace Technology and Management*, *2*(1), 41–46. doi:10.5028/jatm.2010.02014146

Silva, G., Rufino, S. C., & Iha, K. (2013). Green Propellants: Oxidizers. *Journal of AerospaceTechnology and Management*, *5*(2), 139–144. doi:10.5028/jatm.v5i2.229

Silverstein, R. M., Bassler, G. C., & Morril, T. C. (1981). *Spectrometric identification of organic compounds. 5.* New York: John Wiley & Sons.

Sinditskii, V. P., Egorshev, V. Y., Levshenkov, A. I., & Serushkin, V. V. (2006). Combustion of Ammonium Dinitramide, Part 1: Burning Behavior. *Journal of Propulsion and Power*, *22*(4), 769–776. doi:10.2514/1.17950

Sinditskii, V. P., Egorshev, V. Y., Levshenkov, A. I., & Serushkin, V. V. (2006). Combustion of Ammonium Dinitramide, Part 2: Combustion Mechanism. *Journal of Propulsion and Power*, *22*(4), 777–785. doi:10.2514/1.17955

Sinditskii, V. P., Egorshev, V. Y., Serushkin, V. V., & Levshenkov, A. I. (2002) Chemical peculiarities of combustion of solid propellant oxidizers. In *Proc. of the 8th Int. Workshop on Rocket Propulsion: Present and Future*, Pozzuoli, Italy, June 16–20 (pp. 1-20).

Sinditskii, V. P., Egorshev, V. Yu., Serushkin, V. V., Filatov, S. A., & Chernyi, A. N. (2012). Combustion Mechanism of Energetic Binders with Nitramines. *Int. J. Energy Mater. Chem. Propuls.*, *11*(5), 427–449.

Sinditskii, V. P., Filatov, S. A., Kolesov, V. I., Kapranov, K. O., Asachenko, A. F., Nechaev, M. S., & Shishov, N. I. et al. (2015). Combustion behavior and physico-chemical properties of dihydroxylammonium 5, 5-bistetrazole-1, 1-diolate (TKX-50). *Thermochimica Acta*, *614*, 85–92. doi:10.1016/j.tca.2015.06.019

Sinditskii, V. P., Serushkin, V. V., Filatov, S. A., & Egorshev, V. Y. (2002). Flame structure of hydrazinium nitroformate. *International Journal of Energetic Materials and Chemical Propulsion*, *5*(1-6), 576–586. doi:10.1615/IntJEnergeticMaterialsChemProp.v5.i1-6.600

Singh, G., Felix, S. P., & Soni, P. (2003). Studies on energetic compounds part 28: Thermolysis of HMX and its plastic bonded explosives containing Estane. *Thermochimica Acta*, *399*(1-2), 153–165. doi:10.1016/S0040-6031(02)00460-4

Singh, G., Felix, S. P., & Soni, P. (2005). Studies on energetic compounds part 31: Thermolysis and kinetics of RDX and some of its plastic bonded explosives. *Thermochimica Acta*, *426*(1-2), 131–139. doi:10.1016/j.tca.2004.07.013

Singh, H. (1995). High energy material research and development in India. *Journal of Propulsion and Power*, *11*(4), 848–855. doi:10.2514/3.23910

Skidmore, C. B., Phillips, D. S., & Crane, N. B. (1997). *Microscopical examination of plastic-bonded explosives* (Technical Report LA-UR-97-2807). Los Alamos: Los Alamos National Laboratory.

Smith, A. L. (1979). *Applied infrared spectroscopy*. New York: John Wiley & Sons.

Smith, P. D., & Hetherington, J. G. (1994). *Blast and Ballistic Loading of Structures* (1st ed.). Burlington: Butterworth-Heinemann.

Sopena, C., Diéguez, P. M., Sáinz, D., Urroz, J. C., Guelbenzu, E., & Gandía, L. M. (2010). Conversion of a commercial spark ignition engine to run on hydrogen: Performance comparison using hydrogen and gasoline. *International Journal of Hydrogen Energy*, *35*(3), 1420–1429. doi:10.1016/j.ijhydene.2009.11.090

Stroustrup, B. (1994). *A History of C++: 1979– 1991*. New Jersey: AT&T Bell Laboratories.

Sutton, G. P., & Biblarz, O. (2010). Rocket Propulsion Elements (8th ed.). Wiley.

Sutton, G. P., & Biblarz, O. (2017). *Rocket Propulsion Elements*. John Wiley & Sons.

Swisdak, M.M., Jr. (1994). Simplified Kingery Airblast Calculations. Arlington: Naval Surface Warfere Center.

Tachibana, T., Hirata, K., Nishida, H., & Osada, H. (1991). Effect of ozone on combustion of compression ignition engines. *Combustion and Flame*, *85*(3-4), 515–519. doi:10.1016/0010-2180(91)90154-4

Tagliabue, C. (2015). Burning Behavior of ADN/AN based Solid Rocket Propellants [MSc. Thesis]. Politecnico di Milano, Milan, Italy.

Tagliabue, C., Weiser, V., Imiolek, A., Bohn, A. M., Heintz, T., & Gettwert, V. (2016) Burning Behavior of AN/ADN Propellants. In *Proceedings of the 47th International Annual Conference of ICT*, Karlsruhe, Germany.

Tang, C.-J., Lee, Y.-J., Kudva, G., & Litzinger, T. A. (1999). A study the gas-phase chemical strutcture during CO_2 laser assisted combustion of HMX. *Combustion and Flame*, *117*(1-2), 170–188. doi:10.1016/S0010-2180(98)00094-7

Tang, C., Zhang, Y., & Huang, Z. (2014). Progress in combustion investigations of hydrogen enriched hydrocarbons. *Renewable & Sustainable Energy Reviews*, *30*, 195–216. doi:10.1016/j.rser.2013.10.005

Tarzhanov, V. I. (1998). *Fast Initiation of Explosives*. Special Regimes of Detonation.

Thepenier, J. (2001). Advanced technologies available for future solid propellant grains. *Acta Austronautica*, *48*(5), 245-255.

Thompson, R. (1989). RF Sensitivity of Electroexplosives to Pulsed Sources and the Effect of Thermal Time Constants, Explosives and Pyrotechnics. *The Newsletter of Explosives, Pyrotechnics and Their Devices*.

Thompson, D. G., Olinger, B., & Deluca, R. (2005). The effect of pressing parameters on the mechanical properties of plastic bonded explosives. *Propellants, Explosives, Pyrotechnics*, *30*(6), 391–396. doi:10.1002/prep.200500030

Thormählen, P., & Anflo, K. (2012). Low-temperature operational and storable ammonium dinitramide based liquid monopropellant blends. *World Intellectual Property Organization*, *WO2012166046*, A2.

Thornton, J. R. (1967). Salts of nitroform with hydrazines. U.S. Patents, US3,297,747.

Tompa, A. S., & Boswell, R. F. (2000). Thermal stability of a plastic bonded explosive. *Thermochimica Acta, 357-358*, 169–175. doi:10.1016/S0040-6031(00)00386-5

Tummers, M. J., van der Heijden, A. E. D. M., & van Veen, E. H. (2012). Selection of burning rate modifiers for hydrazinium nitroformate. *Combustion and Flame*, *159*(2), 882–886. doi:10.1016/j.combustflame.2011.08.010

Tuzun, F. N., & Uysal, B. Z. (2005). The Effect of ammonium nitrate, coarse/fine ammonium nitrate ratio, plasticizer, bonding agent, and Fe2O3 content on ballistic and mechanical properties of hydroxyl terminated polybutadiene based composite propellants containing 20% AP. *Journal of ASTM International*, 2(6), 233–245.

Umholtz, P. D. (1999). The History of Solid Rocket Propulsion and Aerojet. (*AIAA Paper* 99–2927).

United States. (1979). Department of the Army. *Military Explosives*. Washington, DC. (TM 9-1300-214 C2).

UNODA. (2011). *International Ammunition Technical Guideline* (2nd ed.). New York: United Nations SaferGuard.

Urbański, J., Andrzej Skup, A., & Cameron, G. G. (1977). *Handbook of analysis of synthetic polymers and plastics*. New York: John Wiley & Sons.

Urbanski, T. (1967). *Chemistry and technology of explosives* (Vol. 3). Warszawa: PWN Polish Scientific Publishers.

Urbanski, T. (1984). *Chemistry and Technology of Explosives* (Vol. 4). Great Britain: Pergamon Press.

Urgessa, G. (2010). "Vibration Properties of Beams Using Frequency-Domain System Identification Methods." *Journal of Vibration and Control* 17(9): 1287–1294. Retrieved October 21, 2014 from http://jvc.sagepub.com/cgi/doi/10.1177/1077546310378431

Urgessa, G., and Maji, A. (2010). Dynamic Response of Retrofitted Masonry Walls for Blast Loading. *Journal of Engineering Mechanics-Asce, 136*(7), 858–864.

Urgessa, G. (2009). Finite Element Analysis of Composite Hardened Walls Subjected to Blast Loads. *American Journal of Engineering and Applied Science*, 2(4), 804–811. doi:10.3844/ajeassp.2009.804.811

Valença, U. S. V. (1998). *Curso de emprego de explosivos industriais na engenharia. Unpublished apostille,* Instituto Militar de Engenharia, Rio de Janeiro.

Varde, K. (1981) Combustion characteristics of small spark ignition engines using hydrogen supplemented fuel mixture (SAE paper 810921).

Vasiliev, A. P., Kudriavtcev, V. M., & Kuznetcsov, V. A. (1993). *Theory and calculos of liquid propellant rocket motors.*

Verhelst, S. (2014). Recent progress in the use of hydrogen as a fuel for internal combustion engines. *International Journal of Hydrogen Energy*, *39*(2), 1071–1085. doi:10.1016/j.ijhydene.2013.10.102

Verma, G., Prasad, R. K., Agarwal, R. A., Jain, S., & Agarwal, A. K. (2016, August). Experimental investigations of combustion, performance and emission characteristics of a hydrogen enriched natural gas fueled prototype spark ignition engine. *Fuel*, *178*, 209–217. doi:10.1016/j.fuel.2016.03.022

Vilar, W. D. (2005). *Química e Tecnologia dos Poliuretanos* (3rd ed.). Rio de Janeiro.

Vilyunov, V. N., & Zarko, V. E. (1989). *Ignition of Solids*. New York: Elsevier Science Publishers.

Volk, F., & Bathelt, H. (1997). Influence of energetic materials on the energy-output of gun propellants. *Propellants, Explosives, Pyrotechnics*, *22*(3), 120–124. doi:10.1002/prep.19970220305

Vörde, C., & Skifs, H. (2005). Method of producing salts of dinitramidic acid. *World Intellectual Property Organization, WO2005070823*, A1.

Vyazovkin, S., Burnham, A. K., Criado, J. M., Pérez-Maqueda, L. A., Popescu, C., & Sbirrazzuoli, N. (2011). ICTAC Kinetics Committee recommendations for performing kinetic computations on thermal analysis data. *Thermochimica Acta*, *520*(1-2), 1–19. doi:10.1016/j.tca.2011.03.034

Wang, H. K., Chen, K. S., & Lin, Y. C. (2013). Emission reductions of carbonyl compounds in a heavy-duty diesel engine supplemented with H2/O2 fuel. *Aerosol and Air Quality Research-Index*, *13*, 1790–1795.

Wang, J., Huang, Z., Tang, C., Miao, H., & Wang, X. (2009). Numerical study of the effect of hydrogen addition on methane-air mixtures combustion. *International Journal of Hydrogen Energy*, *34*(2), 1084–1096. doi:10.1016/j.ijhydene.2008.11.010

Wang, J., Huang, Z., Tang, C., & Zheng, J. (2010). Effect of hydrogen addition on early flame growth of lean burn natural gas-air mixtures. *International Journal of Hydrogen Energy*, *35*(13), 7246–7252. doi:10.1016/j.ijhydene.2010.01.004

Wang, S., Ji, C., & Zhang, B. (2010). Effects of hydrogen addition and cylinder cutoff on combustion and emissions performance of a spark-ignited gasoline engine under a low operating condition. *Energy*, *35*(12), 4754–4760. doi:10.1016/j.energy.2010.09.015

Wang, S., Ji, C., Zhang, B., Cong, X., & Liu, X. (2016). Effect of CO2 dilution on combustion and emissions characteristics of the hydrogen-enriched gasoline engine. *Energy*, *96*, 118–126. doi:10.1016/j.energy.2015.12.017

Wang, S., Ji, C., Zhang, B., & Liu, X. (2013). Emissions performance of a hybrid hydrogen e gasoline engine-powered passenger car under the new European driving cycle. *Fuel*, *106*, 873–875. doi:10.1016/j.fuel.2013.01.011

Wang, Z. H., Yang, L., Li, B., Li, Z. S., Sun, Z. W., Aldén, M., & Konnov, A. A. et al. (2012). Investigation of combustion enhancement by ozone additive in CH4/air flames using direct laminar burning velocity measurements and kinetic simulations. *Combustion and Flame*, *159*(1), 120–129. doi:10.1016/j.combustflame.2011.06.017

Wanninger, P., Wild, R., Kleinschmidt, E., & Spath, H. (1996). *Pressable explosives granular product and pressed explosive charge.* U.S. Patents, US5,547,526.

Wegener, G., Brandt, M., Duda, L., & Weisbeck, M. (2001). Trends in industrial catalysis in the polyurethane industry. *Applied Catalysis A, General*, *221*(1-2), 303–335. doi:10.1016/S0926-860X(01)00910-3

Weiser, W., Franzin, A., De Luca, L. T., Fischer, S., Gettwert, V., & Kelzenberg, S., Knapp, S., Raab, A., & Roth, E. (2012). Burning Behavior of ADN Solid Propellants Filled with Aluminum and Alane. In Proceedings of 12-IWCP, Politecnico di Milan, Milan, Italy, June 9-10.

Weng, W., Nilsson, E., Ehn, A., Zhu, J., Zhou, Y., Wang, Z., & Cen, K. et al. (2015). Investigation of formaldehyde enhancement by ozone addition in CH4/air premixed flames. *Combustion and Flame*, *162*(4), 1284–1293. doi:10.1016/j.combustflame.2014.10.021

Wenhui, W., Zhanning, J., & Jianwu, D. 1994. Surface characterization of the coupling interaction on HMX particles by FTIR DR spectroscopy. Energetic materials: analysis, characterization and techniques. In *Proceedings of the International Annual Conference of ICT* (pp. 85-1 – 85-12) Karlsruhe.

Wewerka, E. M., Loughran, D. E., & Williams, M. J. (1976). The effects of long term storage at elevated temperatures on small cylinders of PBX 9501 (Report, Tech. LA-6302-MS). Los Alamos: Los Alamos Scientific Laboratory.

Whiete, C. M., Steeper, R. R., & Lutz, A. E. (2006). The hydrogen-fueled internal combustion engine: A technical review. *International Journal of Hydrogen Energy*, *31*(10), 1292–1305. doi:10.1016/j.ijhydene.2005.12.001

Wiedemanna, C., Homeistera, M., Oswalda, M., Stabrotha, S., Klinkradb, H., & Vörsmanna, P. (2009). Additional historical solid rocket motor burns. *Acta Astronautica, 64*(11-12), 1276–1285. doi:10.1016/j.actaastro.2009.01.011

Willcox, M., Brewster, M., Tang, K., Steward, D., & Kuznetsov, I. (2007). Solid rocket motor internal ballistic simulation using three-dimensional grain burnback. *Journal of Propulsion and Power, 23*(3), 575–584. doi:10.2514/1.22971

Williams, F.A. (1985). Combustion Theory (2nd ed.). MA: The Benjamin/Cummings Publishing Co.

Wingborg, N. (2012). Status of ADN-Based Solid Propellant Development. In Proceedings of 12-IWCP, Politecnico di Milan, Milan, Italy, June 9-10.

Wingborg, N., Andreasson, S., de Flon, J., Johnsson, M., Liljedahl, M., Oscarson, C., & Wanhatalo, M. et al. (2010). Development of ADN-Based Minimum Smoke Propellants. *AIAA Paper*, 2010–6586.

Wingborg, N., & Calabro, M. (2016). Green Solid Propellants for Launchers. In *Proceedings of the Space Propulsion conference*, Rome, Italy, May 2-6.

Wingerden, K. V., Bjerketved, T. D., & Bakke, J. R. (1999). *Detonation in Pipes and in the Open*. Porsgrunn-Norway, Christian Michelsen Research and GexCon.

Won, S. H., Jiang, B., Diévart, P., Sohn, C. H., & Ju, Y. (2015). Self-sustaining n-heptane cool diffusion flames activated by ozone. *Proceedings of the Combustion Institute, 35*(1), 881–888. doi:10.1016/j.proci.2014.05.021

Wylie, C. R. (1975). *Advanced Engineering Mathematics*. Tokyo: McGraw-Hill.

Wynn, J. T. W., & Musker, A. J. (2015). Hydrogen peroxide catalyst. *World Intellectual Property Organization, WO2015033165*, A1.

Yan, Q.-L., Zhao, F.-Q., Kuo, K. K., Zhang, X.-H., Zeman, S., & DeLuca, L. T. (2016). Catalytic Effects of Nano Additives on Decomposition and Combustion of RDX-, HMX-, and AP-Based Energetic Compositions. *Progress in Energy and Combustion Science, 57*, 75–136. doi:10.1016/j.pecs.2016.08.002

Yildirim, C., & Aksel, M. (2005). Numerical simulation of the grain burnback in solid propellant rocket motor. In *Proceedings of the 41th AIAA/ASME/SAE/SEE Joint Propulsion Conference & Exhibit,* Tucson, Arizona, USA. AIAA. doi:10.2514/6.2005-4160

York, J. L., Stubbs, H. E., & Tek, M. R. (1953). The mechanism of disintegration of liquid sheets. Trans. ASME, 1953, 1279-1286.

Yoshihito, S., & Teruo, S. (2005) Effect of hydrogen and gasoline-mixed combustion on spark ignition engine (SAE paper 2005-08-0503).

Yuasa, S., Shimada, O., Imamura, T., Tamura, T., & Yamamoto, K. (1999). A technique for Improving the performance of Hybrid Rocket Engines. In *Proceedings of the 35th AIAA/ASME/SAE/ASEE Joint Propulsion Conference and Exhibit*, Los Angeles, CA. doi:10.2514/6.1999-2322

Zattera, A. J., Bianchi, O., Zeni, M., & Ferreira, C. A. (2005). Caracterização de resíduos de copolímeros de etileno-acetato de Vinila – EVA. *Polímeros: Ciência e Tecnologia, 15*(1), 73–78.

Zee, F. W. M., Mul, J. M., & Hordijk, A. C. (1994). *Method of preparing hydrazine nitroform*. World Intellectual Property Organization.

Zhang, W. R., Lowe, C., & Smith, R. (2009). Depth profiling of coil coating using step-scan photoacoustic FTIR. *Progress in Organic Coatings, 65*(4), 469–476. doi:10.1016/j.porgcoat.2009.04.005

Zhang, Y., Zhu, M., Zhang, Z., Shang, R., & Zhang, D. (2016). Ozone effect on the flammability limit and near-limit combustion of syngas/air flames with N2, CO2, and H2O dilutions. *Fuel, 186*, 414–421. doi:10.1016/j.fuel.2016.08.094

Zhao, C. F., & Chen, J. Y. (2013). Damage Mechanism and Mode of Square Reinforced Concrete Slab Subjected to Blast Loading. *Theoretical and Applied Fracture Mechanics, 63–64*, 54–62. doi:10.1016/j.tafmec.2013.03.006

Zhen, X., Wang, Y., Xu, S., & Zhu, Y. (2013). Study of knock in a high compression ratio spark ignition methanol engine by multi-dimensional simulation. *Energy, 50*, 150–159. doi:10.1016/j.energy.2012.09.062

Zhou, J. H., Cheung, C. S., & Leung, C. W. (2014). Combustion, performance, regulated and unregulated emissions of a diesel engine with hydrogen addition. *Applied Energy, 126*, 1–12. doi:10.1016/j.apenergy.2014.03.089

Zitrin, S. (1998). Forensic Investigation of Explosions. In *Analysis of Explosives by Infrared Spectrometry and Mass Spectrometry* (pp. 267–314). Taylor & Francis Ltd.

Zyl, G. J., & Keyser, R. (1996). Development of a composite propellant with a low Pressure Exponent suitable for Nozzle-less Booster Motors. In *Proc. 27th Intl Ann. Conf. ICT.,* Karlsruhe, Germany (pp. 20/1-20/11).

About the Contributors

Igor Assovskiy is Head of Laboratory of Physics of Solid Propellants Combustion, Semenov Institute of Chemical Physics RAS; Professor of Chemical Physics, National Research Nuclear University "MEPhl", Moscow, Russia.

Manfred A. Bohn has studied Chemistry and Physics at University of Karlsruhe, Germany, where he has got his PhD in Physical Chemistry. He is working more than 25 years at Fraunhofer ICT, Pfinztal, Germany in the field of stability, compatibility and ageing of energetic materials, which comprise gun propellants, rocket propellants, high explosive charges, all types of energetic components, gas generators, gelled propellants and improvised explosives. Besides experimental work also modelling the data is his interest. Stabilizer consumption, molar mass degradation of polymers, mass loss, chemical decomposition analysed by transition state quantum mechanics, atomistic simulation and modelling belong to these activities. Monitoring of ageing is a newer special topic. Numerous publications and presentation on international conferences and symposia document parts of this work. Several memberships in organising and advisory committees: Heat Flow Calorimetry Symposium, NC-Symposium, NTREM conference, KISHEM-Symposium. IPS-Seminar. He is involved in several international collaborations organized in the frame of European Defence Agency (EDA) and a member in the Custodian Nations Group (CNG) of NATO AC/326, SG1.

Fernando Costa holds a degree in Mechanical and Weapon Engineering from the IME (1985), a Master's Degree in Space Science from INPE (1990) and a Masters in Aerospace Engineering from the University of Michigan (1993 and 1995). He is currently Full Researcher, Substitute Head of the Associated Laboratory of Combustion and Propulsion of INPE and Coordinator of the PROPSAT Project to update the INPE propulsion test infrastructure. He works in research and development, consulting, project coordination and postgraduate teaching in the areas of combustion and aerospace propulsion.

Luigi DeLuca has for about forty years, been a professor of Aerospace Propulsion at Politecnico di Milano, Milan, Italy. Since his retirement in the fall of 2014, he has been teaching as Visiting Professor in several foreign universities. His scientific activity has mostly been devoted to fundamental combustion problems of solid-phase energetic materials, from both the experimental and theoretical viewpoints, with specific attention to: steady burning rates and temperature profiles, radiant ignition, static and dynamic extinction, nonsteady burning, linear and nonlinear intrinsic burning stability, radiation assisted/augmented burning, deflagration limits, transient flame modeling, and so on. In the more recent years, interests have moved to combustion of innovative high-energy condensed materials, nanoenergetics for propulsion, performance of metallized formulations, agglomeration and aggregation, dual metal formulations, quasi-steady regression rates, solid and hybrid rocket motors, space launchers, in-space propulsion.

Dimitri de Silva has a Master in Mechanical Engineering with emphasis on toast biomass production process.

Volker Gettwert is the head of the solid rocket propellant and fire protection group in the department of energetic material at the Fraunhofer Institute for Chemical Technology. He received his PhD in chemistry from the University of Ulm. After spending time abroad as a postdoctoral research fellow at Kent State University, USA, he moved back to Germany and began working at the Fraunhofer Institute for Chemical Technology in the field of fire protection systems for construction in 2001. Since 2012 he expands his professional career in the area of solid rocket propellants and gas generators for civil and military applications.

Elizabeth Mattos graduated in Chemical Engineering from University of Mogi das Cruzes (1989), master's degree in Plasma Physics from Instituto Tecnológico de Aeronáutica (2001) and PhD in Aeronautical Engineering and Mechanical Engineering from Instituto Tecnológico de Aeronáutica (2007). It is currently is researcher at the Departamento de Ciência e Tecnologia Aeroespacial (DCTA) - Instituto de Aeronáutica e Espaço (IAE) and permanent professor at the Instituto Tecnológico de Aeronáutica (ITA). He has experience in Chemical Engineering with emphasis in Organic Process, acting on the following topics: synthesis of energetic materials, plastic explosives, development methodologies by FT-IR using transmission techniques, reflection (DRIFT, MIC, ATR and UATR) Photoacoustic detection and the identification and quantification of energy materials, thermal analysis, chromatography (HPLC).

Guilherme Lourenço Mejia graduated in Aerospace Engineering at Aeronautics Institute of Technology – ITA (São José dos Campos, Brazil, 2012) and has Master Degree in Research and Development in Mechanics at Ecole Nationale Superieure Mecanique et d'Aerotechnique – ENSMA (Poitiers, France, 2012), where he worked on experiments of flame propagation, deflagration and detonation. Currently, he is a Doctorate student at Aeronautical and Mechanical Engineering Program, Chemistry Department, ITA. His main research topics are combustion, aerospace propulsion, and solid rocket motors, with emphasis on mathematical problem modeling, simulation and numerical programing. He also has worked as Product Development Engineer in the aerospace industry since 2013.

Fausto Batista Mendonca is a PhD Student at the Aeronautics Institute of Technology - ITA (São José dos Campos, Brazil). He has more than ten years of experience in military target analysis. Currently, he is working with researches having blast against reinforced concrete structures and vulnerability of structures to deal with blast dynamic load. He was at George Mason University as a visiting PhD student sponsored by FAB (Brazilian Air Force) and the C4I & Cyber Center for more than one year.

Leopoldo Rocco Jr. received his Bachelor Degree in Mechanical Engineering from Fundação Armando Alvares Penteado (FAAP) (São Paulo, Brazil, 1983) and Business Administration from Mackenzie University (São Paulo, Brazil, 1990). He has Master Degree at Instituto Tecnológico de Aeronáutica - ITA (São José dos Campos, Brazil, 2006) in Aeronautical and Mechanical Engineering Program, with emphasis on injectors like double swirl (bi-propellants) for liquid fuel and oxidant, used in liquid rocket engines and Doctoral Degree at Instituto Tecnológico de Aeronáutica - ITA (São José dos Campos, Brazil, 2013) in Aeronautical and Mechanical Engineering Program, with emphasis on hybrid rocket motors working with combustion, energetic materials, thermal protection of the combustion chamber and nozzle. He is a member of the National Council for Scientific and Technological Development (Conselho Nacional de Desenvolvimento Científico e Tecnológico - CNPq) Research Groups like: Energetic Materials; Projects, Manufacturing processes and Strategic Materials Defense Area; - Chemical Propulsion. Acting directly at industries like Bertel Metallurgical Industry Ltda., he works with managements, personal, commercial, quality (ISO 9001-2008) and work safety. Manages the company FLOWTEST Engineering and Research Ltda, which operates in Research and Development R & D.

Roberta Jachura Rocha received her Bachelor's Degree in Chemical Engineering from Universidade de São Paulo - USP (São Paulo, Brazil, 2008). She has Doctoral Degree at Instituto Tecnológico de Aeronáutica - ITA (São José dos Campos, Brazil, 2015) in Aeronautical and Mechanical Engineering Program, with emphasis on organic synthesis of polymeric matrices applied to energetic materials. She has experience in combustion, propellant formulations, chemical propulsion of solid and hybrid rocket motors, theoretical chemistry and defence materials. She also works in research projects on paints and industrial coatings.

Claudio Tagliabue is a research engineer at Fraunhofer-Institut für Chemische Technologie (ICT) in the department of energetic materials where he gained some experience in solid rocket propellants. He graduated in Aeronautical Engineering at Politecnico di Milano in 2015.

Girum Urgessa is an Associate Professor of Civil Engineering and an Affiliate Faculty of the C4I & Cyber Center at George Mason University. He received his MS and PhD from the University of New Mexico, and his BS from Addis Ababa University. Prior to joining George Mason University in 2007, he was a research engineer in the Applied Science Division of Weidlinger Associates, and a structural engineer at Dekker/Perich/Sabatini both in Albuquerque, NM. His research interests include dynamic response of structures subjected to extreme loads (blast and impact), the use of fiber reinforced polymer composites in structural design and retrofit applications, and structural stability of temporary structures. Dr. Urgessa is currently a member of the ASCE/SEI's technical committee on dynamic effects, subcommittee on retrofit of structures under dynamic loads. He was a voting member of ACI's technical committee on blast and impact load effects (ACI 370) and TRB's technical committee on field testing and non-destructive evaluation of transportation structures (AFF 40). Dr. Urgessa is a licensed professional engineer.

Index

A

B

C

D

E

F

G

H

I

K

L

M

N

O

P

Q

R

S

T

U

V

www.ingramcontent.com/pod-product-compliance
Lightning Source LLC
LaVergne TN
LVHW060302011026
849161LV00011B/2759